A Textbook of ELECTRODYNAMICS

A Textbook of
ELECTRODYNAMICS

Dr. Mohit Kumar Sharma
M.Sc., M.Phil., Ph.D. (Physics)
Young Scientist Awardee (ISCA, 2014)

National Postdoctoral Fellow, SERB, DST, New Delhi
Amity Center for Astronomy & Astrophysics
Amity Institute of Applied Sciences
Amity University, Noida 201313, India

Prof. (Dr.) Suresh Chandra
President, Physical Sciences (2021)
Indian Science Congress Association (Govt. of India),
AvH Fellow (Germany), FRAS (London), FMAS (Pune)

Deputy Director & Professor
Amity Center for Astronomy & Astrophysics
Amity Institute of Applied Sciences
Amity Universityi, Noida 201313, India

TECHSAR™
NEW DELHI

Published and Printed by
Techsar Pvt. Ltd.
4435–36/7, Ansari Road, Daryaganj
New Delhi-110 002 (India)
E-mail: tech_info@techsarworld.com
Website: www.techsarworld.com

ISBN 978-93-90620-68-5

Reprint 2025

To my Grandchildren
who belong to future

Suresh Chandra

Preface

Electrodynamics is a core subject being taught to graduate students and post-graduate students in India and abroad. In this Textbook of Electrodynamics, we have discussed about Curvilinear Coordinates, Vectors, Electrostatics, Electrostatic Field in Dielectrics, Energy in Electrostatic Fields, Magnetostatics, Electromagnetic Induction, Maxwell's Field Equations, Dispersion and Scattering, Electromagnetic Waves, Wave-guides and Antennas, Retarded Systems, Principle of Relativity, Charged Particle in Fields, Plasma State, and Waves in Fluid Plasma. Prof. Chandra happened to teach this subject at DDU-Gorakhpur University, Gorakhpur (U.P.), S.R.T.M. University, Nanded (Maharashtra), S.M.V.D. University, Katra (J&K), L.P.U. Phagwara (Punjab), and Amity University, Noida. On persistent's demand from students and valuable advice from friends and colleagues, this book has been prepared.

While preparing the manuscript, we have consulted various books and texts, and have got their help. We are grateful to the authors and publishers of those books and texts. For this book, we have been helped, advised and encouraged by our friends working in various institutions/universities in India as well as abroad, and by our friends in the personal life. We are heartily thankful to all of them. Prof. Chandra would like to thank his students whom he taught the subject and got feedback for the improvement of the manuscript. We would greatly appreciate comments and suggestions from the students, our friend teachers and others, which can be sent also to either

mohitkumarsharma32@yahoo.in
suresh492000@yahoo.co.in; schandra2@amity.edu

We are grateful to Hon'ble Dr. Ashok K. Chauhan, Founder President, Amity University, Hon'ble Dr. Atul Chauhan, Chancellor Amity University, Uttar Pradesh, Prof. Balvinder Shukla, Vice-Chancellor, Amity University Uttar Pradesh, Prof. Sunita Rattan, Director, Amity Institute for Applied Sciences, and Prof. S.N. Shukla, Dept of Applied Physics, Amity University, Noida for their moral support and encouragement.

Suresh Chandra is specially thankful to his wife Mrs. Purnima Sharma for her valuable cooperation in his life and for sharing a major part of the responsibility of family affairs, so that he could spend his time for this book. Mohit Kumar Sharma is thankful to his wife Mrs. Shalini Sharma for her valuable cooperation. We are thankful to Dr. Monika Sharma and Mrs. Shalini Sharma for their help in preparation of the manuscript of this textbook. We are deeply grateful to our family members who always have been a source of inspiration and happiness for us.

Mohit Kumar Sharma
Suresh Chandra

CONTENTS

Chapter 1

Curvilinear Coordinates

In science and technology, we have been using several orthogonal coordinate systems, for example, Cartesian coordinates (x, y, z), spherical polar coordinates (r, θ, ϕ), cylindrical polar coordinates (r, ϕ, z) etc., depending on the nature of task. Choice of a particular type of coordinate system often helps in the treatment of task. For example, for the motion of a particle on the surface of a sphere, use of the spherical polar coordinates with origin as the center of the sphere makes the treatment convenient. Here, the path of a particle moving on the surface of a sphere of radius b can be written as $r = b$. This equation is obviously simple as the corresponding equation in Cartesian coordinates would be $\sqrt{(x^2 + y^2 + z^2)} = b$, where we have three variables instead of one. In the present chapter, we have discussed about the curvilinear coordinates whose special cases are these coordinate systems.

1. Relations between some coordinate systems

Coordinates of a point given in one system can be transformed into its coordinates in another system. Here, we consider only three coordinate systems, *viz.*, Cartesian coordinates (x, y, z), spherical coordinates (r, θ, ϕ) and cylindrical coordinates (r, ϕ, z).

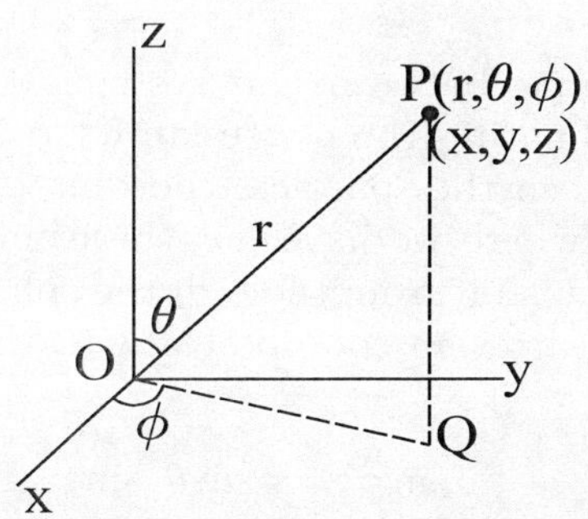

Figure 1: Cartesian as well as spherical polar coordinates having origin at the common point O.

(i) Coordinates of a point in the spherical system $(r,\ \theta,\ \phi)$ can be transformed into the coordinates of the point in the Cartesian system $(x,\ y,\ z)$ through the relations (Figure 1).

$$x = r\sin\theta\ \cos\phi \qquad y = r\sin\theta\ \sin\phi \qquad z = r\cos\theta \tag{1.1}$$

The inverse transformation is through the relations

$$r = (x^2 + y^2 + z^2)^{1/2} \qquad \theta = \tan^{-1}\left(\sqrt{x^2+y^2}/z\right) \qquad \phi = \tan^{-1}(y/x)$$

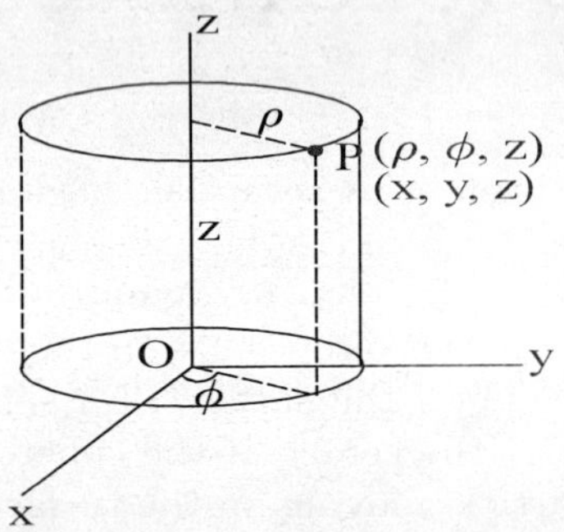

Figure 2: Cartesian as well as cylindrical polar coordinates having origin at the common point O.

(ii) Coordinates of a point in the cylindrical system $(\rho,\ \phi,\ z)$ can be transformed into the coordinates of the point in Cartesian system $(x,\ y,\ z)$ through the relations (Figure 2).

$$x = \rho\cos\phi \qquad y = \rho\sin\phi \qquad z = z \tag{1.2}$$

The inverse transformation is through the relations

$$\rho = (x^2 + y^2)^{1/2} \qquad \phi = \tan^{-1}(y/x) \qquad z = z$$

(iii) Similar transformation equations between Cartesian system and any other system can also be obtained.

(iv) Transformation relations between two systems, other than the Cartesian one, can be obtained by eliminating the Cartesian coordinates. For example, the transformation relations between the spherical coordinates $(r,\ \theta,\ \phi)$ and the cylindrical coordinates $(\rho,\ \phi',\ z)$ (where we have put the prime on ϕ in the cylindrical coordinates in order to distinguish it from the ϕ in the spherical coordinates) can be obtained by eliminating the Cartesian coordinates with the help of equations (1.1) and (1.2). Thus, we have

$$\rho\cos\phi' = r\sin\theta\ \cos\phi \qquad \rho\sin\phi' = r\sin\theta\ \sin\phi \qquad z = r\cos\theta \tag{1.3}$$

Equation (1.3) can be rearranged as

$$\rho = r\sin\theta \qquad \phi' = \phi \qquad z = r\cos\theta \tag{1.4}$$

Equation (1.4) is the equation for transformation from spherical coordinates to the cylindrical ones. The inverse transformation equation can be obtained as

$$r = \sqrt{\rho^2 + z^2} \qquad \theta = \tan^{-1}(\rho/z) \qquad \phi = \phi' \tag{1.5}$$

Expressions for vector operators (*i.e.*, gradient, divergence, curl, Laplacian etc.) depend on the coordinate system used. Therefore, instead of dealing with the expressions for the operators written in various coordinate systems, we intend to introduce a coordinate system, called the 'orthogonal curvilinear coordinate system', whose special cases would be all the above coordinate systems.

2. Orthogonal curvilinear coordinate system

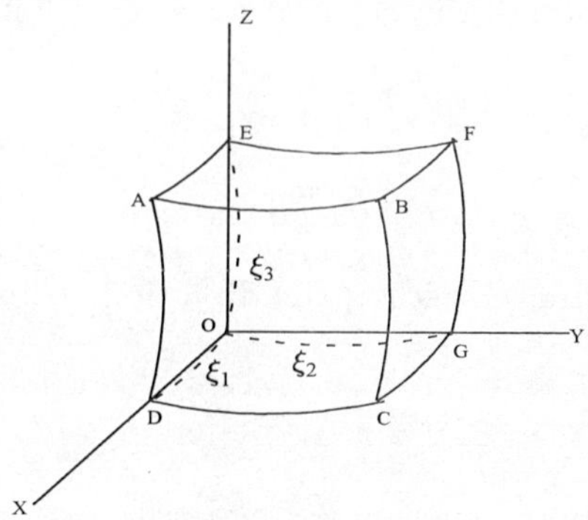

Figure 3: Imagination of the surfaces of an orthogonal curvilinear coordinate system.

The orthogonal curvilinear coordinate system consists of a three-fold family of surfaces whose equations are (Figure 3)

$$\xi_1(x, y, z) = \text{const} \qquad \xi_2(x, y, z) = \text{const} \qquad \xi_3(x, y, z) = \text{const} \tag{1.6}$$

Here, ξ_1, ξ_2, ξ_3 are orthogonal curvilinear coordinates, and can be expressed as functions of the Cartesian coordinates x, y, z. In many cases, it is more convenient to express x, y, z in terms of ξ_1, ξ_2, ξ_3 as

$$x = x(\xi_1, \xi_2, \xi_3) \qquad y = y(\xi_1, \xi_2, \xi_3) \qquad z = z(\xi_1, \xi_2, \xi_3)$$

The lines of intersection of these surfaces (given in equation 1.6) constitute three families of lines, which are in general curve in shape. The lines passing through the origin O are known as the axes ξ_1, ξ_2 and ξ_3 of the orthogonal curvilinear coordinate system. Thus, the axis ξ_1 is the line of intersection of the surfaces $\xi_2 = 0$ and $\xi_3 = 0$; the axis ξ_2 is the line of intersection of surfaces $\xi_3 = 0$ and $\xi_1 = 0$; and the axis ξ_3 is the line of intersection of surfaces $\xi_1 = 0$ and $\xi_2 = 0$. The axes ξ_1, ξ_2 and ξ_3 are orthogonal to one another.

Let $\hat{\xi_1}$, $\hat{\xi_2}$, $\hat{\xi_3}$ be the unit vectors, each tangent to the corresponding coordinate line of the curvilinear coordinate system. Thus, $\hat{\xi_1}$, $\hat{\xi_2}$, $\hat{\xi_3}$ constitute a new trio of unit vector in terms of which the vector quantities may be expressed. For example, the force $\vec{F}$ can be expressed as

$$\vec{F} = F_1\,\hat{\xi}_1 + F_2\,\hat{\xi}_2 + F_3\,\hat{\xi}_3 = \sum_{i=1}^{3} F_i\,\hat{\xi}_i$$

where F_1, F_2 and F_3 are the components of the force acting along the curvilinear coordinate axes.

Let α_1, β_1, γ_1 be the direction cosines of $\hat{\xi}_1$ relative to the Cartesian coordinate system. Thus, we have

$$\hat{\xi}_1 = \alpha_1\,\hat{\imath} + \beta_1\,\hat{\jmath} + \gamma_1\,\hat{k} \tag{1.7}$$

Similarly, let α_2, β_2, γ_2 be the direction cosines of $\hat{\xi}_2$, and α_3, β_3, γ_3 of $\hat{\xi}_3$, then we have

$$\hat{\xi}_2 = \alpha_2\,\hat{\imath} + \beta_2\,\hat{\jmath} + \gamma_2\,\hat{k} \tag{1.8}$$

$$\hat{\xi}_3 = \alpha_3\,\hat{\imath} + \beta_3\,\hat{\jmath} + \gamma_3\,\hat{k} \tag{1.9}$$

For the orthogonal coordinate system, the direction cosines obviously satisfy the relation

$$\alpha_i\alpha_j + \beta_i\beta_j + \gamma_i\gamma_j = \delta_{ij} \tag{1.10}$$

where δ_{ij} is the Kronecker delta function defined as

$$\delta_{ij} = \begin{cases} 1 & \text{when} \quad i = j \\ 0 & \text{when} \quad i \neq j \end{cases} \tag{1.11}$$

By multiplying equations.(1.7), (1.8) and (1.9) by α_1, α_2 and α_3, respectively, and adding the products, we get

$$\hat{\imath} = \alpha_1\,\hat{\xi}_1 + \alpha_2\,\hat{\xi}_2 + \alpha_3\,\hat{\xi}_3 \tag{1.12}$$

Similarly, we can get

$$\hat{\jmath} = \beta_1\,\hat{\xi}_1 + \beta_2\,\hat{\xi}_2 + \beta_3\,\hat{\xi}_3 \tag{1.13}$$

$$\hat{k} = \gamma_1\,\hat{\xi}_1 + \gamma_2\,\hat{\xi}_2 + \gamma_3\,\hat{\xi}_3 \tag{1.14}$$

The other relations satisfied by the direction cosines are

$$\alpha_1^2 + \alpha_2^2 + \alpha_3^2 = 1 = \beta_1^2 + \beta_2^2 + \beta_3^2 = \gamma_1^2 + \gamma_2^2 + \gamma_3^2$$

and

$$\alpha_1\beta_1 + \alpha_2\beta_2 + \alpha_3\beta_3 = 0 = \beta_1\gamma_1 + \beta_2\gamma_2 + \beta_3\gamma_3 = \gamma_1\alpha_1 + \gamma_2\alpha_2 + \gamma_3\alpha_3$$

3. Scale factors

Suppose, in the orthogonal coordinate system (ξ_1, ξ_2, ξ_3) the coordinate ξ_i is changed by $\mathrm{d}\xi_i$. This change in the coordinate produces a displacement $h_i\mathrm{d}\xi_i$ along the line of the coordinate ξ_i. The scale factor h_i may depend on the space coordinates, and therefore may vary from point to point in space. The line element $\vec{\mathrm{d}s}$ is therefore

$$\vec{\mathrm{d}s} = h_1\ \mathrm{d}\xi_1\ \hat{\xi_1} + h_2\ \mathrm{d}\xi_2\ \hat{\xi_2} + h_3\ \mathrm{d}\xi_3\ \hat{\xi_3} = \sum_{i=1}^{3} h_i\ \mathrm{d}\xi_i\ \hat{\xi_i}$$

so that

$$(\mathrm{d}s)^2 = \sum_{i=1}^{3}\sum_{j=1}^{3} h_i h_j \mathrm{d}\xi_i\ \mathrm{d}\xi_j\ \hat{\xi_i}\ .\ \hat{\xi_j} \tag{1.15}$$

However, the line element in Cartesian coordinates is given by

$$(\mathrm{d}s)^2 = (\mathrm{d}x)^2 + (\mathrm{d}y)^2 + (\mathrm{d}z)^2 \tag{1.16}$$

From equations (1.15) and (1.16), we get

$$(\mathrm{d}x)^2 + (\mathrm{d}y)^2 + (\mathrm{d}z)^2 = \sum_{i=1}^{3}\sum_{j=1}^{3} h_i h_j\ \mathrm{d}\xi_i\ \mathrm{d}\xi_j\ \hat{\xi_i}\ .\ \hat{\xi_j} = \sum_{i=1}^{3} h_i^2(\mathrm{d}\xi_i)^2 \tag{1.17}$$

where we used the property of orthogonality of curvilinear coordinates. The expressions for the scale factors can be obtained in the following manner. Suppose, the Cartesian coordinates x, y, z are expressed in terms of the curvilinear coordinates ξ_1, ξ_2, ξ_3

$$x = x(\xi_1, \xi_2, \xi_3) \qquad y = y(\xi_1, \xi_2, \xi_3) \qquad z = z(\xi_1, \xi_2, \xi_3)$$

On differentiation, we get

$$\mathrm{d}x = \frac{\partial x}{\partial \xi_1}\ \mathrm{d}\xi_1 + \frac{\partial x}{\partial \xi_2}\ \mathrm{d}\xi_2 + \frac{\partial x}{\partial \xi_3}\ \mathrm{d}\xi_3 = \sum_{i=1}^{3} \frac{\partial x}{\partial \xi_i}\ \mathrm{d}\xi_i$$

Thus, we have

$$(\mathrm{d}x)^2 = \sum_{i=1}^{3}\sum_{j=1}^{3} \frac{\partial x}{\partial \xi_i}\frac{\partial x}{\partial \xi_j}\ \mathrm{d}\xi_i\ \mathrm{d}\xi_j$$

Similarly, we get

$$(\mathrm{d}y)^2 = \sum_{i=1}^{3}\sum_{j=1}^{3} \frac{\partial y}{\partial \xi_i}\frac{\partial y}{\partial \xi_j}\ \mathrm{d}\xi_i\ \mathrm{d}\xi_j \qquad \text{and} \qquad (\mathrm{d}z)^2 = \sum_{i=1}^{3}\sum_{j=1}^{3} \frac{\partial z}{\partial \xi_i}\frac{\partial z}{\partial \xi_j}\ \mathrm{d}\xi_i\ \mathrm{d}\xi_j$$

Thus, we have

$$(\mathrm{d}x)^2 + (\mathrm{d}y)^2 + (\mathrm{d}z)^2 = \sum_{i=1}^{3}\sum_{j=1}^{3}\left(\frac{\partial x}{\partial \xi_i}\frac{\partial x}{\partial \xi_j} + \frac{\partial y}{\partial \xi_i}\frac{\partial y}{\partial \xi_j} + \frac{\partial z}{\partial \xi_i}\frac{\partial z}{\partial \xi_j}\right)\mathrm{d}\xi_i \mathrm{d}\xi_j \tag{1.18}$$

On comparing equations (1.17) and (1.18), we get

$$h_i h_j \; \hat{\xi}_i \, . \, \hat{\xi}_j = \frac{\partial x}{\partial \xi_i}\frac{\partial x}{\partial \xi_j} + \frac{\partial y}{\partial \xi_i}\frac{\partial y}{\partial \xi_j} + \frac{\partial z}{\partial \xi_i}\frac{\partial z}{\partial \xi_j}$$

and therefore,

$$h_i^2 = \left(\frac{\partial x}{\partial \xi_i}\right)^2 + \left(\frac{\partial y}{\partial \xi_i}\right)^2 + \left(\frac{\partial z}{\partial \xi_i}\right)^2 \tag{1.19}$$

where $i = 1, 2, 3$. Equation (1.19) is the expression for the scale factors h_i where the Cartesian coordinates are expressed in terms of the curvilinear coordinates.

On the other side, the curvilinear coordinates ξ_1, ξ_2, ξ_3 may be expressed in terms of the Cartesian coordinates x, y, z

$$\xi_1 = \xi_1(x, y, z) \qquad \xi_2 = \xi_2(x, y, z) \qquad \xi_3 = \xi_3(x, y, z)$$

On differentiation, we get

$$\mathrm{d}\xi_1 = \frac{\partial \xi_1}{\partial x}\,\mathrm{d}x + \frac{\partial \xi_1}{\partial y}\,\mathrm{d}y + \frac{\partial \xi_1}{\partial z}\,\mathrm{d}z$$

Thus, we have

$$(\mathrm{d}\xi_1)^2 = \left(\frac{\partial \xi_1}{\partial x}\right)^2(\mathrm{d}x)^2 + \left(\frac{\partial \xi_1}{\partial y}\right)^2(\mathrm{d}y)^2 + \left(\frac{\partial \xi_1}{\partial z}\right)^2(\mathrm{d}z)^2 + 2\,\frac{\partial \xi_1}{\partial x}\frac{\partial \xi_1}{\partial y}\mathrm{d}x\,\mathrm{d}y$$
$$+2\,\frac{\partial \xi_1}{\partial y}\frac{\partial \xi_1}{\partial z}\,\mathrm{d}y\,\mathrm{d}z + 2\,\frac{\partial \xi_1}{\partial z}\frac{\partial \xi_1}{\partial x}\,\mathrm{d}z\,\mathrm{d}x \tag{1.20}$$

Similarly, we get

$$(\mathrm{d}\xi_2)^2 = \left(\frac{\partial \xi_2}{\partial x}\right)^2(\mathrm{d}x)^2 + \left(\frac{\partial \xi_2}{\partial y}\right)^2(\mathrm{d}y)^2 + \left(\frac{\partial \xi_2}{\partial z}\right)^2(\mathrm{d}z)^2 + 2\,\frac{\partial \xi_2}{\partial x}\frac{\partial \xi_2}{\partial y}\mathrm{d}x\,\mathrm{d}y$$
$$+2\,\frac{\partial \xi_2}{\partial y}\frac{\partial \xi_2}{\partial z}\,\mathrm{d}y\,\mathrm{d}z + 2\,\frac{\partial \xi_2}{\partial z}\frac{\partial \xi_2}{\partial x}\,\mathrm{d}z\,\mathrm{d}x \tag{1.21}$$

and

$$(\mathrm{d}\xi_3)^2 = \left(\frac{\partial \xi_3}{\partial x}\right)^2(\mathrm{d}x)^2 + \left(\frac{\partial \xi_3}{\partial y}\right)^2(\mathrm{d}y)^2 + \left(\frac{\partial \xi_3}{\partial z}\right)^2(\mathrm{d}z)^2 + 2\,\frac{\partial \xi_3}{\partial x}\frac{\partial \xi_3}{\partial y}\,\mathrm{d}x\,\mathrm{d}y$$
$$+2\,\frac{\partial \xi_3}{\partial y}\frac{\partial \xi_3}{\partial z}\,\mathrm{d}y\,\mathrm{d}z + 2\,\frac{\partial \xi_3}{\partial z}\frac{\partial \xi_3}{\partial x}\,\mathrm{d}z\,\mathrm{d}x \tag{1.22}$$

Using equations (1.20), (1.21) and (1.22) in (1.17) and comparing the coefficients of $(\mathrm{d}x)^2$, $(\mathrm{d}y)^2$ and $(\mathrm{d}z)^2$ separately, we get

$$1 = h_1^2\left(\frac{\partial \xi_1}{\partial x}\right)^2 + h_2^2\left(\frac{\partial \xi_2}{\partial x}\right)^2 + h_3^2\left(\frac{\partial \xi_3}{\partial x}\right)^2 \tag{1.23}$$

$$1 = h_1^2\left(\frac{\partial\xi_1}{\partial y}\right)^2 + h_2^2\left(\frac{\partial\xi_2}{\partial y}\right)^2 + h_3^2\left(\frac{\partial\xi_3}{\partial y}\right)^2 \tag{1.24}$$

$$1 = h_1^2\left(\frac{\partial\xi_1}{\partial z}\right)^2 + h_2^2\left(\frac{\partial\xi_2}{\partial z}\right)^2 + h_3^2\left(\frac{\partial\xi_3}{\partial z}\right)^2 \tag{1.25}$$

On adding equations (1.23), (1.24) and (1.25), we get

$$3 = h_1^2\left[\left(\frac{\partial\xi_1}{\partial x}\right)^2 + \left(\frac{\partial\xi_1}{\partial y}\right)^2 + \left(\frac{\partial\xi_1}{\partial z}\right)^2\right] + h_2^2\left[\left(\frac{\partial\xi_2}{\partial x}\right)^2 + \left(\frac{\partial\xi_2}{\partial y}\right)^2 + \left(\frac{\partial\xi_2}{\partial z}\right)^2\right]$$
$$+ h_3^2\left[\left(\frac{\partial\xi_3}{\partial x}\right)^2 + \left(\frac{\partial\xi_3}{\partial y}\right)^2 + \left(\frac{\partial\xi_3}{\partial z}\right)^2\right]$$

This equation is satisfied when

$$h_i^2 = \left[\left(\frac{\partial\xi_i}{\partial x}\right)^2 + \left(\frac{\partial\xi_i}{\partial y}\right)^2 + \left(\frac{\partial\xi_i}{\partial z}\right)^2\right]^{-1} \tag{1.26}$$

where $i = 1, 2, 3$. Equation (1.26) is another expression for the scale factors h_i where curvilinear coordinates are expressed in terms of the Cartesian coordinates.

Example 1: Find the scale factors for spherical coordinate system (r, θ, ϕ).

Solution: For the spherical coordinates $\xi_1 = r$, $\xi_2 = \theta$ and $\xi_3 = \phi$. The transformation equations between the spherical and Cartesian coordinates are

$$x = r\sin\theta\ \cos\phi \qquad y = r\sin\theta\ \sin\phi \qquad z = r\cos\theta \tag{1.27}$$

The scale factors are

$$h_i^2 = \left(\frac{\partial x}{\partial\xi_i}\right)^2 + \left(\frac{\partial y}{\partial\xi_i}\right)^2 + \left(\frac{\partial z}{\partial\xi_i}\right)^2$$

where $i = 1, 2, 3$. Thus,

$$h_r^2 = \left(\frac{\partial x}{\partial r}\right)^2 + \left(\frac{\partial y}{\partial r}\right)^2 + \left(\frac{\partial z}{\partial r}\right)^2 \tag{1.28}$$

$$h_\theta^2 = \left(\frac{\partial x}{\partial\theta}\right)^2 + \left(\frac{\partial y}{\partial\theta}\right)^2 + \left(\frac{\partial z}{\partial\theta}\right)^2 \tag{1.29}$$

$$h_\phi^2 = \left(\frac{\partial x}{\partial\phi}\right)^2 + \left(\frac{\partial y}{\partial\phi}\right)^2 + \left(\frac{\partial z}{\partial\phi}\right)^2 \tag{1.30}$$

Using equation (1.27) in (1.28), (1.29) and (1.30), we get

$$h_r^2 = (\sin\theta\ \cos\phi)^2 + (\sin\theta\ \sin\phi)^2 + (\cos\theta)^2 = 1$$

$$h_\theta^2 = (r\cos\theta\ \cos\phi)^2 + (r\cos\theta\ \sin\phi)^2 + (-r\sin\theta)^2 = r^2$$

$$h_\phi^2 = (-r\sin\theta\ \sin\phi)^2 + (r\sin\theta\ \cos\phi)^2 = r^2\ \sin^2\theta$$

Hence, the scale factors for spherical coordinates are

$$h_r = 1 \qquad h_\theta = r \qquad h_\phi = r\sin\theta$$

Example 2: Find the scale factors for cylindrical coordinate system (ρ, ϕ, z).

Solution: For the cylindrical coordinates $\xi_1 = \rho$, $\xi_2 = \phi$ and $\xi_3 = z$. The transformation equations between the cylindrical and Cartesian coordinates are

$$x = \rho\cos\phi \qquad y = \rho\sin\phi \qquad z = z \tag{1.31}$$

The scale factors are

$$h_i^2 = \left(\frac{\partial x}{\partial \xi_i}\right)^2 + \left(\frac{\partial y}{\partial \xi_i}\right)^2 + \left(\frac{\partial z}{\partial \xi_i}\right)^2$$

where $i = 1, 2, 3$. Thus, we have

$$h_\rho^2 = \left(\frac{\partial x}{\partial \rho}\right)^2 + \left(\frac{\partial y}{\partial \rho}\right)^2 + \left(\frac{\partial z}{\partial \rho}\right)^2 \tag{1.32}$$

$$h_\phi^2 = \left(\frac{\partial x}{\partial \phi}\right)^2 + \left(\frac{\partial y}{\partial \phi}\right)^2 + \left(\frac{\partial z}{\partial \phi}\right)^2 \tag{1.33}$$

$$h_z^2 = \left(\frac{\partial x}{\partial z}\right)^2 + \left(\frac{\partial y}{\partial z}\right)^2 + \left(\frac{\partial z}{\partial z}\right)^2 \tag{1.34}$$

Using equation (1.27) in (1.32), (1.33) and (1.34), we get

$$h_\rho^2 = (\cos\phi)^2 + (\sin\phi)^2 = 1$$

$$h_\phi^2 = (-\rho\sin\phi)^2 + (\rho\cos\phi)^2 = \rho^2$$

$$h_z^2 = 1$$

Hence, the scale factors for cylindrical coordinates are

$$h_\rho = 1 \qquad h_\phi = \rho \qquad h_z = 1$$

3.1 Volume element

In the orthogonal curvilinear coordinate system (ξ_1, ξ_2, ξ_3) with scale factors h_1, h_2, h_3 the displacement elements are $h_1\,\mathrm{d}\xi_1$, $h_2\,\mathrm{d}\xi_2$, $h_3\,\mathrm{d}\xi_3$, respectively, along the mutually perpendicular axes. Thus, the volume-element is

$$\mathrm{d}v = h_1\mathrm{d}\xi_1\; h_2\mathrm{d}\xi_2\; h_3\mathrm{d}\xi_3 = h_1h_2h_3\mathrm{d}\xi_1\mathrm{d}\xi_2\mathrm{d}\xi_3$$

(i) In spherical coordinates $\mathrm{d}v = r^2\sin\theta\;\mathrm{d}r\;\mathrm{d}\theta\;\mathrm{d}\phi$

(ii) In cylindrical coordinates $\mathrm{d}v = \rho\;\mathrm{d}\rho\;\mathrm{d}\phi\;\mathrm{d}z$

(iii) In Cartesian coordinates $\mathrm{d}v = \mathrm{d}x\;\mathrm{d}y\;\mathrm{d}z$

4. Gradient of scalar field

Gradient of a scalar field $\Psi(\xi_1, \xi_2, \xi_3)$ is defined as the maximum rate of increase of the field. Thus, its magnitude is the value of the maximum rate of increase and its direction is that of the maximum rate of increase. So, the gradient of a scalar quantity is a vector quantity and is given by $\nabla\Psi$. Hence, the components of $\nabla\Psi$ are perpendicular to the surfaces ξ_1 = constant, ξ_2 = constant and ξ_3 = constant. The rate of increase of Ψ in the direction normal to the surfaces ξ_1 = constant, ξ_2 = constant and ξ_3 = constant, respectively, are

$$(\nabla\Psi)_1 = \frac{1}{h_1}\frac{\partial\Psi}{\partial\xi_1} \qquad (\nabla\Psi)_2 = \frac{1}{h_2}\frac{\partial\Psi}{\partial\xi_2} \qquad (\nabla\Psi)_3 = \frac{1}{h_3}\frac{\partial\Psi}{\partial\xi_3}$$

Therefore, gradient of the scalar field Ψ, denoted by $\nabla\Psi$, is

$$\text{grad } \Psi = \nabla\Psi = \frac{\hat{\xi_1}}{h_1}\frac{\partial\Psi}{\partial\xi_1} + \frac{\hat{\xi_2}}{h_2}\frac{\partial\Psi}{\partial\xi_2} + \frac{\hat{\xi_3}}{h_3}\frac{\partial\Psi}{\partial\xi_3} = \sum_{i=1}^{3}\frac{\hat{\xi_i}}{h_i}\frac{\partial\Psi}{\partial\xi_i}$$

4.1 Special cases

(i) In spherical coordinates

$$\begin{array}{lll} \xi_1 = r & \xi_2 = \theta & \xi_3 = \phi \\ h_r = 1 & h_\theta = r & h_\phi = r\ \sin\theta \\ \hat{\xi_1} = \hat{e_r} & \hat{\xi_2} = \hat{e_\theta} & \hat{\xi_3} = \hat{e_\phi} \end{array}$$

Therefore, we have

$$\text{grad } \Psi = \nabla\Psi = \hat{e_r}\frac{\partial\Psi}{\partial r} + \frac{\hat{e_\theta}}{r}\frac{\partial\Psi}{\partial\theta} + \frac{\hat{e_\phi}}{r\ \sin\theta}\frac{\partial\Psi}{\partial\phi}$$

(ii) In cylindrical coordinates

$$\begin{array}{lll} \xi_1 = \rho & \xi_2 = \phi & \xi_3 = z \\ h_r = 1 & h_\phi = \rho & h_z = 1 \\ \hat{\xi_1} = \hat{e_r} & \hat{\xi_2} = \hat{e_\phi} & \hat{\xi_3} = \hat{e_z} \end{array}$$

Therefore, we have

$$\text{grad } \Psi = \nabla\Psi = \hat{e_r}\frac{\partial\Psi}{\partial\rho} + \frac{\hat{e_\phi}}{\rho}\frac{\partial\Psi}{\partial\phi} + \hat{e_z}\frac{\partial\Psi}{\partial z}$$

(iii) In Cartesian coordinates

$$\begin{array}{lll} \xi_1 = x & \xi_2 = y & \xi_3 = z \\ h_x = 1 & h_y = 1 & h_z = 1 \\ \hat{\xi_1} = \hat{i} & \hat{\xi_2} = \hat{j} & \hat{\xi_3} = \hat{k} \end{array}$$

Therefore, we have

$$\text{grad } \Psi = \nabla\Psi = \hat{i}\frac{\partial\Psi}{\partial x} + \hat{j}\frac{\partial\Psi}{\partial y} + \hat{k}\frac{\partial\Psi}{\partial z}$$

5. Divergence of vector field

To derive an expression for the divergence of a vector field in the orthogonal curvilinear coordinates, we consider a volume element of sides $h_1 d\xi_1$, $h_2 d\xi_2$ and $h_3 d\xi_3$ parallel to the axes ξ_1, ξ_2 and ξ_3, respectively (Figure 4). Let the vector field $\vec{F}$ be at the mid-point P(ξ_1, ξ_2, ξ_3) of the volume element, then we have

$$\vec{F} = F_1\,\hat{\xi_1} + F_2\,\hat{\xi_2} + F_3\,\hat{\xi_3} \tag{1.35}$$

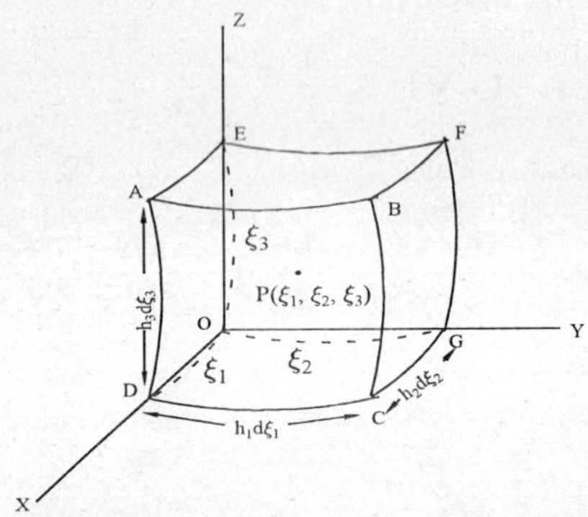

Figure 4: Divergence of a vector field.

where F_1, F_2 and F_3 are the components of the vector field $\vec{F}$ along the coordinate axes. Since the dimensions of the volume element are small, the field at the mid-point of a plane may be considered as the average of the filed over the plane. Therefore, the flux at P parallel to ξ_1-axis is

$$= F_1\,(h_2\,d\xi_2\,h_3\,d\xi_3) = F_1 h_2 h_3\,d\xi_2\,d\xi_3 \tag{1.36}$$

The rate of change of flux along ξ_1-axis is

$$= \frac{1}{h_1}\,\frac{\partial}{\partial \xi_1}(F_1 h_2 h_3\,d\xi_2\,d\xi_3) \tag{1.37}$$

Let the flux, parallel to ξ_1-axis, enter at the face OEFG and go out at the face ABCD. Therefore, the flux entering at the face OEFG is

$$\begin{aligned} &= F_1 h_2 h_3\,d\xi_2\,d\xi_3 - \frac{1}{h_1}\,\frac{\partial}{\partial \xi_1}(F_1 h_2 h_3\,d\xi_2\,d\xi_3)\left(\frac{1}{2}\,h_1\,d\xi_1\right) \\ &= F_1 h_2 h_3\,d\xi_2\,d\xi_3 - \frac{1}{2}\,d\xi_1\,d\xi_2\,d\xi_3\,\frac{\partial}{\partial \xi_1}(F_1 h_2 h_3) \end{aligned} \tag{1.38}$$

and the flux outgoing at the face ABCD is

$$\begin{aligned} &= F_1 h_2 h_3\,d\xi_2\,d\xi_3 + \frac{1}{h_1}\,\frac{\partial}{\partial \xi_1}(F_1 h_2 h_3\,d\xi_2\,d\xi_3)\left(\frac{1}{2}\,h_1\,d\xi_1\right) \\ &= F_1 h_2 h_3\,d\xi_2\,d\xi_3 + \frac{1}{2}\,d\xi_1\,d\xi_2\,d\xi_3\,\frac{\partial}{\partial \xi_1}(F_1 h_2 h_3) \end{aligned} \tag{1.39}$$

Therefore, the net outgoing flux along ξ_1-axis is

$$
\begin{aligned}
&= F_1 h_2 h_3 \, d\xi_2 \, d\xi_3 + \frac{1}{2} \, d\xi_1 \, d\xi_2 \, d\xi_3 \, \frac{\partial}{\partial \xi_1}(F_1 h_2 h_3) \\
&\quad - \Big\{ F_1 h_2 h_3 \, d\xi_2 \, d\xi_3 - \frac{1}{2} \, d\xi_1 \, d\xi_2 \, d\xi_3 \, \frac{\partial}{\partial \xi_1}(F_1 h_2 h_3) \Big\} \\
&= d\xi_1 \, d\xi_2 \, d\xi_3 \, \frac{\partial}{\partial \xi_1}(F_1 h_2 h_3)
\end{aligned} \tag{1.40}
$$

Similarly, the net outgoing fluxes along ξ_2-axis and ξ_3-axis, respectively, are

$$d\xi_1 \, d\xi_2 \, d\xi_3 \, \frac{\partial}{\partial \xi_2}(F_2 h_3 h_1) \tag{1.41}$$

and

$$d\xi_1 \, d\xi_2 \, d\xi_3 \, \frac{\partial}{\partial \xi_3}(F_3 h_1 h_2) \tag{1.42}$$

Thus, the outgoing flux over the volume element is obtained by adding equations (1.40), (1.41) and (1.42), we get

$$= d\xi_1 \, d\xi_2 \, d\xi_3 \Big[\frac{\partial}{\partial \xi_1}(F_1 h_2 h_3) + \frac{\partial}{\partial \xi_2}(F_2 h_3 h_1) + \frac{\partial}{\partial \xi_3}(F_3 h_1 h_2) \Big]$$

The net outgoing flux per unit volume, defined as the divergence of the field $\vec{F}$ is

$$
\begin{aligned}
\text{div } \vec{F} &\doteq \frac{d\xi_1 \, d\xi_2 \, d\xi_3}{h_1 h_2 h_3 \, d\xi_1 \, d\xi_2 \, d\xi_3} \Big[\frac{\partial}{\partial \xi_1}(F_1 h_2 h_3) + \frac{\partial}{\partial \xi_2}(F_2 h_3 h_1) + \frac{\partial}{\partial \xi_3}(F_3 h_1 h_2) \Big] \\
&= \frac{1}{h_1 h_2 h_3} \Big[\frac{\partial}{\partial \xi_1}(F_1 h_2 h_3) + \frac{\partial}{\partial \xi_2}(F_2 h_3 h_1) + \frac{\partial}{\partial \xi_3}(F_3 h_1 h_2) \Big] \\
&= \frac{1}{h_1 h_2 h_3} \sum_{i=1}^{3} \frac{\partial}{\partial \xi_i} \Big(\frac{F_i h_1 h_2 h_3}{h_i} \Big)
\end{aligned}
$$

5.1 Special cases

(i) In spherical coordinates, divergence of the vector $\vec{F} = F_r \, \hat{e_r} + F_\theta \, \hat{e_\theta} + F_\phi \, \hat{e_\phi}$ is

$$\text{div } \vec{F} = \frac{1}{r^2 \sin\theta} \Big[\frac{\partial}{\partial r}(F_r r^2 \sin\theta) + \frac{\partial}{\partial \theta}(F_\theta r \sin\theta) + \frac{\partial}{\partial \phi}(F_\phi r) \Big]$$

(ii) In cylindrical coordinates, divergence of the vector $\vec{F} = F_\rho \, \hat{e_\rho} + F_\phi \, \hat{e_\phi} + F_z \, \hat{e_z}$ is

$$\text{div } \vec{F} = \frac{1}{\rho} \Big[\frac{\partial}{\partial \rho}(F_\rho \rho) + \frac{\partial}{\partial \phi}(F_\phi) + \frac{\partial}{\partial z}(F_z \rho) \Big]$$

(iii) In Cartesian coordinates, divergence of the vector $\vec{F} = F_x \hat{i} + F_y \hat{j} + F_z \hat{k}$ is

$$\text{div } \vec{F} = \frac{\partial F_x}{\partial x} + \frac{\partial F_y}{\partial y} + \frac{\partial F_z}{\partial z}$$

6. Curl of vector field

To derive the expression for the curl (also called the rotation) in the orthogonal curvilinear coordinates, let us consider an area element of sides $h_2\ \mathrm{d}\xi_2$ and $h_3\ \mathrm{d}\xi_3$ which is perpendicular to ξ_1-axis (Figure 5). If the vector field $\vec{F}$ is at the mid-point P of the area-element, its components along the axes ξ_2 and ξ_3 are F_2 and F_3, respectively. Since the dimensions of the area element are small, the field at the mid-point of a line may be considered as the average of the filed over the line. Hence, the line integral at P, parallel to ξ_2-axis, is

$$= F_2 h_2\ \mathrm{d}\xi_2$$

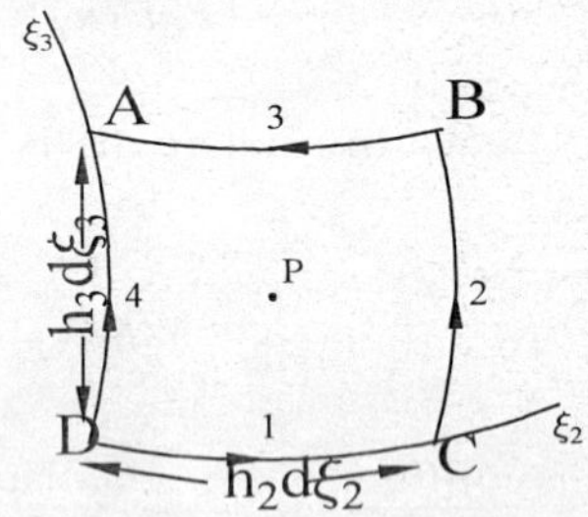

Figure 5: Curl of a vector field.

The rate of change of this line integral along the ξ_3-axis, is

$$= \frac{1}{h_3}\frac{\partial}{\partial\xi_3}(F_2\ h_2\ \mathrm{d}\xi_2) = \frac{\mathrm{d}\xi_2}{h_3}\frac{\partial}{\partial\xi_3}(F_2\ h_2)$$

Thus, the line integral along DC is

$$= F_2 h_2\ \mathrm{d}\xi_2 - \frac{\mathrm{d}\xi_2}{h_3}\ \frac{\partial}{\partial\xi_3}(F_2\ h_2)\left(\frac{1}{2}\ h_3\ \mathrm{d}\xi_3\right)$$

$$= F_2 h_2\ \mathrm{d}\xi_2 - \frac{1}{2}\ \mathrm{d}\xi_2\ \mathrm{d}\xi_3\ \frac{\partial}{\partial\xi_3}(F_2\ h_2)$$

and along AB is

$$= F_2 h_2\ \mathrm{d}\xi_2 + \frac{\mathrm{d}\xi_2}{h_3}\ \frac{\partial}{\partial\xi_3}(F_2 h_2)\left(\frac{1}{2}\ h_3\ \mathrm{d}\xi_3\right)$$

$$= F_2 h_2\ \mathrm{d}\xi_2 + \frac{1}{2}\ \mathrm{d}\xi_2\ \mathrm{d}\xi_3\ \frac{\partial}{\partial\xi_3}(F_2 h_2)$$

Now, the line integral at P, parallel to ξ_3-axis is

$$= F_3 h_3\ \mathrm{d}\xi_3$$

The rate of change of this line integral along the ξ_2-axis, is

$$= \frac{1}{h_2}\frac{\partial}{\partial \xi_2}(F_3 h_3 \ d\xi_3) = \frac{d\xi_3}{h_2}\frac{\partial}{\partial \xi_2}(F_3 h_3)$$

Thus, the line integral along DA is

$$= F_3 h_3 \ d\xi_3 - \frac{d\xi_3}{h_2} \ \frac{\partial}{\partial \xi_2}(F_3 h_3)\left(\frac{1}{2} h_2 \ d\xi_2\right)$$

$$= F_3 h_3 \ d\xi_3 - \frac{1}{2} \ d\xi_2 \ d\xi_3 \frac{\partial}{\partial \xi_2}(F_3 h_3)$$

and along CB is

$$= F_3 h_3 \ d\xi_3 + \frac{d\xi_3}{h_2} \ \frac{\partial}{\partial \xi_2}(F_3 h_3)\left(\frac{1}{2} \ h_2 \ d\xi_2\right)$$

$$= F_3 h_3 \ d\xi_3 + \frac{1}{2} \ d\xi_2 \ d\xi_3 \frac{\partial}{\partial \xi_2}(F_3 h_3)$$

Hence, the net circulation integral along the path DCBAD (*i.e.*, in the anticlockwise direction) is

$$= \left[F_2 h_2 \ d\xi_2 - \frac{1}{2} \ d\xi_2 \ d\xi_3 \frac{\partial}{\partial \xi_3}(F_2 h_2)\right] + \left[F_3 h_3 \ d\xi_3 + \frac{1}{2} \ d\xi_2 \ d\xi_3 \frac{\partial}{\partial \xi_2}(F_3 h_3)\right]$$

$$-\left[F_2 h_2 \ d\xi_2 + \frac{1}{2} \ d\xi_2 \ d\xi_3 \frac{\partial}{\partial \xi_3}(F_2 \ h_2)\right] - \left[F_3 h_3 \ d\xi_3 - \frac{1}{2} \ d\xi_2 \ d\xi_3 \frac{\partial}{\partial \xi_2}(F_3 h_3)\right]$$

$$= d\xi_2 \ d\xi_3 \left[\frac{\partial}{\partial \xi_2}(F_3 h_3) - \frac{\partial}{\partial \xi_3}(F_2 h_2)\right]$$

This circulation integral per unit surface area is defined as ξ_1-component of the curl of $\vec{F}$. Therefore,

$$(\text{curl } \vec{F})_{\xi_1} = \frac{d\xi_2 \ d\xi_3}{h_2 \ d\xi_2 \ h_3 \ d\xi_3}\left[\frac{\partial}{\partial \xi_2}(F_3 h_3) - \frac{\partial}{\partial \xi_3}(F_2 h_2)\right]$$

$$= \frac{1}{h_2 h_3}\left[\frac{\partial}{\partial \xi_2}(F_3 h_3) - \frac{\partial}{\partial \xi_3}(F_2 h_2)\right]$$

Similarly, ξ_2- and ξ_3-components of the curl of $\vec{F}$ can be obtained as

$$(\text{curl } \vec{F})_{\xi_2} = \frac{1}{h_3 \ h_1}\left[\frac{\partial}{\partial \xi_3}(F_1 h_1) - \frac{\partial}{\partial \xi_1}(F_3 h_3)\right]$$

and

$$(\text{curl } \vec{F})_{\xi_3} = \frac{1}{h_1 h_2}\left[\frac{\partial}{\partial \xi_1}(F_2 h_2) - \frac{\partial}{\partial \xi_2}(F_1 h_1)\right]$$

Therefore, curl of $\vec{F}$ is

$$\text{curl } \vec{F} = \frac{\hat{\xi_1}}{h_2h_3}\left[\frac{\partial}{\partial\xi_2}(F_3h_3) - \frac{\partial}{\partial\xi_3}(F_2h_2)\right] + \frac{\hat{\xi_2}}{h_3h_1}\left[\frac{\partial}{\partial\xi_3}(F_1h_1) - \frac{\partial}{\partial\xi_1}(F_3h_3)\right]$$

$$+\frac{\hat{\xi_3}}{h_1h_2}\left[\frac{\partial}{\partial\xi_1}(F_2h_2) - \frac{\partial}{\partial\xi_2}(F_1h_1)\right]$$

$$= \frac{1}{h_1h_2h_3}\begin{bmatrix} h_1\hat{\xi_1} & h_2\hat{\xi_2} & h_3\hat{\xi_3} \\ \frac{\partial}{\partial\xi_1} & \frac{\partial}{\partial\xi_2} & \frac{\partial}{\partial\xi_3} \\ h_1F_1 & h_2F_2 & h_3F_3 \end{bmatrix}$$

6.1 Special cases

(i) In spherical coordinates, curl of the vector $\vec{F} = F_r\hat{e_r} + F_\theta\hat{e_\theta} + F_\phi\hat{e_\phi}$ is

$$\text{curl } \vec{F} = \nabla \times \vec{F} = \frac{1}{r^2 \sin\theta}\begin{bmatrix} \hat{e_r} & r\,\hat{e_\theta} & r\,\sin\theta\,\hat{e_\phi} \\ \frac{\partial}{\partial r} & \frac{\partial}{\partial\theta} & \frac{\partial}{\partial\phi} \\ F_r & r\,F_\theta & r\,\sin\theta\,F_\phi \end{bmatrix}$$

(ii) In cylindrical coordinates, curl of the vector $\vec{F} = F_\rho\,\hat{e_\rho} + F_\phi\,\hat{e_\phi} + F_z\,\hat{e_z}$ is

$$\text{curl } \vec{F} = \nabla \times \vec{F} = \frac{1}{\rho}\begin{bmatrix} \hat{e_\rho} & \rho\hat{e_\phi} & \hat{e_z} \\ \frac{\partial}{\partial\rho} & \frac{\partial}{\partial\phi} & \frac{\partial}{\partial z} \\ F_\rho & \rho\,F_\phi & F_z \end{bmatrix}$$

(iii) In Cartesian coordinates, curl of the vector $\vec{F} = F_x\hat{\imath} + F_y\hat{\jmath} + F_z\hat{k}$ is

$$\text{curl } \vec{F} = \nabla \times \vec{F} = \begin{bmatrix} \hat{\imath} & \hat{\jmath} & \hat{k} \\ \frac{\partial}{\partial x} & \frac{\partial}{\partial y} & \frac{\partial}{\partial z} \\ F_x & F_y & F_z \end{bmatrix}$$

7. Laplacian operator

The Laplacian operator ∇^2 is defined as

$$\nabla^2\Psi = \nabla.\nabla\Psi = \text{div}(\text{grad }\Psi)$$

The expression for the Laplacian operator in the orthogonal curvilinear coordinates can be derived in the following manner. We know that the gradient of a scalar Ψ is

$$\text{grad } \Psi = \nabla\Psi = \frac{\hat{\xi}_1}{h_1}\frac{\partial\Psi}{\partial\xi_1} + \frac{\hat{\xi}_2}{h_2}\frac{\partial\Psi}{\partial\xi_2} + \frac{\hat{\xi}_3}{h_3}\frac{\partial\Psi}{\partial\xi_3}$$

The divergence of a vector field $\vec{F}$ is

$$\text{div } \vec{F} = \frac{1}{h_1h_2h_3}\Big[\frac{\partial}{\partial\xi_1}(F_1h_2h_3) + \frac{\partial}{\partial\xi_2}(F_2h_3h_1) + \frac{\partial}{\partial\xi_3}(F_3h_1h_2)\Big]$$

Hence, the divergence of the vector field grad Ψ is obtained by replacing $\vec{F}$ by grad Ψ

$$\text{div(grad } \Psi) = \frac{1}{h_1h_2h_3}\Big[\frac{\partial}{\partial\xi_1}\Big(\frac{1}{h_1}\frac{\partial\Psi}{\partial\xi_1}h_2h_3\Big) + \frac{\partial}{\partial\xi_2}\Big(\frac{1}{h_2}\frac{\partial\Psi}{\partial\xi_2}h_3h_1\Big) + \frac{\partial}{\partial\xi_3}\Big(\frac{1}{h_3}\frac{\partial\Psi}{\partial\xi_3}h_1h_2\Big)\Big]$$

$$= \frac{1}{h_1h_2h_3}\Big[\frac{\partial}{\partial\xi_1}\Big(\frac{h_2h_3}{h_1}\ \frac{\partial\Psi}{\partial\xi_1}\Big) + \frac{\partial}{\partial\xi_2}\Big(\frac{h_3\ h_1}{h_2}\ \frac{\partial\Psi}{\partial\xi_2}\Big) + \frac{\partial}{\partial\xi_3}\Big(\frac{h_1h_2}{h_3}\ \frac{\partial\Psi}{\partial\xi_3}\Big)\Big]$$

Thus,

$$\nabla^2\Psi = \frac{1}{h_1h_2h_3}\sum_{i=1}^{3}\frac{\partial}{\partial\xi_i}\Big(\frac{h_1h_2h_3}{h_i^2}\frac{\partial\Psi}{\partial\xi_i}\Big)$$

7.1 Special cases

(i) In spherical coordinates

$$\nabla^2\Psi = \frac{1}{r^2\sin\theta}\Big[\frac{\partial}{\partial r}\Big(r^2\sin\theta\ \frac{\partial\Psi}{\partial r}\Big) + \frac{\partial}{\partial\theta}\Big(\sin\theta\ \frac{\partial\Psi}{\partial\theta}\Big) + \frac{\partial}{\partial\phi}\Big(\frac{1}{\sin\theta}\ \frac{\partial\Psi}{\partial\phi}\Big)\Big]$$

$$= \frac{1}{r^2\sin\theta}\Big[\sin\theta\ \frac{\partial}{\partial r}\Big(r^2\ \frac{\partial\Psi}{\partial r}\Big) + \frac{\partial}{\partial\theta}\Big(\sin\theta\ \frac{\partial\Psi}{\partial\theta}\Big) + \frac{1}{\sin\theta}\ \frac{\partial^2\Psi}{\partial\phi^2}\Big]$$

$$= \frac{1}{r^2}\Big[\frac{\partial}{\partial r}\Big(r^2\ \frac{\partial\Psi}{\partial r}\Big) + \frac{1}{\sin\theta}\frac{\partial}{\partial\theta}\Big(\sin\theta\ \frac{\partial\Psi}{\partial\theta}\Big) + \frac{1}{\sin^2\theta}\ \frac{\partial^2\Psi}{\partial\phi^2}\Big]$$

(ii) In cylindrical coordinates

$$\nabla^2\Psi = \frac{1}{\rho}\Big[\frac{\partial}{\partial\rho}\Big(\rho\ \frac{\partial\Psi}{\partial\rho}\Big) + \frac{\partial}{\partial\phi}\Big(\frac{1}{\rho}\ \frac{\partial\Psi}{\partial\phi}\Big) + \frac{\partial}{\partial z}\Big(\rho\ \frac{\partial\Psi}{\partial z}\Big)\Big]$$

$$= \frac{1}{\rho}\frac{\partial}{\partial\rho}\Big(\rho\ \frac{\partial\Psi}{\partial\rho}\Big) + \frac{1}{\rho^2}\frac{\partial^2\Psi}{\partial\phi^2} + \frac{\partial^2\Psi}{\partial z^2}$$

(iii) In Cartesian coordinates

$$\nabla^2\Psi = \frac{\partial^2\Psi}{\partial x^2} + \frac{\partial^2\Psi}{\partial y^2} + \frac{\partial^2\Psi}{\partial z^2}$$

Example 3: Show that curl(grad Ψ) = 0.

Solution: We know

$$\text{grad } \Psi = \nabla\Psi = \frac{\hat{\xi}_1}{h_1}\frac{\partial \Psi}{\partial \xi_1} + \frac{\hat{\xi}_2}{h_2}\frac{\partial \Psi}{\partial \xi_2} + \frac{\hat{\xi}_3}{h_3}\frac{\partial \Psi}{\partial \xi_3}$$

and

$$\text{curl } \vec{F} = \frac{\hat{\xi}_1}{h_2 h_3}\Big[\frac{\partial}{\partial \xi_2}(F_3 h_3) - \frac{\partial}{\partial \xi_3}(F_2 h_2)\Big] + \frac{\hat{\xi}_2}{h_3 h_1}\Big[\frac{\partial}{\partial \xi_3}(F_1 h_1) - \frac{\partial}{\partial \xi_1}(F_3 h_3)\Big] + \frac{\hat{\xi}_3}{h_1 h_2}\Big[\frac{\partial}{\partial \xi_1}(F_2 h_2) - \frac{\partial}{\partial \xi_2}(F_1 h_1)\Big] \tag{1.43}$$

where

$$\vec{F} = F_1\,\hat{\xi}_1 + F_2\,\hat{\xi}_2 + F_3\,\hat{\xi}_3$$

On substituting grad Ψ instead of $\vec{F}$ in equation (1.43), we get

$$\begin{aligned}\text{curl(grad } \Psi) &= \frac{\hat{\xi}_1}{h_2 h_3}\Big[\frac{\partial}{\partial \xi_2}\Big(h_3\frac{1}{h_3}\frac{\partial \Psi}{\partial \xi_3}\Big) - \frac{\partial}{\partial \xi_3}\Big(h_2\frac{1}{h_2}\frac{\partial \Psi}{\partial \xi_2}\Big)\Big] \\ &\quad + \frac{\hat{\xi}_2}{h_3 h_1}\Big[\frac{\partial}{\partial \xi_3}\Big(h_1\frac{1}{h_1}\frac{\partial \Psi}{\partial \xi_1}\Big) - \frac{\partial}{\partial \xi_1}\Big(h_3\,\frac{1}{h_3}\,\frac{\partial \Psi}{\partial \xi_3}\Big)\Big] \\ &\quad + \frac{\hat{\xi}_3}{h_1 h_2}\Big[\frac{\partial}{\partial \xi_1}\Big(h_2\,\frac{1}{h_2}\,\frac{\partial \Psi}{\partial \xi_2}\Big) - \frac{\partial}{\partial \xi_2}\Big(h_1\,\frac{1}{h_1}\,\frac{\partial \Psi}{\partial \xi_1}\Big)\Big] \\ &= \frac{\hat{\xi}_1}{h_2 h_3}\Big[\frac{\partial^2 \Psi}{\partial \xi_2\,\partial \xi_3} - \frac{\partial^2 \Psi}{\partial \xi_3\,\partial \xi_2}\Big] + \frac{\hat{\xi}_2}{h_3\,h_1}\Big[\frac{\partial^2 \Psi}{\partial \xi_3\,\partial \xi_1} - \frac{\partial^2 \Psi}{\partial \xi_1\,\partial \xi_3}\Big] \\ &\quad + \frac{\hat{\xi}_3}{h_1 h_2}\Big[\frac{\partial^2 \Psi}{\partial \xi_1\,\partial \xi_2} - \frac{\partial^2 \Psi}{\partial \xi_2\,\partial \xi_1}\Big] = 0\end{aligned}$$

Example 4: Show that div(curl $\vec{F}$) = 0.

Solution: We know

$$\text{div } \vec{F} = \frac{1}{h_1 h_2 h_3}\Big[\frac{\partial}{\partial \xi_1}(F_1 h_2 h_3) + \frac{\partial}{\partial \xi_2}(F_2 h_3 h_1) + \frac{\partial}{\partial \xi_3}(F_3 h_1 h_2)\Big] \tag{1.44}$$

where

$$\vec{F} = F_1\,\hat{\xi}_1 + F_2\,\hat{\xi}_2 + F_3\,\hat{\xi}_3$$

Now,

$$\text{curl } \vec{A} = \frac{\hat{\xi}_1}{h_2 h_3}\Big[\frac{\partial}{\partial \xi_2}(A_3 h_3) - \frac{\partial}{\partial \xi_3}(A_2 h_2)\Big] + \frac{\hat{\xi}_2}{h_3 h_1}\Big[\frac{\partial}{\partial \xi_3}(A_1 h_1) - \frac{\partial}{\partial \xi_1}(A_3 h_3)\Big] + \frac{\hat{\xi}_3}{h_1 h_2}\Big[\frac{\partial}{\partial \xi_1}(A_2 h_2) - \frac{\partial}{\partial \xi_2}(A_1 h_1)\Big]$$

On substituting curl $\vec{A}$ instead of $\vec{F}$ in equation (1.44), we get

$$\text{div}(\text{curl } \vec{A}) = \frac{1}{h_1h_2h_3}\Big[\frac{\partial}{\partial\xi_1}\Big(h_2h_3\frac{1}{h_2h_3}\Big\{\frac{\partial}{\partial\xi_2}(A_3h_3) - \frac{\partial}{\partial\xi_3}(A_2h_2)\Big\}\Big)$$
$$+\frac{\partial}{\partial\xi_2}\Big(h_3h_1\frac{1}{h_3h_1}\Big\{\frac{\partial}{\partial\xi_3}(A_1h_1) - \frac{\partial}{\partial\xi_1}(A_3h_3)\Big\}\Big)$$
$$+\frac{\partial}{\partial\xi_3}\Big(h_1h_2\frac{1}{h_1h_2}\Big\{\frac{\partial}{\partial\xi_1}(A_2h_2) - \frac{\partial}{\partial\xi_1}(A_1h_1)\Big\}\Big)\Big]$$
$$= \frac{1}{h_1h_2h_3}\Big[\Big\{\frac{\partial^2}{\partial\xi_1\partial\xi_2}(A_3h_3) - \frac{\partial^2}{\partial\xi_1\partial\xi_3}(A_2h_2)\Big\} + \Big\{\frac{\partial^2}{\partial\xi_2\partial\xi_3}(A_1h_1) - \frac{\partial^2}{\partial\xi_2\partial\xi_1}(A_3h_3)\Big\}$$
$$+\Big\{\frac{\partial^2}{\partial\xi_3\partial\xi_1}(A_2h_2) - \frac{\partial^2}{\partial\xi_3\partial\xi_2}(A_1h_1)\Big\}\Big] = 0$$

Review Questions

1. Derive expressions for scale factors in orthogonal curvilinear coordinates. Find the scale factors for: (i) spherical coordinates, (ii) cylindrical coordinates.
2. Derive the expressions for: (i) gradient of a scalar field, (ii) divergence of a vector field, (iii) curl of a vector field and (iv) Laplacian operator in orthogonal curvilinear coordinates.
3. Prove the relations

   (i) curl(grad Ψ) = 0

   (ii) div(curl $\vec{F}$) = 0
4. Write notes on the following:

   (i) Scale factors in orthogonal curvilinear coordinates

   (ii) Gradient of a scalar field in orthogonal curvilinear coordinates

   (iii) Divergence of a vector field in orthogonal curvilinear coordinates

   (iv) Curl of a vector field in orthogonal curvilinear coordinates

   (v) Laplacian operator in orthogonal curvilinear coordinates

Multiple-Choice Questions

1. In which coordinate system the 1 is one of the scale factors?

   A. Cylindrical system only B. Spherical system only

   C. Cartesian system only D. In all the three systems

   Ans. D

2. In which coordinate system all the three scale factors are equal?
   A. Cylindrical system only
   B. Spherical system only
   C. Cartesian system only
   D. In all the three systems

   Ans. C

3. In which coordinate system only two scale factors are equal?
   A. Cylindrical system only
   B. Spherical system only
   C. Cartesian system only
   D. In all the three systems

   Ans. A

4. In which coordinate system, the scale factors do not depend on coordinates?
   A. Cylindrical system only
   B. Spherical system only
   C. Cartesian system only
   D. In all the three systems

   Ans. C

5. In which coordinate system, two scale factors depend on coordinates?
   A. Cylindrical system only
   B. Spherical system only
   C. Cartesian system only
   D. In all the three systems

   Ans. B

6. In which coordinate system, only one scale factor depends on coordinates?
   A. Cylindrical system only
   B. Spherical system only
   C. Cartesian system only
   D. In all the three systems

   Ans. A

Chapter 2

Vectors

The concept of vectors is very helpful in treatment of many physical phenomena in science and technology. A vector represents a physical quantity which has both the direction as well as magnitude. A physical quantity which has only magnitude is known as a scalar. Vector quantities follow some mathematical operations, such as addition, subtraction and multiplication. There is no direct operation for division of one vector by other. Examples of vector quantities are displacement, velocity, acceleration, momentum, force, angular momentum, torque, electric field, magnetic field, electric dipole moment and so on. A vector quantity in this book is denoted by putting an arrow on it. For example, $\vec{r}$ denotes a vector quantity. Magnitude of this vector quantity is denoted by r or by $|\vec{r}|$. A unit vector is that whose magnitude is 1. A unit vector parallel to $\vec{r}$ is denoted by $\hat{r}$, which we read as 'r hat' or 'r caret'. Thus, we can write $\vec{r} = r\,\hat{r} = \hat{r}\,r$.

As mentioned earlier, a vector quantity has direction as well as magnitude. These features are independent of coordinate system used. In most of the discussion in this chapter, we have considered Cartesian coordinates, so that a vector $\vec{A}$ is expressed as $\vec{A} = A_x\,\hat{\imath} + A_y\,\hat{\jmath} + A_z\,\hat{k}$, where A_x, A_y, A_z are respectively known as the x, y, z components of $\vec{A}$. Here, $\hat{\imath}$, $\hat{\jmath}$, $\hat{k}$ denote unit vectors along positive directions of x, y, z axes respectively. The magnitude of $\vec{A}$ is $A = |\vec{A}| = \sqrt{A_x^2 + A_y^2 + A_z^2}$.

1. Addition of vectors

Suppose, $\vec{A}$ and $\vec{B}$ are two vectors. Their addition is expressed as

$$\vec{R} = \vec{A} + \vec{B}$$

For $\vec{A} = A_x\,\hat{\imath} + A_y\,\hat{\jmath} + A_z\,\hat{k}$ and $\vec{B} = B_x\,\hat{\imath} + B_y\,\hat{\jmath} + B_z\,\hat{k}$, we have

$$\vec{R} = \vec{A} + \vec{B} = (A_x + B_x)\,\hat{\imath} + (A_y + B_y)\,\hat{\jmath} + (A_z + B_z)\,\hat{k}$$

When vectors represent physical quantities, two vectors can be added only when they have the same dimensions. For example, a vector representing velocity cannot be added to another vector representing acceleration. (The same is true for subtraction of one vector from other also.)

(i) Commutative law: $\vec{A}+\vec{B}=\vec{B}+\vec{A}$

(ii) Associative law: $\vec{A}+(\vec{B}+\vec{C})=(\vec{A}+\vec{B})+\vec{C}$

(iii) Distribution law: $k(\vec{A}+\vec{B}+\vec{C})=k\vec{A}+k\vec{B}+k\vec{C}$

2. Subtraction of vectors

Suppose, $\vec{A}$ and $\vec{B}$ are two vectors. The subtraction of $\vec{B}$ from $\vec{A}$ is expressed as

$$\vec{R}=\vec{A}-\vec{B}$$

For $\vec{A}=A_x\,\hat{\imath}+A_y\,\hat{\jmath}+A_z\,\hat{k}$ and $\vec{B}=B_x\,\hat{\imath}+B_y\,\hat{\jmath}+B_z\,\hat{k}$, we have

$$\vec{R}=\vec{A}-\vec{B}=(A_x-B_x)\,\hat{\imath}+(A_y-B_y)\,\hat{\jmath}+(A_z-B_z)\,\hat{k}$$

It is obvious that $(\vec{A}-\vec{B})\neq(\vec{B}-\vec{A})$, showing that subtraction of one vector from other is not commutative.

3. Product of vectors

We have seen that addition and subtraction of vectors are vectors. But, in case of product of two vectors, the resultant may be either a scalar quantity or a vector quantity. Accordingly, we have two types of multiplication of two vectors: (i) Scalar product, and (ii) Vector product. In case of scalar product of two vectors, the resultant is a scalar quantity whereas in case of vector product, the resultant is a vector quantity. For multiplication of two vectors, the vectors may have different dimensions.

3.1 Scalar product

Scalar product of two vectors $\vec{A}$ and $\vec{B}$ is expressed as

$$\vec{A}\,.\,\vec{B}=AB\cos\theta$$

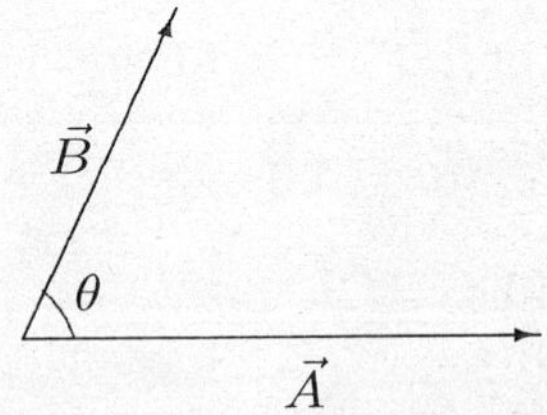

Figure 1: Two vectors $\vec{A}$ and $\vec{B}$ inclined at an angle θ.

where A and B are the magnitudes of $\vec{A}$ and $\vec{B}$, respectively, and θ the angle between two vectors. The $\vec{A}\,.\,\vec{B}$ is read as $\vec{A}$ dot $\vec{B}$. As we use a dot symbol between the two vectors, the product is also known as dot product. The resultant of this product is a scalar quantity, and therefore it is also known as scalar product. As we have to take cosine of angle between two vectors, it does not matter whether the angle θ is measured in clockwise direction or in anticlockwise direction. Thus, we have

$$\vec{A} \,.\, \vec{B} = \vec{B} \,.\, \vec{A}$$

It shows that scalar multiplication of two vectors is commutative. That is, the positions of two vectors can be interchanged in scalar product. Now, we have $\vec{A}.\vec{A} = A^2$. From scalar product, for unit vectors $\hat{\imath}$, $\hat{\jmath}$, $\hat{k}$ along x, y, z axes, respectively, we get

$$\hat{\imath} \,.\, \hat{\imath} = 1 = \hat{\jmath} \,.\, \hat{\jmath} = \hat{k} \,.\, \hat{k} \qquad \text{and} \qquad \hat{\imath} \,.\, \hat{\jmath} = 0 = \hat{\jmath} \,.\, \hat{k} = \hat{k} \,.\, \hat{\imath}$$

For $\vec{A} = A_x\hat{\imath} + A_y\hat{\jmath} + A_z\hat{k}$ and $\vec{B} = B_x\hat{\imath} + B_y\hat{\jmath} + B_z\hat{k}$, we have

$$\vec{A} \,.\, \vec{B} = A_xB_x + A_yB_y + A_zB_z$$

3.2 Vector product

Vector product of two vectors $\vec{A}$ and $\vec{B}$ is expressed as

$$\vec{A} \times \vec{B} = AB \sin\theta \; \hat{n}$$

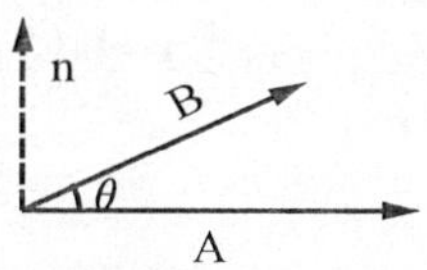

where A and B are magnitudes of $\vec{A}$ and $\vec{B}$, respectively, and θ the angle between two vectors and $\hat{n}$ is a unit vector perpendicular to the plane containing both $\vec{A}$ and $\vec{B}$. The direction of $\hat{n}$ is decided with the help of right hand rule. When fingers of right hand are stretched from $\vec{A}$ towards $\vec{B}$, the direction of thumb is the direction of $\hat{n}$. The $\vec{A} \times \vec{B}$ is read as $\vec{A}$ cross $\vec{B}$. As we use a cross symbol between two vectors, the product is also known as cross product. The resultant of this product is vector quantity and therefore, it is also known as vector product. When we calculate $\vec{B} \times \vec{A}$, the direction of $\hat{n}$ is reversed. Thus, we have

$$\vec{A} \times \vec{B} = -\vec{B} \times \vec{A}$$

From vector product, for unit vectors $\hat{\imath}$, $\hat{\jmath}$, $\hat{k}$ along x, y, z axes, respectively, we get

$$\hat{\imath} \times \hat{\imath} = 0 = \hat{\jmath} \times \hat{\jmath} = \hat{k} \times \hat{k}$$

and

$$\hat{\imath} \times \hat{\jmath} = \hat{k} \qquad \hat{\jmath} \times \hat{k} = \hat{\imath} \qquad \hat{k} \times \hat{\imath} = \hat{\jmath}$$

For $\vec{A} = A_x \,\hat{\imath} + A_y \,\hat{\jmath} + A_z \,\hat{k}$ and $\vec{B} = B_x \,\hat{\imath} + B_y \,\hat{\jmath} + B_z \,\hat{k}$, we have

$$\vec{A} \times \vec{B} = (A_x \,\hat{\imath} + A_y \,\hat{\jmath} + A_z \,\hat{k}) \times (B_x \,\hat{\imath} + B_y \,\hat{\jmath} + B_z \,\hat{k})$$

$$= \hat{\imath}\ (A_yB_z - A_zB_y) - \hat{\jmath}\ (A_xB_z - A_zB_x) + \hat{k}\ (A_xB_y - A_yB_x)$$

It can be expressed in the determinant form as

$$\vec{A} \times \vec{B} = \begin{vmatrix} \hat{\imath} & \hat{\jmath} & \hat{k} \\ A_x & A_y & A_z \\ B_x & B_y & B_z \end{vmatrix}$$

Now, we have $\vec{A} \times \vec{A} = 0$.

3.3 Scalar triple product

Three vectors can be multiplied in two ways. One of them is a scalar triple product whose resultant is scalar quantity. Scalar triple product of three vectors $\vec{A}$, $\vec{B}$ and $\vec{C}$, represented by $[\vec{A}\ \vec{B}\ \vec{C}]$, is expressed as $\vec{A}\ .\ (\vec{B} \times \vec{C})$. For $\vec{A} = A_x\ \hat{\imath} + A_y\ \hat{\jmath} + A_z\ \hat{k}$, $\vec{B} = B_x\ \hat{\imath} + B_y\ \hat{\jmath} + B_z\ \hat{k}$ and $\vec{C} = C_x\ \hat{\imath} + C_y\ \hat{\jmath} + C_z\ \hat{k}$, $[\vec{A}\ \vec{B}\ \vec{C}]$ can be expressed by the following determinant.

$$[\vec{A}\ \vec{B}\ \vec{C}] = \begin{vmatrix} A_x & A_y & A_z \\ B_x & B_y & B_z \\ C_x & C_y & C_z \end{vmatrix}$$

$$= A_x(B_yC_z - B_zC_y) - A_y(B_xC_z - B_zC_x) + A_z(B_xC_y - B_zC_y)$$

3.4 Vector triple product

Three vectors can be multiplied such that the product is a vector quantity. The vector triple product of three vectors $\vec{A}$, $\vec{B}$ and $\vec{C}$ is expressed as

$$\vec{A} \times (\vec{B} \times \vec{C}) = \vec{B}\ (\vec{A}\ .\ \vec{C}) - \vec{C}\ (\vec{A}\ .\ \vec{B})$$

4. Vector derivative

Suppose, $\vec{A} = A_x\ \hat{\imath} + A_y\ \hat{\jmath} + A_z\ \hat{k}$. Then we have

$$\frac{\mathrm{d}\vec{A}}{\mathrm{d}t} = \frac{A_x}{\mathrm{d}t}\ \hat{\imath} + \frac{A_y}{\mathrm{d}t}\ \hat{\jmath} + \frac{A_z}{\mathrm{d}t}\ \hat{k}$$

4.1 Velocity

The velocity $\vec{v}$ of a body is defined as the rate of change of displacement $\vec{r}$ of body with respect to time. That is

$$\vec{v} = \frac{\mathrm{d}\vec{r}}{\mathrm{d}t}$$

If we have $\vec{r} = x\ \hat{\imath} + y\ \hat{\jmath} + z\ \hat{k}$ and $\vec{v} = v_x\ \hat{\imath} + v_y\ \hat{\jmath} + v_z\ \hat{k}$, we get

$$v_x\ \hat{\imath} + v_y\ \hat{\jmath} + v_z\ \hat{k} = \frac{\mathrm{d}}{\mathrm{d}t}(x\ \hat{\imath} + y\ \hat{\jmath} + z\ \hat{k})$$

and therefore, on comparing the coefficients of $\hat{\imath}$, $\hat{\jmath}$, $\hat{k}$ on the two sides of the relation, we have

$$v_x = \frac{\mathrm{d}x}{\mathrm{d}t} \qquad v_y = \frac{\mathrm{d}y}{\mathrm{d}t} \qquad v_z = \frac{\mathrm{d}z}{\mathrm{d}t}$$

Here, v_x, v_y, v_z are respectively known as the x, y, z components of the velocity.

4.2 Acceleration

The acceleration $\vec{a}$ of a body is defined as the rate of change of velocity $\vec{v}$ of body with respect to time. That is

$$\vec{a} = \frac{\mathrm{d}\vec{v}}{\mathrm{d}t}$$

If we have $\vec{v} = v_x\ \hat{\imath} + v_y\ \hat{\jmath} + v_z\ \hat{k}$ and $\vec{a} = a_x\ \hat{\imath} + a_y\ \hat{\jmath} + a_z\ \hat{k}$, we have

$$a_x\ \hat{\imath} + a_y\ \hat{\jmath} + a_z\ \hat{k} = \frac{\mathrm{d}}{\mathrm{d}t}(v_x\ \hat{\imath} + v_y\ \hat{\jmath} + v_z\ \hat{k})$$

and therefore, on comparing the coefficients of $\hat{\imath}$, $\hat{\jmath}$, $\hat{k}$ on the two sides of this relation, we have

$$a_x = \frac{\mathrm{d}v_x}{\mathrm{d}t} \qquad a_y = \frac{\mathrm{d}v_y}{\mathrm{d}t} \qquad a_z = \frac{\mathrm{d}v_z}{\mathrm{d}t}$$

Here, a_x, a_y, a_z are respectively known as the x, y, z components of the acceleration.

4.3 Differentiation

$$\frac{\mathrm{d}(\vec{A}\ \pm\ \vec{B})}{\mathrm{d}x} = \frac{\mathrm{d}\vec{A}}{\mathrm{d}x} \pm \frac{\mathrm{d}\vec{B}}{\partial x}$$

$$\frac{\mathrm{d}(\vec{A}\ .\ \vec{B})}{\mathrm{d}x} = \vec{A}\ .\ \frac{\mathrm{d}\vec{B}}{\mathrm{d}x} + \frac{\mathrm{d}\vec{A}}{\partial x}\ .\ \vec{B}$$

$$\frac{\mathrm{d}(\vec{A}\ \times\ \vec{B})}{\mathrm{d}x} = \vec{A}\ \times\ \frac{\mathrm{d}\vec{B}}{\mathrm{d}x} + \frac{\mathrm{d}\vec{A}}{\partial x}\ \times\ \vec{B}$$

4.4 Partial differentiation

$$\frac{\partial(\vec{A} \pm \vec{B})}{\partial x} = \frac{\partial\vec{A}}{\partial x} \pm \frac{\partial\vec{B}}{\partial x}$$

$$\frac{\partial(\vec{A}\ .\ \vec{B})}{\partial x} = \vec{A}\ .\ \frac{\partial\vec{B}}{\partial x} + \frac{\partial\vec{A}}{\partial x}\ .\ \vec{B}$$

$$\frac{\partial(\vec{A}\ \times\ \vec{B})}{\partial x} = \vec{A}\ \times\ \frac{\partial\vec{B}}{\partial x} + \frac{\partial\vec{A}}{\partial x}\ \times\ \vec{B}$$

5. Del operator

In Cartesian coordinates del operator ∇ is expressed as

$$\nabla \equiv \hat{\imath}\,\frac{\partial}{\partial x} + \hat{\jmath}\,\frac{\partial}{\partial y} + \hat{k}\,\frac{\partial}{\partial z}$$

5.1 Gradient of scalar field

Gradient of a scalar field Ψ, denoted by $\nabla\Psi$, is

$$\text{grad}\Psi = \nabla\Psi = \hat{\imath}\,\frac{\partial\Psi}{\partial x} + \hat{\jmath}\,\frac{\partial\Psi}{\partial y} + \hat{k}\,\frac{\partial\Psi}{\partial z}$$

First product rule for gradient

$$\nabla(\phi\Psi) = (\nabla\phi)\Psi + \phi(\nabla\Psi)$$

Second product rule for gradient

$$\nabla(\vec{F}\,.\,\vec{G}) = (\vec{F}\,.\,\nabla)\vec{G} + \vec{F}\times(\nabla\times\vec{G}) + (\vec{G}\,.\,\nabla)\vec{F} + \vec{G}\times(\nabla\times\vec{F})$$

5.2 Divergence of vector field

Divergence of vector field $\vec{F} = F_x\,\hat{\imath} + F_y\,\hat{\jmath} + F_z\,\hat{k}$, denoted by $\nabla\,.\,\vec{F}$, is

$$\text{div}\vec{F} = \nabla\,.\,\vec{F} = \frac{\partial F_x}{\partial x} + \frac{\partial F_y}{\partial y} + \frac{\partial F_z}{\partial z}$$

For position vector $\vec{r} = x\,\hat{\imath} + y\,\hat{\jmath} + z\,\hat{k}$, we have $\nabla\,.\,\vec{r} = 3$.

First product rule for divergence

$$\nabla\,.\,(\phi\vec{F}) = (\nabla\phi)\,.\,\vec{F} + \phi(\nabla\,.\,\vec{F})$$

Second product rule for divergence

$$\nabla\,.\,(\vec{F}\times\vec{G}) = (\nabla\times\vec{F})\,.\,\vec{G} - (\nabla\times\vec{G})\,.\,\vec{F}$$

5.3 Curl of vector field

Curl of vector field $\vec{F} = F_x\,\hat{\imath} + F_y\,\hat{\jmath} + F_z\,\hat{k}$, denoted by $\nabla\times\vec{F}$, is

$$\text{curl}\,\vec{F} = \begin{bmatrix} \hat{\imath} & \hat{\jmath} & \hat{k} \\ \frac{\partial}{\partial x} & \frac{\partial}{\partial y} & \frac{\partial}{\partial z} \\ F_x & F_y & F_z \end{bmatrix}$$

For position vector $\vec{r} = x\,\hat{\imath} + y\,\hat{\jmath} + z\,\hat{k}$, we have $\nabla\times\vec{r} = 0$.

First product rule for curl

$$\nabla \times (\phi \vec{F}) = (\nabla \phi) \times \vec{F} + \phi(\nabla \times \vec{F})$$

Second product rule for curl

$$\nabla \times (\vec{F} \times \vec{G}) = (\nabla \cdot \vec{G})\vec{F} - (\nabla \cdot \vec{F})\vec{G} + (\vec{G} \cdot \nabla)\vec{F} - (\vec{F} \cdot \nabla)\vec{G}$$

5.4 Laplacian operator

(i) Laplacian operator ∇^2 of scalar field Ψ is defined as

$$\nabla^2 \Psi = \nabla \cdot \nabla \Psi = \text{div}(\text{grad}\Psi)$$

(ii) Laplacian operator ∇^2 of vector field $\vec{F}$ is defined as

$$\nabla^2 \vec{F} = \text{grad}(\text{div}\vec{F}) - \text{curl}(\text{curl}\vec{F})$$

Example 1: Show that

(i) $\text{curl (curl } \vec{A}) = \text{grad (div } \vec{A}) - \nabla^2 \vec{A}$

(ii) $\text{div (grad } f) = \nabla^2 f = \dfrac{\partial^2 f}{\partial x^2} + \dfrac{\partial^2 f}{\partial y^2} + \dfrac{\partial^2 f}{\partial z^2}$

(iii) $\text{curl (grad } f) = \nabla \times (\nabla f) = 0$

(iv) $\text{div (curl}\vec{A}) = \nabla \cdot (\nabla \times \vec{A}) = 0$

(v) $\nabla \left(\dfrac{1}{r} \right) = -\dfrac{\vec{r}}{r^3}$ (vi) $\nabla \times \left(\dfrac{\vec{r}}{r^3} \right) = 0$

5.5 Formulae involving ∇ operator

(i) $\nabla(\phi \pm \psi) = \nabla\phi \pm \nabla\psi$

(ii) $\nabla \cdot (\vec{A} \pm \vec{B}) = \nabla \cdot \vec{A} \pm \nabla \cdot \vec{B}$

(iii) $\nabla \times (\vec{A} \pm \vec{B}) = \nabla \times \vec{A} \pm \nabla \times \vec{B}$

(iv) $\nabla \cdot (\phi \vec{A}) = \nabla\phi \cdot \vec{A} + \phi\, (\nabla \cdot \vec{A})$

(v) $\nabla \times (\phi \vec{A}) = \nabla\phi \times \vec{A} + \phi\, (\nabla \times \vec{A})$

(vi) $\nabla \cdot (\vec{A} \times \vec{B}) = \vec{B}.(\nabla \times \vec{A}) - \vec{A} \cdot (\nabla \times \vec{B})$

(viii) $\nabla\, (\vec{A} \cdot \vec{B}) = (\vec{B} \cdot \nabla)\vec{A} + (\vec{A} \cdot \nabla)\vec{B} + \vec{B} \times (\nabla \times \vec{A}) + \vec{A} \times (\nabla \times \vec{B})$

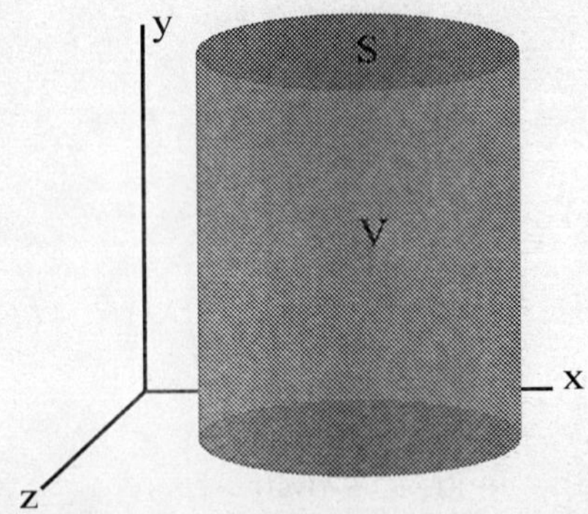

5.6 Gauss' divergence theorem

For a closed surface, Gauss' divergence theorem relates surface integral taken over a closed surface to the volume integral taken over the volume enclosed by the surface. Its mathematical expression is as the following.

$$\int_V \text{div}\vec{F}\ \text{d}V = \oint_S \vec{F}\ .\ \hat{n}\ \text{d}S \qquad \text{or} \qquad \int_V \nabla\ .\ \vec{F}\ \text{d}V = \oint_S \vec{F}\ .\ \hat{n}\ \text{d}S$$

where V is volume enclosed by closed surface S. $\hat{n}$ denotes unit vector normal to the surface in the outward direction.

5.7 Stokes' theorem

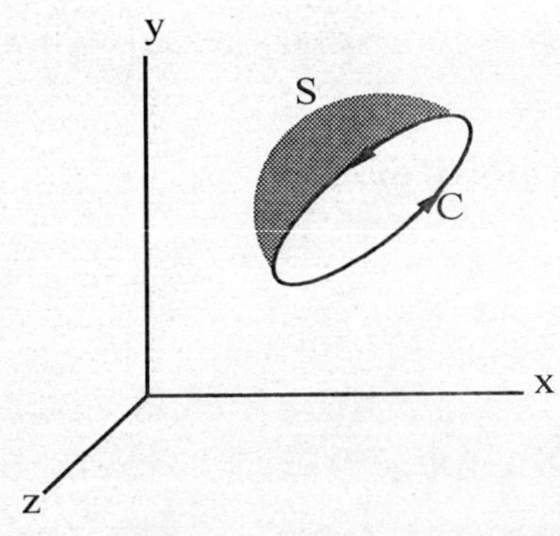

For an open surface, Stokes' theorem relates line integral taken along the boundary of surface enclosed by the line. Its mathematical expression is as the following.

$$\int_S \text{curl}\vec{F}\ .\ \hat{n}\ \text{d}S = \oint_C \vec{F}\ .\ \text{d}\vec{\ell} \qquad \text{or} \qquad \int_S (\nabla \times \vec{F})\ .\ \hat{n}\ \text{d}S = \oint_C \vec{F}\ .\ \text{d}\vec{\ell}$$

where S is surface bounded by the line C. $\hat{n}$ denotes unit vector normal to the surface in the outward direction.

5.8 Green's theorem

For a closed surface, Green's theorem relates surface integral taken over a closed surface to the volume integral taken over the volume enclosed by the surface. Its mathematical expression is as the following.

$$\int_V (\psi \nabla^2 \phi - \phi \nabla^2 \psi)\ \mathrm{d}V = \oint_S (\psi \nabla \phi - \phi \nabla \psi).\hat{n}\ \mathrm{d}S$$

Example 2: For a scalar field ϕ and vector field $\vec{A}$, prove the identity $\mathrm{div}(\phi\vec{A}) = \phi\ \mathrm{div}\vec{A} + \vec{A}\ .\ \mathrm{grad}\phi$.

Solution: Consider a vector field $\vec{A} = A_x\ \hat{i} + A_y\ \hat{j} + A_z\ \hat{k}$. We have

$$\mathrm{div}\vec{A} = \frac{\partial A_x}{\partial x} + \frac{\partial A_y}{\partial y} + \frac{\partial A_z}{\partial z} \qquad \text{and} \qquad \mathrm{grad}\phi = \frac{\partial \phi}{\partial x}\ \hat{i} + \frac{\partial \phi}{\partial y}\ \hat{j} + \frac{\partial \phi}{\partial z}\ \hat{k}$$

Now, we have

$$\mathrm{div}(\phi\vec{A}) = \frac{\partial(\phi A_x)}{\partial x} + \frac{\partial(\phi A_y)}{\partial y} + \frac{\partial(\phi A_z)}{\partial z}$$

$$= \phi\Big[\frac{\partial A_x}{\partial x} + \frac{\partial A_y}{\partial y} + \frac{\partial A_z}{\partial z}\Big] + A_x\ \frac{\partial \phi}{\partial x} + A_y\ \frac{\partial \phi}{\partial y} + A_z\ \frac{\partial \phi}{\partial z}$$

$$= \phi\ \mathrm{div}\vec{A} + [A_x\hat{i} + A_y\hat{j} + A_z\hat{k}].\Big[\frac{\partial \phi}{\partial x}\ \hat{i} + \frac{\partial \phi}{\partial y}\ \hat{j} + \frac{\partial \phi}{\partial z}\ \hat{k}\Big]$$

$$= \phi\ \mathrm{div}\vec{A} + \vec{A}.\mathrm{grad}\phi$$

Note:

Integrating

$$\nabla\ .\ (f\vec{A}) = f(\nabla\ .\ \vec{A}) + \vec{A}\ .\ (\nabla f)$$

over a volume, we have

$$\int_V \nabla\ .\ (f\vec{A})\ \mathrm{d}V = \int_V f(\nabla\ .\ \vec{A})\ \mathrm{d}V + \int_V \vec{A}\ .\ (\nabla f)\ \mathrm{d}V$$

Invoking the Gauss' divergence theorem, we have

$$\int_S (f\vec{A})\ .\ \mathrm{d}\vec{S} = \int_V f(\nabla\ .\ \vec{A})\ \mathrm{d}V + \int_V \vec{A}\ .\ (\nabla f)\ \mathrm{d}V$$

$$\int_V f(\nabla\ .\ \vec{A})\ \mathrm{d}V = \int_S (f\vec{A})\ .\ \mathrm{d}\vec{S} - \int_V \vec{A}\ .\ (\nabla f)\ \mathrm{d}V$$

Review Questions

1. Define addition of two vectors. Can two vectors having different dimensions be added? With the help of two vectors, prove that addition of two vectors is commutative.

2. Define subtraction of one vector from the other. Suppose, $\vec{A}$ and $\vec{B}$ are two vectors having different dimensions. Can we calculate $(\vec{A}-\vec{B})$? With the help of two vectors, prove that subtraction is not commutative.

3. Show that addition of two vectors is commutative whereas the subtraction of one vector from the other is not.

4. For two vectors $\vec{A} = A_x\,\hat{\imath} + A_y\,\hat{\jmath} + A_z\,\hat{k}$ and $\vec{B} = B_x\,\hat{\imath} + B_y\,\hat{\jmath} + B_z\,\hat{k}$, show that $k(\vec{A}+\vec{B}) = k\vec{A} + k\vec{B}$.

5. For two vectors $\vec{A}$ and $\vec{B}$ show that $\vec{A}+\vec{B} = \vec{B}+\vec{A}$.

6. For three vectors $\vec{A}$, $\vec{B}$, $\vec{C}$ show that $\vec{A}+(\vec{B}+\vec{C}) = (\vec{A}+\vec{B})+\vec{C}$.

7. For two vectors $\vec{A} = 2\,\hat{\imath} + 3\,\hat{\jmath} + 2\,\hat{k}$ and $\vec{B} = -\hat{\imath} + 2\,\hat{\jmath} + 3\,\hat{k}$, calculate $2\vec{A} - \vec{B}$ and $2\vec{A} + 3\vec{B}$.

8. Define velocity and acceleration of a body. The displacement of a body is expressed as

$$\vec{r} = (t^2+2)\hat{\imath} + (t+2)^2\hat{\jmath} + (t-1)^2\hat{k}$$

where t denotes the time.At time $t = 2$, calculate (i) magnitude of displacement, (ii) velocity and its magnitude, and (iii) acceleration and its magnitude.

9. Define scalar multiplication of two vectors. With the help of two vectors, prove that scalar multiplication is commutative.

10. Obtain the angle between the vectors $\vec{A} = 4\,\hat{\imath} + 2\,\hat{\jmath} + \hat{k}$ and $\vec{B} = 2\,\hat{\imath} - \hat{\jmath} + 2\,\hat{k}$.

11. Define vector multiplication of two vectors. With the help of two vectors, prove that vector multiplication is not commutative.

12. Define the scalar triple product of three vectors. For three vectors, $\vec{A} = 2\,\hat{\imath} - 2\,\hat{\jmath} + 5\,\hat{k}$, $\vec{B} = -\hat{\imath} + 2\,\hat{\jmath} - 3\,\hat{k}$, $\vec{C} = 3\,\hat{\imath} + \hat{\jmath} + 4\,\hat{k}$, calculate $\vec{A} \,.\, (\vec{B}\times\vec{C})$.

13. Define the vector triple product of three vectors. For three vectors, $\vec{A} = \hat{\imath} + 2\,\hat{\jmath} + 3\,\hat{k}$, $\vec{B} = \hat{\imath} - \hat{\jmath} + 3\,\hat{k}$, $\vec{C} = 3\,\hat{\imath} - \hat{\jmath} - 4\,\hat{k}$, calculate $\vec{A}\times(\vec{B}\times\vec{C})$.

14. Write expressions for: (i) gradient of scalar field, (ii) divergence of vector field, (iii) curl of vector field and (iv) Laplacian operator in Cartesian coordinates.

15. For scalar field Ψ and vector field $\vec{F}$, prove following relations.

    (i) $\text{curl}(\text{grad }\Psi) = 0$

    (ii) $\text{div}(\text{curl }\vec{F}) = 0$

16. For a scalar field ϕ and vector field $\vec{A}$, prove the identity
    $\text{div}(\phi\vec{A}) = \phi\ \text{div}\vec{A} + \vec{A}\ .\ \text{grad}\phi$.

17. Write notes on the following

    (i) Addition of two vectors
    (ii) Subtraction of one vector from other
    (iii) Multiplication of two vectors
    (iv) Velocity
    (v) Acceleration
    (vi) Scalar triple product of vectors
    (vii) Vector triple product of vectors
    (viii) Gradient of scalar field
    (ix) Divergence of vector field
    (x) Curl of vector field
    (xi) Laplacian operator
    (xii) Gauss' divergence theorem
    (xiii) Stokes' theorem

Multiple-Choice Questions

1. The value of $\vec{A}\ .\ \vec{A}$ is

   A. 0 B. A^2 C. A D. $\sqrt{A}$

   Ans. B

2. The value of $\vec{A} \times \vec{A}$ is

   A. 0 B. A^2 C. A D. $\sqrt{A}$

   Ans. A

3. Which of the following relations is not satisfied by vectors?

   A. $\vec{A}\ .\ \vec{B} = \vec{B}\ .\ \vec{A}$ B. $\vec{A}\ .\ (\vec{B} \times \vec{C}) = (\vec{A} \times \vec{B}).\vec{C}$
   C. $\vec{A} \times \vec{B} = \vec{B} \times \vec{A}$ D. $\vec{A} + (\vec{B} + \vec{C}) = (\vec{A} + \vec{B}) + \vec{C}$

   Ans. C

4. For unit vectors $\hat{\imath}$, $\hat{\jmath}$, $\hat{k}$ in Cartesian coordinates, which of the following relations is wrong.

   A. $\hat{\imath}\ .\ \hat{\jmath} = \hat{k}$ B. $\hat{\imath} \times \hat{\jmath} = \hat{k}$ C. $\hat{\imath}\ .\ \hat{\imath} = 1$ D. $\hat{\imath} \times \hat{\imath} = 0$

   Ans. A

5. When $\vec{A} = 2\hat{i} + 2\hat{j} - \hat{k}$ and $\vec{B} = \hat{i} + 2\hat{j} - 3\hat{k}$, then $\vec{A} \times \vec{B}$ is

A. $-4\hat{i} - 5\hat{j} + 2\hat{k}$ B. $-4\hat{i} + 5\hat{j} + 2\hat{k}$ C. $2\hat{i} + 4\hat{j} + 6\hat{k}$ D. $4\hat{i} - 5\hat{j} - 2\hat{k}$

Ans. B

6. Which of the following relations is wrong?

A. $\vec{A} \times (\vec{B} \times \vec{C}) = \vec{B}(\vec{A} \cdot \vec{C}) - \vec{C}(\vec{A} \cdot \vec{B})$ B. $\vec{A} \cdot \vec{B} = \vec{B} \cdot \vec{A}$

C. $\vec{A} \cdot (\vec{B} \times \vec{C}) = (\vec{A} \times \vec{B}) \cdot \vec{C}$ D. $\vec{A} \times \vec{B} = \vec{B} \times \vec{A}$

Ans. D

7. For scalar quantities ϕ and λ, and vector quantity $\vec{A}$, which of the following relations is possible.

A. $\nabla\phi = \lambda\phi$ B. $\nabla \cdot \vec{A} = \lambda\vec{A}$ C. $\nabla \times \vec{A} = \lambda\vec{A}$ D. None of them

Ans. C

8. For a scalar quantity ϕ and vector quantity $\vec{A}$, which of the following relations is wrong?

A. $\nabla\times(\nabla\phi) = 0$ B. $\nabla \cdot (\nabla\times\vec{A}) = 0$ C. $\nabla\times(\nabla\times\vec{A}) = 0$ D. $\nabla \cdot (\nabla\phi) = \nabla^2\phi$

Ans. C

9. Gradient of scalar field ϕ can be expressed as

A. $\nabla\phi$ B. $\nabla \cdot \phi$ C. $\nabla \times \phi$ D. $\nabla^2\phi$

Ans. A

10. Divergence of vector field $\vec{A}$ can be expressed as

A. $\nabla\vec{A}$ B. $\nabla \cdot \vec{A}$ C. $\nabla \times \vec{A}$ D. $\nabla^2\vec{A}$

Ans. B

11. Curl of vector field $\vec{A}$ can be expressed as

A. $\nabla\vec{A}$ B. $\nabla \cdot \vec{A}$ C. $\nabla \times \vec{A}$ D. $\nabla^2\vec{A}$

Ans. C

12. Laplacian of scalar field ϕ can be expressed as

A. $\nabla\phi$ B. $\nabla \cdot \phi$ C. $\nabla \times \phi$ D. $\nabla^2\phi$

Ans. D

13. Gradient of scalar field $\phi = x^2 + y^2 + z^2$ is

A. $2(x\,\hat{i} + y\,\hat{j} + z\,\hat{k})$ B. $2(x + y + z)$

C. $x\,\hat{i} + y\,\hat{j} + z\,\hat{k}$ D. $x^2 + y^2 + z^2$

Ans. A

14. Divergence of vector field $\vec{A} = x^2\ \hat{\imath} + y^2\ \hat{\jmath} + z^2\ \hat{k}$ is

 A. $2(x\ \hat{\imath} + y\ \hat{\jmath} + z\ \hat{k}\ )$ B. $2(x + y + z)$

 C. $x\ \hat{\imath} + y\ \hat{\jmath} + z\ \hat{k}$ D. $x^2 + y^2 + z^2$

Ans. B

15. Curl of vector field $\vec{A} = x^2\ \hat{\imath} + y^2\ \hat{\jmath} + z^2\ \hat{k}$ is

 A. $2(x\ \hat{\imath} + y\ \hat{\jmath} + z\ \hat{k}\ )$ B. $2(x + y + z)$

 C. $x\ \hat{\imath} + y\ \hat{\jmath} + z\ \hat{k}$ D. 0

Ans. D

16. Laplacian of scalar field $\phi = x^2 + y^2 + z^2$ is

 A. $2(x\ \hat{\imath} + y\ \hat{\jmath} + z\ \hat{k}\ )$ B. 6 C. $x\ \hat{\imath} + y\ \hat{\jmath} + z\ \hat{k}$ D. 2

Ans. B

17. Gauss' divergence theorem relates

 A. line & volume integrals B. surface & volume integrals

 C. line & surface integrals D. None of these

Ans. B

18. Stokes' theorem relates

 A. line & volume integrals B. surface & volume integrals

 C. line & surface integrals D. None of these

Ans. C

19. Which statement is wrong for the Stokes' theorem?

 A. Applicable to open surface

 B. Relates surface integral and line integral

 C. Relates surface integral and volume integral

Ans. C

20. Which statement is correct for the Stokes' theorem?

 A. Applicable to close surface

 B. Relates surface integral and line integral

 C. Relates surface integral and volume integral

 D. Relates line integral and volume integral

Ans. B

21. Which statement is wrong for the Gauss' divergence theorem?

 A. Applicable to close surface

 B. Relates surface integral and line integral

 C. Relates surface integral and volume integral

Ans. B

22. Which statement is correct for the Gauss' divergence theorem?
    A. Applicable to open surface
    B. Relates surface integral and line integral
    C. Relates surface integral and volume integral
    D. Relates line integral and volume integral

Ans. C

23. Which is equal to zero?
    A. Curl of gradient of scalar
    B. Divergence of gradient of scalar
    C. Gradient of divergence of vector

Ans. A

24. Which is equal to the Laplacian?
    A. Divergence of curl
    B. Divergence of gradient
    C. Gradient of divergence

Ans. B

25. Which is vector quantity?
    A. Divergence of curl of vector
    B. Divergence of gradient of scalar
    C. Gradient of divergence of vector

Ans. C

26. Which is scalar quantity?
    A. Divergence of curl of vector
    B. Curl of gradient of scalar
    C. Gradient of divergence of vector

Ans. A

27. Which statement is correct for the Stokes' theorem?
    A. Applicable to close surface
    B. Relates surface integral and line integral
    C. Relates surface integral and volume integral
    D. Relates line integral and volume integral

Ans. B

28. Which statement is wrong for the Gauss' divergence theorem?
    A. Applicable to close surface
    B. Relates surface integral and line integral
    C. Relates surface integral and volume integral

Ans. B

Chapter 3

Electrostatics

Electrostatics is a branch of science which deals with forces, fields, potentials produced and experienced by stationary charges. Electric charges are source of interactions between them and these charges can have positive and negative values. By convention, charge of electron is taken negative and the charge of proton is taken positive, with both having magnitude $e = 1.602 \times 10^{-19}$ C. From experiments, it is well established that all free charges are integral multiples of e. We have adopted M.K.S.I. (meter-kilogram-second-international) system of units where Coulomb is the unit of charge. It is well known that charge is basically discrete rather than continuous. However, charge densities which refer to continuous charge distribution are also considered. Experimentally it is found that total charge of an isolated system does not change with time; conservation of charge. Total charge is defined as algebraic sum of positive and negative charges in a system. In this chapter, we have discussed about forces, fields and potentials produced and experienced by stationary charges.

1. Electric field

An electric field is created by electric charge, or by time-varying magnetic field. Electric field surrounds an electric charge, and exerts force on other charges in the field, attracting or repelling them.

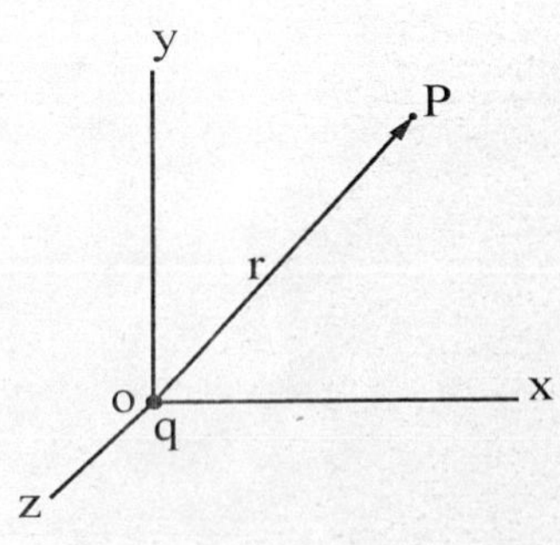

Figure 1: Electric field of point-charge

Suppose a charge q is placed at the origin of coordinate system (Figure 1). The electric field intensity $\vec{E}$ developed at a position $\vec{r}$ in vacuum by the charge is

$$\vec{E}(\vec{r}) = \frac{1}{4\pi\epsilon_0} \frac{q}{r^2} \hat{r}$$

where ϵ_0 is called the permittivity of free space and

$$\frac{1}{4\pi\epsilon_0} = 8.987552 \times 10^9 \ \text{N m}^2/\text{C}^2$$

The force applied by the electric field on charge q' placed at the position $\vec{r}$ is

$$\vec{F}_{q\,q'} = q' \, \vec{E}(\vec{r}) = \frac{1}{4\pi\epsilon_0} \frac{q\,q'}{r^2} \hat{r}$$

This is the force applied by the charge q placed at the origin on the charge q' placed at the position $\vec{r}$. The force between two charges is known as the **Coulomb force**. Similarly, the force applied by the charge q' placed at the position $\vec{r}$ on the charge q placed at the origin is

$$\vec{F}_{q'\,q} = \frac{1}{4\pi\epsilon_0} \frac{q\,q'}{r^2} (-\hat{r})$$

Therefore,

$$\vec{F}_{q\,q'} = -\vec{F}_{q'\,q}$$

Two charges having the same kind of charges (both positive or both negative) would repel to each other and two charges having opposite kind of charges (one positive and one negative) would attract each other. The electric field intensity as a vector quantity having SI unit volts per meter (V/m), is exactly equivalent to newtons per coulomb (N/C).

Example 1. Calculate electric force between an electron and a proton, separated by 0.5 Å.

2. Gauss' law

Let us consider a single point charge q, situated at the origin. The electric fieid produced is

$$\vec{E}(\vec{r}) = \frac{1}{4\pi\epsilon_0} \frac{q}{r^2} \hat{r} \tag{3.1}$$

The **flux** of $\vec{E}$ through a surface S is

$$\Phi_E \equiv \int_S \vec{E} \, . \, \mathrm{d}\vec{S} \tag{3.2}$$

In the case of a point charge q at the origin, the flux of $\vec{E}$ through a spherical surface of radius r is ($\mathrm{d}\vec{S} = r^2 \sin\theta \ \mathrm{d}\theta \ \mathrm{d}\phi \ \hat{r}$)

$$\int_S \vec{E} \, . \, \mathrm{d}\vec{S} = \int_S \frac{1}{4\pi\epsilon_0} \left(\frac{q}{r^2} \hat{r} \right) . \ (r^2 \sin\theta \ \mathrm{d}\theta \ \mathrm{d}\phi \ \hat{r})$$

$$= \frac{q}{\epsilon_0} \tag{3.3}$$

For any closed surface, we have

$$\int_S \vec{E} \cdot d\vec{S} = \frac{Q_{enc}}{\epsilon_0} \tag{3.4}$$

where Q_{enc} is the total charge enclosed within the surface. This is integral form of Gauss' law.

It states that total electric flux over a closed surface is total charge enclosed within the surface divided by ϵ_0.

Example 2. One electron and one proton are enclosed within a closed surface. Calculate total flux passing through the surface.

Example 3. How much is total flux in SI unit, passing through a closed surface enclosing 2 protons and 1 electron?

Total charge enclosed may be expressed as

$$Q_{enc} = \int_V \rho \, dV$$

where ρ is volume charge density. Thus, we have

$$\int_S \vec{E} \cdot d\vec{S} = \frac{1}{\epsilon_0} \int_V \rho \, dV$$

By applying the Gauss' divergence theorem and using expression for Q_{enc}, we get

$$\int_V (\nabla \cdot \vec{E}) \, dV = \frac{1}{\epsilon_0} \int_V \rho \, dV \qquad \text{or} \qquad \int_V \left[\nabla \cdot \vec{E} - \frac{\rho}{\epsilon_0}\right] dV = 0$$

Since it is valid for arbitrary volume, we have

$$\nabla \cdot \vec{E} - \frac{\rho}{\epsilon_0} = 0 \qquad \text{or} \qquad \nabla \cdot \vec{E} = \frac{\rho}{\epsilon_0}$$

This is differential form of Gauss' law.

We have

$$\vec{E}(\vec{r}) = \frac{1}{4\pi\epsilon_0} \frac{q}{r^2} \hat{r}$$

Electric field is conservative field. We want to calculate the line integral of this field from some point A to some other point B

$$\int_A^B \vec{E} \cdot d\vec{\ell}$$

The value of integral does not depend on path, and depends only on the end-points. In spherical polar coordinates, $d\vec{\ell} = dr \, \hat{r} + r \, d\theta \, \hat{\theta} + r \sin\theta \, d\phi \, \hat{\phi}$, so

$$\vec{E} \cdot d\vec{\ell} = \frac{1}{4\pi\epsilon_0} \frac{q}{r^2} \, dr$$

Therefore,

$$\int_A^B \vec{E} \, . \, \mathrm{d}\vec{\ell} = \frac{1}{4\pi\epsilon_0} \int_A^B \frac{q}{r^2} \, \mathrm{d}r$$

$$= \frac{q}{4\pi\epsilon_0} \Big(\frac{1}{r_A} - \frac{1}{r_B} \Big) \tag{3.5}$$

where r_A and r_B are distances of points A and B, respectively, from the origin of coordinates system. The integral around a closed path is evidently zero (for that $r_A = r_B$):

$$\oint \vec{E} \, . \, \mathrm{d}\vec{\ell} = 0 \tag{3.6}$$

On applying the Stokes' theorem,

$$\int_S (\nabla \times \vec{E}) \, . \, \mathrm{d}\vec{S} = 0$$

Since it is valid for arbitrary surface, we have

$$\nabla \times \vec{E} = 0 \tag{3.7}$$

We know that the curl of gradient of scalar is zero. Therefore, we can write

$$\vec{E} = -\nabla\phi$$

where ϕ is **electrostatic potential**. The negative sign comes to have proper sign of field and work done by the field.

Example 4. Electrostatic potential ϕ between two charges q_1 and q_2, separated by a distance r is given by

$$\phi = \frac{1}{4\pi\epsilon_0} \, \frac{q_1 \, q_2}{r}$$

Calculate the electrostatic force between the charges.

Solution. The electrostatic force $\vec{F}$ between the charges is

$$\vec{F} = -\nabla\phi$$

We are dealing in space. In spherical polar coordinates, we have

$$\nabla \equiv \hat{r} \, \frac{\partial}{\partial r} + \frac{\hat{\theta}}{r} \, \frac{\partial}{\partial \theta} + \frac{\hat{\phi}}{r \, \sin\theta} \, \frac{\partial}{\partial \phi}$$

Since the potential ϕ is not depending on the coordinates θ and ϕ coordinates, and therefore,

$$\vec{F} = -\nabla\phi = -\hat{r} \, \frac{\partial}{\partial r} \Big(\frac{1}{4\pi\epsilon_0} \, \frac{q_1 \, q_2}{r} \Big) = \frac{1}{4\pi\epsilon_0} \, \frac{q_1 \, q_2}{r^2} \, \hat{r}$$

2.1 Poisson's equation and Laplace's equation

The electric field can be written as the gradient of a scalar potential.

$$\vec{E} = -\nabla\phi$$

We have

$$\nabla \cdot \vec{E} = \frac{\rho}{\epsilon_0} \qquad \text{and} \qquad \text{curl } \vec{E} = 0$$

Now, $\nabla \cdot \vec{E} = \nabla \cdot (-\nabla\phi) = -\nabla^2\phi$, so, apart from that persistent minus sign, the divergence of $\vec{E}$ is the Laplacian of ϕ. Gauss' law, then, says

$$\nabla^2\phi = -\frac{\rho}{\epsilon_0}$$

This is known as **Poisson's equation**. It is a linear inhomogeneous equation.

In the regions where there is no charge, so $\rho = 0$, Poisson's equation reduces to **Laplace's equation**,

$$\nabla^2\phi = 0$$

It is a linear homogeneous equation.

Note: For distributed electric charges, electric potential can be calculated easily as addition of scalar quantities is much easier than that of vector quantities. Then negative gradient of potential would give electric field due to distributed electric charges.

3. Potential and electric field due to electric dipole

Consider an electric dipole having charges $+q$ and $-q$, and length $2a$ placed along z-axis (Figure 2), so that electric dipole moment of dipole is $p = 2aq$. Suppose, the charge $+q$ is situated at a point with coordinates (0, 0, a) and charge $-q$ is situated at a point with coordinates (0, 0, $-a$).

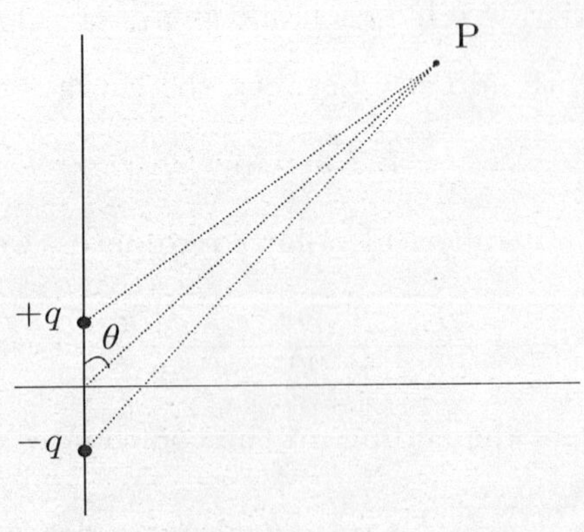

Figure 2: Electric dipole of length $2a$ placed along z-axis.

Let us have a point P (x, y, z) at a distance r from origin so that $r = \sqrt{x^2 + y^2 + z^2}$. The electric potential due to electric dipole at point P is

$$\phi(\vec{r}) = \frac{1}{4\pi\epsilon_0}\left[\frac{q}{\sqrt{x^2 + y^2 + (z-a)^2}} - \frac{q}{\sqrt{x^2 + y^2 + (z+a)^2}}\right]$$

We take that size $2a$ of electric dipole is much smaller than the distance r of P from origin, *i.e.*, $a << r$. Then we have

$$\frac{1}{\sqrt{x^2 + y^2 + (z-a)^2}} = \frac{1}{r}\left[1 - \frac{2az}{r^2}\right]^{-1/2} = \frac{1}{r}\left[1 + \frac{az}{r^2}\right]$$

Here, higher order (smaller) terms are neglected. Similarly, we have

$$\frac{1}{\sqrt{x^2 + y^2 + (z+a)^2}} = \frac{1}{r}\left[1 - \frac{az}{r^2}\right]$$

Thus, we have

$$\phi(\vec{r}) = \frac{1}{4\pi\epsilon_0}\left[\frac{q}{r}\left(1 + \frac{az}{r^2}\right) - \frac{q}{r}\left(1 - \frac{az}{r^2}\right)\right] = \frac{1}{4\pi\epsilon_0}\frac{pz}{r^3}$$

where $p = 2aq$ is electric dipole moment of electric dipole. If θ is angle between $\vec{r}$ and z-axis, then $z = r\cos\theta$ and we have

$$\phi(\vec{r}) = \frac{1}{4\pi\epsilon_0}\frac{p\cos\theta}{r^2}$$

When the point P lies on the axis of electric dipole, such that $\theta = 0$ and we have

$$\phi(\vec{r}) = \frac{1}{4\pi\epsilon_0}\frac{p}{r^2}$$

Note that electric potential due to electric dipole varies as r^{-2} in contrast to the potential of a point-charge which varies as r^{-1}.

In spherical polar coordinates (r, θ, ϕ), we have

$$\nabla \equiv \hat{r}\frac{\partial}{\partial r} + \frac{\hat{\theta}}{r}\frac{\partial}{\partial \theta} + \frac{\hat{\phi}}{r\sin\theta}\frac{\partial}{\partial \phi}$$

Thus, components of electric field due to electric dipole are

$$E_r = -\frac{\partial \phi}{\partial r} = \frac{1}{4\pi\epsilon_0}\frac{2p\cos\theta}{r^3}$$

$$E_\theta = -\frac{1}{r}\frac{\partial \phi}{\partial \theta} = \frac{1}{4\pi\epsilon_0}\frac{p\sin\theta}{r^3}$$

$$E_\phi = -\frac{1}{r\sin\theta}\frac{\partial \phi}{\partial \phi} = 0$$

4. Potential and electric field due to linear electric quadrupole

Consider linear quadrupole having charges $+q$, $-2q$, and $+q$, and length $2a$ placed along z-axis (Figure 3), so that electric dipole moment is zero. Suppose, the charge $+q$ is situated at a point with coordinates $(0, 0, a)$, charge $-2q$ is situated at a point with coordinates $(0, 0, 0)$, and charge $+q$ is situated at a point with coordinates $(0, 0, -a)$.

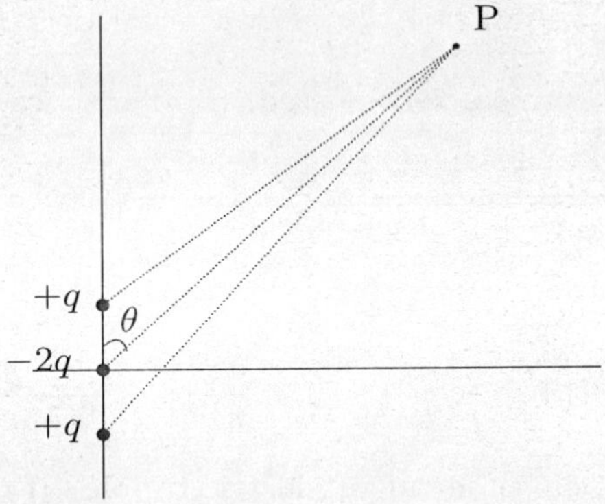

Figure 3: Linear electric quadrupole of length $2a$ placed along z-axis.

Let us have a point P (x, y, z) at a distance r from origin so that $r = \sqrt{x^2 + y^2 + z^2}$. The electric potential due to linear electric quadrupole at point P is

$$\phi(\vec{r}) = \frac{1}{4\pi\epsilon_0}\Big[\frac{q}{\sqrt{x^2 + y^2 + (z-a)^2}} - \frac{2q}{\sqrt{x^2 + y^2 + z^2}} + \frac{q}{\sqrt{x^2 + y^2 + (z+a)^2}}\Big]$$

We take that size $2a$ of electric quadrupole is much smaller than distance r of P from origin, *i.e*, $a << r$. Then we have

$$\frac{1}{\sqrt{x^2 + y^2 + (z-a)^2}} = \frac{1}{r}\Big[1 - \frac{2az}{r^2} + \frac{a^2}{r^2}\Big]^{-1/2} = \frac{1}{r}\Big[1 + \frac{az}{r^2} - \frac{1}{2}\frac{a^2}{r^2} + \frac{3}{2}\frac{a^2z^2}{r^4}\Big]$$

Here, higher order (smaller terms) are neglected. Similarly, we have

$$\frac{1}{\sqrt{x^2 + y^2 + (z+a)^2}} = \frac{1}{r}\Big[1 + \frac{2az}{r^2} + \frac{a^2}{r^2}\Big]^{-1/2} = \frac{1}{r}\Big[1 - \frac{az}{r^2} - \frac{1}{2}\frac{a^2}{r^2} + \frac{3}{2}\frac{a^2z^2}{r^4}\Big]$$

Thus, we have

$$\phi(\vec{r}) = \frac{1}{4\pi\epsilon_0}\Big[\frac{q}{r}\Big(1 + \frac{az}{r^2} - \frac{1}{2}\frac{a^2}{r^2} + \frac{3}{2}\frac{a^2z^2}{r^4}\Big) - \frac{2q}{r} + \frac{q}{r}\Big(1 - \frac{az}{r^2} - \frac{1}{2}\frac{a^2}{r^2} + \frac{3}{2}\frac{a^2z^2}{r^4}\Big)\Big]$$

$$= \frac{qa^2}{4\pi\epsilon_0}\Big(\frac{3z^2}{r^5} - \frac{1}{r^3}\Big)$$

If θ is angle between $\vec{r}$ and z-axis, then $z = r\cos\theta$ and we have

$$\phi(\vec{r}) = \frac{qa^2}{4\pi\epsilon_0}\Big(\frac{3\cos^2\theta - 1}{r^3}\Big)$$

Note that electric potential due to linear electric quadrupole varies as r^{-3} in contrast to the potential of electric dipole which varies as r^{-2}.

In spherical polar coordinates $(r,\ \theta,\ \psi)$, gradient has polar components

$$\nabla \equiv \Big(\frac{\partial}{\partial r},\ \frac{1}{r}\frac{\partial}{\partial \theta},\ \frac{1}{r\sin\theta}\frac{\partial}{\partial \psi}\Big)$$

Thus, components of electric field due to linear electric quadrupole are

$$E_r = -\frac{\partial \phi}{\partial r} = \frac{3qa^2}{4\pi\epsilon_0}\Big(\frac{3\cos^2\theta - 1}{r^4}\Big)$$

$$E_\theta = -\frac{1}{r}\frac{\partial \phi}{\partial \theta} = \frac{qa^2}{4\pi\epsilon_0}\Big(\frac{3\sin 2\theta}{r^4}\Big)$$

$$E_\psi = -\frac{1}{r\sin\theta}\frac{\partial \phi}{\partial \psi} = 0$$

Review Questions

1. State and describe Coulomb's law between two point-charges, a group of discrete charges, and between a charge and continuous distribution of charge.
2. Describe electric field due to a charge, a group of discrete charges, and continuous distribution of charge.
3. What is flux of vector field?
4. State and derive Gauss' law.
5. Calculate potential and electric field due to electric dipole.
6. Calculate potential and electric field due to linear electric quadrupole.
7. Write notes on the following:

   (i) Coulomb's law
   (ii) Electric field
   (iii) Flux of vector field
   (iv) Gauss' law
   (v) Electric potential due to electric dipole
   (vi) Electric potential due to electric linear quadrupole

Multiple-Choice Questions

1. Law describing force between static charges was given by

   A. Coulomb B. Faraday C. Gauss D. Stokes

   Ans. A

2. The statement that total electric flux through any closed surface surrounding a charge q is q/ϵ_0 was given by

   A. Coulomb B. Faraday C. Gauss D. Stokes

   Ans. C

3. With convention that work done by electric field on positive charge is positive, the relation between electric field $\vec{E}$ and electrostatic potential ϕ is

   A. $\vec{E} = \nabla\phi$ B. $\vec{E} = -\nabla\phi$ C. $\vec{E} = \nabla \times \phi$ D. $\vec{E} = -\nabla \times \phi$

   Ans. B

4. On a stationary charge, both electric and magnetic fields are applied simultaneously. Due to which field the charge would start moving?

   A. Electric field only

   B. Magnetic field only

   C. Both electric and magnetic fields

   D. Neither electric field nor magnetic field

   Ans. A

5. Which statement is wrong?

   A. Poisson's equation is linear homogeneous.

   B. Poisson's equation is linear inhomogeneous.

   C. Laplace's equation is linear homogeneous.

   Ans. A

6. Which statement is wrong?

   A. Poisson's equation is linear inhomogeneous.

   B. Laplace's equation is linear inhomogeneous.

   C. Laplace's equation is linear homogeneous.

   Ans. B

7. Which statement is correct?

   A. Poisson's equation is linear homogeneous.

   B. Poisson's equation is linear inhomogeneous.

   C. Laplace's equation is linear inhomogeneous.

   Ans. B

8. Which statement is correct?

   A. Poisson's equation is linear inhomogeneous.

   B. Laplace's equation is linear inhomogeneous.

   C. Poisson's equation is linear homogeneous.

Ans. C

9. How many Newton is the force between an electron and a proton in hydrogen atom? Given, $(1/4\pi\epsilon_0) = 8.99 \times 10^9$ N m^2/C^2.

   A. Zero B. 9.2 e-8 C. Infinite D. 1.0 e-17

Ans. B

10. How much is total flux in SI unit, passing through a closed surface enclosing 2 protons and 1 electron? Given, $(1/4\pi\epsilon_0) = 8.99 \times 10^9$ N m^2/C^2.

   A. Zero B. 1.6 e-19 C. 4.8 e-19 D. 1.81 e-8

Ans. D

11. How much is total flux in SI unit, passing through a closed surface enclosing 2 neutrons, 3 protons and 2 electrons? Given, $(1/4\pi\epsilon_0) = 8.99 \times 10^9$ N m^2/C^2.

   A. 11.2 e -19 B. 1.6 e-19 C. 8.0 e-19 D. 1.81 e-8

Ans. D

Chapter 4

Electrostatic Field in Dielectrics

In the preceding chapter, we have discussed about the electric potential and field due to isolated charge, electric dipole and linear electric quadrupole. In a material medium, such as insulators, positive and negative charges get separated when an external electric field is applied. Such a charge separation leads to a dipole field of many dipoles. Electrostatics in a dielectric medium is of great interest and is discussed in this chapter. In this chapter, we have also discussed electric field energy due to discrete and continuous charge distributions in a material medium.

1. Dielectric polarization

Contrary to conductors, in an insulator, electrons are tightly bound to atoms; there are no free electrons or the number of free electrons is too low. When a potential difference is applied across an insulator, though there is no flow of electric current, but the behaviour of insulator in the field may be modified and may play important role in science and technology. The insulators whose behaviour gets modified on application of electric field are known as **dielectrics**. When the change in the behaviour of dielectric is independent of the direction of applied field, the dielectric is called isotropic. On the other hand, if the change in behaviour of dielectric depends on the direction of applied field, the dielectric is called anisotropic.

When a dielectric is placed in an external electric field, the field exerts a force on each charged particle in the dielectric. Positive charges are pushed towards one direction (direction of electric field)[1] and negative charges are pushed in opposite direction. As a result, positive and negative parts of each molecule have relative displacement in opposite directions from their equilibrium positions. The overall effect is a displacement of entire positive charge relative to negative charge in the dielectric. The phenomenon of relative displacement of charges is called polarization and the dielectric is said polarized. Molecules of dielectrics may be of two kinds:

1. polar molecules

2. non-polar molecules

[1]Direction of applied electric field is from the positive pole towards the negative pole.

Positive charges of nuclei of atoms of molecule may be considered as concentrated at a point, say, center of positive charges. Similarly, negative charges of electrons revolving around nuclei of atoms of molecule may be considered as concentrated at a point, say, center of negative charges. When the center of positive charges coincides with the center of negative charges in a molecule, the molecule has no permanent electric dipole moment, and is called non-polar molecule. When the center of positive charges does not coincide with the center of negative charges in a molecule, the molecule has permanent electric dipole moment, and is called polar molecule. A polar molecule in an applied electric field exhibits both permanent and induced electric dipole moments whereas a non-polar molecule exhibits only induced electric dipole moment.

In absence of external electric field, polar or non-polar molecules are aligned in a random manner such that total electric dipole moment of dielectric is zero. When either of these types of molecules is placed in an applied electric field, non-polar molecules have induced electric dipoles whereas polar molecules have resultant of permanent electric dipoles and induced electric dipoles. The orientation of induced dipoles or of permanent dipoles in an external electric field align along the field lines. Thus, resultant electric dipole moment gets modified, in either type of molecules, after application of electric field.

Definition of dielectric polarization:

In a homogeneous isotropic dielectric, the polarization depends directly on the net electric field $\vec{E}$ in the medium. Electric field is applied such that the plane AB is positive and plane CD is negative, as shown in Figure 1. All the molecules are polarized to the same extent. The induced charges within the body of medium neutralize each other except at the surface of end faces (close to AB and CD planes) in Figure 1.

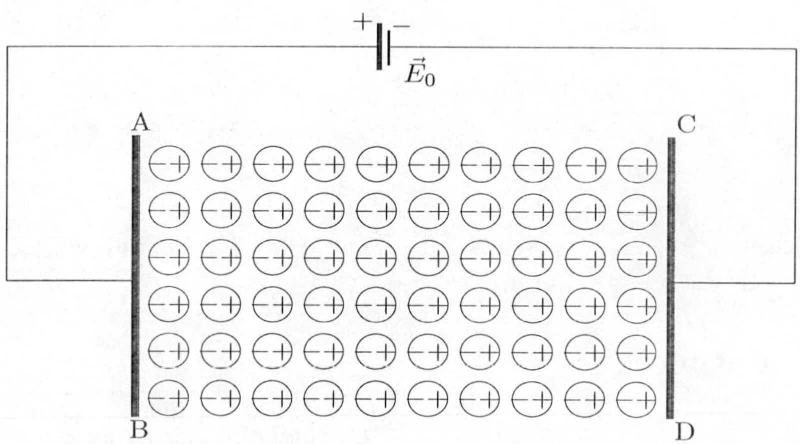

Figure 1: Dielectric material placed in external electric field $\vec{E}_0$ applied on the planes AB and CD.

The electric field due to induced charges is opposite to the external (applied) field $\vec{E}_0$. Hence, the resultant field $\vec{E}$ is less than $\vec{E}_0$. Provided $\vec{E}$ is not too large, average electric dipole moment $\vec{p}$ induced in each molecule of dielectric is proportional to $\vec{E}$.

$$\vec{p} \propto \vec{E} \qquad \text{or} \qquad \vec{p} = \alpha\vec{E}$$

where α is a constant, called **molecular polarizibility**. Suppose, there are n molecules per unit volume in dielectric and average electric dipole moment $\vec{p}$ of each molecule is independent of others, electric dipole moment per unit volume is $n\vec{p}$ or $n\alpha\vec{E}$, and is represented by $\vec{P}$. That is,

$$\vec{P} = n\vec{p} = n\alpha\vec{E}$$

This electric dipole moment per unit volume is known as polarization $\vec{P}$.

Consider a rectangular block of dielectric material of length ℓ (Figure 2). Let A be the area of cross section (face ABCD and face EFGH) of block perpendicular to the direction of applied (external) electric field, which is along the length of slab. Due to electric field (Figure 2), negative charge $-q'$ develops on the face ABCD and positive charge $+q'$ develops on the face EFGH. The induced electric dipole moment is $q'\ell$.

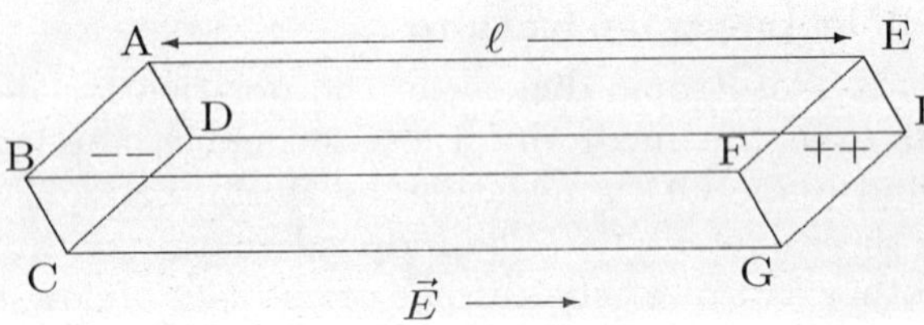

Figure 2: Dielectric material of length ℓ.

For polarization P, total electric dipole moment of dielectric is $A\ell P$. Therefore, we have

$$A\ell P = q'\ell \qquad \text{or} \qquad P = \frac{q'}{A}$$

Thus, polarization may be defined as surface charge per unit area appearing at faces perpendicular to the direction of applied electric field.

Electrical susceptibility

Experimental observations have found that, provided electric field $\vec{E}$ is not too large, in a large number of dielectrics, polarization $\vec{P}$ is proportional to $\vec{E}$.

$$\vec{P} \propto \vec{E} \qquad \text{or} \qquad \vec{P} = k\epsilon_0\vec{E}$$

where k is dimensionless constant called **electrical susceptibility** of dielectric and ϵ_0 is permittivity of free space.

2. Dielectric constant (relative permittivity)

Capacity of a parallel plate capacitor increases when the gap between the plates is filled with a dielectric medium. The increase in capacity of capacitor with a dielectric compared to the capacity of the same capacitor with vacuum between the plates, measures a constant known as **dielectric constant** or **relative permittivity**, ϵ_r.

$$\epsilon_r = \frac{\text{capacity with dielectric present}}{\text{capacity with vacuum}} = \frac{\epsilon}{\epsilon_0}$$

where ϵ and ϵ_0 are absolute permittivities of dielectric and vacuum, respectively. The dielectric constant may be defined as the ratio of absolute permittivity of dielectric to that of free space. Thus, $\epsilon = \epsilon_r \epsilon_0$ is absolute permittivity of dielectric.

The dielectric constant may be defined in terms of electric field intensities as the following. On application of electric field $\vec{E}_0$, when dielectric is polarized, the charges appearing on the opposite surfaces perpendicular to the field are opposite to polarities applied on the plates. Thus, the field in dielectric decreases to $\vec{E}$. The relative permittivity of dielectric may also be defined as

$$\epsilon_r = \frac{\text{field in vacuum}}{\text{field in dielectric}} = \frac{\vec{E}_0}{\vec{E}}$$

The electric field in vacuum due a charge q at a distance r is

$$\vec{E}_0 = \frac{1}{4\pi\epsilon_0} \frac{q\,\vec{r}}{r^3}$$

where $\vec{r}$ is vector from the charge to the point. The electric field in dielectric is

$$\vec{E} = \frac{1}{4\pi\epsilon} \frac{q\,\vec{r}}{r^3}$$

Electric displacement

Since the electric intensity depends on medium, we define another electrical vector which depends only on magnitude of charge and not on the medium. The vector is known as **electric displacement**, $\vec{D}$, given by

$$\vec{D} = \frac{1}{4\pi} \frac{q\,\vec{r}}{r^3}$$

so that $\vec{D} = \epsilon \vec{E}$.

3. External field of dielectric medium

When a dielectric is placed in an external electric field, it becomes polarized and acquires electrical dipole moment $\vec{P}$ per unit volume. We wish to calculate electric field intensity produced in space at an external point due to these elementary dipoles.

Consider a finite volume of dielectric having polarization $\vec{P}(\vec{r})$. The polarization gives rise to an electric field. We want to calculate potential and field at a point Q with position vector $\vec{r}\,'$ situated outside the dielectric.

Suppose, the dielectric is divided into small elementary volumes. Let us consider volume-element dV at point A with position vector $\vec{r}$. Electric dipole moment of element is $\vec{P}(\vec{r})$ dV. Potential at Q due to volume-element is

$$d\phi(\vec{r}\,') = \frac{1}{4\pi\epsilon_0} \frac{\vec{P}(\vec{r})\ dV(\vec{r}\,' - \vec{r})}{|\vec{r}\,' - \vec{r}|^3}$$

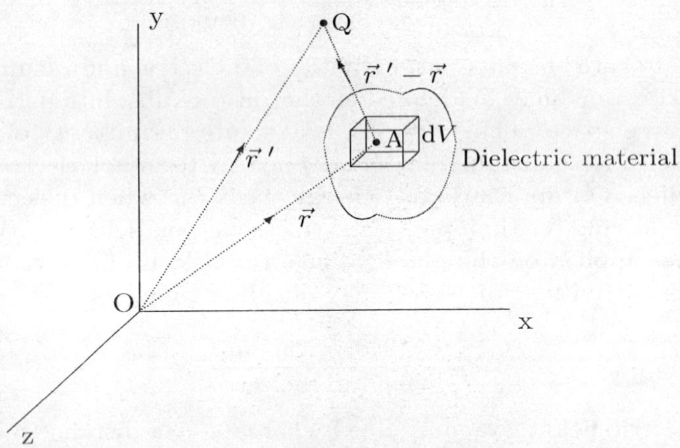

Figure 3: Dielectric material with external point Q.

Hence, the potential due to dielectric body at outside point Q is

$$\phi(\vec{r}\,') = \frac{1}{4\pi\epsilon_0} \int_V \frac{\vec{P}(\vec{r})\ dV(\vec{r}\,' - \vec{r})}{|\vec{r}\,' - \vec{r}|^3} = \frac{1}{4\pi\epsilon_0} \int_V \frac{\vec{P}(\vec{r})\ dV(\vec{r}\,' - \vec{r})^0}{|\vec{r}\,' - \vec{r}|^2}$$

where $(\vec{r}\,' - \vec{r})^0$ denotes unit vector directed towards Q from P. It is obviously equal to

$$(\vec{r}\,' - \vec{r})^0 = \frac{(x' - x)\ \hat{i} + (y' - y)\ \hat{j} + (z' - z)\ \hat{k}}{\sqrt{(x' - x)^2 + (y' - y)^2 + (z' - z)^2}}$$

where $(x',\ y',\ z')$ are Cartesian coordinates of point Q and $(x,\ y,\ z)$ are Cartesian coordinates of point A. Further,

$$(\vec{r}\,' - \vec{r})^0 = -\Big[\hat{i}\ \frac{\partial}{\partial x}(r' - r) + \hat{j}\ \frac{\partial}{\partial y}(r' - r) + \hat{k}\ \frac{\partial}{\partial z}(r' - r)\Big]$$

and

$$\vec{P}(\vec{r}) = P_x(x, y, z)\ \hat{i} + P_y(x, y, z)\ \hat{j} + P_z(x, y, z)\ \hat{k}$$

Substituting these values, we get

$$\phi(\vec{r}\,') = -\frac{1}{4\pi\epsilon_0}\Big[\int_V \frac{P_x}{(r' - r)^2}\ \frac{\partial}{\partial x}(r' - r)\ dV + \int_V \frac{P_y}{(r' - r)^2}\ \frac{\partial}{\partial y}(r' - r)\ dV$$

$$+\int_V \frac{P_z}{(r'-r)^2}\frac{\partial}{\partial z}(r'-r)\ \mathrm{d}V\Big] = \phi_1+\phi_2+\phi_3$$

Now, we consider three parts separately. It is obvious that

$$\frac{1}{|\vec{r}\,'-\vec{r}|^2}\frac{\partial}{\partial x}(r'-r) = -\frac{\partial}{\partial x}\Big(\frac{1}{|\vec{r}\,'-\vec{r}|}\Big) \qquad \text{and} \qquad \mathrm{d}V = \mathrm{d}x\ \mathrm{d}y\ \mathrm{d}z$$

Thus,

$$\phi_1 = \frac{1}{4\pi\epsilon_0}\iiint P_x\frac{\partial}{\partial x}\Big(\frac{1}{|\vec{r}\,'-\vec{r}|}\Big)\mathrm{d}x\ \mathrm{d}y\ \mathrm{d}z$$

Let us first integrate ϕ_1 with respect to x. If we consider integration along x-axis only, it would be simply a contribution of a straight rod of material area of cross section $\mathrm{d}y\mathrm{d}z$. If such element cuts the surface of body at x_1 and x_2 coordinates of respective points, then x integral would be

$$\int_{x_1}^{x_2} P_x\frac{\partial}{\partial x}\Big(\frac{1}{|\vec{r}\,'-\vec{r}|}\Big)\mathrm{d}x = \Big[\frac{P_x}{|\vec{r}\,'-\vec{r}|}\Big]_{x=x_2} - \Big[\frac{P_x}{|\vec{r}\,'-\vec{r}|}\Big]_{x=x_1} - \int_{x_1}^{x_2}\frac{1}{|\vec{r}\,'-\vec{r}|}\frac{\partial P_x}{\partial x}\ \mathrm{d}x$$

Thus,

$$\phi_1 = \frac{1}{4\pi\epsilon_0}\Big\{\iint\Big[\frac{P_x}{|\vec{r}\,'-\vec{r}|}\Big]_{x=x_2}\mathrm{d}y\mathrm{d}z - \iint\Big[\frac{P_x}{|\vec{r}\,'-\vec{r}|}\Big]_{x=x_1}\mathrm{d}y\mathrm{d}z - \iiint\frac{1}{|\vec{r}\,'-\vec{r}|}\frac{\partial P_x}{\partial x}\mathrm{d}x\mathrm{d}y\mathrm{d}z\Big\}$$

If $\hat{n}$ is unit vector in outward direction and normal at area-element $\mathrm{d}\vec{A}$, then

$$\mathrm{d}y\mathrm{d}z = \hat{n}\ .\ \mathrm{d}\vec{A}_2 = n_x\mathrm{d}A_2 \qquad \text{and} \qquad \mathrm{d}y\mathrm{d}z = \hat{n}\ .\ \mathrm{d}\vec{A}_1 = -n_x\mathrm{d}A_1$$

Since at x_1 which is on the part S_1, the x-component of $\hat{n}$, *i.e.*, n_x has negative value. Now the integral for ϕ_1 can be written as

$$\phi_1 = \frac{1}{4\pi\epsilon_0}\Big\{\int_{S_2}\frac{P_x n_x}{|\vec{r}\,'-\vec{r}|}\ \mathrm{d}A_2 + \int_{S_1}\frac{P_x n_x}{|\vec{r}\,'-\vec{r}|}\ \mathrm{d}A_1 - \int_V\frac{1}{|\vec{r}\,'-\vec{r}|}\frac{\partial P_x}{\partial x}\ \mathrm{d}V\Big\}$$

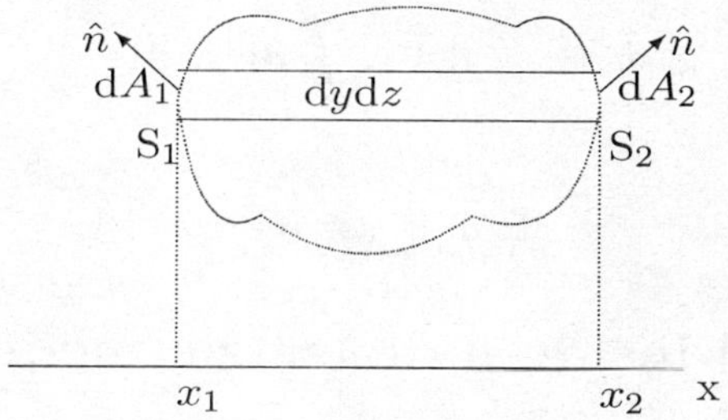

Figure 4: Integration with respect to x.

Since two integrals over S_1 and S_2 cover whole boundary S, we have

$$\phi_1 = \frac{1}{4\pi\epsilon_0}\int_S\frac{P_x n_x}{|\vec{r}\,'-\vec{r}|}\ \mathrm{d}A - \frac{1}{4\pi\epsilon_0}\int_V\frac{1}{|\vec{r}\,'-\vec{r}|}\frac{\partial P_x}{\partial x}\ \mathrm{d}V$$

Similar integrals for ϕ_2 and ϕ_3 can be written and hence value of ϕ is

$$\phi(\vec{r}\,') = \frac{1}{4\pi\epsilon_0}\Big[\int_S \frac{P_x n_x + P_y n_y + P_z n_z}{|\vec{r}\,' - \vec{r}|}\,\mathrm{d}A - \int_V \frac{1}{|\vec{r}\,' - \vec{r}|}\Big(\frac{\partial P_x}{\partial x} + \frac{\partial P_y}{\partial y} + \frac{\partial P_z}{\partial z}\Big)\,\mathrm{d}V\Big]$$

or

$$\phi(\vec{r}\,') = \frac{1}{4\pi\epsilon_0}\int_S \frac{\vec{P}\,.\,\hat{n}}{|\vec{r}\,' - \vec{r}|}\,\mathrm{d}A + \frac{1}{4\pi\epsilon_0}\int_V \frac{-\mathrm{div}\vec{P}}{|\vec{r}\,' - \vec{r}|}\,\mathrm{d}V \tag{4.1}$$

This equation obviously shows that potential at a point $\vec{r}\,'$ due to dielectric body is the same as if it were replaced by a system of bound charges in empty space. A part of these bound charges appears on the dielectric surface as a surface density $\sigma'_p = \vec{P}\,.\,\hat{n}$, and in the dielectric volume as a volume density $\rho'_p = -\mathrm{div}\vec{P}$. Therefore,

$$\phi(\vec{r}\,') = \frac{1}{4\pi\epsilon_0}\int_S \frac{\sigma'_p}{|\vec{r}\,' - \vec{r}|}\,\mathrm{d}A + \frac{1}{4\pi\epsilon_0}\int_V \frac{\rho'_p}{|\vec{r}\,' - \vec{r}|}\,\mathrm{d}V \tag{4.2}$$

In case of homogeneous and isotropic dielectric medium, where the polarization of each molecule is similar, the volume distribution of polarized charge is zero because the positive and negative charges of dipoles neutralize each other. Hence the second term does not contribute anything to the potential, *i.e.*, $\mathrm{div}\vec{P} = 0$. The potential is due to surface distribution of charges only. The intensity of electric field due to polarized volume may be expressed as

$$\vec{E}(\vec{r}\,') = -\mathrm{grad}[\phi(\vec{r}\,')] = \frac{1}{4\pi\epsilon_0}\Big[\int_S \frac{(\vec{r} - \vec{r}\,')\sigma'_p}{|(\vec{r} - \vec{r}\,')|^3}\,\mathrm{d}A + \int_V \frac{(\vec{r} - \vec{r}\,')\rho'_p}{|(\vec{r} - \vec{r}\,')|^3}\,\mathrm{d}V\Big]$$

This provides outer field due to polarization. Total bound charge of a dielectric body of volume V and complete boundary surface S is zero as the total charge is

$$q'_{\text{total}} = \oint_S \sigma'_p\,\mathrm{d}A + \int_V \rho'_p\,\mathrm{d}V = \oint_S \vec{P}\,.\,\hat{n}\,\mathrm{d}A - \int_V \mathrm{div}\vec{P}\,\mathrm{d}V$$

From Gauss' divergence theorem, we have

$$\oint_S \vec{P}\,.\,\hat{n}\,\mathrm{d}A = \int_V \mathrm{div}\vec{P}\,\mathrm{d}V$$

Therefore,

$$q'_{\text{total}} = 0$$

4. Electric field inside a dielectric medium

In the preceding section, we have calculated electric field intensity due to a polarized dielectric at an external point. Now, we intend to calculate field at an internal point, *i.e.*, within the dielectric material. Calculation of electric field intensity at an internal point poses a problem which is not encountered in the calculation of field at an external point. Actual field inside an insulator is complicated on the microscopic level, *i.e.*, on atom to atom basis. Very close an electron, the field is extremely large whereas a short distance away it may be small or it points in totally different direction. Moreover, an instant later, as the atoms move about, the field may change entirely. Thus, it is very difficult to calculate microscopic field. Instead, we ignore all such effects and concentrate on the macroscopic field, defined as an average field over regions that are large enough to contain many thousands of atoms, so as to ignore variations on atomic scale but not washing out any significant large scale variations of the field.

For getting $\vec{E}$ of polarization charges of medium, we make use the property that electrostatic field in a dielectric must have the same basic properties that are applicable to electric field $\vec{E}_0$ in vacuum; in particular, $\vec{E}$ is a conservative field, and thus, derivable from a scalar potential. That is $\vec{E} = -\text{grad}\phi$. Therefore (we know that curl of gradient of scalar is zero),

$$\nabla \times \vec{E} = 0$$

Using Stokes' theorem

$$\int_S (\nabla \times \vec{E}) \, . \, \hat{n} \, \mathrm{d}S = \oint_C \vec{E} \, . \, \mathrm{d}\vec{\ell}$$

above equation can be expressed as

$$\oint_C \vec{E} \, . \, \mathrm{d}\vec{\ell} = 0$$

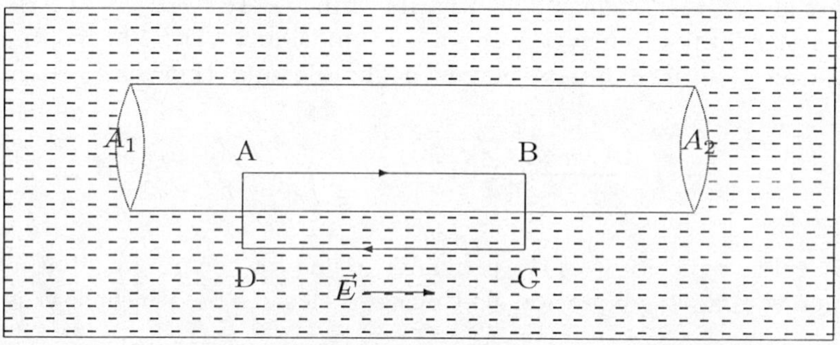

Figure 5: Cavity within dielectric having axis along the direction of applied electric field.

Let us apply this equation on path ABCD shown in Figure 5, where part AB lies in a cylindrical shaped cavity cut out of dielectric, and part CD lies in the dielectric. Since the parts AD and BC can be made arbitrarily small, line integral reduces to

$$\vec{E}_v \, . \, \vec{\ell} - \vec{E}_d \, . \, \vec{\ell} = 0$$

where $\vec{E}_v$ and $\vec{E}_d$ are electric fields in vacuum and dielectric, respectively. It gives

$$\vec{E}_{vt} = \vec{E}_{dt} \tag{4.3}$$

where $\vec{E}_{vt}$ and $\vec{E}_{dt}$ are tangential components of electric fields. Equation (4.3) is valid regardless of the orientation of cylinder-shaped cavity. If the cylinder is oriented along the direction of $\vec{E}$, then $\vec{E}_{dt} = \vec{E}_d$; furthermore by the symmetry, the field in the cavity is along the direction of cylinder, that is $\vec{E}_{vt} = \vec{E}_v$. Thus, for isotropic dielectric, we conclude as the following.

The electric field in a dielectric is equal to the electric field inside a cylinder-shaped cavity in dielectric provided the cavity-axis is oriented parallel to the direction of electric field.

Hence, the problem of calculation of electric field inside a dielectric reduces to the calculation of electric field inside a cylinder-shaped cavity in the dielectric. But, the dielectric field in the cavity is an external field and thus, may be determined with the help of result of previous article. Similar to previous article, here also we assume that polarization of dielectric is a given function $\vec{P}(x, y, z)$ and we calculate the potential and electric field due to this polarization. Considering the field point $\vec{r}\,'$ at the center of cavity and using equation (4.1)

$$\phi(\vec{r}\,') = \frac{1}{4\pi\epsilon_0}\int_{V_0-V_1}\frac{\rho_p'(x,y,z)}{|\vec{r}\,'-\vec{r}|}\,\mathrm{d}V + \frac{1}{4\pi\epsilon_0}\int_{A_0+A'}\frac{\sigma_p'(x,y,z)}{|\vec{r}\,'-\vec{r}|}\,\mathrm{d}A \tag{4.4}$$

where $(V_0 - V_1)$ is the volume of dielectric excluding cylinder, A_0 is exterior of dielectric and $A' = A_1 + A_2 + A_c$. But, from Figure 5, it can be seen that $\sigma_p' = 0$ on the cylindrical surface A_c of cylinder ($\sigma_p' = \vec{P}\,.\,\hat{n} = P\ell\cos 90° = 0$ over curved surface). As the cylinder is made arbitrarily thin so $A_1 = A_2 = 0$. Hence, only A_0 remains in the limit of surface integral of equation (4.4). The volume integral excludes the cavity as its contribution is negligible and therefore, equation (4.4) reduces to

$$\phi(\vec{r}\,') = \frac{1}{4\pi\epsilon_0}\int_{V_0}\frac{\rho_p'(x,y,z)}{|\vec{r}\,'-\vec{r}|}\,\mathrm{d}V + \frac{1}{4\pi\epsilon_0}\int_{A_0}\frac{\sigma_p'(x,y,z)}{|\vec{r}\,'-\vec{r}|}\,\mathrm{d}A$$

which is identical to equation (4.1) of previous article for the potential at an external point. In other words, this equation will provide electric field intensity as $\vec{E}(\vec{r}\,') = -\mathrm{grad}\phi(\vec{r}\,')$, giving

$$\vec{E}(\vec{r}\,') = \frac{1}{4\pi\epsilon_0}\int_{V_0}\frac{(\vec{r}-\vec{r}\,')\rho_p'}{|(\vec{r}-\vec{r}\,')|^3}\,\mathrm{d}V + \frac{1}{4\pi\epsilon_0}\int_{A_0}\frac{(\vec{r}-\vec{r}\,')\sigma_p'}{|(\vec{r}-\vec{r}\,')|^3}\,\mathrm{d}A$$

which gives the contribution of dielectric medium to the electric field at $\vec{r}\,'$, irrespective of whether $\vec{r}\,'$ is inside or outside the medium.

Example 1: Show that electric field inside a dielectric depends on the shape of cavity outside of dielectric.

Solution: Suppose a cavity is formed in a dielectric having center at O and axis along the direction of applied electric field. At O, field in the cavity is

$$\vec{E}_c = \vec{E}_1 + \vec{E}_2$$

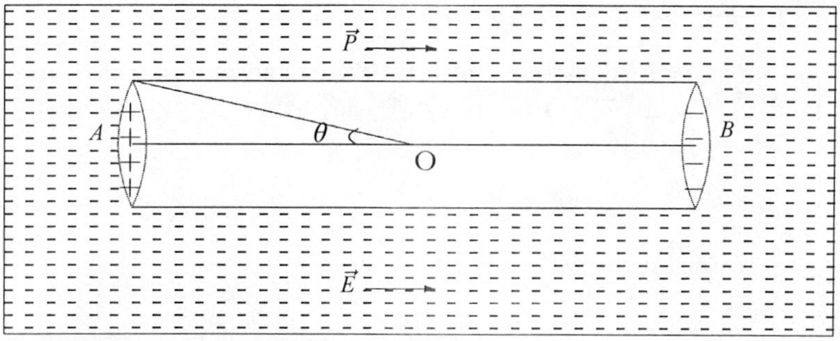

Figure 6: Cavity within dielectric having center at O and axis along the direction of applied field.

where the field $\vec{E}_1$ is due to macroscopic field $\vec{E}$ in the dielectric. For a very small cavity, contribution due to macroscopic field at O is the same as that at O in absence of cavity, and thus, $\vec{E}_1 = \vec{E}$. Field $\vec{E}_2$ at O due to polarization charges over the walls of cavity. As there is no charge induced on curved surface ($\sigma'_p = \vec{P} \cdot \hat{n} = 0$, as the angle is 90°), only induced charges on two circular ends A and B contribute to the field at O. The field at O due to one of the end charges is

$$\vec{E}_2 = \frac{1}{4\pi\epsilon_0} \int \frac{\sigma'_p \vec{r}}{r^3} \, dA = \frac{1}{4\pi\epsilon_0} \sigma'_p \, \hat{n} \int d\omega$$

where $\int d\omega$ is solid angle subtended by the face area at O and is given by

$$\int d\omega = 2\pi(1 - \cos\theta)$$

so that the field at O due to both end charges is

$$\vec{E}_2 = \frac{1}{4\pi\epsilon_0} \sigma'_p \, \hat{n} \, 2\pi(1 - \cos\theta) \times 2$$

Putting polarization $\vec{P} = \sigma'_p \, \hat{n}$, we have

$$\vec{E}_2 = \frac{\vec{P}}{\epsilon_0} (1 - \cos\theta)$$

Therefore

$$\vec{E}_c = \vec{E} + \frac{\vec{P}}{\epsilon_0} (1 - \cos\theta)$$

showing that $\vec{E}_c$ depends on θ, *i.e.*, on the shape of cavity.

Example 2: In a dielectric having uniform and fixed polarization $\vec{P}$, cavities are cut: (i) long cylinder-shaped cavity, (ii) a narrow slit at right angle to its polarization, and

(iii) spherical cavity. Find electric field at the center of each cavity due to charge produced on the surface of cavity. Hence, write expression for net field strength within the cavity for applied electric field $\vec{E}$.

Solution: Let us consider a dielectric slab of infinite extent polarized along x-axis due to applied field $\vec{E}$ (Figure 7). One cavity of above three types is cut in that slab at a time. The net field in side cavity, $\vec{E}_c$ is

$$\vec{E}_c = \vec{E} + \vec{E}_2$$

where the field due to induced charges is given by

$$\vec{E}_2 = \frac{\vec{P}}{\epsilon_0}(1 - \cos\theta)$$

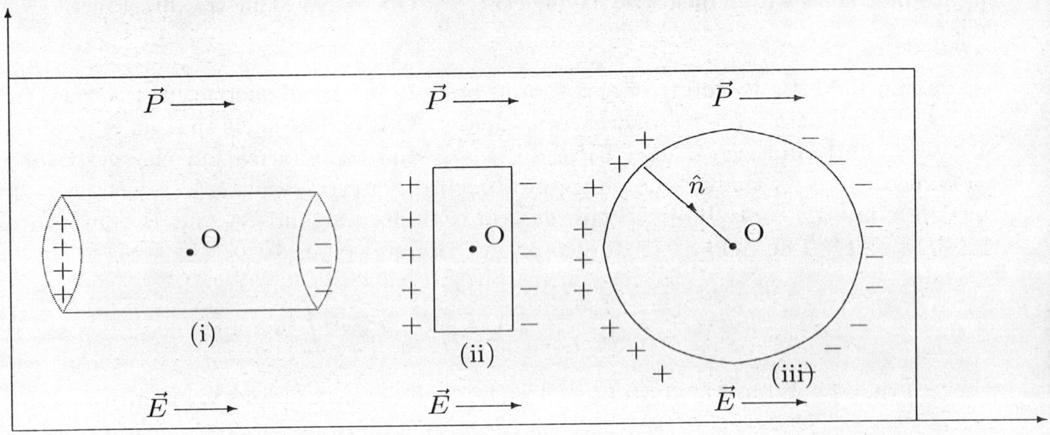

Figure 7: Cavities of various shapes within dielectric of infinite extent having center at O.

(i) For a cylinder-shaped cavity: When the side surfaces are very small, it approaches to zero in the limit and cavity has only length, such that $\theta = 0°$, then we have

$$\vec{E}_c = \vec{E} + \frac{\vec{P}}{\epsilon_0}(1 - \cos 0°) = \vec{E}$$

That is the field inside the cavity is same as the field in the dielectric provided the axis of cavity is along the electric field $\vec{E}$.

(ii) For disc (or narrow slit at right angle to polarization) type cavity: In this case, $\theta = \pi/2$ and therefore, the field inside the cavity is

$$\vec{E}_c = \vec{E} + \frac{\vec{P}}{\epsilon_0}(1 - \cos 90°) = \vec{E} + \frac{\vec{P}}{\epsilon_0}$$

(iii) For a spherical cavity: Because equation (4.4) is not valid in this type of cavity, we calculate electric field at the center of cavity due to induced charges using basic relation

$$\vec{E_2} = \frac{1}{4\pi\epsilon_0}\int \frac{\sigma'_p \vec{r}\,'}{r'^3}\,\mathrm{d}A = \frac{1}{4\pi\epsilon_0}\int \frac{\vec{P}\,.\,\hat{n}}{r'^2}\,\hat{n}\,\mathrm{d}A$$

$$= \frac{1}{4\pi\epsilon_0}\int \frac{\vec{P}\cos\theta}{r'^2}\,n\,\mathrm{d}A$$

By symmetry, the component of field from each surface element is parallel to $\vec{P}$ or $\vec{E}$. That is,

$$\vec{E_2} = \frac{1}{4\pi\epsilon_0}\int \frac{\vec{P}\cos\theta}{r^2}\,\cos\theta\,\mathrm{d}A$$

$$= \frac{1}{4\pi\epsilon_0}\int_0^{\pi} \frac{\vec{P}\cos\theta}{r^2}\,\cos\theta\,(2\pi r^2\sin\theta\,\mathrm{d}\theta)$$

where $\mathrm{d}A = 2\pi r^2\sin\theta\,\mathrm{d}\theta$ is the area of ring strip taken over the surface of spherical cavity. Therefore,

$$\vec{E_2} = \frac{\vec{P}}{2\epsilon_0}\int_0^{\pi}\cos^2\theta\,\sin\theta\,\mathrm{d}\theta = \frac{\vec{P}}{3\epsilon_0}$$

Hence, the net field within cavity is

$$\vec{E_c} = \vec{E} + \frac{\vec{P}}{3\epsilon_0}$$

5. Relation between electric displacement $\vec{D}$, electric field $\vec{E}$ and polarization $\vec{P}$

(1) **Electric intensity $\vec{E}$** : The electric intensity $\vec{E}$ at any point in electric field is numerically equal to the force experienced by a unit positive charge placed at that point. The direction of $\vec{E}$ is the same as that of the field.

(2) **Dielectric polarization $\vec{P}$** : Suppose, dielectric material is placed between two oppositely charged metal plates, each of surface area A and separated by a distance ℓ. The dielectric gets polarized. The distorted atoms due to relative displacement of positive and negative charges of atom are called electric dipoles. The electric dipole moment per unit volume of dielectric is called dielectric polarization, denoted by $\vec{P}$. It is a vector lying parallel to the applied electric field[2]. Suppose, $-q'$ and $+q'$ induced surface charges appear near positively and negatively charged metal plates. The electric dipole moment is $q'l$ and thus, the polarization is

$$P = \frac{q'\ell}{A\ell} = \frac{q'}{A}$$

(3) **Dielectric displacement $\vec{D}$** : When no dielectric is present, Gauss' law gives

$$\oint \vec{E}\,.\,\mathrm{d}\vec{S} = \frac{q}{\epsilon_0} \qquad E_0 A = \frac{q}{\epsilon_0} \qquad E_0 = \frac{q}{\epsilon_0 A} \tag{4.5}$$

[2]Direction of electric dipole moment is from negative charge towards positive charge.

where E_0 is electric field in free space and q is free charge. When dielectric is present, induced surface charge, say, $-q'$ appears, so that Gaussian surface encloses $(q-q')$ charge, and electric field is modified to $\vec{E}$, such that

$$\vec{E} = \frac{\vec{E}_0}{\epsilon_r}$$

where ϵ_r is relative permittivity of dielectric medium. So that

$$E = \frac{q}{\epsilon_r \epsilon_0 A} = \frac{q}{\epsilon A}$$

where $\epsilon = \epsilon_r \epsilon_0$ is permittivity of dielectric medium. Now, Gauss' law gives

$$\oint \vec{E} \cdot d\vec{S} = \frac{(q-q')}{\epsilon_0} \qquad EA = \frac{(q-q')}{\epsilon_0} \qquad \frac{q}{A} = \epsilon_0 E + \frac{q'}{A} \tag{4.6}$$

In the quantity on left q/A is called electric displacement, denoted by D. Thus,

$$D = \epsilon_0 E + P \tag{4.7}$$

This relation in vector form is expressed as

$$\vec{D} = \epsilon_0 \vec{E} + \vec{P}$$

Relation between electrical susceptibility and relative permittivity

Using expressions in Equation (4.7), we have

$$\epsilon E = \epsilon_0 E + k\epsilon_0 E \qquad \epsilon_r = 1 + k \tag{4.8}$$

Relation between ρ and ρ'

A relation can also be obtained between ρ and ρ' as the following.

$$\rho' = -\nabla \cdot \vec{P} = -\nabla \cdot \epsilon_0(\epsilon_r - 1)\vec{E} = -\nabla \cdot \epsilon_0 \epsilon_r \vec{E}\left(1 - \frac{1}{\epsilon_r}\right)$$

$$= -\nabla \cdot \vec{D}\left(1 - \frac{1}{\epsilon_r}\right) = -\rho\left(1 - \frac{1}{\epsilon_r}\right)$$

As $\epsilon_r > 1$, we have $\rho' < -\rho$, *i.e.*, the bound charge density is always opposite in sign and less in magnitude compared to free charge density.

6. Point charge in dielectric fluid

Consider a point charge +q located in a homogeneous and isotropic dielectric medium of infinite extent. The dielectric medium is supposed to be linear and characterized by dielectric constant ϵ_r. Due to the point charge +q, the medium is polarized radially, *i.e.*, the centers of negative charges of constituent molecules of the medium are pulled towards the charge +q and that of positive charges are pushed away. We know that the electric field due to a point charge +q is radial when it is placed in vacuum. In the dielectric also the field is radial. This is not changed due to the presence of medium because at each point $\vec{E}$, $\vec{D}$ and $\vec{P}$ are parallel to each other. From the symmetry of problem, $\vec{E}$, $\vec{D}$ and $\vec{P}$ depend only on the distance from the point charge and not on any angular coordinate. Considering +q at the origin of coordinate system, let us draw a spherical surface of radius r as shown in Figure 8. Applying Gauss' law to this spherical surface, we get

$$\oint \vec{E} \,.\, d\vec{S} = \frac{q}{\epsilon_0} \qquad \oint \epsilon_0 \vec{E} \,.\, \hat{n}\, dS = q \qquad \oint \vec{D} \,.\, \hat{n}\, dS = q$$

$$D\, 4\pi r^2 = q \qquad D = \frac{q}{4\pi r^2} \qquad \vec{D} = \frac{q\vec{r}}{4\pi r^3}$$

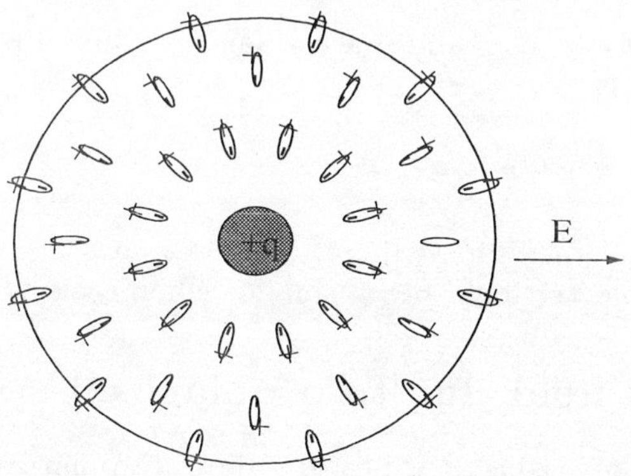

Figure 8: A point charge +q is situated at the origin coinciding with the center of spherical surface.

We know that electric field $\vec{E}$ is expressed as $\vec{E} = \vec{D}/\epsilon$, where $\vec{D}$ is electric displacement and ϵ is permittivity of medium. Thus,

$$\vec{E} = \frac{\vec{D}}{\epsilon_r \epsilon_0} = \frac{q\vec{r}}{4\pi r^3 \epsilon_r \epsilon_0}$$

showing that the electric field in dielectric is smaller by a factor ϵ_r as compared to the electric field for free space. As $\vec{D} = \epsilon_0 \vec{E} + \vec{P}$, the polarization is

$$\vec{P} = \vec{D} - \epsilon_0 \vec{E} = \epsilon_r \epsilon_0 \vec{E} - \epsilon_0 \vec{E} = (\epsilon_r - 1)\epsilon_0 \vec{E} = \frac{(\epsilon_r - 1) q\vec{r}}{4\pi r^3 \epsilon_r}$$

Let us try to understand why dielectric has weakened the electric field. The electric field has its origin in all of the charges whether free or bound. The free charge is just the point charge +q. The bound charge is developed from two contributions:

(i) due to volume density $\rho_P = -\nabla \,.\, \vec{P}$

(ii) due to surface density $\sigma_P = -\vec{P} \,.\, \hat{n}$ on the surface of dielectric in contact with the point charge +q.

Taking divergence of $\vec{P}$, we have

$$\nabla \cdot \vec{P} = \text{div}\left[\frac{(\epsilon_r - 1)q\vec{r}}{4\pi r^3 \epsilon_r}\right] = \frac{(\epsilon_r - 1)q}{4\pi\epsilon_r} \text{ div}\left[\frac{\vec{r}}{r^3}\right] = 0$$

Therefore, $\rho_P = 0$, *i.e.*, there is no volume density of bound charge in this case. Thus, at the macroscopic scale, in the volume, +q is only charge. However, at the molecular scale, it may not be true. Consider a radius b which will be made to approach to zero. Then, total surface bound charge is

$$Q_P = \lim_{b \to 0} 4\pi b^2 (-\vec{P} \cdot \hat{n})_{r=b}$$

$$= -\lim_{b \to 0} 4\pi b^2 \left(\frac{(\epsilon_r - 1)q\vec{r}}{4\pi r^3 \epsilon_r} \cdot \hat{n}\right)_{r=b} = -\frac{(\epsilon_r - 1)q}{\epsilon_r}$$

Here, we have used $\vec{r} \cdot \hat{n} = r$, as $\vec{r}$ and $\hat{n}$ are along the same direction. Total charge within Gaussian surface is

$$q + Q_P = q - \frac{(\epsilon_r - 1)q}{\epsilon_r} = \frac{q}{\epsilon_r}$$

It shows that presence of medium reduces the point charge q to a value q/ϵ_r. It explains the reason for electric field being smaller when dielectric is present.

7. Potential and field due to a polarized sphere

Consider a solid sphere of radius R which is uniformly polarized with polarization $\vec{P}$. Let us calculate potential and field inside and outside the sphere.

We consider a thin cylindrical column (Figure 9 (i)) with edge surface area δs. The component of $\vec{P}$ in the direction of δs is $\vec{P}\cos\theta$, where θ is angle between the directions of $\vec{P}$ and δs. Thus, the charge on the upper edge surface area is

$$+\vec{P}\cos\theta\ \delta s$$

while on the lower edge surface area of this column is

$$-\vec{P}\cos\theta\ \delta s$$

For calculating surface charge on the upper half of sphere, we consider a cone as shown in Figure 9 (ii), with ring area on the surface as $2\pi R^2 \sin\theta\ d\theta$. As the sphere is uniformly polarized, $\vec{P}$ is same everywhere, so that total charge on upper half surface of the sphere is

$$\int_0^{\pi/2} (2\pi R^2 \sin\theta\ d\theta)(P\cos\theta) = \pi R^2 P \int_0^{\pi/2} \sin 2\theta\ d\theta = \pi R^2 P$$

Similarly, total surface charge on lower half of this sphere is $-\pi R^2 P$. Thus, total charge on the polarized surface is zero. To calculate electrical properties, we consider this sphere as composed of two identical spheres, one having a uniform

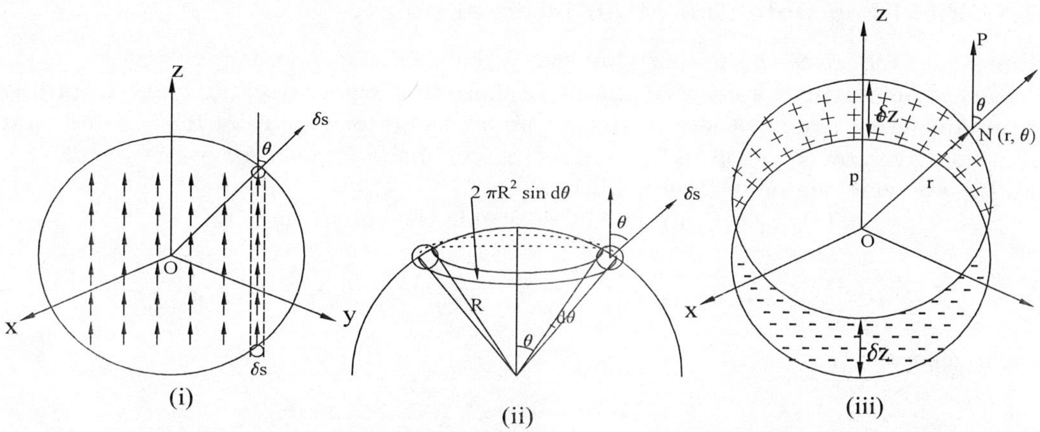

Figure 9: (i) A sphere with cylindrical element having base area δs. (ii) A cone cut out of sphere.

positive charge of volume density $\vec{P}/\delta z$ and the other having a uniform positive charge of volume density $-\vec{P}/\delta z$ with their centers separated by a distance δz as shown in Figure 9 (iii). In the overlap region, net charge density is obviously zero.

Total charge inside the sphere is $\pm(4\pi R^3/3)(\vec{P}/\delta z)$. For calculation of field at an external point, we consider these charges to be concentrated at their centers, thus constituting an electric dipole of moment

$$\vec{p} = \left(\frac{4\pi R^3}{3}\right)\frac{\vec{P}}{\delta z}\,\delta z = \frac{4\pi R^3}{3}\,\vec{P}$$

In order to calculate field and potential at an external point N $(r,\ \theta)$ we use expression of potential and field due to a dipole and they are obtained by putting $\vec{p} = (4\pi R^3/3)\ \vec{P}$. The potential is

$$\phi_0 = \frac{1}{4\pi\epsilon_0}\frac{p\cos\theta}{r^2} = \frac{PR^3\cos\theta}{3\epsilon_0 r^2}$$

Electrical field is

$$(E_r)_0 = \frac{1}{4\pi\epsilon_0}\frac{2p\cos\theta}{r^3} = \frac{2PR^3\cos\theta}{3\epsilon_0 r^3}$$

and

$$(E_\theta)_0 = \frac{1}{4\pi\epsilon_0}\frac{p\sin\theta}{r^3} = \frac{PR^3\sin\theta}{3\epsilon_0 r^3}$$

If this point lies on the surface of polarized sphere, then $r = R$, so that

$$\phi_0 = \frac{PR^3\cos\theta}{3\epsilon_0 R^2} = \frac{PR\cos\theta}{3\epsilon_0}$$

But, $R\cos\theta$ is component of R along z-axis, *i.e.*, z-coordinate of point N. Thus, we have

$$\phi_0 = \frac{Pz}{3\epsilon_0} \tag{4.9}$$

It shows that potential at an external point depends on its z-coordinate.

7.1 Field and potential at an internal point:

Equation (4.9) gives the potential at every point on the boundary of region, inside which ϕ must satisfy Laplace's equation because this region is charge free. According to uniqueness theorem, there can exist one and only one solution for ϕ_m and that must be given by equation (4.9). Therefore, ϕ_{in} may be taken as given by equation (4.9), satisfying the boundary condition.

As $(E_x)_{in} = (E_y)_{in} = 0$, we have field inside the sphere as

$$(E_z)_{in} = -\frac{\partial \phi_{in}}{\partial z} = -\frac{\partial}{\partial z}\left(\frac{Pz}{3\epsilon_0}\right) = -\frac{P}{3\epsilon_0}$$

or in general

$$\vec{E_{in}} = -\frac{\vec{P}}{3\epsilon_0}$$

It remains constant throughout and points opposite to the direction of polarization (Fig. 10).

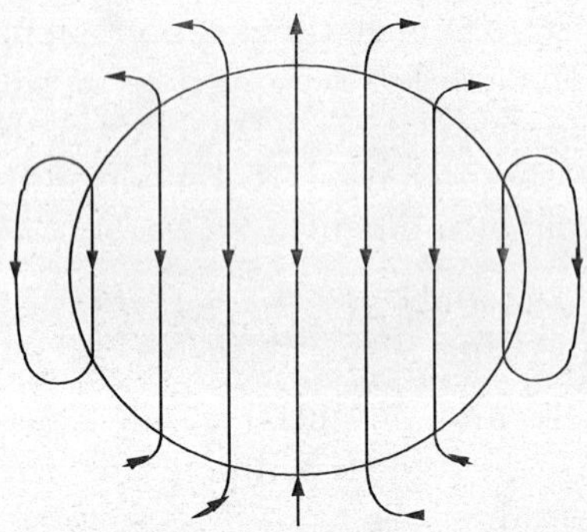

Figure 10: Field inside and outside a polarized sphere.

Example 3: A metal sphere of radius A carrying charge Q is surrounded, out to radius B, by linear dielectric of permittivity ϵ. Calculate potential at the center (relative to infinity). Compute the polarization and hence the surface and volume bound charges.

Solution: Free charge Q on metal sphere is distributed spherically symmetric. The displacement for all point $r > A$ can be expressed as

$$\vec{D} = \frac{Q}{4\pi r^2}\,\hat{r}$$

where $\hat{r}$ denotes unit vector. For all points $r < A$, *i.e.*, inside the metal sphere,

$$\vec{E} = \vec{P} = \vec{D} = 0$$

Now, there are two regions for consideration:

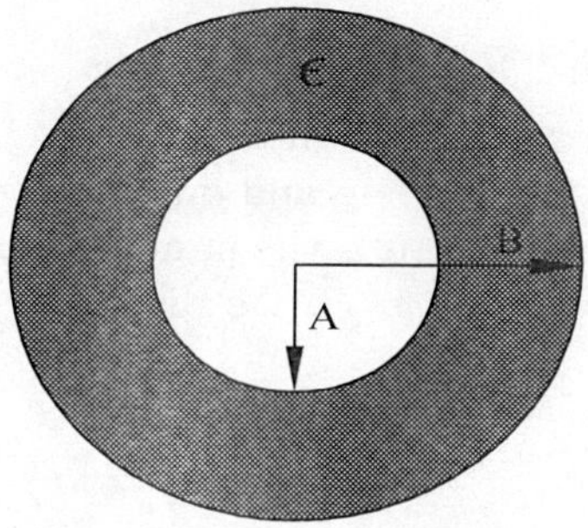

Figure 11: Metal sphere of radius A carrying charge Q is surrounded, out to radius B, by linear dielectric of permittivity ϵ.

(i) $A < r < B$ which is linear dielectric material of permittivity ϵ.

(ii) $r > B$ which is free space of permittivity ϵ_0.

For these regions, we can write electric field on using the expression for $\vec{D}$. That is,

$$\vec{E} = \begin{cases} \frac{1}{4\pi\epsilon}\frac{Q}{r^2}\,\hat{r} & \text{for } A < r < B \\ \frac{1}{4\pi\epsilon_0}\frac{Q}{r^2}\,\hat{r} & \text{for } r > B \end{cases}$$

Therefore, potential at the center is

$$V = -\int_{\infty}^{A} \vec{E}\,.\,d\vec{\ell} = -\int_{\infty}^{B} \frac{1}{4\pi\epsilon_0}\frac{Q}{r^2}\,dr - \int_{B}^{A} \frac{1}{4\pi\epsilon}\frac{Q}{r^2}\,dr$$

$$= -\frac{Q}{4\pi\epsilon_0}\Big(-\frac{1}{r}\Big)_{\infty}^{B} - \frac{Q}{4\pi\epsilon}\Big(-\frac{1}{r}\Big)_{B}^{A} = \frac{Q}{4\pi}\Big[\frac{1}{\epsilon_0 B} + \frac{1}{\epsilon A} - \frac{1}{\epsilon B}\Big]$$

Polarization in dielectric is

$$\vec{P} = k\epsilon_0\vec{E}$$

where k is susceptibility of dielectric. Thus,

$$\vec{P} = k\epsilon_0\frac{1}{4\pi\epsilon}\frac{Q}{r^2}\,\hat{r}$$

where $\vec{E}$ has been written for $A < r < B$. Therefore, bound charges that appear on the surface are

$$\sigma'_p = \vec{P}\,.\,\hat{n} = \begin{cases} \frac{k\epsilon_0 Q}{4\pi\epsilon B^2} & \text{at the outer surface} \\ -\frac{k\epsilon_0 Q}{4\pi\epsilon A^2} & \text{at the inner surface} \end{cases}$$

Example 4: A dielectric sphere is placed in a uniform electrostatic field. Draw a neat diagram showing a distortion produced in the field. Calculate the electric field outside and inside the dielectric sphere.

Solution: (i) Resultant field inside the sphere

Suppose a dielectric sphere is placed in a uniform electric field $\vec{E}_0$. The sphere is polarized uniformly so that field inside and outside the sphere is not equal to $\vec{E}_0$. The resultant field $\vec{E}$ inside the sphere is the sum of uniform applied field $\vec{E}_0$ and a field $\vec{E}_{in}$ generated by the polarization $\vec{P}$. As discussed earlier, the field developed inside the sphere is

$$\vec{E}_{in} = -\frac{\vec{P}}{3\epsilon_0}$$

Thus, electric field inside the sphere is

$$\vec{E} = \vec{E}_0 + \vec{E}_{in} = \vec{E}_0 - \frac{\vec{P}}{3\epsilon_0}$$

Using the relation $\vec{P} = \epsilon_0(\epsilon_r - 1)\vec{E}$, where ϵ_r is dielectric constant of material of sphere, we get

$$\vec{E} = \vec{E}_0 + \vec{E}_{in} = \vec{E}_0 - \frac{\epsilon_0(\epsilon_r - 1)\vec{E}}{3\epsilon_0} \quad \text{or} \quad \vec{E} = \left(\frac{3}{\epsilon_r + 2}\right)\vec{E}_0$$

It gives the field inside the dielectric. Since $\epsilon_r > 1$, it implies that $\vec{E} < \vec{E}_0$. That is, the resultant field inside the dielectric is less than the applied field $\vec{E}_0$. It is along the direction of $\vec{E}_0$ and is uniform within the sphere.

The polarization is therefore

$$\vec{P} = \frac{3\epsilon_0(\epsilon_r - 1)}{\epsilon_r + 2}\vec{E}_0$$

(ii) Resultant field outside the dielectric sphere

Field outside dielectric sphere, at a point not very far from it (at the off points, field is $\vec{E}_0$) is the sum of uniform applied field $\vec{E}_0$ and $\vec{E}_{out}$ due to polarization of dielectric. From earlier discussion, we note that external field components are

$$(E_r)_{out} = \frac{2PR^3\cos\theta}{3\epsilon_0 r^3} \quad \text{and} \quad (E_\theta)_{out} = \frac{2PR^3\sin\theta}{3\epsilon_0 r^3}$$

where R is radius of sphere, r and θ specify the distance and direction of the point from the center of sphere. Thus,

$$\vec{E}_{out} = (E_r)_{out}\,\hat{r}' + (E_\theta)_{out}\,\hat{\theta}'$$

where $\hat{r}'$ and $\hat{\theta}'$ are unit vectors in the directions of r increasing and θ increasing. Thus, we have

$$\vec{E}_{out} = \frac{2PR^3\cos\theta}{3\epsilon_0 r^3}\,\hat{r}' + \frac{PR^3\sin\theta}{3\epsilon_0 r^3}\,\hat{\theta}'$$

so that resultant field outside the dielectric sphere at a point (r, θ) is

$$\vec{E} = \vec{E}_0 + \vec{E}_{out}$$

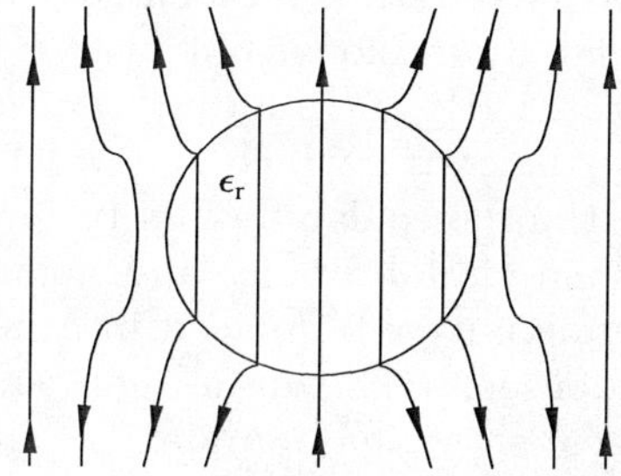

Figure 12: Electric field inside and outside dielectric sphere.

$$= \vec{E}_0 + \frac{2PR^3 \cos\theta}{3\epsilon_0 r^3}\,\hat{r}' + \frac{PR^3 \sin\theta}{3\epsilon_0 r^3}\,\hat{\theta}'$$

Using the value of P, we get

$$\vec{E} = \vec{E}_0 + \frac{2(\epsilon_r - 1)R^3 E_0 \cos\theta}{(\epsilon_r + 2)r^3}\,\hat{r}' + \frac{(\epsilon_r - 1)R^3 E_0 \sin\theta}{(\epsilon_r + 2)r^3}\,\hat{\theta}'$$

8. Molecular field in a dielectric (Claussius-Mossotti relation) (dispersion in liquids and solids)

Let us examine the nature of dielectric and see how the electric field, responsible for polarizing the molecule, is related to the macroscopic electric field, which is known as the molecular field or local field. This molecular field is produced by all the external sources and by all the polarized molecules in the dielectric except the molecule under consideration. The molecular field $\vec{E}_{in}$ may be calculated in the following manner.

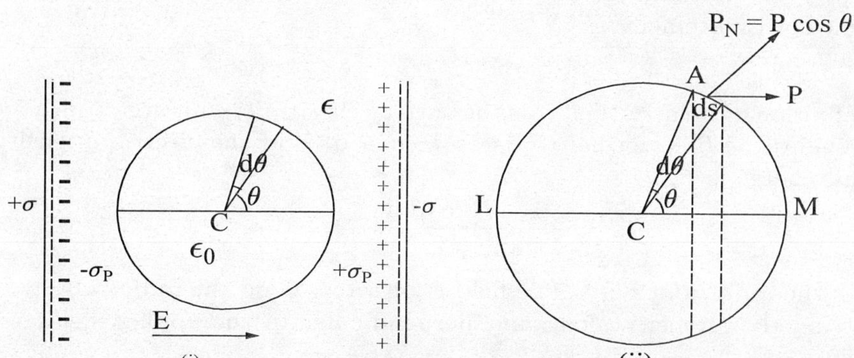

Figure 13: (i) Spherical cavity in dielectric placed in a uniform electric field. (ii) Spherical cavity.

Let us cut a spherical cavity of radius r (such that its dimensions are very large as compared to the molecular dimensions, but are very small as compared to the volume of dielectric) (Figure 13 (i)) surrounding the point at which the molecular field is to be calculated. The dielectric, which is now left, is treated as a continuum. Except the molecule where the molecular field is to be calculated, other molecules in the cavity are treated as individual dipoles and not as a continuum.

Let the dielectric of permittivity ϵ be placed in a uniform electric field between two parallel plates (of a condenser) as shown in Figure 13 (i). The dotted lines (close to the plates of the condenser) show the boundary of the dielectric. Let the surface density of real charges on the capacitor plates be σ, and the surface of dielectric has the polarized charges of surface density σ_P.

The field $\vec{E}_{in}$ experienced by the molecule of dielectric at the center C of the cavity is given by

$$\vec{E}_{in} = \vec{E}_1 + \vec{E}_2 + \vec{E}_3 + \vec{E}_4$$

where

(i) $\vec{E}_1$ is the field between two plates with no dielectric, so that $\vec{E}_1 = \vec{\sigma}/\epsilon_0$.

(ii) $\vec{E}_2$ is the field due to polarized charges on the plane surface of dielectric facing the capacitor plates and is expressed as $\vec{E}_2 = \vec{\sigma}_P/\epsilon_0$.

(iii) $\vec{E}_3$ is the field due to polarized charges on the surface of the cavity, to be calculated.

(iv) $\vec{E}_4$ is the field due to permanent dipoles. But, in present case, for non-polar isotropic dielectrics $\vec{E}_4 = 0$.

Thus,

$$\vec{E}_{in} = \frac{\vec{\sigma}}{\epsilon_0} - \frac{\vec{\sigma}_P}{\epsilon_0} + \vec{E}_3$$

Evaluation of $\vec{E}_3$: Consider a small element area ds on the surface of cavity of an angular width dθ and at an angle θ with the direction of field $\vec{E}$ (Figure 13 (i)). The vector $\vec{P}$ shows the direction of displacement at the center of ds. The normal component of displacement is

$$\vec{P}_N = \vec{P}\cos\theta$$

By definition of polarization, it is the surface charge per unit area. Such a charge on ds provides the flux normal to ds as $\vec{P}\cos\theta$ ds and the electric intensity at C due to this charge is

$$\frac{1}{4\pi\epsilon_0}\frac{\vec{P}\cos\theta\, \mathrm{d}s}{r^2}$$

where r is the radius of cavity. The field is directed along the radius CA.

Resolving the intensity along and perpendicular to the applied field, we have components

$$\frac{\vec{P}\cos\theta\, \mathrm{d}s}{4\pi\epsilon_0 r^2}\cos\theta \qquad \text{along the field}$$

$$\frac{\vec{P}\cos\theta \; ds}{4\pi\epsilon_0 r^2} \sin\theta \qquad \text{perpendicular to field}$$

If the area ds be taken round through 2π radians about LM, it describes a ring, whose surface area is $2\pi r \sin\theta \;.\; r \; d\theta = 2\pi r^2 \sin\theta \; d\theta$. Therefore, intensity at C due to ring in the direction of field is

$$\frac{\vec{P}\cos^2\theta}{4\pi\epsilon_0 r^2} 2\pi r^2 \sin\theta \; d\theta = \frac{\vec{P}\cos^2\theta\sin\theta}{2\epsilon_0} \, d\theta$$

while the normal components of intensity due to ring cancel each other. Integrating the intensity at C due to charges on the surface of cavity, we have

$$\vec{E}_3 = \frac{\vec{P}}{2\epsilon_0}\int_0^{\pi} \cos^2\theta\sin\theta \; d\theta = \frac{\vec{P}}{2\epsilon_0} \times \frac{2}{3} = \frac{\vec{P}}{3\epsilon_0}$$

Therefore, total intensity or $\vec{E}_{in}$ at C is

$$\vec{E}_{in} = \frac{\vec{\sigma}}{\epsilon_0} - \frac{\vec{\sigma}_P}{\epsilon_0} + \frac{\vec{P}}{3\epsilon_0}$$

The resultant intensity $\vec{E}$ between the plates is

$$\vec{E} = \frac{\vec{\sigma}}{\epsilon_0} - \frac{\vec{\sigma}_P}{\epsilon_0}$$

Thus, we have

$$\vec{E}_{in} = \vec{E} + \frac{\vec{P}}{3\epsilon_0}$$

Further, we know that

$$\vec{D} = \epsilon\vec{E} = \epsilon_0\vec{E} + \vec{P}$$

so that

$$\vec{E} = \frac{\vec{P}}{\epsilon - \epsilon_0}$$

Using this value of $\vec{E}$, we have

$$\vec{E}_{in} = \frac{\vec{P}}{\epsilon - \epsilon_0} + \frac{\vec{P}}{3\epsilon_0} = \frac{\vec{P}}{3\epsilon_0}\left(\frac{\epsilon + 2\epsilon_0}{\epsilon - \epsilon_0}\right)$$

If N be the number of molecules per unit volume and α the molecular polarizability, then polarization $\vec{P}$ is defined as electric moment per unit volume. That is

$$\vec{P} = N\alpha\vec{E}_{in}$$

Thus, we have

$$\frac{\vec{P}}{N\alpha} = \frac{\vec{P}}{3\epsilon_0}\left(\frac{\epsilon + 2\epsilon_0}{\epsilon - \epsilon_0}\right) \qquad \frac{N\alpha}{3\epsilon_0} = \frac{\epsilon - \epsilon_0}{\epsilon + 2\epsilon_0} \qquad \frac{N\alpha}{3\epsilon_0} = \frac{\epsilon_r - 1}{\epsilon_r + 2}$$

where $\epsilon_r = \epsilon/\epsilon_0$ is relative permittivity or dielectric constant. This relation is well known **Claussius-Mossotti relation.**

8.1 Validity of Claussius-Mossotti relation:

The number of molecules per unit volume is proportional to the density of gas. If molecular polarizability α is taken constant, the number n is proportional to $(\epsilon_r - 1)/(\epsilon_r + 2)$. Also the constant α is proportional to the cubic of radius of molecule. Hence, finding ϵ_r experimentally and calculating n from the density at definite temperature and pressure, the value of radius of molecule of dielectric may be estimated. The value, so obtained, fairly agrees with the values obtained by other methods and thus, proves the validity of Claussius-Mossotti relation. It is true only in case of mono-atomic gases and weak solutions. The experimental and theoretical values disagree in case of strong solutions and solids. It is due to the fact that in these cases, the interaction forces among molecules are sufficiently large, which has not been considered.

8.2 Molar polarizibility

Multiplying Claussius-Mossotti relation by $M^\star/d$ on both sides, where $M^\star$ is molecular weight and d the density of dielectric, we get

$$\frac{\epsilon_r - 1}{\epsilon_r + 2}\frac{M^\star}{d} = \frac{\alpha}{3\epsilon_0}\frac{nM^\star}{d}$$

We know that $(nM^\star/d) = N$ is Avogadro number. Thus,

$$\frac{N\alpha}{3\epsilon_0} = \frac{\epsilon_r - 1}{\epsilon_r + 2}\frac{M^\star}{d}$$

where $N\alpha$ is known as **molar polarizability**.

9. Polarization of polar molecules - Debye's formula

In case of non-polar molecules, we have found that molecular polarizability α does not depend on temperature. But, it is not the case with polar molecules. If polar molecules are subjected to electric field then in addition to induced polarization, the permanent polarization of molecules is also present. In absence of electric field, the permanent dipoles point in random directions so that the net polarization is zero. On application of electric field, the permanent dipoles are forced to orient in the direction of applied field. But, the orientations along the field direction are continuously disturbed by thermal agitation. Therefore, an equilibrium is attained and the dipoles make zero to π angles with the field direction. The polarization of such permanent dipoles may therefore be resolved in the direction of field and average value of $P_p \cos\theta$ may be found (Figure 14). Thus, we have $\vec{P}_p$ the permanent polarization and $\vec{P}_i$ the induced polarization. Therefore, total polarization is

$$\vec{P}_t = \vec{P}_p + \vec{P}_i$$

If $\vec{E}_{in}$ is inner field acting on dielectric molecules, the potential energy of permanent dipoles per unit volume of moment $\vec{P}_p$ is

$$\vec{E}_{in} \cdot \vec{P}_p = E_{in} P_p \cos\theta$$

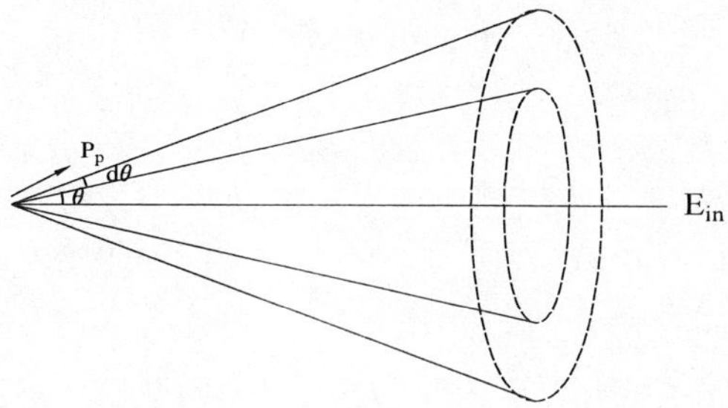

Figure 14: Permanent dipole of polar molecule making an angle θ with the field direction.

Calculation of average value of $P_p \cos\theta$

Following Boltzmann statistical method (kinetic theory of gases), the probability that a molecule is found in an energy state W, at temperature T is proportional to $\exp(-W/kT)$, where $k = R/N$ is Boltzmann constant, R and N are gas constant and Avogadro number, respectively. In the present case, the potential energy of dipoles per unit volume is $-\vec{E}_{in} \cdot \vec{P}_p$. Thus, the probability of finding dipoles at an angle θ with the field direction is

$$P_\theta \propto \mathrm{e}^{\vec{E}_{in}\cdot\vec{P}_p/kT} \qquad \text{or} \qquad P_\theta \propto \mathrm{e}^{E_{in}P_p\cos\theta/kT}$$

The number of dipoles at an orientation is proportional to its probability. Hence, if N_θ is the number of electric dipoles per unit volume at an angle θ and $N_{\pi/2}$ at $\theta = \pi/2$, we have

$$N_\theta \propto \mathrm{e}^{E_{in}P_p\cos\theta/kT} \qquad \text{and} \qquad N_{\pi/2} \propto 1$$

Thus, we have

$$\frac{N_\theta}{N_{\pi/2}} = \mathrm{e}^{E_{in}P_p\cos\theta/kT} \qquad \text{or} \qquad N_\theta = N_{\pi/2}\,\mathrm{e}^{E_{in}P_p\cos\theta/kT}$$

On differentiation with respect to θ, we get

$$\mathrm{d}N_\theta = -N_{\pi/2}\,\mathrm{e}^{E_{in}P_p\cos\theta/kT}\,\frac{E_{in}P_p}{kT}\,\sin\theta\,\mathrm{d}\theta$$

The average moment per unit volume in the direction of field is

$$\langle P_p\cos\theta\rangle = \int_0^\pi P_p\cos\theta\,\mathrm{d}N_\theta \Big/ \int_0^\pi \mathrm{d}N_\theta$$

Substituting the expression for $\mathrm{d}N_\theta$, we get

$$\langle P_p\cos\theta\rangle = \frac{E_{in}P_p^2}{kT}N_{\pi/2}\int_0^\pi \cos\theta\sin\theta\,\mathrm{e}^{E_{in}P_p\cos\theta/kT}\,\mathrm{d}\theta \Big/$$

$$\frac{E_{in}P_p}{kT}N_{\pi/2}\int_0^\pi \sin\theta\ \mathrm{e}^{E_{in}P_p\cos\theta/kT}\ \mathrm{d}\theta$$

$$= P_p\int_0^\pi \cos\theta\sin\theta\ \mathrm{e}^{u\cos\theta}\ \mathrm{d}\theta\Big/\int_0^\pi \sin\theta\ \mathrm{e}^{u\cos\theta}\ \mathrm{d}\theta$$

where we have put $u = E_{in}P_p/kT$. Now, we substitute $t = u\cos\theta$, so that $\mathrm{d}t = -u\sin\theta\ \mathrm{d}\theta$, and we have

$$\langle P_p\cos\theta\rangle = P_p\int_{-u}^{u}\mathrm{e}^t(t/u)\ \mathrm{d}t\Big/\int_{-u}^{u}\mathrm{e}^t\ \mathrm{d}t$$

$$= \frac{P_p[t\ \mathrm{e}^t - \mathrm{e}^t]_{-u}^{u}}{u[\mathrm{e}^t]_{-u}^{u}} = \frac{P_p}{u}\left[\frac{u(\mathrm{e}^u + \mathrm{e}^{-u})}{\mathrm{e}^u - \mathrm{e}^{-u}} - \frac{\mathrm{e}^u - \mathrm{e}^{-u}}{\mathrm{e}^u - \mathrm{e}^{-u}}\right]$$

$$= P_p\left[\coth u - \frac{1}{u}\right] = P_p\left[\coth\left(\frac{E_{in}P_p}{kT}\right) - \frac{kT}{E_{in}P_p}\right]$$

This is known as **Langevin equation.** When the inner field E_{in} is small and temperature T is large, we have

$$\frac{E_{in}P_p}{kT} << 1$$

Then on expanding exponential in the Langevin equation and we have

$$\langle P_p\cos\theta\rangle = P_p\left[\frac{\cosh u}{\sinh u} - \frac{1}{u}\right] = P_p\left[\frac{1 + \frac{u^2}{2!} + \frac{u^4}{4!} + \ldots}{u + \frac{u^3}{3!} + \ldots} - \frac{1}{u}\right]$$

$$= P_p\left[\left(1 + \frac{u^2}{2!} + \ldots\right)\left(u + \frac{u^3}{3!} + \ldots\right)^{-1} - \frac{1}{u}\right]$$

$$= \frac{P_p}{u}\left[\left(1 + \frac{u^2}{2!} + \ldots\right)\left(1 + \frac{u^2}{3!} + \ldots\right)^{-1} - 1\right]$$

$$= \frac{P_p}{u}\left[\left(1 + \frac{u^2}{2} - \frac{u^2}{6} + \ldots\right) - 1\right] = \frac{P_p}{u}\times\frac{u^2}{3} = \frac{P_p u}{3}$$

Using the value of u, we get

$$\langle P_p\cos\theta\rangle = \frac{P_p^2E_{in}}{3kT}$$

Therefore, net polarization $\vec{P_t}$, in case of polar molecules, is

$$\vec{P_t} = \vec{P_i} + \langle P_p\cos\theta\rangle = \vec{P_i} + \frac{P_p^2\vec{E_{in}}}{3kT}$$

But, induced polarization is $\vec{P_i} = \alpha_i\vec{E_{in}}$, so that

$$\vec{P_t} = \alpha_i\vec{E_{in}} + \frac{P_p^2\vec{E_{in}}}{3kT} = \left(\alpha_i + \frac{P_p^2}{3kT}\right)\vec{E_{in}} = \alpha_t\vec{E_{in}}$$

where total polarizibility per unit volume α_t is

$$\alpha_t = \alpha_i + \frac{P_p^2}{3kT}$$

Thus, per unit volume , total polarizibility α_t is sum of induced polarizibility α_i and permanent polarizibility $P_p^2/3kT$.

Study of molecular structure and Debye's relation

Claussius-Mossotti relation, modified for polar molecules, shows that polarizability per unit volume α_t is temperature dependent. This modification was utilized by Debye to study the molecular structure. The modified Claussius-Mossotti relation is

$$\frac{\epsilon_r - 1}{\epsilon_r + 2} = \frac{\alpha_t}{3\epsilon_0} \qquad \frac{\epsilon_r - 1}{\epsilon_r + 2} = \frac{1}{3\epsilon_0}\left[\alpha_i + \frac{P_p^2}{3kT}\right] \qquad \frac{\epsilon_r - 1}{\epsilon_r + 2} = \frac{\alpha_i}{3\epsilon_0} + \frac{P_p^2}{9\epsilon_0 kT}$$

This equation represents straight line, $y = mx + c$ with

$$y = \frac{\epsilon_r - 1}{\epsilon_r + 2} \qquad x = \frac{1}{T} \qquad m = \frac{P_p^2}{9\epsilon_0 k} \qquad c = \frac{\alpha_i}{3\epsilon_0}$$

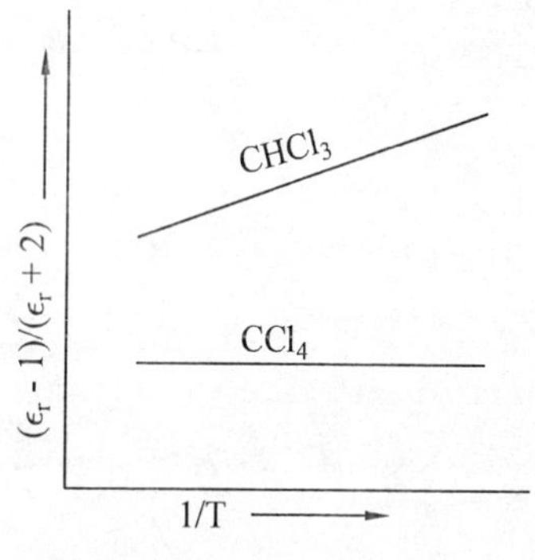

Figure 15: Variation of $(\epsilon_r - 1)/(\epsilon_r + 2)$ versus $1/T$ for CCl_4 and $CHCl_3$.

For non-polar molecules $P_p = 0$ and therefore $m = 0$. That is, the straight line is parallel to x-axis. The plat of $(\epsilon_r - 1)/(\epsilon_r + 2)$ versus $1/T$ is straight line and the gradient m depends on the molecules in dielectric. Figure 15 shows the qualitative plots for CCl_4 and $CHCl_3$.

(i) The line for CCl_4 is parallel to x-axis and shows that its molecules are non-polar. It implies the chlorine atoms in CCl_4 are symmetrically placed with respect to carbon atom, so that the center of negative charges coincides with that of positive charges and no permanent electric dipoles are present.

(ii) The line for $CHCl_3$ is inclined and shows that its molecules are polar. The polarity appears due to charge asymmetry of molecule. The permanent dipole moment per unit volume is calculated from the slope of line. The dipole moment is then calculated by dividing it by the number of molecules per unit volume n. Value of α_i is determined from the intercept on y-axis.

The study has been extended to a large number of molecules of mono-atomic gases, like Argon, and homonuclear diatomic molecules, like H_2. They do not show charge asymmetry and are therefore, non-polar. But, for molecule made up of different atoms, the charge asymmetry appears and the molecules act like permanent dipoles. The examples are NH_3, H_2O, etc.

Example 5: The polarizability of NH_3 molecule is found approximately by the measurement of dielectric constant as 2.53×10^{-39} Coulomb^2-m/N and 1.95×10^{-39} Coulomb^2-m/N at 315 K and 468 K, respectively. Calculate for each temperature the polarizability due to deformation of molecules.

Solution: Total molecular polarizability α_t per unit volume is

$$\alpha_t = \alpha_i + \frac{P_p^2}{3kT}$$

We have $\alpha_t = 2.53 \times 10^{-39}$ Coulomb^2-m/N at $T = 315$ K and $\alpha_t = 1.95 \times 10^{-39}$ Coulomb^2-m/N at $T = 468$ K. Substituting values, we have

$$2.53 \times 10^{-39} = \alpha_i + \frac{P_p^2}{3k\ 315} \qquad 1.95 \times 10^{-39} = \alpha_i + \frac{P_p^2}{3k\ 468}$$

Therefore,

$$\frac{2.53 \times 10^{-39} - \alpha_i}{1.95 \times 10^{-39} - \alpha_i} = \frac{468}{315}$$

On solving, we have $\alpha_i = 7.613 \times 10^{-40}$ Coulomb^2-m/N. This gives value of deformation polarizability. The polarizability due to permanent dipole at 315 K is

$$\alpha_0 = (2.53 - 0.7613) \times 10^{-39} = 1.769 \times 10^{-39} \text{Coulomb}^2\text{-m/N}$$

and at 468 K

$$\alpha_0 = (1.95 - 0.7613) \times 10^{-39} = 1.189 \times 10^{-39} \text{Coulomb}^2\text{-m/N}$$

Review Questions

1. What is difference between insulator and dielectric?
2. Define dielectric polarization.
3. Find relation between electric displacement $\vec{D}$, electric field $\vec{E}$ and polarization $\vec{P}$.
4. Show that potential due to a polarized medium can be expressed in terms of a sum of volume and surface integrals. Discuss physical meaning of the two.
5. Derive Claussius-Mossotti relation. Discuss validity of relation.
6. What is dielectric? Explain difference between isotropic and anisotropic dielectrics.
7. What are polar molecule and non-polar molecules? How does polarization of polar molecules depend on temperature? Explain with the help of Debye's modification of Claussius-Mossotti relation, the method to study molecular structure.

8. Derive expression for effective field inside a dielectric medium placed in an external electric field.

9. Define polarizability of a material and derive Claussius-Mossotti relation connecting their quantities with dielectric constant of material.

10. A dielectric sphere is placed in a uniform electrostatic field. Draw a neat diagram showing a distortion produced in the field. Calculate electric field outside and inside the dielectric sphere.

11. What are polar and non-polar molecules of a dielectric? Obtain Langevin's expression for polarizability when a gaseous dielectric is subjected to an electrostatic field. What is the effect of induced polarization?

12. What is orientation of molecules in a dielectric when (i) no external electric field is applied, and (ii) external electric field is applied?

13. What is polarization in dielectric? Explain.

14. What is electric displacement in dielectric? Explain.

15. For dielectric medium, describe electric field intensity, electric displacement and polarization. Obtain relation between them.

16. For dielectric medium, show that bound charge density is always opposite in sign and less in magnitude compared to free charge densities.

17. If electric field in an infinite dielectric is $\vec{E}$, show that electric field inside a needle shaped cavity in the dielectric is $\vec{E}$.

18. If electric field in an infinite dielectric is $\vec{E}$, show that electric field inside a narrow slit at right angle to polarization in the dielectric is $\vec{E} + \vec{P}/\epsilon_0$.

19. If electric field in an infinite dielectric is $\vec{E}$, show that electric field inside a spherical cavity in the dielectric is $\vec{E} + \vec{P}/3\epsilon_0$.

20. A point charge is placed in dialectic having infinite extent. Show that presence of medium reduces the electric field strength. Calculate this reduction.

21. A sphere consisting of linear dielectric is placed in homogeneous electric field. Show that electric field inside the sphere is smaller than the applied field.

22. Write notes on the following:

    (i) Dielectric material

    (ii) Dielectric polarization

    (iii) Polar and non-polar molecules

    (iv) Dielectric constant

    (v) Electrical susceptibility

    (vi) Electric displacement

Multiple-Choice Questions

1. Dielectric is a form of

   A. Conductor B. Semiconductor C. Insulator D. Superconductor

   Ans. C

2. Relation between electric displacement $\vec{D}$, electric field $\vec{E}$ and polarization $\vec{P}$ is

   A. $\vec{D} = \epsilon_0\vec{E} - \vec{P}$ B. $\vec{D} = \epsilon_0\vec{E} + \vec{P}$ C. $\vec{E} = \epsilon_0\vec{D} - \vec{P}$ D. $\vec{E} = \epsilon_0\vec{D} + \vec{P}$

   Ans. B

3. ϵ and ϵ_0 are absolute permittivities of dielectric medium and vacuum, respectively. The dielectric constant ϵ_r is

   A. $\epsilon + \epsilon_0$ B. $\epsilon - \epsilon_0$ C. $\epsilon\epsilon_0$ D. ϵ/ϵ_0

   Ans. D

4. $\vec{E}$ and $\vec{E}_0$ are electric fields when dielectric and vacuum, respectively, are present between the plates of capacitor. The dielectric constant ϵ_r is

   A. $\vec{E}_0 + \vec{E}$ B. $\vec{E}_0 - \vec{E}$ C. $\vec{E}_0\vec{E}$ D. $\vec{E}_0/\vec{E}$

   Ans. D

5. Curl of displacement vector $\vec{D}$ vanishes when electrostatic field is

   A. non-solenoidal and dielectric is not isotropic and homogeneous

   B. solenoidal and dielectric is not isotropic and homogeneous

   C non-solenoidal and dielectric is isotropic and homogeneous

   D. solenoidal and dielectric is isotropic and homogeneous

   Ans. D

6. In a dielectric, as compared to free charge density, the bound charge density is always

   A. opposite in sign and more in magnitude

   B. opposite in sign and less in magnitude

   C. same in sign and more in magnitude

   D. same in sign and less in magnitude

   Ans. B

7. The straight line, $y = mx + c$, for CCl_4 with

   $$y = \frac{\epsilon_r - 1}{\epsilon_r + 2} \qquad x = \frac{1}{T} \qquad m = \frac{P_p^2}{9\epsilon_0 k} \qquad c = \frac{\alpha_i}{3\epsilon_0}$$

   is

   A. parallel to x-axis B. parallel to y-axis

   C. inclined to x-axis D. may be parallel to x-axis or y-axis

   Ans. A

8. The straight line, $y = mx + c$, for $CHCl_3$ with

$$y = \frac{\epsilon_r - 1}{\epsilon_r + 2} \qquad x = \frac{1}{T} \qquad m = \frac{P_p^2}{9\epsilon_0 k} \qquad c = \frac{\alpha_i}{3\epsilon_0}$$

is

A. parallel to x-axis B. parallel to y-axis

C. inclined to x-axis D. may be parallel to x-axis or y-axis

Ans. C

9. A point charge +q is located at origin of coordinate system in homogeneous and isotropic dielectric medium of infinite extent. The electric field due to the point charge is

A. along x-axis B. along y-axis C. along z-axis D. radial

Ans. D

10. A point charge +q is located at origin of coordinate system in vacuum. The electric field due to the point charge is

A. along x-axis B. along y-axis C. along z-axis D. radial

Ans. D

11. Suppose, electric field in free space is $\vec{E}$. With the presence of dielectric medium with permittivity ϵ and dielectric constant ϵ_r, the electric field is modified to

A. $\vec{E}/\epsilon_0$ B. $\vec{E}/\epsilon$ C. $\vec{E}/\epsilon_r$ D. $\vec{E}/1.4$

Ans. C

12. Suppose, electric field in a dielectric medium with permittivity ϵ and dielectric constant ϵ_r is $\vec{E}$,. In vacuum (after removal of dielectric medium), the electric field is modified to

A. $\epsilon_0\ \vec{E}$ B. $\epsilon\ \vec{E}$ C. $\epsilon_r\ \vec{E}$ D. $1.4\ \vec{E}$

Ans. C

Chapter 5

Energy in Electrostatic Fields

For finding the solution of mechanical problems, energy consideration plays vital role. For example, in classical mechanics, we can easily find the path followed by a particle under investigation, where Hamiltonian represents total energy of system. In the present case, system is made up of static charges and therefore there is no kinetic part of energy and the entire energy of the system is in the form of potential energy. Further, we are concerned particularly with the potential energy, so called electrostatic energy, that arises from electrical interactions of charges.

In order to account for electrical interactions, we have to consider Coulomb's law which explains the action at a distance. For giving the action at a distance a physical meaning, we consider the presence of the field. Thus, the electrostatic phenomenon is explained either (i) by the direct interaction of the fields produced by the charges, or (ii) by the direct interaction of charges. To explain the phenomenon by the former way, it is essential to consider the field in free space and in presence of dielectric substances. We proceed to evaluate energy and force due to charges in terms of fields. So, let us consider field energy in free space and in dielectrics.

1. Work and energy in electrostatics

Consider a stationary configuration of source charges, so that the electric field developed is stationary, and we want to move a test charge Q from a point A to another point B (Figure 1).

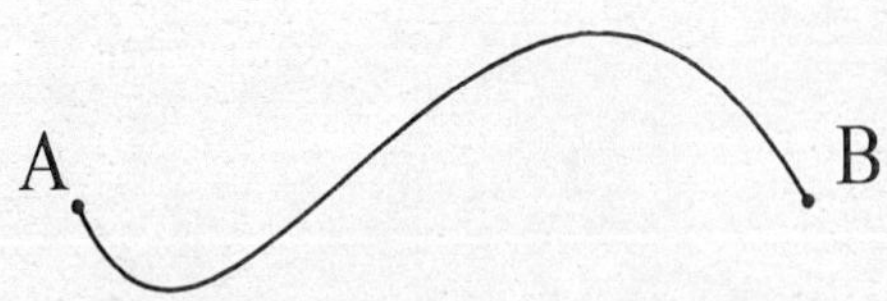

Figure 1: Movement of test charge from a point A to another point B.

Let us calculate the amount of work done in moving the charge Q from point A to point B against the electric field in the region. At any point along the path, the force $\vec{F}$ due to electric field on the charge is $\vec{F} = Q\,\vec{E}$, where $\vec{E}$ is the electric field intensity at the position of the charge Q. The force applied by us is $-Q\vec{E}$. (Here, we are interested in the minimum force to do the job.) The work done is therefore

$$W = \int_A^B \vec{F} \cdot d\vec{\ell} = -Q \int_A^B \vec{E} \cdot d\vec{\ell}$$

We know that

$$\vec{E} = -\nabla\phi \qquad \text{or} \qquad \vec{E} = -\frac{\partial\phi}{\partial\ell}\,\hat{\ell}$$

where ϕ electric potential. Therefore,

$$W = Q \int_A^B d\phi = Q\,[\phi]_A^B = Q\,[\phi(B) - \phi(A)]$$

It shows that the work done is independent of the path followed between the points A and B. It is because the electrostatic force is conservative. Considering $Q = 1$, we have

$$[\phi(B) - \phi(A)] = W$$

It implies that, the potential difference between the points B and A is equal to the work done on a particle carrying unit charge for moving from position A to position B. Thus, for moving a charge Q from infinity to a point at distance r from the source of field, the amount of work done is

$$W = Q[\phi(r) - \phi(\infty)]$$

By convention, the potential at infinity (at infinite distance) is taken zero (that is, the reference point is taken at infinity), and we have

$$W = Q\,\phi(r) \tag{5.1}$$

Suppose, the electric potential is developed by a charge q placed at the origin of coordinate system, the electric potential at the distance r is

$$\phi(r) = \frac{1}{4\pi\epsilon_0}\,\frac{q}{r}$$

so that

$$W = \frac{1}{4\pi\epsilon_0}\,\frac{q\,Q}{r}$$

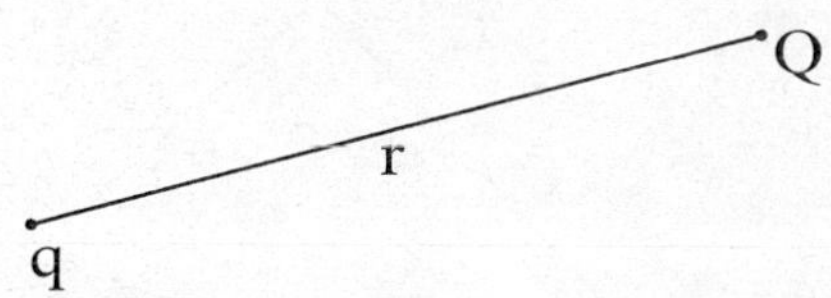

Figure 2: Charges q and Q separated by a distance r.

Thus, W is the amount of work done in arranging the charges q and Q (Figure 2) (of the same kind, so that the force between the charges is repulsive), separated by a distance r. This amount of work done is stored in the form of potential energy.

1.1 Energy of a point charge distribution

Let us consider a collection of point charges and estimate the amount of work done for assembling the collection. We bring the charges one by one from far away, so that there is no interaction between them. In absence of all charges, there is no field, and bringing of the first charge, q_1 (Figure 3), requires no work to be done.

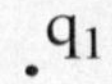

Figure 3: Single charge q_1 placed in space.

Now, the electric field created by the charge q_1 is present in the field, and we bring a charge q_2 in the system (Figure 4). According to equation (5.1), the amount of work required in making the arrangement of charges q_1 and q_2 separated by a distance r_{12} is

$$W_2 = \frac{1}{4\pi\epsilon_0}\left(\frac{q_1\, q_2}{r_{12}}\right)$$

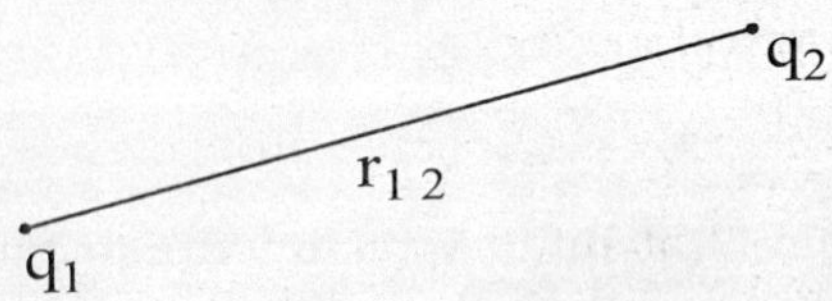

Figure 4: Two charges q_1 and q_2 placed in space.

The positions of the charges q_1 and q_2 are now fixed, and the electric field created by both the charges, simultaneously, is present. Now, we bring a charge q_3 in the system (Figure 5). This requires an additional amount of work

$$W_3 = \frac{1}{4\pi\epsilon_0}\,\frac{q_1\, q_3}{r_{13}} + \frac{1}{4\pi\epsilon_0}\,\frac{q_2\, q_3}{r_{23}}$$

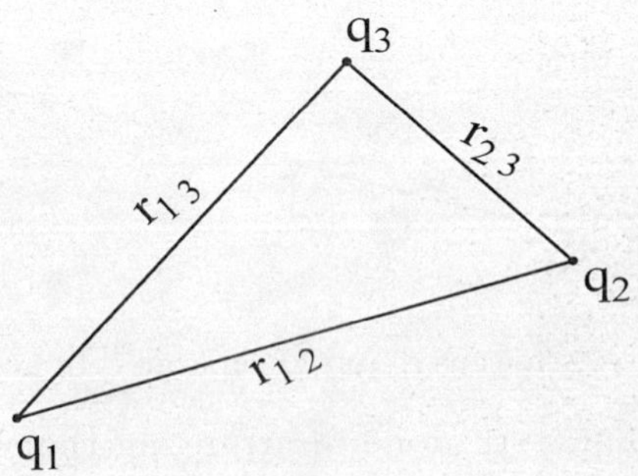

Figure 5: Three charges q_1, q_2 and q_3 placed in space.

The positions of the charges q_1, q_2 and q_3 are now fixed, and the electric field created by the three charges, simultaneously, is present. Now, we bring a charge q_4 in the system (Figure 6). This requires an additional amount of work.

$$W_4 = \frac{1}{4\pi\epsilon_0}\frac{q_1\, q_4}{r_{14}} + \frac{1}{4\pi\epsilon_0}\frac{q_2\, q_4}{r_{24}} + \frac{1}{4\pi\epsilon_0}\frac{q_3\, q_4}{r_{34}}$$

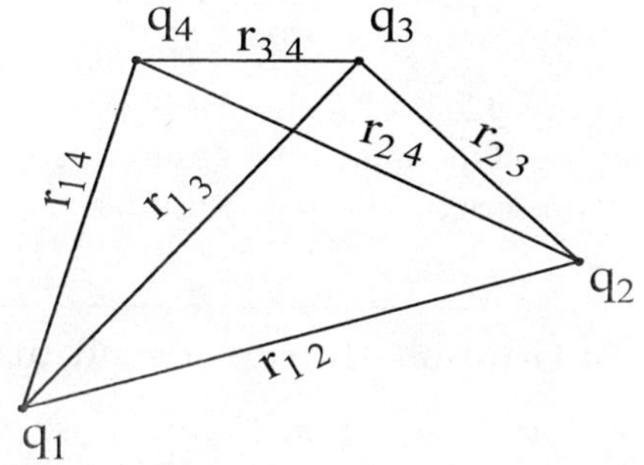

Figure 6: Four charges q_1, q_2, q_3 and q_4 placed in space.

The total work required to assemble the system of these four charges is

$$\begin{aligned} W &= W_2 + W_3 + W_4 \\ &= \frac{1}{4\pi\epsilon_0}\left(\frac{q_1q_2}{r_{12}} + \frac{q_1q_3}{r_{13}} + \frac{q_1q_4}{r_{14}} + \frac{q_2q_3}{r_{23}} + \frac{q_2q_4}{r_{24}} + \frac{q_3q_4}{r_{34}}\right) \\ &= \frac{1}{4\pi\epsilon_0}\sum_{i=1}^{3}\sum_{j>i}^{4}\frac{q_iq_j}{r_{ij}} = \frac{1}{4\pi\epsilon_0}\sum_{i=1}^{4}\sum_{j>i}^{4}\frac{q_iq_j}{r_{ij}} \end{aligned}$$

This rule can be generalized for n charged particles brought in a system:

$$W = \frac{1}{4\pi\epsilon_0}\sum_{i=1}^{n}\sum_{j>i}^{n}\frac{q_iq_j}{r_{ij}}$$

The stipulation $j > i$ is to remind you not to count the same pair twice. We can count each pair twice, as the following:

$$W = \frac{1}{8\pi\epsilon_0}\sum_{i=1}^{n}\sum_{j\neq i}^{n}\frac{q_iq_j}{r_{ij}}$$

where, we avoid $i = j$, as a charged particle cannot attract to itself.

Let us now pull out the factor q_i:

$$W = \frac{1}{2}\sum_{i=1}^{n} q_i\left(\frac{1}{4\pi\epsilon_0}\sum_{j\neq i}^{n}\frac{q_j}{r_{ij}}\right)$$

The term in parentheses is the potential at point r_i (the position of charge q_i) due to all the other charges. That is, all of the charges, except the ones on which he others are acting. Thus,

$$W = \frac{1}{2}\sum_{i=1}^{n} q_i V(r_i) \tag{5.2}$$

where

$$V(r_i) = \frac{1}{4\pi\epsilon_0}\sum_{j\neq i}^{n} \frac{q_j}{r_{ij}}$$

is the potential at point r_i. The W is the total work required to assemble a configuration of n point charges. This is also the amount of work that would be received back if the system is dismantled. This is amount of potential energy stored in the configuration.

1.2 Energy of a continuous charge distribution

For a volume charge density ρ, equation (5.2) can be expressed as

$$W = \frac{1}{2}\int_v \rho\ V\ \mathrm{d}\tau \tag{5.3}$$

where $\mathrm{d}\tau$ is volume element, and v denotes the volume, in order to distinguish it from the electric potential V. This result can be rewritten by expressing charge density ρ and potential V in terms of the electric field $\vec{E}$. From the Gauss's law, we have

$$\rho = \epsilon_0\ \nabla\ .\ \vec{E} \qquad \text{so} \qquad W = \frac{\epsilon_0}{2}\int_v (\nabla\ .\ \vec{E})\ V\ \mathrm{d}\tau$$

Using the relation,

$$\int_v f(\nabla\ .\ \vec{A})\ \mathrm{d}\tau = \int_S (f\vec{A})\ .\ \mathrm{d}\vec{S} - \int_v \vec{A}\ .\ (\nabla f)\ \mathrm{d}\tau$$

we have,

$$W = \frac{\epsilon_0}{2}\left[\int_S (V\vec{E})\ .\ \mathrm{d}\vec{S} - \int_v \vec{E}\ .\ (\nabla V)\ \mathrm{d}\tau\right]$$

But, $\vec{E} = -\nabla V$, so

$$W = \frac{\epsilon_0}{2}\left[\int_S V\vec{E}\ .\ \mathrm{d}\vec{S} + \int_v E^2\ \mathrm{d}\tau\right] \tag{5.4}$$

It is now clear that we should integrate over the region where the charge is located. At large distances from the charge, E varies as $1/r^2$ and V varies as $1/r$, while the surface area varies as r^2; roughly speaking, then, the surface integral varies as $1/r$.

Equation (5.4) gives the correct energy W, whatever volume we use (as long as it encloses all the charge), but the contribution from the volume integral goes up, and that of the surface integral goes down, when we consider larger and larger volumes. If we integrate overall space, then the surface integral goes to zero, and we are left with

$$W = \frac{\epsilon_0}{2}\int_v E^2\ \mathrm{d}\tau \qquad \text{(all space)}$$

2. Thermodynamic interpretation of electrostatic energy

The electrostatic energy is expressed as

$$U = \frac{1}{2}\int_V \epsilon\ E^2\ \mathrm{d}V = \frac{1}{2}\int_V \vec{E}\ .\ \vec{D}\ \mathrm{d}V$$

It is equal to the amount of work done in assembling the system. We now want to identify increment in work (hence in electrostatic energy) with a thermodynamic property of system for having thermodynamical interpretation of electrostatic energy. Let us see whether the increase in work $\mathrm{d}W$ can be related to the change in internal energy $\mathrm{d}U$.

The first law of thermodynamics is

$$\mathrm{d}Q = \mathrm{d}U + \mathrm{d}W \tag{5.5}$$

with

$$\mathrm{d}Q = T\ \mathrm{d}S$$

where T and S are absolute temperature and entropy of the system, respectively, $\mathrm{d}Q$ the amount of heat added to the system, $\mathrm{d}U$ the change in internal energy, $\mathrm{d}S$ the change in entropy and $\mathrm{d}W$ the work done by the system. Consider that the process is adiabatic so that $\mathrm{d}Q = 0$. Then equation (5.5) gives

$$0 = \mathrm{d}U + \mathrm{d}W \qquad \text{or} \qquad \mathrm{d}U = -\mathrm{d}W$$

The negative sign shows that work is done on the system. It shows that change in work can be identified in terms of the change in internal energy of system.

It is not yet the case of electrostatic energy, as in an adiabatic process, the temperature changes and therefore dielectric constants also change, as they depend on temperature. But, in the derivation of the expression for electrostatic energy, we have assumed the dielectric constants to be constant. Therefore, we must consider an isothermal process, and not adiabatic one. For having isothermal situation, the dielectric material must be kept in contact with a heat bath which can exchange heat to maintain constant temperature.

Under isothermal conditions, which can obtain the maximum work from a system is Helmholtz free energy F of the system, and not the total energy. Helmholtz free energy of a system is expressed as

$$F = U - TS \qquad \text{therefore,} \qquad \mathrm{d}F = \mathrm{d}U - T\mathrm{d}S - S\mathrm{d}T$$

Using expression for $\mathrm{d}U$, we have

$$\mathrm{d}F = \mathrm{d}Q - \mathrm{d}W - T\mathrm{d}S - S\mathrm{d}T$$

Using expression for $\mathrm{d}Q$, we have

$$\mathrm{d}F = T\mathrm{d}S - \mathrm{d}W - T\mathrm{d}S - S\mathrm{d}T = -\mathrm{d}W - S\mathrm{d}T$$

For an isothermal process $\mathrm{d}T = 0$ so that $\mathrm{d}F = -\mathrm{d}W = \mathrm{d}U$. Thus, the change in electrostatic energy is identical to the change in Helmholtz free energy of system. In other words, electrostatic energy forms part of free energy of system. This energy represents the maximum work which can be extracted at a later instant from the electrostatic field.

3. Concept of free energy and Thomson's theorem

For most of the dielectrics, the dielectric constant is temperature dependent. A dielectric absorbs electrical energy and consequently its temperature increases. Thus, in order to have an isothermal process, for polarization of dielectrics, it is essential to keep a heat sink to absorb the liberated heat. In this manner, total electric field energy is used not only in doing external work, but a part of it is also converted into heat. On thermodynamical grounds, the maximum energy convertible into work is known as free energy. Therefore, the expression $U = \frac{1}{2}\int \vec{E} \cdot \vec{D}\ \mathrm{d}V$ represents free energy in the presence of dielectric (of course for the substances, which have temperature dependent dielectric properties, the free energy represents total energy).

Statement of Thomson's theorem may be as the following.

Free energy of system in a dielectric medium is in equilibrium state.

The theorem may be proved in the following manner. Let us consider a virtual process in which charges in equilibrium on a number of conductors are displaced by an infinitesimal amount along the conductor surfaces of constant potential in such a manner that total charge remains unchanged. Then the variation δU in free energy is

$$\delta U = \delta\Big(\frac{1}{2}\int_V \epsilon\ E^2\ \mathrm{d}V\Big) = \frac{1}{2}\int_V \epsilon\ \delta(E^2)\ \mathrm{d}V$$

$$= \int_V \epsilon\ \vec{E} \cdot \delta\vec{E}\ \mathrm{d}V = \int_V \vec{E} \cdot \delta\vec{D}\ \mathrm{d}V = \int_V (-\nabla\phi) \cdot \delta\vec{D}\ \mathrm{d}V$$

Here, we have used $\vec{D} = \epsilon\ \vec{E}$ and $\vec{E} = -\nabla\phi$. Using vector identity $\mathrm{div}(\phi\ \delta\vec{D}) = \phi\ \mathrm{div}\ \delta\vec{D} + \delta\vec{D} \cdot \nabla\phi$, we have

$$\delta U = \int_V [\phi\ \mathrm{div}\ \delta\vec{D} - \mathrm{div}(\phi\ \delta\vec{D})]\ \mathrm{d}V$$

For charge density ρ, we have

$$\rho = \nabla \cdot \vec{D} \qquad \text{so that} \qquad \delta\rho = \nabla \cdot \delta\vec{D}$$

and therefore

$$\delta U = \int_V \phi\ \delta\rho\ \mathrm{d}V - \int_V \mathrm{div}(\phi\ \delta\vec{D})\ \mathrm{d}V$$

Using Gauss' divergence theorem for second integral on right side, we have

$$\delta U = \int_V \phi\ \delta\rho\ \mathrm{d}V - \int_S \phi\ \delta\vec{D} \cdot \mathrm{d}\vec{S}$$

For arbitrary volume V which includes all charges of system, we can consider the bounding surface $\vec{S}$ at a large distance from the charge distribution. At a large distance, we have

$$\phi \propto \frac{1}{r} \qquad \delta D \propto \frac{1}{r^2} \qquad \mathrm{d}S \propto r^2$$

so that

$$\int_S (\phi\ \delta\vec{D})\ .\ \mathrm{d}\vec{S} = \frac{1}{r}$$

and for $r \to \infty$, we have

$$\int_S (\phi\ \delta\vec{D})\ .\ \mathrm{d}\vec{S} = 0$$

Thus, we have

$$\delta U = \int_V \phi\ \delta\rho\ \mathrm{d}V$$

Writing it as a sum of individual integrals over elements of system, we have

$$\delta U = \sum_i \int_{V_i} \phi_i\ \delta\rho_i\ \mathrm{d}V_i$$

If potential ϕ_i is constant over each conductor, we get

$$\delta U = \sum_i \phi_i \int_{V_i} \delta\rho_i\ \mathrm{d}V_i$$

The integral

$$\int_{V_i} \delta\rho_i\ \mathrm{d}V_i$$

represents change in charge q_i on the conductor, *i.e.*, δq_i. The charge has been assumed constant in this virtual process, so that

$$\delta q_i = \int_{V_i} \delta\rho_i\ \mathrm{d}V_i = 0$$

leading to

$$\delta U = 0$$

or U is extremum. We have got this proof taking the equilibrium condition in which each conductor surface is equipotential. Thus, the equilibrium that each conductor be at equipotential is equivalent to make the free energy an extremum. The requirement is usually known as Thomson's theorem.

Example 1: In a parallel plate condenser, each plate is of length ℓ and width a; the separation between the plates is d. The capacitor is filled with a dielectric of permittivity ϵ. The plates are maintained at a constant potential difference ϕ and the dielectric block is withdrawn along the length until only its length x remains between the plates. Calculate the force tending to pull the block back into the plates.

Solution: If ϕ is the potential difference between the plates then electric field existing between them is

$$E = \frac{\phi}{d}$$

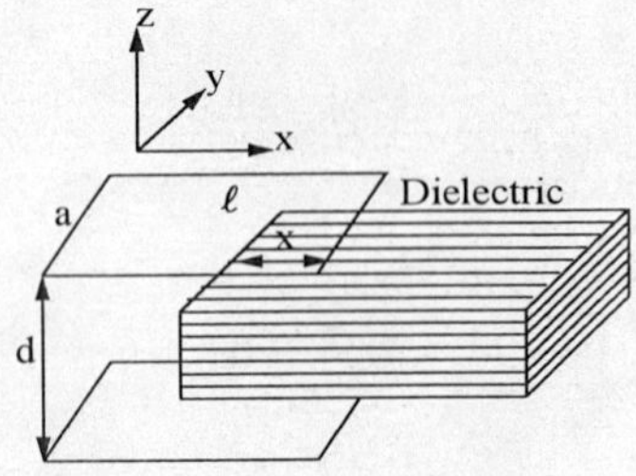

Figure 7: Dielectric of length x is between the plates of parallel plate condenser.

and the energy of system is

$$U = \frac{1}{2}\int_V \epsilon\ E^2\ dV$$

where integration extends over the entire region of the system where $\vec{E} \neq 0$.

Neglecting the fringing effects over the edge of capacitor, we can write from equation for the region of length $(\ell - x)$ without dielectric and of length x with dielectric, as

$$U = \frac{1}{2}\int_0^a \epsilon\left(\frac{\phi}{d}\right)^2 x\ dy + \frac{1}{2}\int_0^a \epsilon_0\left(\frac{\phi}{d}\right)^2 (\ell - x)\ dy$$

$$= \frac{1}{2}\ \epsilon\left(\frac{\phi}{d}\right)^2 xa + \frac{1}{2}\ \epsilon_0\left(\frac{\phi}{d}\right)^2 (\ell - x)a$$

Therefore, the force that brings back the block into the plates is

$$F_x = -\left(\frac{\partial U}{\partial x}\right)_\phi$$

$$= -\frac{1}{2}\ \epsilon\left(\frac{\phi}{d}\right)^2 a - \frac{1}{2}\ \epsilon_0\left(\frac{\phi}{d}\right)^2 (0 - 1)a = -\frac{1}{2}\ (\epsilon - \epsilon_0)\left(\frac{\phi}{d}\right)^2 a$$

which is in the direction of x-decreasing (as $\epsilon > \epsilon_0$), that is, towards the left into the plates. It may be mentioned here that it is the fringing effect which is responsible for pulling the dielectric into the capacitor.

4. Volume forces in electrostatic field

When a dielectric body is placed in an external electromagnetic field, it is subjected to a driving force. We shall show that force per unit volume or volume force acting on the dielectric body arises due to (i) electric intensity in vacuum, and (ii) inhomogeneity of electrostatic field.

The force per unit volume that acts on a dielectric body when placed in an external electrostatic field can be calculated in terms of changes in the free energy U expressed as

$$U = \frac{1}{2}\int_V \vec{E}\ .\ \vec{D}\ dV \tag{5.6}$$

Let us consider a virtual process in which the physical coordinates of charges and dielectric are given a virtual displacement $\delta\vec{r}$ at each point in space but where no external charges are added so that equation (5.6) holds. Then consider unit volume of dielectric in an electromagnetic field and let it be displaced through a small virtual displacement $\delta\vec{r}$. If $\vec{F}$ is the necessary force per unit volume for the displacement $\delta\vec{r}$, then the change in energy is

$$\delta U = -\int \vec{F}\,.\,\delta\vec{r}\,\mathrm{d}V$$

where magnitude of $\delta\vec{r}$ is quite arbitrary so that the force $\vec{F}$ is truly a volume force. The rate of change of displacement is

$$\frac{\delta\vec{r}}{\delta t} = \vec{u}$$

where $\vec{u}$ is an arbitrary velocity field within dielectric, then the rate at which the energy is lost by the field is

$$\frac{\delta U}{\delta t} = -\int \vec{F}\,.\,\vec{u}\,\mathrm{d}V \tag{5.7}$$

where the volume force $\vec{F}$ is to be expressed in terms of suitable physical quantities.

The change in energy δU is attributed to (i) the change in true charge distribution $\delta\rho$ and (ii) the change in dielectric $\delta\epsilon$. This change can be found by the energy relation in presence of dielectric. That is

$$U = \frac{1}{2}\int_V \vec{E}\,.\,\vec{D}\,\mathrm{d}V \qquad \text{so that} \qquad \delta U = \frac{1}{2}\int_V \delta\vec{E}\,.\,\vec{D}\,\mathrm{d}V + \frac{1}{2}\int_V \vec{E}\,.\,\delta\vec{D}\,\mathrm{d}V \tag{5.8}$$

From $\vec{E} = \vec{D}/\epsilon = \vec{D}/\epsilon_0\epsilon_r$, we have

$$\delta\vec{E} = \frac{1}{\epsilon_0\epsilon_r}\,\delta\vec{D} - \frac{\vec{D}}{\epsilon_0\epsilon_r^2}\,\delta\epsilon_r$$

Therefore,

$$\delta\vec{E}\,.\,\vec{D} = \frac{1}{\epsilon_0\epsilon_r}\vec{D}\,.\,\delta\vec{D} - \frac{D^2}{\epsilon_0\epsilon_r^2}\,\delta\epsilon_r$$

$$= \vec{E}\,.\,\delta\vec{D} - \epsilon_0 E^2\delta\epsilon_r$$

Using this relation in equation (5.8), we have

$$\delta U = \frac{1}{2}\int_V (\vec{E}\,.\,\delta\vec{D} - \epsilon_0E^2\delta\epsilon_r)\,\mathrm{d}V + \frac{1}{2}\int_V \vec{E}\,.\,\delta\vec{D}\,\mathrm{d}V$$

$$= \int_V \vec{E}\,.\,\delta\vec{D}\,\mathrm{d}V - \frac{\epsilon_0}{2}\int_V E^2\,\delta\epsilon_r\,\mathrm{d}V \tag{5.9}$$

First term on right side of this equation can be transformed to a suitable form in the following manner.

$$\int_V \vec{E} \cdot \delta\vec{D} \, dV = \int_V (-\nabla\phi) \cdot \delta\vec{D} \, dV$$

Using the identity $\nabla \cdot (\phi \, \delta\vec{D}) = \phi\nabla \cdot \delta\vec{D} + \delta\vec{D} \cdot \nabla\phi$, we have

$$\int_V \vec{E} \cdot \delta\vec{D} \, dV = \int_V \phi\nabla \cdot \delta\vec{D} \, dV - \int_V \nabla \cdot (\phi \, \delta\vec{D}) \, dV$$

Further, we have

$$\nabla \cdot \vec{D} = \rho \qquad \text{so that} \qquad \nabla \cdot \delta\vec{D} = \delta\rho$$

and therefore,

$$\int_V \vec{E} \cdot \delta\vec{D} \, dV = \int_V \phi \, \delta\rho \, dV - \int_V \nabla \cdot (\phi \, \delta\vec{D}) \, dV$$

Using Gauss' divergence theorem for second term on right side, we have

$$\int_V \vec{E} \cdot \delta\vec{D} \, dV = \int_V \phi \, \delta\rho \, dV - \oint_S (\phi \, \delta\vec{D}) \cdot d\vec{S} \tag{5.10}$$

At a large distance, we have

$$\phi \propto \frac{1}{r} \qquad \delta D \propto \frac{1}{r^2} \qquad dS \propto r^2$$

so that

$$\int_S (\phi \, \delta\vec{D}) \cdot d\vec{S} = \frac{1}{r}$$

and for $r \to \infty$, we have

$$\int_S (\phi \, \delta\vec{D}) \cdot d\vec{S} = 0$$

Thus, from equation (5.10), we have

$$\int_V \vec{E} \cdot \delta\vec{D} \, dV = \int_V \phi \, \delta\rho \, dV$$

Using this relation in equation (5.9), we have

$$\delta U = \int_V \phi \, \delta\rho \, dV - \frac{\epsilon_0}{2} \int_V E^2 \, \delta\epsilon_r \, dV$$

Therefore, total rate of change of field energy is

$$\frac{\partial U}{\partial t} = \int_V \phi \, \frac{\partial \rho}{\partial t} \, dV - \frac{\epsilon_0}{2} \int_V E^2 \, \frac{\partial \epsilon_r}{\partial t} \, dV \tag{5.11}$$

Now, we want to evaluate $\partial\rho/\partial t$ and $\partial\epsilon_r/\partial t$ in terms of velocity $\vec{u}$. If ρ and m are charge density and mass density, then the principles of conservation of charge and mass is expressed in terms of equation of continuity as the following:

$$\nabla \cdot (\rho \, \vec{u}) + \frac{\partial \rho}{\partial t} = 0 \qquad \text{for charge} \tag{5.12}$$

$$\nabla \cdot (m\,\vec{u}) + \frac{\partial m}{\partial t} = 0 \qquad \text{for mass} \tag{5.13}$$

In order to find the value of the rate of change of dielectric constant $\partial\epsilon_r/\partial t$, it should be related with the velocity flow. This involves transport of material and the change of ϵ_r, then dielectric constant can be related with the changes in geometry. We can then express

$$\epsilon_r = \epsilon_r(x, y, z, t)$$

so that total time-derivative is

$$\frac{d\epsilon_r}{dt} = \frac{\partial\epsilon_r}{\partial x}\frac{dx}{dt} + \frac{\partial\epsilon_r}{\partial y}\frac{dy}{dt} + \frac{\partial\epsilon_r}{\partial z}\frac{dz}{dt} + \frac{\partial\epsilon_r}{\partial t}\frac{dt}{dt}$$

$$= \left(\hat{i}\,\frac{\partial\epsilon_r}{\partial x} + \hat{j}\,\frac{\partial\epsilon_r}{\partial y} + \hat{k}\,\frac{\partial\epsilon_r}{\partial z}\right)\cdot\left(\hat{i}\,\frac{dx}{dt} + \hat{j}\,\frac{dy}{dt} + \hat{k}\,\frac{dz}{dt}\right) + \frac{\partial\epsilon_r}{\partial t}$$

$$= \nabla\epsilon_r \cdot \vec{u} + \frac{\partial\epsilon_r}{\partial t}$$

Therefore,

$$\frac{\partial\epsilon_r}{\partial t} = \frac{d\epsilon_r}{dt} - \nabla\epsilon_r \cdot \vec{u} \tag{5.14}$$

Similarly, we can have

$$\frac{\partial m}{\partial t} = \frac{dm}{dt} - \nabla m \cdot \vec{u} \qquad \text{and} \qquad \frac{dm}{dt} = \nabla m \cdot \vec{u} + \frac{\partial m}{\partial t} \tag{5.15}$$

Now, with the help of relations, like Claussius-Mossotti relation, which relate dielectric constant with the density of substance, the total time-derivative can be related to each other as the following. We have

$$\frac{d\epsilon_r}{dt} = \frac{d\epsilon_r}{dm}\,\frac{dm}{dt}$$

assuming that the process is isothermal. Putting the value of dm/dt from equation (5.15) on right side, we get

$$\frac{d\epsilon_r}{dt} = \frac{d\epsilon_r}{dm}\left[\nabla m \cdot \vec{u} + \frac{\partial m}{\partial t}\right]$$

Putting the value of $\partial m/\partial t$ from equation (5.13) on right side, we get

$$\frac{d\epsilon_r}{dt} = \frac{d\epsilon_r}{dm}\left[\nabla m \cdot \vec{u} - \nabla \cdot (m\,\vec{u})\right] = -\frac{d\epsilon_r}{dm}\,m\,\nabla \cdot \vec{u} \tag{5.16}$$

Using equation (5.16) in (5.14), we have

$$\frac{\partial\epsilon_r}{\partial t} = -\frac{d\epsilon_r}{dm}\,m\,\nabla \cdot \vec{u} - \nabla\epsilon_r \cdot \vec{u} \tag{5.17}$$

Using values of $\partial\epsilon_r/\partial t$ and $\partial\rho/\partial t$ from equations (5.12) and (5.17) in (5.11), we have

$$\frac{\partial U}{\partial t} = -\int_V \phi\,\nabla\,.\,(\rho\,\vec{u})\,\mathrm{d}V + \frac{\epsilon_0}{2}\int_V E^2\,\frac{\mathrm{d}\epsilon_r}{\mathrm{d}m}\,m\,\nabla\,.\,\vec{u}\,\mathrm{d}V + \frac{\epsilon_0}{2}\int_V E^2\,\nabla\epsilon_r\,.\,\vec{u}\,\mathrm{d}V \quad (5.18)$$

Using vector identity $\mathrm{div}(\phi\,\rho\vec{u}) = \phi\,\mathrm{div}(\rho\vec{u}) + \rho\,\vec{u}\,.\,\mathrm{grad}\phi$, the first term on right side of equation (5.18) becomes

$$-\int_V \phi\,\nabla\,.\,(\rho\,\vec{u})\,\mathrm{d}V = \int_V \rho\,\vec{u}\,.\,\nabla\phi\,\mathrm{d}V - \int_V \mathrm{div}(\phi\,\rho\vec{u})\,\mathrm{d}V$$

Using Gauss' divergence theorem for second term on tight side, we have

$$-\int_V \phi\,\nabla\,.\,(\rho\,\vec{u})\,\mathrm{d}V = \int_V \rho\,\vec{u}\,.\,\nabla\phi\,\mathrm{d}V - \int_S (\phi\,\rho\vec{u})\,.\,\mathrm{d}\vec{S}$$

For the situation $r \to \infty$, second integral on right side of this equation reduces to zero, and using $\vec{E} = -\nabla\phi$, we have

$$-\int_V \phi\,\nabla\,.\,(\rho\,\vec{u})\,\mathrm{d}V = -\int_V \rho\,\vec{u}\,.\,\vec{E}\,\mathrm{d}V \quad (5.19)$$

Using the identity

$$\left(E^2\,\frac{\mathrm{d}\epsilon_r}{\mathrm{d}m}\,m\right)\nabla\,.\,\vec{u} = \nabla\,.\,\left(E^2\,\frac{\mathrm{d}\epsilon_r}{\mathrm{d}m}\,m\vec{u}\right) - \vec{u}\,.\,\nabla\left(E^2\,\frac{\mathrm{d}\epsilon_r}{\mathrm{d}m}\,m\right)$$

second term on right side of equation (5.18) can be expressed as

$$\frac{\epsilon_0}{2}\int_V E^2\,\frac{\mathrm{d}\epsilon_r}{\mathrm{d}m}\,m\,\nabla\,.\,\vec{u}\,\mathrm{d}V = \frac{\epsilon_0}{2}\int_V \nabla\,.\,\left(E^2\,\frac{\mathrm{d}\epsilon_r}{\mathrm{d}m}\,m\vec{u}\right)\mathrm{d}V - \frac{\epsilon_0}{2}\int_V \vec{u}\,.\,\nabla\left(E^2\,\frac{\mathrm{d}\epsilon_r}{\mathrm{d}m}\,m\right)\mathrm{d}V$$

Applying Gauss' divergence theorem for the first integral on right side, we have

$$\frac{\epsilon_0}{2}\int_V E^2\,\frac{\mathrm{d}\epsilon_r}{\mathrm{d}m}\,m\,\nabla\,.\,\vec{u}\,\mathrm{d}V = \frac{\epsilon_0}{2}\oint_S E^2\,\frac{\mathrm{d}\epsilon_r}{\mathrm{d}m}\,m\vec{u}\,.\,\mathrm{d}\vec{S} - \frac{\epsilon_0}{2}\int_V \vec{u}\,.\nabla\left(E^2\,\frac{\mathrm{d}\epsilon_r}{\mathrm{d}m}\,m\right)\mathrm{d}V$$

For the situation $r \to \infty$, first integral on right side of this equation reduces to zero, and we have

$$\frac{\epsilon_0}{2}\int_V E^2\,\frac{\mathrm{d}\epsilon_r}{\mathrm{d}m}\,m\,\nabla\,.\,\vec{u}\,\mathrm{d}V = -\frac{\epsilon_0}{2}\int_V \vec{u}\,.\,\nabla\left(E^2\,\frac{\mathrm{d}\epsilon_r}{\mathrm{d}m}\,m\right)\mathrm{d}V \quad (5.20)$$

Using equations (5.19) and (5.20) in (5.18), we have

$$\frac{\partial U}{\partial t} = -\int_V (\rho\,\vec{u})\,.\,\vec{E}\,\mathrm{d}V - \frac{\epsilon_0}{2}\int_V \vec{u}\,.\,\nabla\left(E^2\,\frac{\mathrm{d}\epsilon_r}{\mathrm{d}m}\,m\right)\mathrm{d}V + \frac{\epsilon_0}{2}\int_V E^2\,\nabla\epsilon_r.\vec{u}\,\mathrm{d}V$$

$$= -\int_V \left[\rho\vec{E} + \frac{\epsilon_0}{2}\nabla\left(E^2\,\frac{\mathrm{d}\epsilon_r}{\mathrm{d}m}\,m\right) - \frac{\epsilon_0}{2}\,E^2\,\nabla\epsilon_r\right]i\,.\,\vec{u}\,\mathrm{d}V$$

Comparing this equation with (5.7), we get the volume force

$$\vec{F} = \rho\vec{E} - \frac{\epsilon_0}{2}\,E^2\,\nabla\epsilon_r + \frac{\epsilon_0}{2}\nabla\left(E^2\,\frac{\mathrm{d}\epsilon_r}{\mathrm{d}m}\,m\right)$$

The three terms on right side can be interpreted as the following:

(i) The first term $\rho\vec{E}$ gives the electrostatic force on true charges.

(ii) The second term appears due to inhomogeneity of dielectric.

(iii) The last term, known as electrostriction term, gives a volume force on a dielectric in an inhomogeneous electric field. The magnitude of this term depends on $\mathrm{d}\epsilon_r/\mathrm{d}m$ and on m. However, as these changes in ordinary field are quite low, the effect of the third term is often neglected.

Example 2: Derive an expression for electrostatic energy of a uniformly charged non-conducting sphere.

Solution: Suppose, a charge Q is distributed uniformly in the volume of non-conducting sphere of radius R. The charge density ρ is

$$\rho = \frac{Q}{4\pi R^3/3}$$

Divide the sphere into concentric spherical shells. Let the sphere of radius r $(r < R)$ has charge q. Its potential at its surface is

$$\frac{1}{4\pi\epsilon_0}\frac{q}{r}$$

We now add a spherical shell of thickness $\mathrm{d}r$ carrying a charge $\mathrm{d}q$. The amount of work done in adding a concentric spherical layer of radius r and thickness $\mathrm{d}r$ carrying a charge $\mathrm{d}q$ is

$$\mathrm{d}U = \text{potential of sphere} \times \text{charge } \mathrm{d}q = \frac{1}{4\pi\epsilon_0}\frac{q}{r}\,\mathrm{d}q$$

For the sphere of radius r, we have

$$q = \frac{4\pi}{3}r^3\rho \qquad \text{and} \qquad \mathrm{d}q = 4\pi r^2\rho\,\mathrm{d}r$$

Using these expressions, we have

$$\mathrm{d}U = \frac{4\pi}{3\epsilon_0}\rho^2 r^4\,\mathrm{d}r$$

Thus, total energy U required for assembling the sphere of radius R is

$$U = \frac{4\pi\rho^2}{3\epsilon_0}\int_0^R r^4\,\mathrm{d}r = \frac{4\pi\rho^2 R^5}{15\epsilon_0}$$

Using expression for charge density ρ, we have

$$U = \frac{3}{5}\frac{Q^2}{4\pi\epsilon_0 R}$$

Example 3: A conducting sphere of radius R is placed in vacuum and carries a charge Q. Calculate total electrostatic energy stored in the surrounding space. What does it signify in case of a point charge?

Solution: A charge given to a conducting sphere resides on the outer surface. So it is the case of surface charge distribution. Electric field exists only outside the sphere. The intensity of electric field $\vec{E}$ at a distance r $(r > R)$ from the center of sphere carrying charge Q is

$$E = \frac{1}{4\pi\epsilon_0} \frac{Q}{r^2}$$

The energy per unit volume at a distance r from the center of sphere is

$$u = \frac{1}{2} \epsilon_0 E^2 = \frac{\epsilon_0}{2} \left(\frac{1}{4\pi\epsilon_0} \frac{Q}{r^2} \right)^2$$

At this distance r, we consider a concentric shell of radius r and thickness dr, so that energy of shell is

$$dU = \text{volume of shell} \times \text{energy density}$$

$$= (4\pi r^2\, dr) \frac{\epsilon_0}{2} \left(\frac{1}{4\pi\epsilon_0} \frac{Q}{r^2} \right)^2 = \frac{Q^2}{8\pi\epsilon_0} \frac{dr}{r^2}$$

Total electrostatic energy U surrounding the sphere is obtained by integrating the expression from $r = R$ to $r = \infty$ as

$$U = \int_R^\infty \frac{Q^2}{8\pi\epsilon_0} \frac{dr}{r^2} = \frac{Q^2}{8\pi\epsilon_0 R}$$

This is energy of field due to charged conducting sphere, also called **self energy** of conducting sphere. It may be mentioned that if $R \to 0$, then U should represent the energy in the field of a point-charge, which comes out to be infinite. It simply points that more compact is the charge distribution, the more is energy required to assemble it otherwise, in real world there is no point-charge (having zero dimension) and hence no infinite electrostatic energy.

Review Questions

1. Drive expression for total electrostatic energy of assembly of charges.
2. How mechanical properties exhibited by electrostatic fields can be explained? Discuss the field theory concepts for free space and dielectric medium.
3. What is thermodynamic interpretation of field energy? Derive an expression for total energy of an electrostatic system. What is effect of temperature dependence of dielectric constant?
4. Explain the thermodynamic interpretation of electrostatic energy.
5. State and prove Thomson's theorem.
6. What is volume force in dielectric in electrostatic field? Derive expression for it.

7. Derive an expression for field energy density in an isotropic and linear dielectric medium as a function of electric field intensity. Also obtain an expression for force on a dielectric body placed in an electric field.

8. Show that energy U within a dielectric in an electric field $\vec{E}$ is

$$U = \frac{1}{2} \int_V \vec{D} \cdot \vec{E} \, dV$$

9. Derive an expression for electrostatic energy of a uniformly charged non-conducting sphere.

10. A conducting sphere of radius R is placed in vacuum and carries a charge Q. Calculate total electrostatic energy stored in the surrounding space. What does it signify in the case of point charge?

11. Write notes on the following:

    (i) Electrostatic energy of assembly of charges

    (ii) Volume force

Multiple-Choice Questions

1. In vacuum, n point charges, $q_1, q_2, \ldots, q_n$, are assembled together. Total electrostatic energy of assembly is

A. $\displaystyle\sum_{i=1}^{n}\sum_{j>i}^{n} \frac{1}{4\pi\epsilon_0} \frac{q_i\, q_j}{r_{ij}}$ B. $\displaystyle\sum_{i=1}^{n-1}\sum_{j>i}^{n} \frac{1}{4\pi\epsilon_0} \frac{q_i\, q_j}{r_{ij}}$

C. $\displaystyle\sum_{i=1}^{n-1}\sum_{j>i}^{n} \frac{1}{4\pi\epsilon_r} \frac{q_i\, q_j}{r_{ij}}$ D. $\displaystyle\sum_{i=1}^{n-1}\sum_{j>i}^{n} \frac{1}{4\pi\epsilon} \frac{q_i\, q_j}{r_{ij}}$

Ans. B

2. For a dielectric medium, energy density is

A. $\epsilon_0\, E^2/2$ B. $\epsilon\, E^2/2$ C. $\epsilon_r\, E^2/2$ D. $\epsilon\, D^2/2$

Ans. B

3. Energy within a dielectric in an electric field $\vec{E}$ and displacement $\vec{D}$ is

A. $\displaystyle\frac{1}{2}\int_V \vec{D} \cdot \vec{E}\, dV$ B. $\displaystyle\int_V \vec{D} \cdot \vec{E}\, dV$ C. $\displaystyle\int_V E^2\, dV$ D. $\displaystyle\int_V D^2\, dV$

Ans. A

Chapter 6

Magnetostatics

We know that a steady current passing through a conductor produces magnetic field around the conductor. If two current-carrying conductors are brought close to each other, the magnetic fields produced by them interact. Such interaction is known as magnetic interaction of steady currents and is studied under magnetostatics. In the present chapter, we have discussed about magnetostatics.

1. Current density

Consider a stream of charges flowing through a conductor. The quantity of charge passing per second through unit area of plane whose normal is along the direction of flow of charges is known as **current density**, denoted by $\vec{J}$. It is a vector quantity as its direction depends on the direction of flow of charges. Thus, total current I flowing through a surface S is obtained by integrating the normal component of current density as the following.

$$I = \int_S \vec{J} \,.\, \mathrm{d}\vec{S}$$

It has been found experimentally that in a metallic conductor, at constant temperature, the current density is linearly proportional to the electric field. It can be expressed as

$$\vec{J} = \sigma \, \vec{E}$$

where proportionality constant σ is known as **electrical conductivity** of material and its unit in SI system is mho per meter. This equation is equivalent to Ohm's law for the flow of electricity.

2. Equation of continuity

Following the law of conservation of charge, the net amount of charge in an isolated system remains constant. For generality, let us assume that the volume charge density is a function of time. The principle of conservation of charge can be stated as the following.

If the net charge crossing a surface bounding a volume is not zero, then the charge density within the volume must change with time in a manner that the time rate of decrease of charge within the volume equals the net rate of flow of charge out of the volume.

This statement of conservation of charge can be expressed in the form of **equation of continuity** which can be derived in the following manner. For the motion of charges, let us consider that the volume charge density ρ is a function of time. The charge q is

$$q = \int_V \rho \, \mathrm{d}V \tag{6.1}$$

where we have considered that the current is extended in space of volume V enclosed by a surface S (Figure 1).

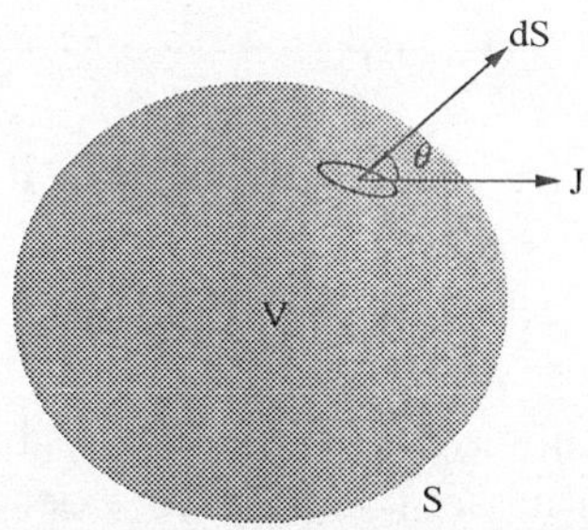

Figure 1: Volume V is enclosed by surface S.

The net amount of charge which crosses a unit area (normal to the direction of charge flow) of a surface in unit time is defined as the **volume current density** $\vec{J}$. Based on a large number of experiments, charge in a closed system is always conserved. Thus, it is a net amount of current flowing outward a closed surface, the charge contained within that volume should decrease at the rate,

$$-\frac{\partial q}{\partial t} = I \tag{6.2}$$

where I is total current flowing through surface S. Minus sign shows that the current is in the direction of negatively charged particles (electrons). For volume current density $\vec{J}$, the total current is

$$I = \int_S \vec{J} \, . \, \mathrm{d}\vec{S} \tag{6.3}$$

From equations (6.2), (6.3) and (6.1), we have

$$\int_S \vec{J} \, . \, \mathrm{d}\vec{S} = -\frac{\partial q}{\partial t} = -\frac{\partial}{\partial t}\int_V \rho \, \mathrm{d}V = -\int_V \frac{\partial \rho}{\partial t} \, \mathrm{d}V$$

We have used the fact that space and time are independent of each other. Using Gauss' divergence theorem for left side, we get

$$\int_V (\nabla \, . \, \vec{J}) \, \mathrm{d}V = -\int_V \frac{\partial \rho}{\partial t} \, \mathrm{d}V \qquad \text{or} \qquad \int_V \left(\nabla \, . \, \vec{J} + \frac{\partial \rho}{\partial t}\right) \mathrm{d}V = 0$$

This equation holds good for arbitrary volume and can only be satisfied when the integrand is equal to zero:

$$\nabla \cdot \vec{J} + \frac{\partial \rho}{\partial t} = 0 \tag{6.4}$$

This is known as **equation of continuity**. It is mathematical expression for the conservation of charge. It states that **total current flowing out of some volume must be equal to the rate of decrease of charge within the volume assuming that charge cannot be created or destroyed, i.e., no sources or sinks are present in that volume.**

In the case of steady currents, charge density at any point within the region remains unchanged with time

$$\frac{\partial \rho}{\partial t} = 0 \qquad \text{so that} \qquad \nabla \cdot \vec{J} = 0$$

It expresses the fact that there is no net outward flux of current density

3. Magnetic induction

Consider a positive charge q moving with velocity $\vec{v}$ in xy-plane at the point P as shown in Figure 2. A magnetic field is said to exist there if this moving charge experiences a deflecting force $\vec{F}$, along z-axis (perpendicular to the xy-plane). This magnetic field described by a vector quantity $\vec{B}$ and is known as **magnetic induction**, lying in the xy-plane, inclined at an angle θ with velocity $\vec{v}$, such that $0 > \theta > \pi$. Say, it is along y-axis. Hence, the deflecting force $\vec{F}$ is perpendicular to the plane containing both the velocity $\vec{v}$ and magnetic induction $\vec{B}$.

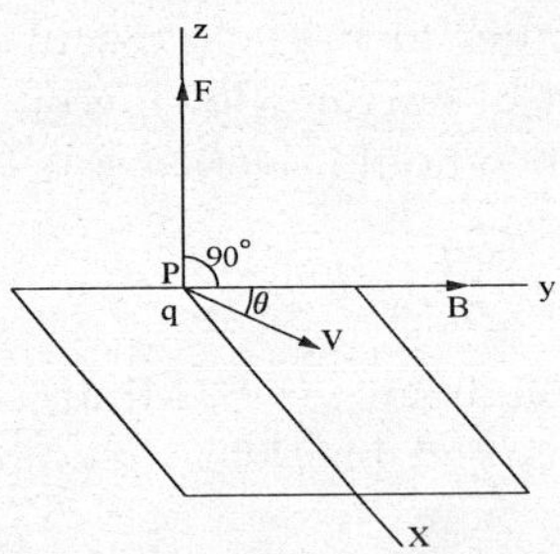

Figure 2: A positive charge q moving with velocity $\vec{v}$ experiences a deflecting force $\vec{F}$ due to magnetic induction $\vec{B}$.

Keeping the magnitude constant, if the direction of $\vec{v}$ through P is varied, it is found that the magnitude of $\vec{F}$ changes, but its direction always remains at right angle to $\vec{v}$. Further, it is found that for a particular orientation of $\vec{v}$, the magnitude of force becomes zero. It is the direction of magnetic induction $\vec{B}$. Thus, if a charge moving in a magnetic field experiences no side-away deflection force, then the direction of motion of charge is the direction of magnetic induction $\vec{B}$.

Now, on moving the charge at right angle to $\vec{B}$, the magnitude of force is found to be maximum. It may be summarized as

(i) when $\vec{v}$ is parallel to $\vec{B}$, magnitude of force $\vec{F}$ is zero.

(ii) when $\vec{v}$ is perpendicular to $\vec{B}$, magnitude of force $\vec{F}$ is maximum.

These observations show that the magnitude of force depends on the angle θ between $\vec{v}$ and $\vec{B}$. Further, it is found that for constant magnitudes of $\vec{v}$ and $\vec{B}$, the variation is as $\sin\theta$. Further, for a given value of θ,

(i) for constant magnitude of $\vec{v}$, magnitude of force is found proportional to the magnitude of $\vec{B}$.

(ii) for constant magnitude of $\vec{B}$, magnitude of force is found proportional to the magnitude of $\vec{v}$.

All above observations can be incorporated by expressing the force as

$$\vec{F} = q\,\vec{v} \times \vec{B}$$

The SI unit of magnetic induction $\vec{B}$ weber/m^2 or Tesla.

3.1 Lorentz force

When electric field $\vec{E}$ and magnetic field $\vec{B}$ both are present simultaneously, in a space, the net force $\vec{F}$ acting on a particle carrying charge q and moving with velocity $\vec{v}$ is

$$\vec{F} = q\,[\vec{E} + \vec{v} \times \vec{B}]$$

This force is known as the **Lorentz force**. In spite of the presence of both the electric and magnetic fields, the Lorentz force can be zero. For the zero Lorentz force, we require

$$\vec{E} = -\vec{v} \times \vec{B} \qquad \text{or} \qquad \vec{E} = \vec{B} \times \vec{v}$$

It shows that the directions for electric and magnetic forces are perpendicular to each other.

Example 1. A charge particle is moving with velocity $\vec{v} = 5\,\hat{\imath} - 3\,\hat{\jmath} + 2\,\hat{k}$ in a magnetic field $\vec{B} = 6\,\hat{\imath} - 2\,\hat{\jmath} + 3\,\hat{k}$. Calculate the electric field $\vec{E}$ which would neutralize the influence of magnetic field.

4. Force on a current element: Ampere's force law

A region is said to be characterized by a magnetic field $\vec{B}$ when a charge q moving with velocity $\vec{v}$ experiences a deflecting force $\vec{F}$, expressed as

$$\vec{F} = q\,\vec{v} \times \vec{B}$$

The current I is expressed as the rate of flow of charge.

$$I = \frac{\mathrm{d}q}{\mathrm{d}t} \qquad \text{and therefore} \qquad \mathrm{d}q = I\,\mathrm{d}t \tag{6.5}$$

The force $d\vec{F}$ experienced by the charge dq moving with velocity $\vec{v}$ is

$$d\vec{F} = dq\, \vec{v} \times \vec{B}$$

Using equation (6.5) here, we get

$$d\vec{F} = I\, dt\, \vec{v} \times \vec{B}$$

Suppose, in time dt, charge dq travels a displacement $d\vec{\ell}$, velocity of charge is

$$\vec{v} = \frac{d\vec{\ell}}{dt} \qquad \text{and therefore} \qquad d\vec{\ell} = \vec{v}\, dt$$

Using this expression, we get

$$d\vec{F} = I\, d\vec{\ell} \times \vec{B}$$

This is known as **Ampere's force law**. It provides deflecting force on current element I flowing through a length element $d\vec{\ell}$ placed in a magnetic field $\vec{B}$.

Example 2: Using Ampere's force law, derive Lorentz force.

Solution: From the Ampere's force law, the deflecting force $d\vec{F}$ on current element $I\, d\vec{\ell}$ placed in a magnetic field $\vec{B}$ is

$$d\vec{F} = I\, d\vec{\ell} \times \vec{B}$$

If $\vec{J}$ is current density (current per unit area) and a is area of cross section of conductor, then

$$I\, d\vec{\ell} = \vec{J}\, a\, d\ell = \vec{J}\, dV$$

where $dV (= a\, d\ell)$ is volume element; we have used the fact that the current density $\vec{J}$ and line element $d\vec{l}$ are along the same direction. Thus, we have

$$d\vec{F} = \vec{J} \times \vec{B}\, dV \qquad \text{or} \qquad \frac{d\vec{F}}{dV} = \vec{J} \times \vec{B}$$

If there are n charged particles per unit volume, each having charge q and moving with velocity $\vec{v}$, then we have

$$\vec{J} = n\, q\, \vec{v}$$

Using this expression, we get

$$\frac{d\vec{F}}{dV} = n\, q\, \vec{v} \times \vec{B} \qquad \text{or} \qquad \frac{1}{n}\frac{d\vec{F}}{dV} = q\, \vec{v} \times \vec{B}$$

Now, $n\, dV$ is total number of charged particles in volume element dV, then the term

$$\frac{1}{n}\frac{d\vec{F}}{dV}$$

represents force on an individual charge. Denoting it by $\vec{F}$, we get

$$\vec{F} = q\, \vec{v} \times \vec{B}$$

This is known as **Lorentz force** and is perpendicular to both the velocity $\vec{v}$ and magnetic induction $\vec{B}$.

5. Magnetic interaction of steady line current: Biot-Savart law

From observations, Ampere found that the force between two elements $d\vec{\ell}_1$ and $d\vec{\ell}_2$, carrying the steady currents I_1 and I_2, respectively (Figure 3), is

(i) proportional to the magnitude of each current.

(ii) inversely proportional to the square of distance between the two elements.

(iii) proportional to the lengths elements and their orientations.

(iv) dependent on the medium in which elements are situated.

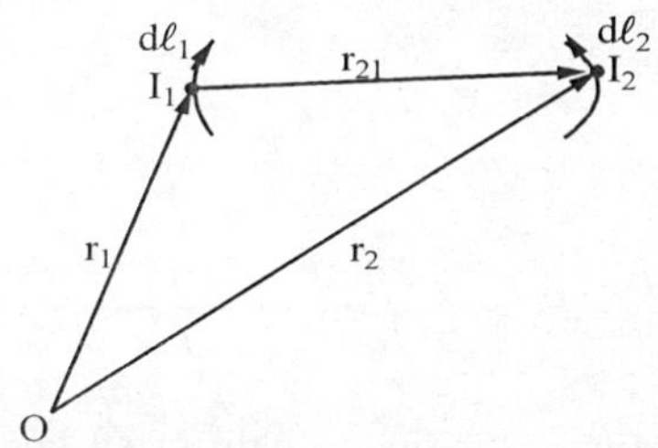

Figure 3: Two elements $d\vec{\ell}_1$ and $d\vec{\ell}_2$ are carrying steady currents I_1 and I_2, respectively.

From these observations, the force exerted on element $d\vec{\ell}_2$ by the element $d\vec{\ell}_1$ can be expressed mathematically as

$$d\vec{F}_{21} = \left(\frac{\mu_0}{4\pi}\right)(I_1 I_2)\left(\frac{1}{r_{21}^2}\right)\left[d\vec{\ell}_2 \times \left(d\vec{\ell}_1 \times \frac{\vec{r}_{21}}{r_{21}}\right)\right]$$

where $\vec{r}_{21} = \vec{r}_2 - \vec{r}_1$, and in this expression, the term

$\left(\frac{\mu_0}{4\pi}\right)$ arises due to factor (iv).

$(I_1 I_2)$ arises due to factor (i).

$\left(\frac{1}{r_{21}^2}\right)$ arises due to factor (ii).

$\left[d\vec{\ell}_2 \times \left(d\vec{\ell}_1 \times \frac{\vec{r}_{21}}{r_{21}}\right)\right]$ arises due to factor (iii).

From this expression for $d\vec{F}_{21}$, the force for whole lengths of conductors can be obtained as the following.

$$\vec{F}_{21} = \frac{\mu_0}{4\pi} \oint_1 \oint_2 \frac{I_2\, d\vec{\ell}_2 \times (I_1\, d\vec{\ell}_1 \times \vec{r}_{21})}{r_{21}^3} \tag{6.6}$$

This equation represents the force exerted on current I_2 by the current I_1. The vectors $d\vec{\ell}_1$ and $d\vec{\ell}_2$ point in the directions of positive current flow. r_{21} is the distance between the two current elements.

In order to determine the directions of force, we first find cross product $(d\vec{\ell}_1 \times \vec{r}_{21})$ and then obtain the cross product of $d\vec{\ell}_2$ with $(d\vec{\ell}_1 \times \vec{r}_{21})$. It may be noted that $d\vec{\ell}_2 \times (d\vec{\ell}_1 \times \vec{r}_{21}) \neq (d\vec{\ell}_2 \times d\vec{\ell}_1) \times \vec{r}_{21}$. By interchanging $I_1 d\vec{\ell}_1$ and $I_2 d\vec{\ell}_2$, replacing $\vec{r}_{21}$ by $\vec{r}_{12}$ in equation (6.6), we can write for the force exerted on current I_1 due to the current I_2 as

$$\vec{F}_{12} = \frac{\mu_0}{4\pi} \oint_2 \oint_1 \frac{I_1\, d\vec{\ell}_1 \times (I_2\, d\vec{\ell}_2 \times \vec{r}_{12})}{r_{12}^3} \tag{6.7}$$

It shows that for current loops, action and reaction are equal and opposite, and we can put

$$\vec{F}_{12} = -\vec{F}_{21}$$

From equations (6.6) and (6.7), we take I_1 and I_2 outside the integrals so that

$$\vec{F}_{21} = -\vec{F}_{12} = \frac{\mu_0}{4\pi}\, I_2 I_1 \oint_1 \oint_2 \frac{d\vec{\ell}_2 \times (d\vec{\ell}_1 \times \vec{r}_{21})}{r_{21}^3} \tag{6.8}$$

Equation (6.8) is mathematical formulation of Ampere's law of force. Let us write equation (6.8) in terms of quantity $\vec{B}$

$$\vec{F}_{21} = I_2 \oint_2 d\vec{\ell}_2 \times \left(\frac{\mu_0}{4\pi}\, I_1 \oint_1 \frac{d\vec{\ell}_1 \times \vec{r}_{21}}{r_{21}^3}\right)$$

$$= I_2 \oint_2 d\vec{\ell}_2 \times \vec{B}_1$$

where

$$\vec{B}_1 = \frac{\mu_0}{4\pi}\, I_1 \oint_1 \frac{d\vec{\ell}_1 \times \vec{r}_{21}}{r_{21}^3} \tag{6.9}$$

can be taken as magnetic field of current I_1 at the position of the element $d\vec{\ell}_2$ of current I_2. The vector $\vec{B}_1$ is called magnetic induction and is expressed in weber/m^2.

Equation (6.9) for $\vec{B}_1$ is called **Biot-Savart law**. On differentiation, we get

$$d\vec{B}_1 = \frac{\mu_0}{4\pi} \frac{I_1 d\vec{\ell}_1 \times \vec{r}_{21}}{r_{21}^3} \tag{6.10}$$

This equation tells that magnetic flux density or magnetic induction $\vec{B}_1$ at a point due to a current element is directed normal to the plane containing the current elements $I_1 d\vec{\ell}_1$ and the line joining the current element to the point (direction of $\vec{r}_{21}$).

If the current I_1 is distributed in space with a current density $\vec{J}_1$, then

$$I_1 = \vec{J}_1 \cdot d\vec{S} \qquad \text{and} \qquad I_1\, d\vec{\ell}_1 = \vec{J}_1 \cdot d\vec{S}\, d\vec{\ell}_1 = \vec{J}_1\, dV_1$$

Equation (6.9) then becomes

$$\vec{B}_1 = \frac{\mu_0}{4\pi} \int_V \frac{\vec{J}_1 \times \vec{r}_{21}}{r_{21}^3} \, \mathrm{d}V_1 \tag{6.11}$$

where the integration is carried over a volume V which includes all the currents.

Example 3: Two current elements $I_1 \, \mathrm{d}\vec{\ell}_1$ and $I_2 \, \mathrm{d}\vec{\ell}_2$ show that $\vec{F}_{12} \neq -\vec{F}_{21}$. Show that when the force of one entire circuit is considered on the other circuit, we have $\vec{F}_{12} = -\vec{F}_{21}$.

Solution: Consider two length elements $\mathrm{d}\vec{\ell}_1$ and $\mathrm{d}\vec{\ell}_2$ of circuits 1 and 2, carrying currents I_1 and I_2, respectively. Force on current element $I_1 \, \mathrm{d}\vec{\ell}_1$ of circuit 1 due to circuit 2, according to Ampere's force law is

$$\mathrm{d}\vec{F}_{12} = I_1 \, \mathrm{d}\vec{\ell}_1 \times \vec{B}_2$$

Following Biot-Savart law, we have

$$\vec{B}_2 = \frac{\mu_0}{4\pi} \, I_2 \oint_2 \frac{\mathrm{d}\vec{\ell}_2 \times \vec{r}_{12}}{r_{12}^3} \tag{6.12}$$

and therefore,

$$\begin{aligned} \mathrm{d}\vec{F}_{12} &= \frac{\mu_0}{4\pi} \, I_1 I_2 \, \mathrm{d}\vec{\ell}_1 \times \oint_2 \frac{\mathrm{d}\vec{\ell}_2 \times \vec{r}_{12}}{r_{12}^3} \\ &= \frac{\mu_0}{4\pi} \, I_1 I_2 \oint_2 \frac{\mathrm{d}\vec{\ell}_1 \times (\mathrm{d}\vec{\ell}_2 \times \vec{r}_{12})}{r_{12}^3} \end{aligned}$$

Using relation $\vec{A} \times (\vec{B} \times \vec{C}) = (\vec{A} \, . \, \vec{C})\vec{B} - (\vec{A} \, . \, \vec{B}) \, \vec{C}$, we have

$$\mathrm{d}\vec{F}_{12} = \frac{\mu_0}{4\pi} \, I_1 I_2 \oint_2 \frac{(\mathrm{d}\vec{\ell}_1 \, . \, \vec{r}_{12}) \, \mathrm{d}\vec{\ell}_2 \; - \; (\mathrm{d}\vec{\ell}_1 \, . \, \mathrm{d}\vec{\ell}_2) \, \vec{r}_{12}}{r_{12}^3} \tag{6.13}$$

Similarly, force on current element $I_2 \, \mathrm{d}\vec{\ell}_2$ of circuit 2 due to circuit 1 is

$$\mathrm{d}\vec{F}_{21} = \frac{\mu_0}{4\pi} \, I_1 I_2 \oint_1 \frac{(\mathrm{d}\vec{\ell}_2 \, . \, \mathrm{d}\vec{\ell}_1) \, \vec{r}_{12} - (\mathrm{d}\vec{\ell}_2 \, . \, \vec{r}_{12}) \, \mathrm{d}\vec{\ell}_1}{r_{12}^3} \tag{6.14}$$

From equations (6.13) and (6.14), it is obvious that $\mathrm{d}\vec{F}_{12} \neq -\mathrm{d}\vec{F}_{21}$.

Now, we consider force due to complete circuit, and from equation (6.13), we get

$$\vec{F}_{12} = \oint_1 \mathrm{d}\vec{F}_{12} = \frac{\mu_0}{4\pi} \, I_1 I_2 \oint_1 \oint_2 \frac{(\mathrm{d}\vec{\ell}_1 \, . \, \vec{r}_{12}) \, \mathrm{d}\vec{\ell}_2 - (\mathrm{d}\vec{\ell}_1 \, . \, \mathrm{d}\vec{\ell}_2) \, \vec{r}_{12}}{r_{12}^3}$$

We can show that

$$\oint_1 \oint_2 \frac{(\mathrm{d}\vec{\ell}_1 \, . \, \vec{r}_{12}) \, \mathrm{d}\vec{\ell}_2}{r_{12}^3} = 0$$

in the following manner. We have

$$\oint_1\oint_2 \frac{(\mathrm{d}\vec{\ell}_1 \,.\, \vec{r}_{12})\,\mathrm{d}\vec{\ell}_2}{r_{12}^3} = \oint_2 \mathrm{d}\vec{\ell}_2 \oint_1 \frac{(\mathrm{d}\vec{\ell}_1 \,.\, \vec{r}_{12})}{r_{12}^3} = \oint_2 \mathrm{d}\vec{\ell}_2 \oint_1 \mathrm{d}\vec{\ell}_1 \,.\, \Big[-\nabla\Big(\frac{1}{r_{12}}\Big)\Big]$$

we have used following known relation

$$\nabla\Big(\frac{1}{r_{12}}\Big) = -\frac{\vec{r}_{12}}{r_{12}^3}$$

Using the Stokes' law

$$\oint \vec{A} \,.\, \mathrm{d}\vec{\ell} = \int_S \nabla \times \vec{A} \,.\, \mathrm{d}\vec{S}$$

we have

$$\oint_1\oint_2 \frac{(\mathrm{d}\vec{\ell}_1 \,.\, \vec{r}_{12})\mathrm{d}\vec{\ell}_2}{r_{12}^3} = -\oint_2 \mathrm{d}\vec{\ell}_2 \int_S \nabla \times \Big[\nabla\Big(\frac{1}{r_{12}}\Big)\Big] \,.\, \mathrm{d}\vec{S}$$

We know that curl of gradient of a scalar is zero. Therefore,

$$\oint_1\oint_2 \frac{(\mathrm{d}\vec{\ell}_1 \,.\, \vec{r}_{12})\,\mathrm{d}\vec{\ell}_2}{r_{12}^3} = 0$$

Thus, we have

$$\vec{F}_{12} = -\frac{\mu_0}{4\pi}\, I_1 I_2 \oint_1\oint_2 \frac{(\mathrm{d}\vec{\ell}_1 \,.\, \mathrm{d}\vec{\ell}_2)\,\vec{r}_{12}}{r_{12}^3} \tag{6.15}$$

Similarly, from equation (6.14), we can have

$$\vec{F}_{21} = \frac{\mu_0}{4\pi}\, I_1 I_2 \oint_1\oint_2 \frac{(\mathrm{d}\vec{\ell}_2 \,.\, \mathrm{d}\vec{\ell}_1)\,\vec{r}_{12}}{r_{12}^3} \tag{6.16}$$

From equations (6.15) and (6.16), we have $\vec{F}_{12} = -\vec{F}_{21}$

6. Magnetic induction due to infinitely long current-carrying straight wire

Consider infinitely long wire carrying current I (Figure 4). Suppose, there is a point P on the line perpendicular to the wire at a distance d from the wire.

(i) Using Biot-Savart law:

In the diagram, the magnetic field $\mathrm{d}\vec{B}(\vec{r})$ at the point P due to current I in element $\mathrm{d}\vec{\ell}\,'$ is

$$\mathrm{d}\vec{B}(\vec{r}) = \frac{\mu_0}{4\pi}\,\frac{I\,\mathrm{d}\vec{\ell}\,' \times \hat{r}}{r^2}$$

The $(\mathrm{d}\vec{\ell'} \times \hat{r})$ points out of the page, and has the magnitude

$$\mathrm{d}\ell' \sin\alpha = \mathrm{d}\ell' \sin(90^\circ + \theta) = \mathrm{d}\ell' \cos\theta$$

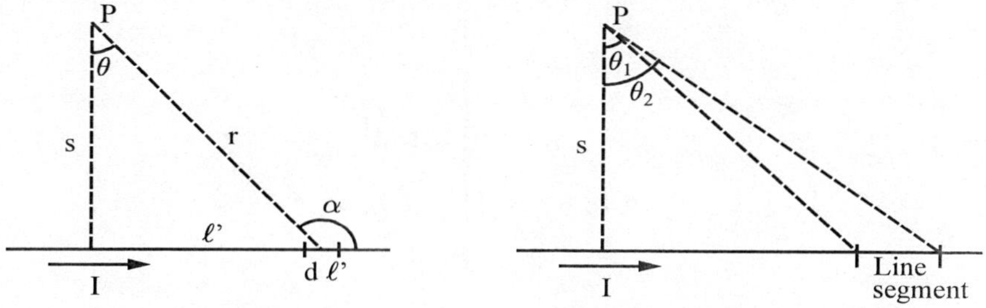

Figure 4:

Also, $\ell' = s\tan\theta$, so

$$d\ell' = \frac{s}{\cos^2\theta}\, d\theta$$

and $s = r\cos\theta$, so

$$\frac{1}{r^2} = \frac{\cos^2\theta}{s^2}$$

Thus,

$$dB(\vec{r}) = \frac{\mu_0 I}{4\pi}\frac{d\ell'\cos\theta}{r^2} = \frac{\mu_0 I}{4\pi}\left(\frac{s}{\cos^2\theta}\, d\theta\right)\cos\theta\left(\frac{\cos^2\theta}{s^2}\right) = \frac{\mu_0 I}{4\pi s}\cos\theta\, d\theta$$

If θ_1 and θ_2 are angles at the two ends of line segment, then the magnetic field

$$B(\vec{r}) = \frac{\mu_0 I}{4\pi s}\int_{\theta_1}^{\theta_2}\cos\theta\, d\theta = \frac{\mu_0 I}{4\pi s}(\sin\theta_2 - \sin\theta_1)$$

In the case of an infinite long wire, $\theta_1 = -\pi/2$ and $\theta_2 = \pi/2$, so we obtain

$$B = \frac{\mu_0 I}{4\pi s}[\sin\pi/2 - \sin(-\pi/2)] = \frac{\mu_0 I}{2\pi s}$$

Consider a current I is flowing through a wire of infinite length. The magnetic at distance s from the wire is

$$B = \frac{\mu_0 I}{2\pi s}$$

As $\vec{B}$ and $d\vec{\ell}$ are along the same direction, the integral of $\vec{B}$ around a circular path of radius s, centered at the wire, is

$$\oint \vec{B}\,.\, d\vec{\ell} = \oint \frac{\mu_0 I}{2\pi s}\, d\ell = \frac{\mu_0 I}{2\pi s}\oint d\ell = \frac{\mu_0 I}{2\pi s}\, 2\pi s = \mu_0 I$$

It may be noted that the result is independent of the distance s from the wire. This equation is called the **Ampere's law** (in integral form). Now, suppose we have a bundle of straight wires. Each wire that passes through the loop contributes $\mu_0 I$, and those outside contribute nothing. The line integral is

$$\oint \vec{B}\,.\, d\vec{\ell} = \mu_0 I_{enc}$$

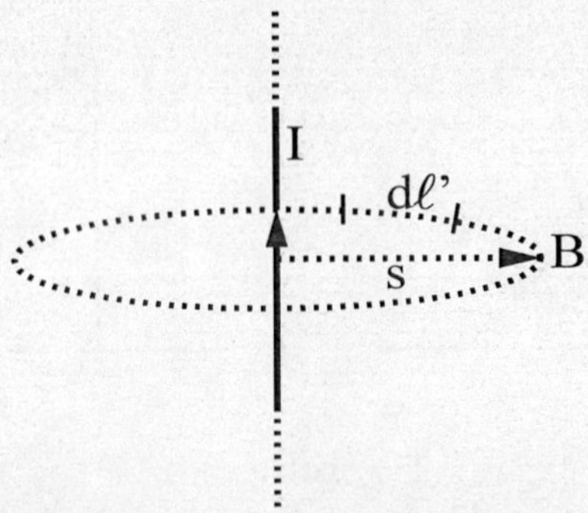

Figure 5: Current in infinite long wire.

where I_{enc} stands for the total current enclosed by the integration path. If the flow of charge is represented by a volume current density $\vec{J}$, the enclosed current is

$$I_{enc} = \int \vec{J} \,.\, \mathrm{d}\vec{S}$$

Therefore, we have

$$\oint \vec{B} \,.\, \mathrm{d}\vec{\ell} = \mu_0 \int \vec{J} \,.\, \mathrm{d}\vec{S}$$

Applying Stokes' theorem, we have

$$\int (\nabla \times \vec{B}) \,.\, \mathrm{d}\vec{S} = \mu_0 \int \vec{J} \,.\, \mathrm{d}\vec{S} \qquad \text{or} \qquad \int \left[(\nabla \times \vec{B}) - \mu_0 \, \vec{J}\right] .\, \mathrm{d}\vec{S} = 0$$

This relation is valid for arbitrary surface, and can only be satisfied when the integrand is equal to zero:

$$(\nabla \times \vec{B}) - \mu_0 \, \vec{J} = 0 \qquad \text{or} \qquad \nabla \times \vec{B} = \mu_0 \, \vec{J}$$

This equation is called the **Ampere's law** (in differential form).

(ii) Using Ampere's law:

If wire is vertical then lines of magnetic induction $\vec{B}$ are concentric circles in horizontal plane. Suppose, P is the point at distance s from wire (Figure 5), then we can draw a circle of lines of magnetic induction through it. On using this circle as the path of integration, we can write Ampere's circuital law as

$$\oint_C \vec{B} \,.\, \mathrm{d}\vec{\ell} = \int \vec{J} \,.\, \mathrm{d}\vec{S} \qquad \text{or} \qquad B 2\pi d = \mu_0 I \qquad \text{or} \qquad B = \frac{\mu_0}{4\pi} \frac{2I}{s}$$

7. Magnetic induction on the axis of a current-carrying circular loop

Consider a circular loop of radius R carrying current I. Let us consider an element $d\vec{\ell}$ of circular loop (Figure 6). Then following Biot-Savart law, magnetic induction due to this element at point P on the axis at distance x from center of loop is

$$d\vec{B} = \frac{\mu_0}{4\pi}\frac{Id\vec{\ell}\times\vec{r}}{r^3} = \frac{\mu_0}{4\pi}\frac{Id\ell\; r\sin 90^\circ\hat{n}}{r^3}$$

The angle between $d\vec{\ell}$ and $\vec{r}$ is 90° and $\hat{n}$ is unit vector in the direction of $d\vec{B}$. The magnetic induction points in a direction perpendicular to both $d\vec{\ell}$ and $\vec{r}$. As shown in Figure 5, $d\vec{B}$ can be resolved into two directions parallel and perpendicular to axis as $d\vec{B}_{\|}$ and $d\vec{B}_{\perp}$, respectively. The perpendicular components when summed up for the whole loop comes out to be zero. Thus, only $d\vec{B}_{\|}$ contributes for the whole induction $\vec{B}$ at the point due to circular loop. That is

$$\vec{B} = \int d\vec{B}_{\|}$$

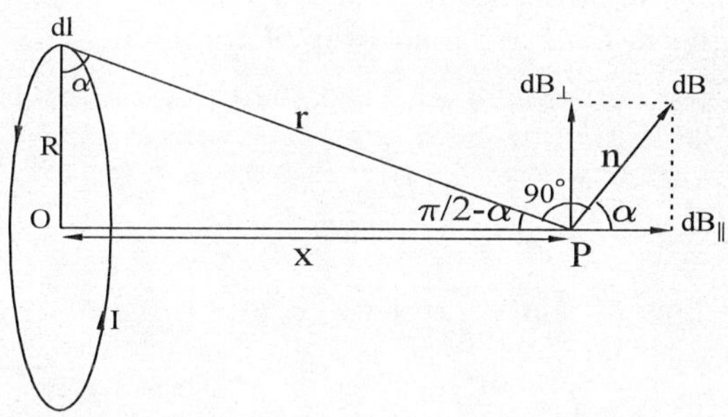

Figure 6: Magnetic induction of circular loop of radius R carries a current I.

The component parallel to axis is

$$dB_{\|} = \hat{i}\;.\;d\vec{B}_{\|}$$

where $\hat{i}$ is the unit vector along the axis of loop. Thus, we have

$$dB_{\|} = \frac{\mu_0}{4\pi}\frac{I\,d\ell}{r^2}\hat{i}\;.\;\hat{n} = \frac{\mu_0}{4\pi}\frac{I\,dl}{r^2}\cos\alpha$$

where α is angle between $\hat{i}$ and $\hat{n}$. Thus, we have

$$B = \frac{\mu_0}{4\pi}\int\frac{I\,dl}{r^2}\cos\alpha$$

We have

$$\sin\left(\frac{\pi}{2}-\alpha\right) = \frac{R}{r} \qquad \text{and therefore} \qquad \cos\alpha = \frac{R}{\sqrt{R^2+x^2}}$$

Using the values of r and $\cos\alpha$, we have

$$B = \frac{\mu_0}{4\pi}\int \frac{I\,d\ell}{R^2+x^2}\frac{R}{\sqrt{R^2+x^2}} = \frac{\mu_0 I}{4\pi}\frac{R}{(R^2+x^2)^{3/2}}\int d\ell$$

$$= \frac{\mu_0 I}{4\pi}\frac{R}{(R^2+x^2)^{3/2}}2\pi R = \frac{\mu_0 I R^2}{2(R^2+x^2)^{3/2}}$$

This is the required result.

For a point far away from the center of coil O, we put $x >> R$ and we have

$$B = \frac{\mu_0 I R^2}{2x^3} = \frac{\mu_0 I A}{2\pi x^3}$$

where $A = \pi R^2$ is area of cross section of loop. As magnetic dipole moment m of loop is $m = IA$ and we have

$$B = \frac{\mu_0 m}{2\pi x^3}$$

Example 4: In the Bohr model of hydrogen atom, the electron circulates around the nucleus along a path of radius 0.5 Å at a frequency of 5×10^{15} Hz. Calculate the magnitude of magnetic field at the center of the orbit.

Solution: From above example, the magnetic field at the center of orbit $(x = 0)$ is

$$B = \frac{\mu_0 I R^2}{2(R^2+x^2)^{3/2}} = \frac{\mu_0 I}{2R}$$

The current is rate of flow of charge and we have

$$I = \frac{dq}{dt} = \frac{e}{1/n} = n\,e$$

Therefore,

$$B = \frac{\mu_0 n e}{2R}$$

Using values, we get

$$B = \frac{(4\pi\times10^{-7})(5\times10^{15})(1.6\times10^{-19})}{2(0.5\times10^{-10})} = 3.2\pi \text{ weber/m}^2$$

8. Magnetic induction for two infinitely long parallel wires carrying currents and separated by a distance

Consider two infinitely long parallel wires separated by a distance d, carrying currents I_1 and I_2 as shown in Figure 7. Wire 1 carrying current I_1 produces magnetic induction B_1 at a distance d as

$$B_1 = \frac{\mu_0}{4\pi}\frac{2I_1}{d}$$

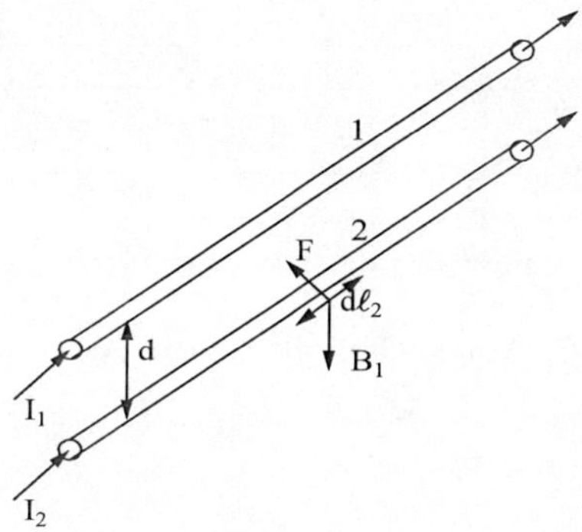

Figure 7: Two infinitely long wires separated by a distance d are carrying currents I_1 and I_2.

The direction of magnetic field B_1 on wire 2 is in downward direction. Due to B_1, wire 2 experiences a side way force $\mathrm{d}F_2$, say for a length $\mathrm{d}l_2$ of this wire

$$\mathrm{d}\vec{F}_2 = I_2\ \mathrm{d}\vec{\ell}_2 \times \vec{B}_1$$

$\mathrm{d}\vec{F}_2$ is in a plane perpendicular to the plane containing $\vec{B}_1$ and $\mathrm{d}\vec{\ell}_2$. That is, it points in the plane of wires and points to the left. The magnitude of force is

$$\mathrm{d}F_2 = I_2\ \mathrm{d}\ell_2 B_1 = \mathrm{d}\ell_2 \frac{\mu_0}{4\pi}\ \frac{2I_1I_2}{d}$$

If we calculate force due to wire 2 on wire 1, it is the same in magnitude but it points to the right (in the plane of wires). Thus, this example signifies that two wires attract each other *i.e.,* action and reaction are equal and opposite.

If the current in two parallel wires is in opposite directions then they repel each other.

9. Magnetic induction inside a long solenoid

Consider a solenoid of length l and radius a having N turns and carrying current i as shown in Figure 8. To calculate magnetic field at the point P situated on the axis, the solenoid may be imagined as divided into a number of narrow equidistant coils and consider one such coil of width $\mathrm{d}x$. The number of turns is $N\ \mathrm{d}x/l$. Let x be the distance of the point P from the center of the coil.

The field at point P due to elementary coil of width $\mathrm{d}x$ carrying current i is

$$\mathrm{d}B = \frac{\mu_0 i a^2}{2(a^2+x^2)^{3/2}}\left(\frac{N}{\ell}\ \mathrm{d}x\right)$$

From Figure 8, we have

$$x = a\cot\theta \qquad \text{and therefore,} \qquad \mathrm{d}x = -a\ \mathrm{cosec}^2\theta\ \mathrm{d}\theta$$

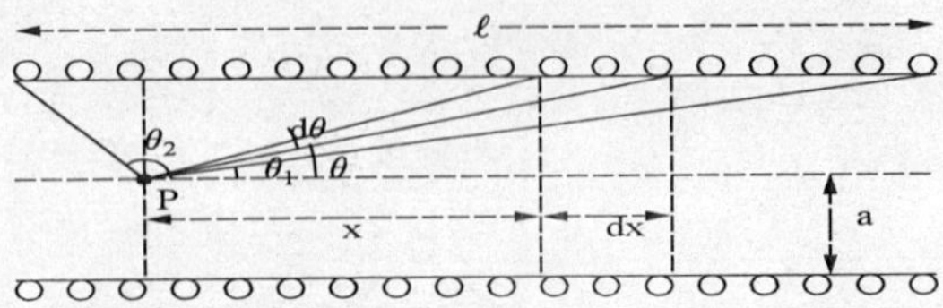

Figure 8: Magnetic induction inside a long solenoid.

Thus, we have

$$B = \int dB = \int_{\theta_2}^{\theta_1} \frac{\mu_0 i a^2}{2(a^2 + a^2 \cot^2 \theta)^{3/2}} \left(\frac{N}{\ell}\right)(-a \text{ cosec}^2\theta \, d\theta)$$

$$= -\frac{\mu_0 Ni}{2\ell} \int_{\theta_2}^{\theta_1} \sin\theta \, d\theta = \frac{\mu_0 Ni}{2\ell}(\cos\theta_1 - \cos\theta_2)$$

When the point P is well inside a very long solenoid, we have $\theta_1 = 0$ and $\theta_2 = \pi$ and

$$B = \frac{\mu_0 Ni}{2\ell}(\cos 0 - \cos\pi) - = \frac{\mu_0 Ni}{\ell}$$

When the point P is at one end of very long solenoid, we have $\theta_1 = 0$ and $\theta_2 = \pi/2$ and

$$B = \frac{\mu_0 Ni}{2\ell}(\cos 0 - \cos\pi/2) = \frac{\mu_0 Ni}{2\ell}$$

That is, the magnetic field at either side of solenoid is half of that well inside the solenoid.

9.1 Field at the center of solenoid of finite length

Let us move the point P at the center of the solenoid so that

$$\cos\theta_1 = \frac{\ell/2}{(a^2 + \ell^2/4)^{1/2}} = \frac{\ell}{(4a^2 + l^2)^{1/2}}$$

and

$$\cos(\pi - \theta_2) = \frac{\ell/2}{(a^2 + \ell^2/4)^{1/2}} \qquad \text{or} \qquad \cos\theta_2 = -\frac{\ell}{(4a^2 + \ell^2)^{1/2}}$$

Using values of $\cos\theta_1$ and $\cos\theta_2$, we get

$$B = \frac{\mu_0 Ni}{2\ell}\left[\frac{\ell}{(4a^2 + \ell^2)^{1/2}} + \frac{\ell}{(4a^2 + \ell^2)^{1/2}}\right] = \frac{\mu_0 Ni}{(4a^2 + \ell^2)^{1/2}}$$

The field of a solenoid is a vector sum of the fields set up by all turns of wire bound around the solenoid.

Example 5: A coil having a shape of square of side $2a$ is carrying a current i. Calculate the magnetic field at the center.

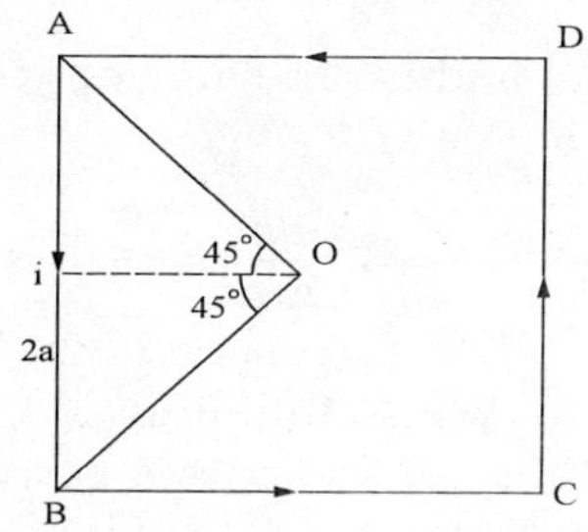

Figure 9: Magnetic induction at the center of square shaped coil.

Solution: Using Biot-Savart law, the magnetic field at the center O due to side AB, carrying a current i is

$$B_1 = \frac{\mu_0 i}{4\pi a}\,(\sin 45^\circ + \sin 45^\circ) = \frac{\mu_0 i}{4\pi a}\,\sqrt{2}$$

Magnetic field at O due to each side is directed outward perpendicular to the plane of paper. Thus, total magnetic induction at O is

$$B = 4B_1 = 4\,\frac{\mu_0 i}{4\pi a}\,\sqrt{2} = \frac{\mu_0 i\sqrt{2}}{\pi a}$$

Example 6: A coil having a shape of equilateral triangle of side a is carrying a current i. Calculate magnetic field at the center of triangle.

Solution: Since each side of equilateral triangle is a, we have

$$\text{AD} = \sqrt{a^2 - (a/2)^2} = \sqrt{3}\;a/2$$

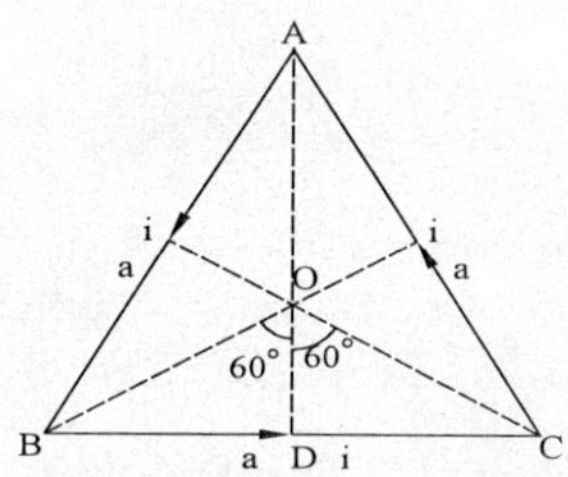

Figure 10: Magnetic induction at the center of equilateral triangle shaped coil.

Thus, we have

$$\text{OD} = \frac{1}{3}\;\text{AD} = a/2\sqrt{3}$$

Magnetic field at O due to current i in arm BC is

$$B_1 = \frac{\mu_0}{4\pi}\,\frac{i}{\text{OD}}(\sin 60^\circ + \sin 60^\circ) = \frac{\mu_0}{4\pi}\,\frac{i2\sqrt{3}}{a}\sqrt{3} = \frac{3\mu_0 i}{2\pi a}$$

The field due to all sides is equal and in the same direction at O. Thus, the magnetic field at O due to equilateral triangle is

$$B = 3B_1 = 3\,\frac{3\mu_0 i}{2\pi a} = \frac{9\mu_0 i}{2\pi a}$$

Example 7: A pair of stationary and infinitely long bent wires are placed in xy plane as shown in Figure 11. Each wire carries a current i. Find the magnetic field at point O, where OB = OF.

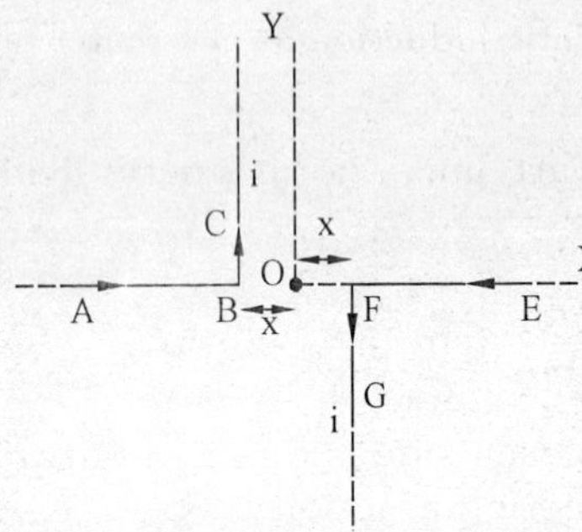

Figure 11: Magnetic induction between a pair of stationary and infinitely long bent wires.

Solution: A pair of stationary and infinitely long bent wires is shown in Figure 11. The field at O due to AB and EF parts of wires is zero, as they are along the same line and is directed in opposite directions. For BC and FG parts of wires, point O is at one end for which magnetic field is

$$B_1 = \frac{\mu_0}{4\pi}\,\frac{i}{\text{OB}}$$

Current flow is such that magnetic field due to BC and FG is down the plane. Thus, the net magnetic field is

$$B = 2B_1 = \frac{\mu_0}{2\pi}\,\frac{i}{x}$$

Example 8: A loop of flexible conducting wire of length l lies in a magnetic field B perpendicular to the plane of loop. Show that when a current i is passed through the loop, it opens out into a circle. Find the tension development in the wire.

Solution: Consider a loop of length l and radius r. The magnetic field is perpendicular to the plane of loop and is directed into the plane of loop. Force $iB\delta l$ is in the direction shown in Figure 12 due to the current in δl. Since it is directed outward on all the current elements of the loop, the loop opens out into a circle. From Figure 12, we write

$$T\sin\frac{\alpha}{2} + T\sin\frac{\alpha}{2} = iB\delta l \qquad \text{or} \qquad 2T\sin\frac{\alpha}{2} = iB\delta l$$

If α is small *i.e.*, δl is small, then

$$2T\,\frac{\alpha}{2} = iB\delta l \qquad \text{or} \qquad T\,\alpha = iB\delta l$$

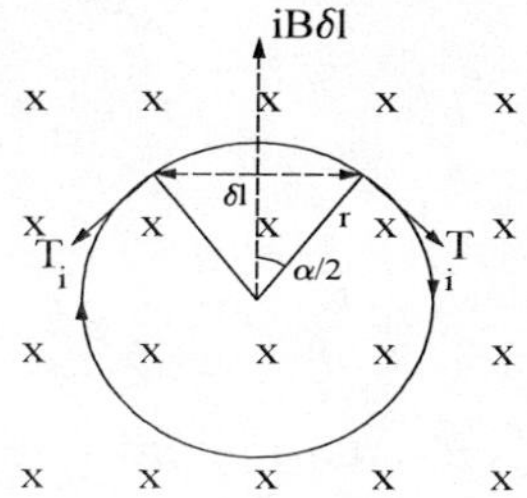

Figure 12: Loop of flexible conducting wire lying in a magnetic field perpendicular to the plane of loop.

Using $\delta\ell = r\alpha$, we get

$$T = iBr = \frac{2\pi riB}{2\pi} = \frac{\ell iB}{2\pi}$$

Here, $\ell = 2\pi r$ is length of loop.

Example 9: A long horizontal wire AB free to move in the vertical plane and carrying current i_1 is in equilibrium at a height d over another parallel long wire CD, which is fixed in a horizontal plane and carries steady current i_2. If wire AB is slightly depressed, it executes simple harmonic motion. Calculate the period of oscillation.

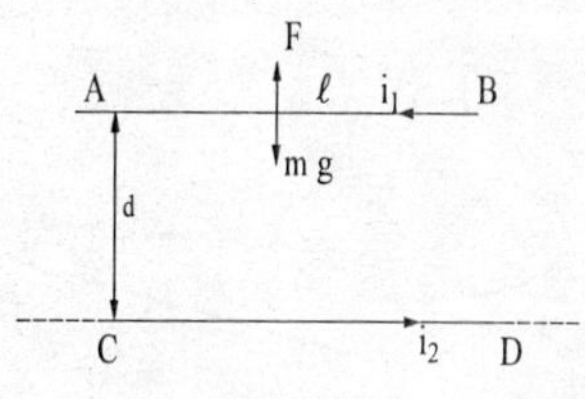

Figure 13: Wire CD fixed in the horizontal plane carries steady current i_2.

Solution: To balance weight mg, the force F on it due to the wire CD must be in upward direction, for which the currents in these two wires must be in opposite directions. Suppose, when separation between the wires is d, the force F and mg are in equilibrium. Then we have

$$F = \frac{\mu_0}{2\pi}\frac{i_1 i_2 \ell}{d} = mg$$

If wire AB is slightly depressed downward by x then the force that would act as restoring force is

$$F' = -\Big(\frac{\mu_0}{2\pi}\frac{i_1 i_2 \ell}{(d-x)} - mg\Big)$$

Using above relation, we get

$$F' = -\Big(\frac{mgd}{(d-x)} - mg\Big) = -\frac{mgx}{(d-x)}$$

Since $x << d$, we have

$$F' = -\frac{mg}{d}\,x$$

The acceleration is

$$\frac{d^2x}{dt^2} = \frac{F'}{m} = -\frac{g}{d}\,x$$

This is equation of simple harmonic motion with angular frequency

$$\omega = \sqrt{\frac{g}{d}}$$

The time-period of simple harmonic oscillations is

$$T = \frac{2\pi}{\omega} = 2\pi\sqrt{\frac{d}{g}}$$

10. Divergence of magnetic induction

The Biot-Savart law is

$$d\vec{B}(\vec{r}) = \frac{\mu_0}{4\pi}\,\frac{\vec{I}(\vec{r}') \times \hat{R}}{R^2}\,d\ell'$$

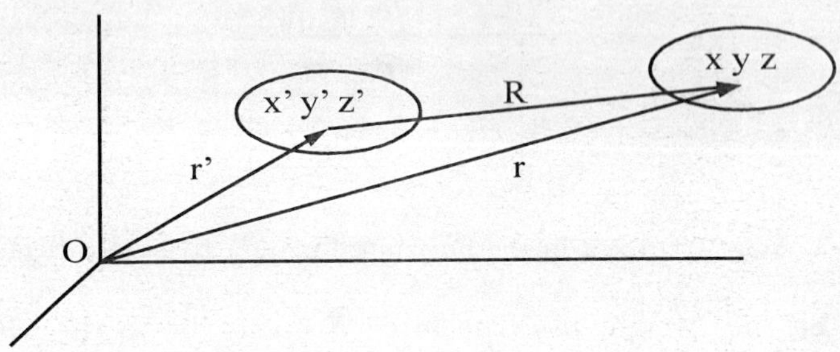

Figure 14:

Here, $\vec{R} = \vec{r} - \vec{r}\,'$. Writing $\vec{I}(\vec{r}\,') = \vec{J}(\vec{r}\,')\ dS'$, where $\vec{J}(\vec{r}\,')$ is current density and dS' is surface element, we have

$$d\vec{B}(\vec{r}) = \frac{\mu_0}{4\pi}\,\frac{\vec{J}(\vec{r}\,') \times \hat{R}}{R^2}\,d\tau'$$

where $d\tau' = dS'\ d\ell'$ is volume element. On integration, we get

$$\vec{B}(\vec{r}) = \frac{\mu_0}{4\pi}\int \frac{\vec{J}(\vec{r}\,') \times \hat{R}}{R^2}\,d\tau' \tag{6.17}$$

This formula gives the magnetic field at a point $\vec{r} = (x, y, z)$ in terms of an integral over the current distribution $\vec{J}(x', y', z')$. We have $d\tau' = dx'\ dy'\ dz'$. The integration is over the primed coordinates; the divergence and the curl of $\vec{B}$ are with respect to the unprimed coordinates. Applying the divergence to equation (6.17), we obtain:

$$\nabla \cdot \vec{B} = \frac{\mu_0}{4\pi} \int \nabla \cdot \left(\frac{\vec{J} \times \hat{R}}{R^2} \right) \mathrm{d}\tau'$$

Invoking product rule number

$$\nabla \cdot \left(\vec{J} \times \frac{\hat{R}}{R^2} \right) = \frac{\hat{R}}{R^2} \cdot (\nabla \times \vec{J}) - \vec{J} \cdot \left(\nabla \times \frac{\hat{R}}{R^2} \right)$$

But, $\nabla \times \vec{J} = 0$, because $\vec{J}$ does not depend on the unprimed variables, while $\nabla \times (\hat{R}/R^2) = 0$. Thus,

$$\nabla \cdot \vec{B} = 0$$

It shows that divergence of magnetic induction is always zero. The magnetic flux through the surface enclosed by the current loop is

$$\Phi = \int_S \vec{B} \cdot \mathrm{d}\vec{S}$$

Using the Gauss' divergence theorem for closed surface, we have

$$\Phi = \int_V (\nabla \cdot \vec{B}) \, \mathrm{d}V = 0$$

It shows that net flux of magnetic induction through any closed surface is always zero.

Example 10: State and prove Ampere's circuital law.

Solution: Ampere's circuital law states that the line integral of magnetic induction $\vec{B}$ around a closed path is equal to μ_0 times total current crossing any surface bounded by the line integral path. The mathematical form is

$$\oint_C \vec{B} \cdot \mathrm{d}\vec{\ell} = \mu_0 \int_S \vec{J} \cdot \mathrm{d}\vec{S} = \mu_0 I$$

We have proved

$$\nabla \times \vec{B} = \mu_0 \vec{J}$$

Integrating it over an open surface S bounded by a closed curve C, we have

$$\oint_S (\nabla \times \vec{B}) \cdot \mathrm{d}\vec{S} = \mu_0 \oint_S \vec{J} \cdot \mathrm{d}\vec{S}$$

According to Stokes' theorem on right side, we get

$$\int_S (\nabla \times \vec{B}) \cdot \mathrm{d}\vec{S} = \oint_C \vec{B} \cdot \mathrm{d}\vec{\ell}$$

Using Stokes' theorem, we have

$$\oint_C \vec{B} \cdot \mathrm{d}\vec{\ell} = \mu_0 \int_S \vec{J} \cdot \mathrm{d}\vec{S} = \mu_0 I$$

It is Ampere's circuital law.

Example 11: (a) Show that magnetic field of hollow circular cylindrical conductor carrying current vanishes everywhere inside a hollow region.

(b) A long straight copper tube having an inner radius R_1 and outer radius R_2 carries a current I. Compute magnetic field at distance r from the axis, where (i) $r < R_1$, (ii) $R_1 < r < R_2$, and (iii) $r > R_2$.

Solution: Cross section of hollow circular cylinder with axis through O, and inner radius R_1 and outer radius R_2 is shown in Figure 15. Suppose, the current through the conductor is I. Let us consider a circular path of radius r around the axis of hollow cylinder.

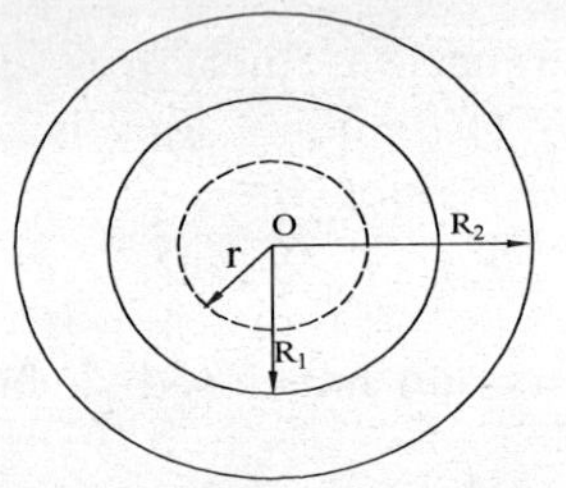

Figure 15: Cross section of hollow circular cylinder with axis through O, and inner radius R_1 and outer radius R_2.

(a) Applying Ampere's law to the path of radius r, we get

$$\oint_C \vec{B} \cdot d\vec{\ell} = \mu_0 I$$

where I is the current enclosed by circular path of radius r. As this does not enclose any current, $i = 0$ so that

$$\oint_C \vec{B} \cdot d\vec{\ell} = 0$$

But

$$\oint_C \vec{B} . d\vec{\ell} = B \oint_C d\ell = B(2\pi r)$$

Therefore, we get

$$B(2\pi r) = 0 \qquad \text{or} \qquad B = 0$$

Thus, the magnetic field of a hollow circular cylindrical conductor carrying current vanishes everywhere inside the hollow cylinder.

(b) (i) For the case $r < R_1$, we have

$$\oint_C \vec{B} \cdot d\vec{\ell} = \mu_0 I'$$

where I' is the current enclosed by the circular path of radius $r < R_1$, which is zero. Therefore,

$$B(2\pi r) = 0 \qquad \text{or} \qquad B = 0$$

(ii) For the case $R_1 < r < R_2$, we have

$$\oint_C \vec{B} \cdot d\vec{\ell} = \mu_0 I'$$

where I' is the current enclosed by the circular path of radius $R_1 < r < R_2$ and is given by

$$I' = \frac{I}{\pi(R_2^2 - R_1^2)} \times \pi(r^2 - R_1^2)$$

where I is total current through the conductor. Therefore,

$$\oint_C \vec{B} \cdot d\vec{\ell} = \mu_0 \frac{I}{\pi(R_2^2 - R_1^2)} \times \pi(r^2 - R_1^2) \qquad B(2\pi r) = \frac{\mu_0 I(r^2 - R_1^2)}{(R_2^2 - R_1^2)}$$

$$B = \frac{\mu_0 I(r^2 - R_1^2)}{2\pi r(R_2^2 - R_1^2)}$$

(iii) For the case $r > R_2$, we have

$$\oint_C \vec{B} \cdot d\vec{\ell} = \mu_0 I'$$

where I' is the current enclosed by the circular path of radius $r > R_2$. which I. Therefore,

$$B(2\pi r) = \mu_0 I \qquad \text{and} \qquad B = \frac{\mu_0 I}{2\pi r}$$

Review Questions

1. What is current density? Derive expression for it.
2. What is magnetic induction? Derive expression for Lorentz force.
3. What is Ampere's force law? Using Ampere's force law, derive Lorentz force.
4. Write the equation of continuity and derive it.
5. Derive expression for magnetic interaction of steady line current.
6. What is Biot-Savart law? Derive it.
7. Derive expression for magnetic induction due to infinitely long straight wire carrying steady current. What happens when the length of wire is finite.
8. Derive expression for magnetic induction on the axis of a circular loop carrying steady current.

9. Derive expression for magnetic induction for two infinitely long parallel wires carrying different steady currents.

10. Derive expression for magnetic induction inside infinitely long solenoid. What happens when the length of solenoid is finite.

11. Prove that gradient of magnetic induction is always zero.

12. Shows that net flux of magnetic induction through any closed surface is always zero.

13. Write notes on the following:

    (i) Current density
    (ii) Magnetic induction
    (iii) Ampere's force law
    (iv) Lorentz force
    (v) Biot-Savart law
    (vi) Equation of continuity

Multiple-Choice Questions

1. For current density $\vec{J}$ and charge density ρ, which is the equation of continuity?

    A. $\nabla \cdot \vec{J} - \dfrac{\partial \rho}{\partial t} = 0$ B. $\nabla \times \vec{J} - \dfrac{\partial \rho}{\partial t} = 0$

    C. $\nabla \cdot \vec{J} + \dfrac{\partial \rho}{\partial t} = 0$ D. $\nabla \times \vec{J} + \dfrac{\partial \rho}{\partial t} = 0$

    Ans. C

2. A positively charged particle is moving in space with velocity $\vec{v}$ where magnetic induction $\vec{B}$, inclined to $\vec{v}$, is present. What is the direction of force on the particle?

    A. parallel to $\vec{v}$ B. in the plane containing both $\vec{v}$ and $\vec{B}$

    C. parallel to $\vec{B}$ D. perpendicular to plane containing both $\vec{v}$ and $\vec{B}$

    Ans. D

3. A positively charged particle is moving along x-axis in a magnetic field working along y-axis. The deflecting force is

    A. along x-axis B. along y-axis C. along z-axis D. in xy-plane

    Ans. C

4. A particle carrying charge q is moving with velocity $\vec{v}$ in a magnetic field $\vec{B}$. The Lorentz force acing on the particle is

   A. $q\,\vec{v}\,.\,\vec{B}$ B. $q\,\vec{v}\times\vec{B}$ C. $q\,\vec{B}\times\vec{v}$ D. $q\,\vec{v}+\vec{B}$

   Ans. B

5. A particle carrying charge q is moving with velocity $\vec{v}$ in a magnetic field $\vec{B}$. The direction of Lorentz force acing on the particle is

   A. along $\vec{v}$ B. perpendicular to both $\vec{v}$ and $\vec{B}$

   C. along $\vec{B}$ D. in the plane having $\vec{v}$ and $\vec{B}$

   Ans. B

6. For magnetic vector potential $\vec{A}$ and electric scalar potential ϕ, the Lorentz condition is

   A. $\nabla\,.\,\vec{A}-\frac{1}{c^2}\,\frac{\partial\phi}{\partial t}=0$ B. $\nabla\,.\,\vec{A}+\frac{1}{c^2}\,\frac{\partial\phi}{\partial t}=0$

   C. $\nabla\times\vec{A}+\frac{1}{c^2}\,\nabla\phi=0$ D. $\nabla\times\vec{A}-\frac{1}{c^2}\,\nabla\phi=0$

   Ans. B

7. Which of the following relations is correct?

   A. $\nabla\times\vec{B}=\vec{J}+\frac{\partial\vec{D}}{\partial t}$ B. $\nabla\times\vec{H}=\vec{J}+\frac{\partial\vec{D}}{\partial t}$

   C. $\nabla\times\vec{B}=\vec{J}+\frac{\partial\vec{E}}{\partial t}$ D. $\nabla\times\vec{H}=\vec{J}+\frac{\partial\vec{E}}{\partial t}$

   Ans. B

8. On a stationary charge, both electric and magnetic fields are applied simultaneously. Due to which field the change would start moving?

   A. Electric field only

   B. Magnetic field only

   C. Both electric and magnetic fields

   D. Neither electric field nor magnetic field

   Ans. A

9. In a magnetic field, under which situation, a charged particle would experience a force?

   A. Particle is stationary

   B. Particle is moving parallel to the direction of magnetic field

   C. Particle is moving opposite to the direction of magnetic field

   D. Particle is moving perpendicular to the direction of magnetic field

   Ans. D

10. Under which situation, the Lorentz force due to electric and magnetic fields acting simultaneously on a charge can be zero?

    A. When the fields are parallel to each other.

    B. When the fields are opposite to each other.

    C. When the fields are perpendicular to each other.

Ans. C

11. Which statement is wrong in case of electric and magnetic fields?

    A. Work done by electric field on a free charge can never be zero.

    B. Work done by magnetic field on a freely moving charge is always zero.

    C. For the Lorentz force to be zero, the electric and magnetic fields are parallel to each other.

    D. For the Lorentz force to be zero, the electric and magnetic fields are perpendicular to each other.

Ans. C

12. Which statement is correct in case of electric and magnetic fields?

    A. Work done by electric field on a free charge can be zero.

    B. Work done by magnetic field on a freely moving charge is always zero.

    C. For the Lorentz force to be zero, the electric and magnetic fields are parallel to each other.

    D. For the Lorentz force to be zero, the electric and magnetic fields are opposite to each other.

Ans. B

13. Conservation of what, the equation of continuity corresponds to?

    A. Energy B. Mass C. Momentum D. Charge

Ans. D

14. Current I is flowing through a wire of infinite length. The magnetic field at a distance s from the wire, varies as

    A. s B. $s * s$ C. $1/s$ D. does not depend on s

Ans. C

15. For force-free magnetic field, the motion of charged particle in the field is

    A. perpendicular to the direction of field.

    B. along the direction of field.

    C. inclined at 45 degree with the direction of field.

Ans. B

16. A charged particle is moving in electric and magnetic fields applied simultaneously. The amount of work done for motion between two points does not depend on the path followed because of

    A. electric field only.

    B. magnetic field only.

    C. electric and magnetic fields both.

    D. neither electric field nor magnetic field.

Ans. C

Chapter 7

Electromagnetic Induction

The fields we have discussed so far are time-independent. Electric field produced by stationary electric charges is known as static electric field or electrostatic field and the field at any point in space is time independent. This electric field is conservative *i.e.*, the line integral of field intensity vector $\vec{E}(\vec{r})$ around any close path in space is equal to zero. We have also discussed that the steady current generates the magnetic field called the static magnetic field or magnetostatic field.

Now, we intend to study the time-dependent fields. There is another kind of electric field which varies with time and is non-conservative. Thus, when current is not steady and varies with time, then it produces a time-varying magnetic field around it. Such time-varying magnetic field $\vec{B}(\vec{r},\ t)$ in its turn produces an electric field $\vec{E}(\vec{r},\ t)$ which also varies with time. Also, time-varying electric fields give rise to magnetic fields $\vec{B}(\vec{r},\ t)$ through displacement currents. Thus, both electric and magnetic fields are interdependent.

1. Faraday's law of induction and Lenz's law

Faraday discovered that if magnetic flux through a closed circuit of a conduction wire is changed, an electric current, called induction current, flows in the circuit and is said to be caused by an electric field which is produced by the varying magnetic field. It is known as induced electric field $\vec{E}(\vec{r},\ t)$, the line integral of which around the contour of the closed circuit is equal to the negative rate of change of magnetic flux Φ of the circuit. That is,

$$\oint_C \vec{E}(\vec{r},\ t)\ .\ \mathrm{d}\vec{\ell} = -\frac{\mathrm{d}\Phi}{\mathrm{d}t}$$

Left side of this equation is called electromotive force (emf) e induced in the circuit. Magnetic flux through the closed circuit is given by the surface integral of the magnetic induction $\vec{B}(\vec{r},\ t)$ over a surface S bounded by the contour C of closed circuit. That is,

$$\Phi = \oint_S \vec{B}(\vec{r},\ t)\ .\ \mathrm{d}\vec{S}$$

Using expression for Φ, we have

$$\mathcal{E} = \oint_C \vec{E}(\vec{r},\ t)\ .\ \mathrm{d}\vec{\ell} = -\frac{\mathrm{d}}{\mathrm{d}t}\oint_S \vec{B}(\vec{r},\ t)\ .\ \mathrm{d}\vec{S}$$

Using Stokes' theorem on left side, and considering space and time are independent on right side, we get

$$\oint_S \nabla \times \vec{E}\ .\ \mathrm{d}\vec{S} = -\oint_S \frac{\partial \vec{B}}{\partial t}\ .\ \mathrm{d}\vec{S} \qquad \text{or} \qquad \oint_S \left[\nabla \times \vec{E} + \frac{\partial \vec{B}}{\partial t}\right]\ .\ \mathrm{d}\vec{S} = 0$$

For arbitrary surface, this equation is valid when the integrand is zero. Thus, we have

$$\nabla \times \vec{E} + \frac{\partial \vec{B}}{\partial t} = 0 \qquad \text{or} \qquad \nabla \times \vec{E} = -\frac{\partial \vec{B}}{\partial t}$$

This is differential form of Faraday's law. The negative sign on right side determines the polarity of induced emf. It opposes the change in the magnetic flux producing it. This fact is known as **Lenz's law** and is a consequence of the law of conservation of energy. It may be noted that not only conducting wire closed path but any closed path in the region has induced emf but a flow of current occurs only if the path is conducting.

2. Vector and scalar potentials

We have $\nabla\ .\ \vec{B} = 0$. We know that divergence of curl of a vector is zero. So, we can write

$$\vec{B} = \text{curl}\ \vec{A}$$

where $\vec{A}$ is known as **vector potential**. $\vec{A}$ is not uniquely defined:

$$\vec{A} \to \vec{A} + \nabla f$$

From Faraday's law, we have

$$\nabla \times \vec{E} = -\frac{\partial \vec{B}}{\partial t}$$

Substituting $\vec{B}$, we have

$$\nabla \times \vec{E} = -\frac{\partial}{\partial t}(\text{curl}\ \vec{A}) \qquad \text{or} \qquad \text{curl}\ \vec{E} + \text{curl}\left[\frac{\partial \vec{A}}{\partial t}\right] = 0$$

$$\text{curl}\left[\vec{E} + \frac{\partial \vec{A}}{\partial t}\right] = 0$$

We know that curl of gradient of a scalar function is zero. Therefore, we have

$$\vec{E} + \frac{\partial \vec{A}}{\partial t} = -\text{grad}\ \phi$$

Here, ϕ is known as the **scalar potential**. Negative sign is used to relate the field and potential. Thus, electric field intensity $\vec{E}$ is expressed as

$$\vec{E} = -\frac{\partial \vec{A}}{\partial t} - \text{ grad } \phi$$

$\vec{A}$ and ϕ are not defined uniquely:

$$\vec{A}' \to \vec{A} + \nabla f \qquad\qquad \phi' \to \phi + \frac{\partial f}{\partial t}$$

For static (time-independent) case, we have

$$\vec{E} = -\nabla\phi$$

3. Gauge invariance

For an arbitrary scalar function f of time and space, let us have a new set of scalar and vector potentials (ϕ', $\vec{A}'$) related to the old set of scalar and vector potentials (ϕ, $\vec{A}$) as

$$\vec{A}' = \vec{A} + \text{ grad } f \qquad\qquad \phi' = \phi - \frac{\partial f}{\partial t} \tag{7.1}$$

The function f is known as gauge function. For the new set of potentials, let us calculate new electric field

$$\vec{E}' = -\frac{\partial \vec{A}'}{\partial t} - \text{ grad } \phi'$$

Using equation (7.1), we get

$$\vec{E}' = -\frac{\partial}{\partial t}(\vec{A} + \text{grad} f) - \text{ grad } \left(\phi - \frac{\partial f}{\partial t}\right)$$

$$= -\frac{\partial \vec{A}}{\partial t} - \frac{\partial}{\partial t}(\text{grad} f) - \text{ grad } \phi + \text{grad } \left(\frac{\partial f}{\partial t}\right)$$

$$= -\frac{\partial \vec{A}}{\partial t} - \text{ grad } \phi = \vec{E}$$

It shows that the electric field intensity does not change.

Let us calculate new magnetic field

$$\vec{B}' = \text{curl } \vec{A}'$$

Using equation (7.1), we get

$$\vec{B}' = \text{curl}(\vec{A} + \text{grad } f)$$

$$= \text{curl } \vec{A} + \text{curl (grad } f)$$

$$= \text{curl } \vec{A} = \vec{B}$$

It shows that the magnetic field intensity does not change. The potential are therefore not uniquely defined, as f is arbitrary function of time and space. The invariance of electric and magnetic field with the variations of vector and scalar potentials expressed by equation (7.1) is known as the Gauge invariance.

Example: Decide if the followings are gauge functions.

(i) $x^2 + y^2 + z^2$, (ii) $e^{i\omega t}$ (iii) $(x^2 + y^2 + z^2)\ e^{i\omega t}$ (iv) 100

3.1 Lorentz force law in potential form

When electric field $\vec{E}$ and magnetic field $\vec{B}$ both are present in a space simultaneously, the net force action a particle carrying charge q and moving with velocity $\vec{v}$ is

$$\vec{F} = q[\vec{E} + \vec{v} \times \vec{B}]$$

This force is known as the **Lorentz force**. We know that the electric and magnetic fields can be expressed in terms of vector potential $\vec{A}$ and scalar potential ϕ as

$$\vec{B} = \nabla \times \vec{A} \qquad \text{and} \qquad \vec{E} = -\frac{\partial \vec{A}}{\partial t} - \nabla\phi$$

Further, in terms of momentum, the force can be expressed as

$$\vec{F} = \frac{d\vec{p}}{dt}$$

Using these relations, we have

$$\frac{d\vec{p}}{dt} = q\Big[\Big(-\frac{\partial \vec{A}}{\partial t} - \nabla\phi\Big) + \vec{v} \times (\nabla \times \vec{A})\Big]$$

We know

$$\nabla(\vec{v} \cdot \vec{A}) = \vec{v} \times (\nabla \times \vec{A}) + (\vec{v} \cdot \nabla)\vec{A}$$

Thus,

$$\frac{d\vec{p}}{dt} = -q\Big[\frac{\partial \vec{A}}{\partial t} + (\vec{v} \cdot \nabla)\vec{A} + \nabla(\phi - \vec{v} \cdot \vec{A})\Big]$$

We have

$$\frac{d\vec{A}}{dt} = \frac{\partial \vec{A}}{\partial t} + (\vec{v} \cdot \nabla)\vec{A}$$

Therefore,

$$\frac{d}{dt}(\vec{p} + q\vec{A}) = -\nabla[q(\phi - \vec{v} \cdot \vec{A})] \tag{7.2}$$

In mechanics, for the motion of a particle whose potential energy U is a specified function of position:

$$\frac{d\vec{p}}{dt} = -\nabla U$$

$(\vec{p} + q\vec{A})$ is called **canonical momentum**, denoted by $\vec{p}$. The part of U is taken by the velocity-dependent quantity $U = q(\phi - \vec{v} \cdot \vec{A})$

3.2 Lorentz gauge

Ampere's law for vacuum is

$$\nabla \times \vec{B} = \frac{1}{c^2}\,\frac{\partial \vec{E}}{\partial t}$$

Using expressions for $\vec{B}$ and $\vec{E}$ in terms of vector and scalar potentials, we get

$$\nabla \times (\nabla \times \vec{A}) = \frac{1}{c^2}\,\frac{\partial}{\partial t}\Big(-\frac{\partial \vec{A}}{\partial t} - \text{ grad } \phi\Big)$$

$$\nabla(\nabla \,.\, \vec{A}) - \nabla^2 \vec{A} = -\frac{1}{c^2}\,\frac{\partial^2 \vec{A}}{\partial t^2} - \frac{1}{c^2}\nabla\Big(\frac{\partial \phi}{\partial t}\Big)$$

$$\frac{1}{c^2}\,\frac{\partial^2 \vec{A}}{\partial t^2} - \nabla^2 \vec{A} = -\nabla\Big(\nabla \,.\, \vec{A} + \frac{1}{c^2}\,\frac{\partial \phi}{\partial t}\Big)$$

Now, $\vec{A}$ and ϕ are not unique. This fact is known as the gauge freedom, which allows us to impose conditions on the vector and scalar potentials. Let us choose the gauge such that

$$\nabla . \vec{A} + \frac{1}{c^2}\,\frac{\partial \phi}{\partial t} = 0 \tag{7.3}$$

This is called the **Lorentz gauge**. On adoption of this gauge, we have

$$\frac{1}{c^2}\,\frac{\partial^2 \vec{A}}{\partial t^2} - \nabla^2 \vec{A} = 0$$

Thus, vector potential satisfies the wave equation. From the Gauss' law for vacuum, we have

$$\nabla \,.\, \vec{E} = 0$$

Using the expression for $\vec{E}$, we have

$$\nabla \,.\, \Big(-\frac{\partial \vec{A}}{\partial t} - \text{ grad } \phi\Big) = 0 \qquad\qquad -\frac{\partial(\nabla \,.\, \vec{A})}{\partial t} - \nabla^2 \phi = 0$$

Using equation (7.3), we have

$$-\frac{\partial}{\partial t}\Big(-\frac{1}{c^2}\,\frac{\partial \phi}{\partial t}\Big) - \nabla^2 \phi = 0 \qquad\qquad \frac{1}{c^2}\,\frac{\partial^2 \phi}{\partial t^2} - \nabla^2 \phi = 0$$

Thus, scalar potential also satisfies the wave equation.

3.3 Coulomb gauge

Ampere's law for steady state is

$$\nabla \times \vec{B} = \mu_0 \vec{J}$$

Using expression for $\vec{B}$, we have

$$\nabla \times (\nabla \times \vec{A}) = \mu_0 \vec{J}$$

$$\nabla(\nabla \cdot \vec{A}) - \nabla^2 \vec{A} = \mu_0 \vec{J}$$

It is now convenient to exploit the non-uniqueness of $\vec{A}$ by choosing

$$\nabla \cdot \vec{A} = 0$$

This choice is known as the **Coulomb gauge**. This is Lorentz gauge for steady sate. Thus, we have vector equation

$$\nabla^2 \vec{A} = -\mu_0 \vec{J}$$

4. Self inductance

When a current flowing in a circuit changes, magnetic field associated with it also changes and that induces an emf (electromotive force) in the circuit. This induced emf (according to Lenz's law) opposes the change (growth or decay) of the current in the circuit. This phenomenon is known as **self induction**. In Figure 1, circuit C is shown in which a current I is flowing. At any instant, the magnetic flux Φ associated with the circuit is proportional to the current flowing though the circuit. That is,

$$\Phi \propto I \qquad \text{or} \qquad \Phi = LI$$

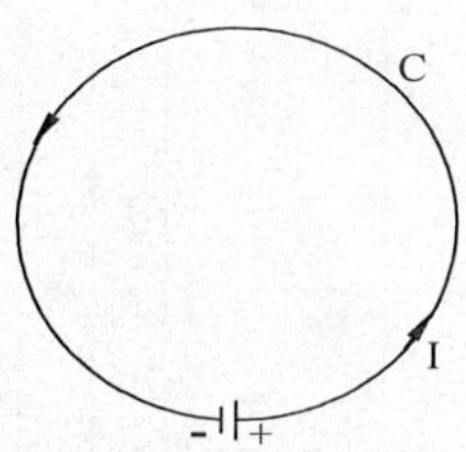

Figure 1: Magnetic flux is associated with the current I flowing through the circuit.

where L is coefficient of self induction. Putting expression for Φ in Faraday's law, we have

$$\mathcal{E} = -\frac{d\Phi}{dt} = -\frac{d(LI)}{dt} = -L\,\frac{dI}{dt}$$

Thus, we have

$$L = \frac{\mathcal{E}}{-dI/dt} = \frac{\text{emf induced}}{\text{rate of decay of current}}$$

The self inductance of a circuit is the emf induced in it when the rate of decay of current in the circuit is unity.

When current in circuit is first switched on, the induced emf (also called back emf) opposes growth of emf. In other words, the current does work against the back emf. If the growing current at any instant is I, then work done against back emf in a short time dt is

$$dW = -\mathcal{E}\, I\, dt$$

Using expression for e, we have

$$dW = L\,\frac{dI}{dt}\, I\, dt = LI\, dI$$

Total work done when the current increases from zero to steady maximum value I_0 is

$$W = L\int_0^{I_0} I\, dI = \frac{LI_0^2}{2} = U$$

where U is the energy stored in the magnetic field.

Example 1: A long cylindrical wire of radius R and length ℓ carries a current i_0 which is distributed uniformly over the cross section of wire. Derive an expression for self inductance of wire associated with the flux inside the wire only.

Solution: Let us consider a cylindrical shell of radius r and thickness dr. If i_0 is current in the wire, then current enclosed by the shell is

$$i' = \left(\frac{i_0}{\pi R^2}\right)\pi r^2 = \frac{i_0 r^2}{R^2}$$

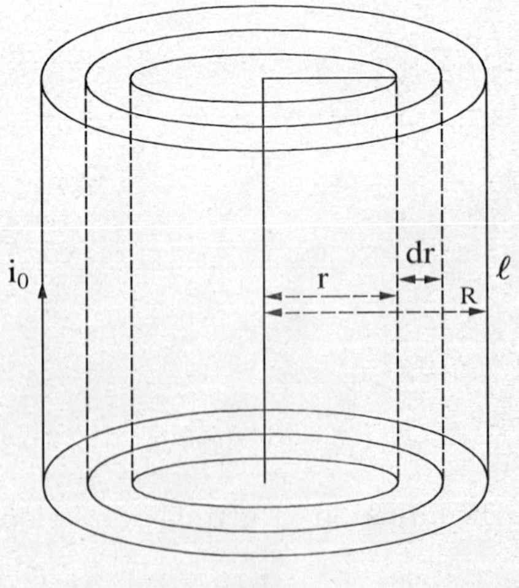

Figure 2: A cylindrical shell of radius r and thickness dr.

Thus, we have

$$\oint \vec{B}\,.\,d\vec{\ell} = \mu_0 i' \qquad B(2\pi r) = \mu_0\,\frac{i_0 r^2}{R^2} \qquad B = \frac{\mu_0 i_0 r}{2\pi R^2}$$

As discussed later on, the energy stored per unit volume in magnetic field is $B^2/2\mu_0$. The energy stored within this shell of length ℓ is

$$dV = \text{volume of shell} \times \text{energy density} = 2\pi r l dr \times \frac{B^2}{2\mu_0}$$

Using expression for B, we have

$$dV = \frac{2\pi r \ell dr}{2\mu_0} \times \left(\frac{\mu_0\, i_0\, r}{2\pi R^2}\right)^2 = \left(\frac{\mu_0\, i_0^2\, \ell}{4\pi R^4}\right) r^3\, dr$$

Total energy stored within the wire of length ℓ and radius R is

$$V = \left(\frac{\mu_0\, i_0^2\, \ell}{4\pi R^4}\right) \int_0^R r^3\, dr = \frac{\mu_0\, i_0^2\, \ell}{16\pi}$$

Equating it with $Li_0^2/2$, we get

$$\frac{\mu_0\, i_0^2\, \ell}{16\pi} = \frac{Li_0^2}{2} \qquad \text{and thus,} \qquad L = \frac{\mu_0\, \ell}{8\pi}$$

5. Mutual inductance

The phenomenon in which changing current in a circuit induces emf in another nearby circuit is called mutual induction. Suppose, I_1 is current flowing in circuit 1 (C_1) producing magnetic induction B_1 in space surrounding C_1. This magnetic field is also linked with circuit 2 (C_2). If S_2 is area of C_2 through which the flux of B_1 passes, then flux linked with C_2 is

$$\Phi_2 = B_1 S_2$$

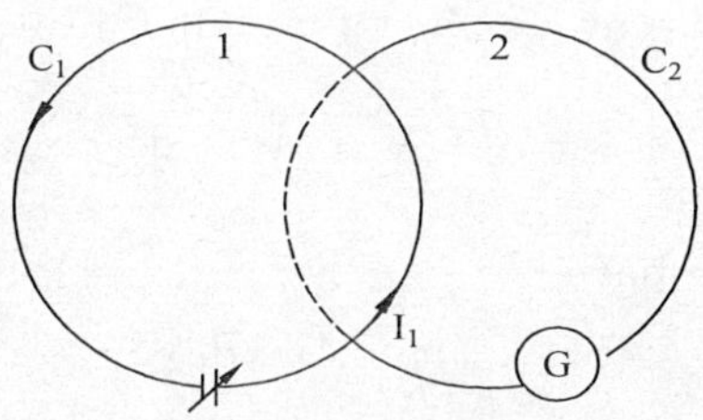

Figure 3: Changing current in a circuit 1 induces emf in another nearby circuit 2.

If the shapes, orientations and positions of two circuits C_1 and C_2 are fixed then

$$\Phi_2 \propto I_1 \qquad \text{or} \qquad \Phi_2 = M_{21} I_1$$

where M_{21} is known as coefficient of mutual induction of circuit C_2 with respect to circuit C_1. Consequently, if I_1 changes, emf induced in C_2 is

$$\mathcal{E}_2 = -\frac{d\Phi_2}{dt} = -\frac{d(M_{21} I_1)}{dt} = -M_{21}\, \frac{dI_1}{dt}$$

$$M_{21} = \frac{\mathcal{E}_2}{-\mathrm{d}I_1/\mathrm{d}t} = \frac{\text{emf induced in C}_2}{\text{rate of decay of current in C}_1}$$

Thus, the coefficient of mutual induction is the emf induced in circuit C_2 when decay rate of current in circuit C_1 is unity.

Similarly, if current I_2 is made to flow in circuit C_2 and we find emf induced in circuit C_1, then we have

$$M_{12} = \frac{\mathcal{E}_1}{-\mathrm{d}I_2/\mathrm{d}t} = \frac{\text{emf induced in C}_1}{\text{rate of decay of current in C}_2}$$

5.1 Reciprocity theorem of mutual inductance

The reciprocity theorem of mutual inductance states that $M_{21} = M_{12}$. Suppose there are two closed coils C_1 and C_2 placed very close to each other. The current I_1 in C_1 links a flux to C_2 expressed as

$$\Phi_{21} = \oint_{S_2} \vec{B}_1 \cdot \mathrm{d}\vec{S}_2$$

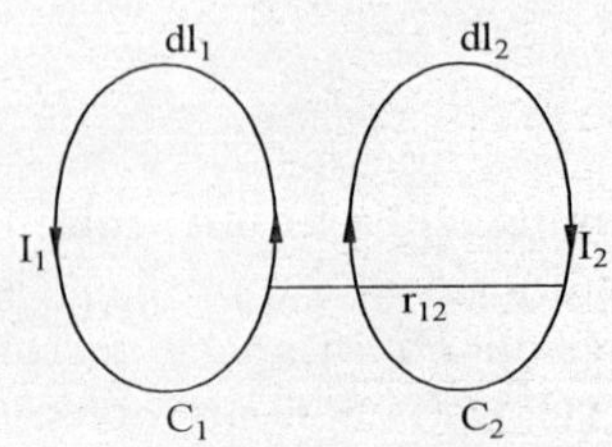

Figure 4: Two circuits C_1 and C_2 carrying currents I_1 and I_2, respectively.

where $\vec{B}_1$ is magnetic field at a point in circuit C_2 due to current I_1 in the circuit C_1. If $\vec{A}_1$ is vector potential at a point of circuit C_2 due to current I_1 in C_1, then

$$\Phi_{21} = \oint_{S_2} (\nabla \times \vec{A}_1) \cdot \mathrm{d}\vec{S}_2$$

Using Stokes' theorem, we have

$$\Phi_{21} = \oint_{C_2} \vec{A}_1 \cdot \mathrm{d}\vec{\ell}_2$$

Using expression for $\vec{A}_1$, we have

$$\Phi_{21} = \oint_{C_2} \left(\frac{\mu_0 I_1}{4\pi} \oint_{C_1} \frac{\mathrm{d}\vec{\ell}_1}{\vec{r}_{12}} \right) \cdot \mathrm{d}\vec{\ell}_2 = \frac{\mu_0 I_1}{4\pi} \oint_{C_2} \oint_{C_1} \frac{\mathrm{d}\vec{\ell}_1 \cdot \mathrm{d}\vec{\ell}_2}{\vec{r}_{12}}$$

or

$$M_{21} = \frac{\Phi_{21}}{I_1} = \frac{\mu_0}{4\pi} \oint_{C_2} \oint_{C_1} \frac{\mathrm{d}\vec{\ell}_1 \cdot \mathrm{d}\vec{\ell}_2}{\vec{r}_{12}}$$

This is known as **Neumann equation**. As it is symmetric with respect to indices 1 and 2, we find that

$$M_{12} = M_{21} = \frac{\Phi_{12}}{I_2} = \frac{\Phi_{21}}{I_1}$$

This is reciprocity relation.

Example 2: A very small circular loop of area a, resistance R and negligible inductance is initially coplanar and concentric with a much larger fixed circuit loop is rotated with angular velocity ω rad/s about a diameter. Calculate (i) flux linked with smaller loop, (ii) induced emf, and (iii) induced current in smaller loop as a function of time.

Solution: (i) Magnetic field produced due to large loop 1 is

$$\vec{B}_1 = \frac{\mu_0 I_1}{2r_1}$$

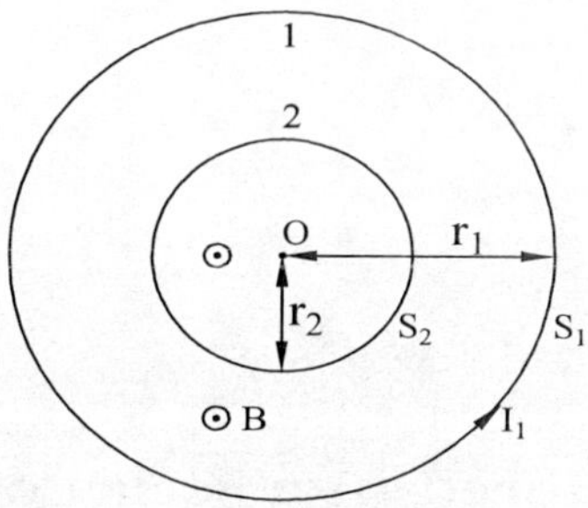

Figure 5: Concentric loops 1 and 2 of of radii r_1 and r_2, respectively. Magnetic field is perpendicular to the plane of loop in upward direction.

It is perpendicular to the plane of loop. It is in the outward direction of the paper in Figure 5. This field is at the center O where the smaller coil is placed. Smaller coil is limited to the region near the center, so flux linked with smaller coil 2 is

$$\Phi_2 = B_1 A_2 \cos\theta$$

As smaller coil 2 is rotating $\theta = \omega t$ so that

$$\Phi_2 = B_1 A_2 \cos\omega t$$

Using value of B_1, we have

$$\Phi_2 = \frac{\mu_0 I_1}{2r_1} A_2 \cos\omega t$$

(ii) Induced emf in coil 2 is

$$\mathcal{E}_2 = -\frac{d\Phi_2}{dt} = -\frac{\mu_0 I_1 \omega}{2r_1} A_2 \sin\omega t$$

(iii) Induced current in coil 2 is

$$i_2 = -\frac{\mathcal{E}_2}{R_2} = \frac{\mu_0 I_1 \omega}{2r_1 R_2} A_2 \sin\omega t$$

Example 3: A long solenoid of diameter D has N turns per unit length. At the center of solenoid a n turn coil of radius r is placed with axis coinciding with the solenoid axis. The current in the solenoid is decreased at a constant rate from $+i$ to $-i$ in t sec. Find the emf induced in the coil. Also find total charge flowing through the coil during this time when the resistance of coil is R.

Solution: Consider solenoid as primary and coil inside as secondary. Then mutual inductance of this system is

$$M = \mu_0 n N(\pi r^2)$$

(i) The emf induced in the coil is

$$\mathcal{E}_2 = -M \frac{\mathrm{d}i_p}{\mathrm{d}t} = -M \frac{-i - (+i)}{t} = M \frac{2i}{t}$$

(ii) Current in the coil is

$$i_2 = \frac{\mathcal{E}_2}{R} = M \frac{2i}{Rt}$$

(iii) Charge induced in the coil is

$$q_2 = i_2\, t = M \frac{2i}{R}$$

6. Magnetic energy stored in a series of inductances

Consider a number of coils connected in series. The initial current in each coil is zero and final equilibrium current in k-th circuit is I_k. Let i_k be the current in kth circuit at time t (instantaneous current). The current $i_k(t)$ in circuit and flux through it $\phi_k(t)$ is some fraction α of their final values I_k and Φ_k, respectively. That is,

$$i_k(t) = \alpha I_k \qquad \text{and} \qquad \phi_k(t) = \alpha \Phi_k$$

The emf induced in kth circuit is

$$\mathcal{E}_k = \frac{\mathrm{d}\phi_k}{\mathrm{d}t} = \frac{\mathrm{d}(\alpha\Phi_k)}{\mathrm{d}t} = \Phi_k \frac{\mathrm{d}\alpha}{\mathrm{d}t}$$

Thus, total work done in time T by the applied source for kth circuit is

$$W_k = \int_0^T \mathcal{E}_k\, i_k(t)\, \mathrm{d}t$$

$$= \int_0^T \Phi_k \frac{\mathrm{d}\alpha}{\mathrm{d}t} \alpha I_k\, \mathrm{d}t = \Phi_k I_k \int_0^1 \alpha\, \mathrm{d}\alpha = \frac{1}{2} \Phi_k I_k$$

On summing for all circuits, magnetic energy U is

$$U = \frac{1}{2} \sum_k \Phi_k I_k$$

The flux linked through the k-th circuit is

$$\Phi_k = L_k I_k + \sum_{j=1}^{k-1} M_{kj} I_j$$

where M_{kj} is mutual inductance between k-th and j-th coils. Thus,

$$U = \frac{1}{2}\sum_k L_k I_k^2 + \frac{1}{2}\sum_k \sum_{j=1}^{k-1} M_{kj} I_j I_k$$

For two coupled circuits (circuit 1 and circuit 2), this equation reduces to

$$U = \frac{1}{2} L_1 I_1^2 + \frac{1}{2} L_2 I_2^2 + M I_1 I_2$$

where $M = M_{12} = M_{21}$. The term MI_1I_2 may either be positive or negative, but total magnetic energy U must be positive or zero for any pair of current values I_1 and I_2.

6.1 Standard form of magnetic energy

For k-th circuit, the flux linked is

$$\phi_k = \int_k \vec{B} \,.\, \mathrm{d}\vec{S}_k$$

Using $\vec{B} = \nabla \times \vec{A}$, we have

$$\phi_k = \int_k (\nabla \times \vec{A}) \,.\, \mathrm{d}\vec{S}_k$$

Using Stokes' theorem, we get

$$\phi_k = \oint_k \vec{A} \,.\, \mathrm{d}\vec{\ell}_k$$

The magnetic energy U is

$$U = \frac{1}{2}\sum_k I_k \phi_k = \frac{1}{2}\sum_k I_k \oint_k \vec{A} \,.\, \mathrm{d}\vec{\ell}_k$$

For more general case, we change summation into integration which can be understood as the following. Suppose, that we do not have current circuits defined by wires, but instead each circuit is a closed path in the medium (assumed to be conducting) that follows a line current density. Then we change

$$\sum_k \oint_k \quad \text{to} \quad \int_V$$

so that for this situation, approximate magnetic energy is

$$U = \frac{1}{2}\int_V (\vec{A} \,.\, \vec{J})\, \mathrm{d}V$$

Using $\vec{J} = \nabla \times \vec{B}/\mu_0$, we get

$$U = \frac{1}{2\mu_0} \int_V [\vec{A} \,.\, (\nabla \times \vec{B})] \text{ d}V$$

Using vector identity

$$\nabla \,.\, (\vec{A} \times \vec{B}) = \vec{B} \,.\, (\nabla \times \vec{A}) - \vec{A} \,.\, (\nabla \times \vec{B})$$

the magnetic energy is

$$U = \frac{1}{2\mu_0} \int_V \vec{B} \,.\, (\nabla \times \vec{A}) \text{ d}V - \frac{1}{2\mu_0} \int_V \nabla \,.\, (\vec{A} \times \vec{B}) \text{ d}V$$

Applying Gauss' divergence theorem to second term on right side, we have

$$U = \frac{1}{2\mu_0} \int_V \vec{B} \,.\, (\nabla \times \vec{A}) \text{ d}V - \frac{1}{2\mu_0} \int_S (\vec{A} \times \vec{B}) \,.\, \text{d}\vec{S}$$

where the surface S bounds volume V. Since, by assumption, none of the current circuits extends to infinity, it is convenient to move the surface S out to very large distance so that all parts of this surface are far from currents. The volume of the system must be increased accordingly. Now $\vec{B}$ falls off at least as fast as $1/r^2$, where r is the distance from an origin near the middle of the current distribution to a characteristic point on the surface S; $\vec{A}$ falls off at least as fast as $1/r$, and the surface area is proportional to r^2. Thus, the contribution of surface integral in this equation falls off as $1/r$ or faster and if S is moved out to infinity, this contribution vanishes.

By dropping the surface integral from equation, and extending the volume term to include all space, we have

$$U = \frac{1}{2} \int_V \vec{B} \,.\, (\nabla \times \vec{A}) \text{ d}V$$

Using $\vec{B} = \nabla \times \vec{A}$, we have

$$U = \frac{1}{2\mu_0} \int_V (\vec{B} \,.\, \vec{B}) \text{ d}V$$

As the integral is extended to include all space and if we take that this magnetic energy is distributed uniformly through the space, then the magnetic energy density (magnetic energy stored per unit volume) can be expressed as

$$u = \frac{1}{2\mu_0} \vec{B} \,.\, \vec{B}$$

For the case of isotropic, magnetic material, it reduces to

$$u = \frac{1}{2} \frac{B^2}{\mu_0}$$

Example 4: A long wire carries a current of uniform density. If i_0 is total current carried by the wire, show that the magnetic energy per unit length stored within the wire is $\mu_0 i_0^2/32\pi$.

Solution: Let us consider a point P within the wire such that $r < R$. Let us consider a cylindrical shell of radius r, thickness dr. If i_0 is current in the wire, then current enclosed by the shell is

$$i' = \left(\frac{i_0}{\pi R^2}\right)\pi r^2 = \frac{i_0 r^2}{R^2}$$

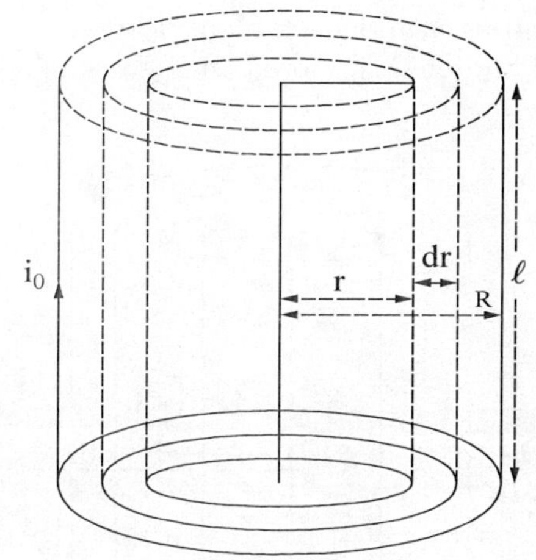

Figure 6: A cylindrical shell of radius r and thickness dr.

Thus, we have

$$\oint \vec{B} \,.\, d\vec{\ell} = \mu_0 i' \qquad B(2\pi r) = \mu_0 \frac{i_0 r^2}{R^2} \qquad B = \frac{\mu_0 i_0 r}{2\pi R^2}$$

Since B is proportional to r, we can take average magnetic field as that at $r = R/2$. Therefore,

$$B_{av} = \frac{\mu_0 i_0 (R/2)}{2\pi R^2} = \frac{\mu_0 i_0}{4\pi R}$$

The magnetic energy density within the wire is

$$u = \frac{B^2}{2\mu_0} = \frac{B_{av}^2}{2\mu_0} = \frac{1}{2\mu_0}\left(\frac{\mu_0 i_0}{4\pi R}\right)^2$$

Total magnetic energy is $= \pi R^2 l u$

$$\text{Total magnetic energy per unit length is} = \pi R^2 u = \frac{\mu_0 i_0^2}{32\pi}$$

Example 5: A circular loop of radius R carries a current i_0. Find magnetic energy density at the center of loop.

Solution: Magnetic field at the center of loop

$$B = \frac{\mu_0 i}{2R}$$

Energy density at the center is

$$u = \frac{B^2}{2\mu_0} = \frac{1}{2\mu_0}\left(\frac{\mu_0 i}{2R}\right)^2 = \frac{\mu_0 i^2}{8R^2}$$

Example 6: A long coaxial cable carries a current i (current flows down the surface of the inner cylinder of radius a and back along the outer cylinder of radius b) as shown in Figure 7. Find the energy stored in a section of length ℓ, and find its inductance.

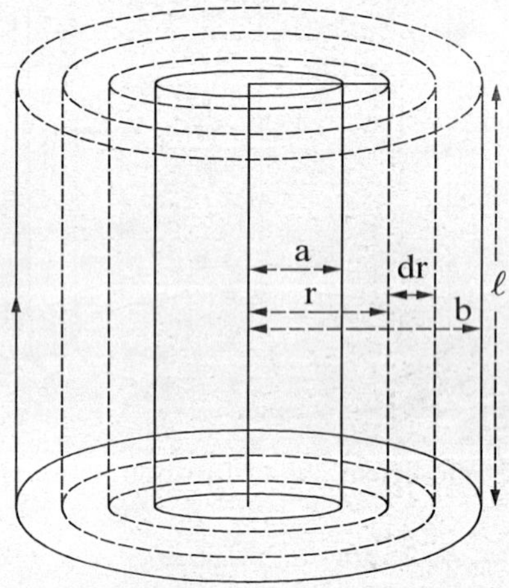

Figure 7: A cylindrical shell of radius r and thickness dr.

Solution: We consider a cylindrical shell of radius r and thickness dr and apply Ampere's law

$$\oint \vec{B} \,.\, d\vec{\ell} = \mu_0 i \qquad\qquad B(2\pi r) = \mu_0 i \qquad\qquad B = \frac{\mu_0 i}{2\pi r}$$

where i is current in the inner conductor. The energy contained in the shell of radius r and thickness dr is

$$dU = \text{energy density} \times \text{volume of shell} = \frac{B^2}{2\mu_0}\,(2\pi r\ell dr)$$

Using expression for B, we get

$$dU = \frac{1}{2\mu_0}\left(\frac{\mu_0 i}{2\pi r}\right)^2 (2\pi r\ell dr) = \frac{\mu_0 i^2 \ell}{4\pi}\,\frac{dr}{r}$$

Thus, total energy stored between coaxial cylinders is

$$U = \frac{\mu_0 i^2 \ell}{4\pi} \int_a^b \frac{\mathrm{d}r}{r} = \frac{\mu_0 i^2 \ell}{4\pi} \ln\left(\frac{b}{a}\right)$$

The inductance is

$$L = \frac{2\,U}{i^2} = \frac{\mu_0 \ell}{2\pi} \ln\left(\frac{b}{a}\right)$$

$$\text{Self inductance per unit length} = \frac{\mu_0}{2\pi} \ln\left(\frac{b}{a}\right)$$

Example 7: Find self inductance of a toroidal coil with rectangular cross section (inner radius a, outer radius b, height h) which carries a total of n turns.

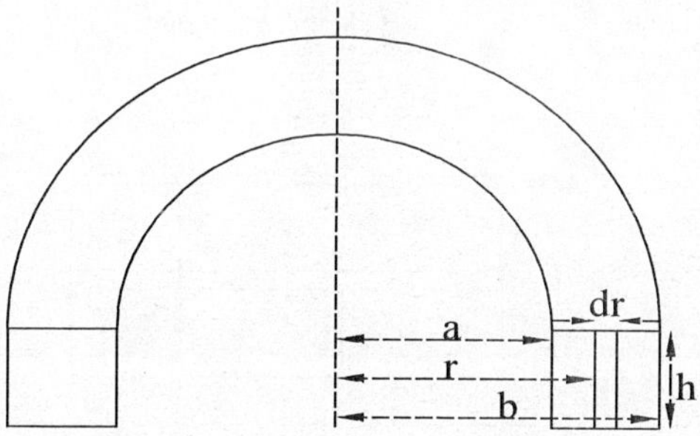

Figure 8: A toroidal coil.

Solution: Suppose, in coil windings the current flowing is i. Then, considering a circular path of radius r and thickness dr, the magnetic field is

$$\oint \vec{B}\,.\,\mathrm{d}\vec{\ell} = \mu_0(ni) \qquad B(2\pi r) = \mu_0 ni \qquad B = \frac{\mu_0 ni}{2\pi r}$$

Flux through cross section h dr of the toroid is

$$\mathrm{d}\Phi = Bh\mathrm{d}r$$

Flux for the toroid through a single turn is

$$\Phi = \int_a^b Bh\mathrm{d}r$$

Using expression for B, we have

$$\Phi = \frac{\mu_0 nih}{2\pi} \int_a^b \frac{\mathrm{d}r}{r} = \frac{\mu_0 nih}{2\pi} \ln\left(\frac{b}{a}\right)$$

Total flux is then

$$n\Phi = \frac{\mu_0 n^2 i h}{2\pi} \ln\left(\frac{b}{a}\right)$$

Self inductance is

$$L = \frac{n\Phi}{i} = \frac{\mu_0 n^2 h}{2\pi} \ln\left(\frac{b}{a}\right)$$

Example 8: An infinitely long straight wire carries a slowly varying current $I(t)$. Calculate the induced electric field.

Solution: Though the problem is time-dependent, we have a time-dependent magnetic field, but because the current is slowly varying, we can have quasi-static approximation and can apply Ampere's law to obtain the magnetic field encircling the wire at a distance r as

$$\oint \vec{B} \cdot \mathrm{d}\vec{\ell} = \mu_0 I \qquad B(2\pi r) = \mu_0 I \qquad B = \frac{\mu_0 I}{2\pi r}$$

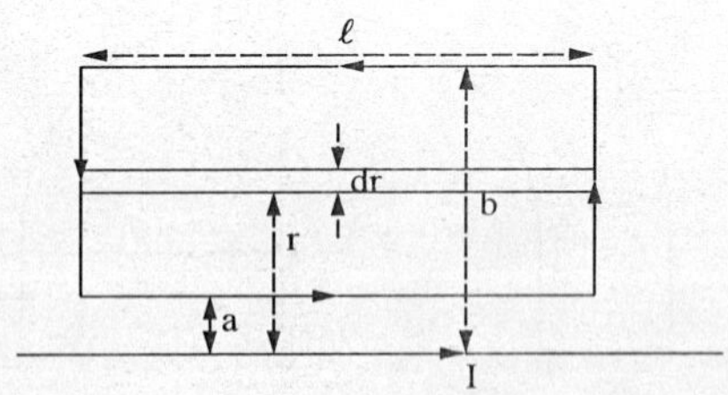

Figure 9: Infinitely long wire carrying slowly varying current I.

Like $\vec{B}$-field of a solenoid, $\vec{E}$ here runs parallel to axis. For rectangular loop shown in Figure 9, Faraday's law gives

$$\oint \vec{E} \cdot \mathrm{d}\vec{\ell} = E(a)\ell - E(b)\ell = -\frac{\mathrm{d}}{\mathrm{d}t}\int_S \vec{B} \cdot \mathrm{d}\vec{S}$$

$$= -\frac{\mathrm{d}}{\mathrm{d}t}\int_S \frac{\mu_0 I}{2\pi r} \cdot \ell \,\mathrm{d}r = -\frac{\mu_0 \ell}{2\pi}\left(\frac{\mathrm{d}I}{\mathrm{d}t}\right)\int_a^b \frac{\mathrm{d}r}{r} = -\frac{\mu_0 \ell}{2\pi}\left(\frac{\mathrm{d}I}{\mathrm{d}t}\right) \ln\left(\frac{b}{a}\right)$$

If $(\mathrm{d}I/\mathrm{d}t) > 0$, a current flows around the loop in counter clockwise direction.

Example 9: A wire PQ of length l is sliding on non-parallel rails making an angle 2α as shown in Figure 10. If the speed with which the wire PQ slides is v and magnetic field $\vec{B}$ is perpendicular to the plane of rails, find an expression for the induced emf as a function of time t.

Solution: Let at any time t, the rod PQ is at a distance x from the end O of nails. At this instant, flux linked with the circuit (formed by nails and rod) is

$$\Phi = B \times (\text{area OAC}) = B \times \left(\frac{1}{2}\ \text{AC} \times \text{OF}\right)$$

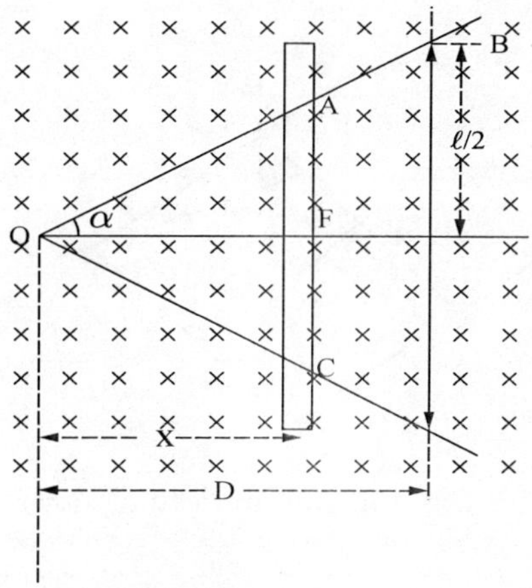

Figure 10: A wire PQ of length l is sliding on non-parallel rails making an angle 2α.

In Figure 10, OF $= x$ and

$$\frac{\ell/2}{D} = \frac{\mathrm{AC}/2}{x} \qquad \text{or} \qquad \mathrm{AC} = \frac{x\ell}{D}$$

Using the value of AC, we have

$$\Phi = \frac{B}{2} \times \frac{x\ell}{D}\, x = \frac{1}{2}\frac{B\ell}{D}\, x^2$$

The emf induced is

$$\mathcal{E} = \frac{\mathrm{d}\Phi}{\mathrm{d}t} = \frac{B\ell}{D}\, x\, \frac{\mathrm{d}x}{\mathrm{d}t}$$

We have $x = vt$ and $v = \mathrm{d}x/\mathrm{d}t$ and therefore,

$$\mathcal{E} = \frac{B\ell}{D}\, vt\, v = 2Bv^2\left(\frac{\ell/2}{D}\right)t = (2Bv^2 \tan\alpha)t$$

It is emf induced in the rod as a function of time.

Example 10: A thin conducting rod of mass m and length ℓ, having electrical resistance R, slides down an incline consisting of two parallel friction-less rails of negligible resistance as shown in Figure 11. The uniform magnetic field is directed upward vertically and the angle of incline is θ. Show that the terminal velocity of the rod is

$$v = \frac{mgR}{B^2\ell^2}\,\frac{\sin\theta}{\cos^2\theta}$$

Solution: Rod of length l, mass m is sliding down the incline with velocity v. Angle between magnetic field B and velocity is $(180 - \theta)$. Induced emf in the rod is

$$\mathcal{E} = B\ell v\cos\theta$$

Induced current is

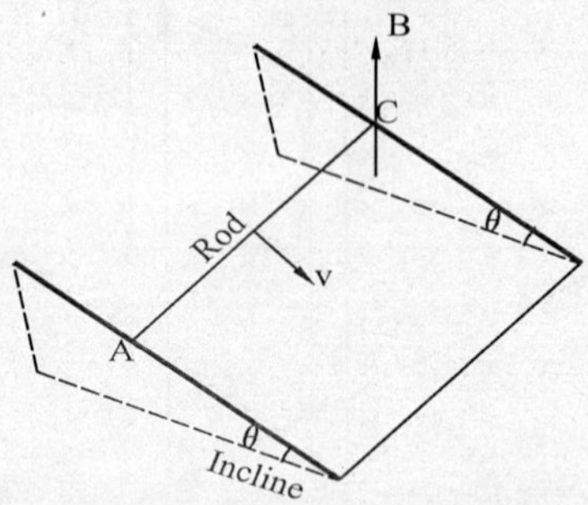

Figure 11: A thin conducting rod sliding down an incline.

$$i = \frac{B\ell v \cos\theta}{R}$$

Magnetic force on rod AC is then

$$F = i\ell B$$

Its component $F\cos\theta$ along the incline should act up the incline so that it balances the component $mg\sin\theta$ of the weight of rod acting along and down the incline, in order to attain equilibrium so that rod slides down the incline with constant terminal velocity. That is, for constant terminal velocity of rod

$$F\cos\theta = mg\sin\theta \qquad i\ell B\cos\theta = mg\sin\theta \qquad \left(\frac{B\ell v\cos\theta}{R}\right)\ell B\cos\theta = mg\sin\theta$$

$$\frac{B^2\ell^2 v\cos^2\theta}{R} = mg\sin\theta \qquad v = \frac{mgR}{B^2\ell^2}\,\frac{\sin\theta}{\cos^\theta}$$

Review Questions

1. What is Faraday's law of induction? Derive differential form of Faraday's law. What is Lenz's law?

2. State and explain Faraday's law of electromagnetic induction.

3. Deduce Faraday's law of electromagnetic induction in the form

$$\nabla \times \vec{E} = -\frac{\partial \vec{B}}{\partial t}$$

4. Obtain an expression for electromotive force for the following cases:

   (i) a fixed loop in a time-dependent magnetic field.

   (ii) a rotating loop in a fixed magnetic field.

5. Explain the meaning of self and mutual induction. Obtain Neumann equation as related to mutual inductance.

6. Derive expression for induced electric field in terms of vector potential.

7. Show that magnetic energy stored in a series of inductances is

$$U = \frac{1}{2\mu_0}\int_V [\vec{B} \,.\, (\nabla \times \vec{A})]\mathrm{d}V$$

8. Define self inductance. A long cylindrical wire of radius R and length l carries current i_0 which is distributed uniformly over the cross section of wire. Derive an expression for self inductance of wire associated with the flux inside the wire only.

9. Define mutual inductance. State and prove reciprocity theorem of mutual inductance.

10. Drive expression for magnetic energy.

11. Write notes on the following:

    (i) Faraday's law of induction
    (ii) Lenz's law
    (iii) Self inductance
    (iv) Mutual inductance
    (v) Reciprocity theorem of mutual inductance

Multiple-Choice Questions

1. Differential form of Faraday's law is

A. $\nabla \times \vec{E} = -\dfrac{\partial \vec{B}}{\partial t}$ B. $\nabla \,.\, \vec{E} = -\dfrac{\partial \vec{B}}{\partial t}$ C. $\nabla \times \vec{E} = \dfrac{\partial \vec{B}}{\partial t}$ D. $\nabla \,.\, \vec{E} = \dfrac{\partial \vec{B}}{\partial t}$

Ans. A

2. Relation between induced electric field $\vec{E}$ and magnetic vector potential $\vec{A}$ and scalar potential ϕ is

A. $\vec{E} = -\nabla \times \phi - \dfrac{\partial \vec{A}}{\partial t}$ B. $\vec{E} = -\nabla\phi - \dfrac{\partial \vec{A}}{\partial t}$

C. $\vec{E} = -\nabla \,.\, \phi - \dfrac{\partial \vec{A}}{\partial t}$ D. $\vec{E} = \nabla\phi + \dfrac{\partial \vec{A}}{\partial t}$

Ans. B

3. Reciprocity theorem of mutual inductance states that

A. $M_{12} = M_{21}$ B. $M_{12} = -M_{21}$ C. $M_{12}^2 = -M_{21}^2$ D. $M_{12}^{1/2} = -M_{21}^{1/2}$

Ans. A

4. For a linear medium, magnetic energy stored per unit volume is

A. $B^2/2\mu_0$ B. $B^2/2\mu_0^2$ C. $\mu_0 B^2/2$ D. $B^2/2$

Ans. A

Chapter 8

Maxwell's Field Equations

For electric field $\vec{E}$, in electrostatics, we have obtained the relations:

$$\nabla \cdot \vec{E} = \frac{\rho}{\epsilon_0} \qquad \text{and} \qquad \nabla \times \vec{E} = 0$$

For magnetic field $\vec{B}$, in magnetostatics, we have obtained the relations:

$$\nabla \cdot \vec{B} = 0 \qquad \text{and} \qquad \nabla \times \vec{B} = \mu_0 \vec{J}$$

After introduction of electromagnetic induction, the equation $\nabla \times \vec{E} = 0$ was modified and the set of equations was obtained as

$$\nabla \cdot \vec{E} = \frac{\rho}{\epsilon_0} \qquad \nabla \cdot \vec{B} = 0 \qquad \nabla \times \vec{E} = -\frac{\partial \vec{B}}{\partial t} \qquad \nabla \times \vec{B} = \mu_0 \vec{J}$$

There is space variation of $\vec{E}$ and $\vec{B}$ both, but the time variation of $\vec{B}$ only. It shows that time variation of $\vec{E}$ is still missing. In this chapter, we shall find time variation of $\vec{E}$ and then will discuss about the Maxwell's equations.

1. Displacement current

We here discuss how Maxwell changed the definition of total current density to adopt the equation of continuity to time dependent fields. We have

$$\nabla \times \vec{B} = \mu_0 \vec{J} \tag{8.1}$$

Taking divergence, we have

$$\nabla \cdot [\nabla \times \vec{B}] = \mu_0 \nabla \cdot \vec{J} \qquad \text{or} \qquad \nabla \cdot \vec{J} = 0$$

Here, we have used the fact that divergence of a curl of vector is zero. Using the equation of continuity, we have

$$\frac{\partial \rho}{\partial t} = 0$$

This is steady condition, *i.e.*, the charge density is not changing with time. It is because of equation (8.1). Therefore, for the time-dependent (changing) fields, equation (8.1) needs to be modified.

Maxwell said that the definition of total current density is incomplete and suggested to add something to $\vec{J}$. Suppose, current density $\vec{J}'$ is added so that

$$\nabla \times \vec{B} = \mu_0 \, (\vec{J} + \vec{J}\,')$$

Taking divergence of this equation, we have

$$\nabla \, . \, [\nabla \times \vec{B}] = \mu_0(\nabla \, . \, \vec{J} + \nabla \, . \, \vec{J}\,') \qquad \text{or} \qquad 0 = \nabla \, . \, \vec{J} + \nabla \, . \, \vec{J}\,'$$

Thus, we have

$$\nabla \, . \, \vec{J}\,' = -\nabla \, . \, \vec{J} = \frac{\partial \rho}{\partial t}$$

We know that $\rho = \epsilon_0 \, \nabla \, . \, \vec{E}$ and therefore,

$$\nabla \, . \, \vec{J}\,' = \epsilon_0 \, \frac{\partial}{\partial t}(\nabla \, . \, \vec{E}) = \epsilon_0 \, \nabla \, . \, \frac{\partial \vec{E}}{\partial t}$$

or

$$\nabla \, . \left(\vec{J}\,' - \epsilon_0 \, \frac{\partial \vec{E}}{\partial t} \right) = 0$$

We have used the fact that space and time are independent of each other. This equation holds good for arbitrary space, therefore we have

$$\left(\vec{J}\,' - \epsilon_0 \, \frac{\partial \vec{E}}{\partial t} \right) = 0 \qquad \text{or} \qquad \vec{J}\,' = \epsilon_0 \, \frac{\partial \vec{E}}{\partial t}$$

Therefore, modified Ampere's law is

$$\nabla \times \vec{B} = \mu_0 \, \vec{J} + \mu_0 \epsilon_0 \, \frac{\partial \vec{E}}{\partial t} \qquad \text{or} \qquad \nabla \times \vec{B} = \mu_0 \, \vec{J} + \frac{1}{c^2} \, \frac{\partial \vec{E}}{\partial t}$$

where $c = 1/\sqrt{\mu_0 \epsilon_0}$ is the speed of light. Thus, time variation of $\vec{E}$ is also obtained.

Example 1: Show that for a conductor subject to electric field $\vec{E} = \vec{E}_0 \cos \omega t$, the displacement current density is negligible compared to conduction current density at frequencies less than 10^{15} Hz.

Solution: We have $\vec{D} = \epsilon \vec{E} = \epsilon \vec{E}_0 \cos \omega t$. Therefore,

$$\vec{J}\,' = \frac{\partial \vec{D}}{\partial t} = -\omega \epsilon \vec{E}_0 \sin \omega t = \omega \epsilon \vec{E}_0 \cos \left(\omega t + \frac{\pi}{2} \right)$$

Following Ohm's law, conduction current density is

$$\vec{J} = \sigma \vec{E} = \sigma \vec{E}_0 \cos \omega t$$

Thus,

$$\frac{\vec{J}\,'}{\vec{J}} = \frac{\omega\epsilon}{\sigma}\,\frac{\cos(\omega t+\pi/2)}{\cos\omega t} \qquad \text{or} \qquad \left|\frac{\vec{J}\,'}{\vec{J}}\right| = \frac{\omega\epsilon}{\sigma} = \frac{2\pi f\epsilon}{\sigma}$$

For a good conductor, we have $\sigma \approx 10^7$ mho/m. Further, $\epsilon \approx 9\times10^{-12}$ A^2 s^4/m^3 kg. Thus, we have

$$\left|\frac{\vec{J}\,'}{\vec{J}}\right| = 5.6\times10^{-18} f$$

Therefore, for $f < 10^{15}$ Hz, $\vec{J}\,' << \vec{J}$ and is negligible.

2. Maxwell's equations (Differential form)

Differential form of four Maxwell's equations is as the following.

$$\nabla\,.\,\vec{E} = \frac{\rho}{\epsilon_0} \qquad \nabla\,.\,\vec{B} = 0 \qquad \nabla\times\vec{E} = -\frac{\partial\vec{B}}{\partial t} \qquad \nabla\times\vec{B} = \mu_0\vec{J} + \frac{1}{c^2}\frac{\partial\vec{E}}{\partial t}$$

where $\vec{D} = \epsilon_0\vec{E}$ and $\vec{B} = \mu_0\vec{H}$ for free space.

3. Maxwell's equations (Integral form)

In the last section, we have obtained differential form of Maxwell's equations. Now, we want to transform them into their integral form.

3.1 First Maxwell's equation

Integrating first Maxwell's equation $\nabla\,.\,\vec{D} = \rho$ over volume V, we have

$$\int_V \nabla\,.\,\vec{D}\,\mathrm{d}V = \int_V \rho\,\mathrm{d}V$$

Using Gauss' divergence theorem for left side, we get

$$\int_S \vec{D}\,.\,\mathrm{d}\vec{S} = \int_V \rho\,\mathrm{d}V = q$$

where q is net charge contained in volume V and S is surface bounding the volume V. This, Maxwell's equation signifies that:

Total electric displacement through the surface enclosing a volume is equal to the total charge within the volume.

3.2 Second Maxwell's equation

Integrating second Maxwell's equation $\nabla.\vec{B} = 0$ over volume V, we have

$$\int_V \nabla \cdot \vec{B} \, \mathrm{d}V = 0$$

Using Gauss' divergence theorem for left side, we get

$$\int_S \vec{B} \cdot \mathrm{d}\vec{S} = 0$$

This Maxwell's equation signifies that

Total outward flux of magnetic induction $\vec{B}$ through any closed surface S is equal to zero.

3.3 Third Maxwell's equation

Integrating third Maxwell's equation $\nabla \times \vec{E} = -\partial\vec{B}/\partial t$ over surface S

$$\int_S (\nabla \times \vec{E}) \cdot \mathrm{d}\vec{S} = -\int_S \frac{\partial \vec{B}}{\partial t} \cdot \mathrm{d}\vec{S} = -\frac{\partial}{\partial t}\int_S \vec{B} \cdot \mathrm{d}\vec{S}$$

We have used the fact that space and time are independent of each other. Applying Stokes' theorem on the left side, we have

$$\oint_S \vec{E} \cdot \mathrm{d}\vec{l} = -\frac{\partial}{\partial t}\int_S \vec{B} \cdot \mathrm{d}\vec{S}$$

This, Maxwell's equation signifies that

Electromotive force around a closed path is equal to the time-derivative of the magnetic flux through any surface bounded by the path.

3.4 Fourth Maxwell's equation

Integrating fourth Maxwell's equation $\nabla \times \vec{H} = \vec{J} + \partial\vec{D}/\partial t$ over surface S, we get

$$\int_S (\nabla \times \vec{H}) \cdot \mathrm{d}\vec{S} = \int_S \left(\vec{J} + \frac{\partial \vec{D}}{\partial t}\right) \cdot \mathrm{d}\vec{S}$$

Applying Stokes' theorem on the left side, we have

$$\oint_S \vec{H} \cdot \mathrm{d}\vec{l} = \int_S \left(\vec{J} + \frac{\partial \vec{D}}{\partial t}\right) \cdot \mathrm{d}\vec{S}$$

This, Maxwell's equation signifies that

Magnetomotive force around a closed path is equal to the conduction current plus time derivative of electric displacement through any surface bounded by the path.

Note: The correspondence of $\vec{B}$ and $\vec{H}$ with $\vec{E}$ and $\vec{D}$ through the Maxwell's (third and fourth) curl equations implies that time varying electric and magnetic fields in empty space are interdependent. That is, a changing electric field is capable to generate a magnetic field and vice versa. We can infer that a time-changing electromagnetic field would propagate energy through empty space with the velocity of light and further that the light is electromagnetic in nature.

Example 2: Using Maxwell's equations

$$\nabla \times \vec{E} = -\frac{\partial \vec{B}}{\partial t} \qquad \text{and} \qquad \nabla \times \vec{B} = \mu_0 \vec{J} + \frac{1}{c^2}\frac{\partial \vec{E}}{\partial t}$$

show that

$$\nabla \cdot \vec{E} = \frac{\rho}{\epsilon_0} \qquad \text{and} \qquad \nabla \cdot \vec{B} = 0$$

Solution: Taking divergence of first curl equation, we have

$$\nabla \cdot (\nabla \times \vec{E}) = -\nabla \cdot \frac{\partial \vec{B}}{\partial t} = -\frac{\partial}{\partial t}(\nabla \cdot \vec{B})$$

We have used the fact that space and time are independent of each other. Since divergence of a curl of vector is zero, we have

$$\frac{\partial}{\partial t}(\nabla \cdot \vec{B}) = 0 \qquad \text{or} \qquad \nabla \cdot \vec{B} = 0$$

Taking divergence of second curl equation, we have

$$\nabla \cdot (\nabla \times \vec{B}) = \mu_0 \nabla \cdot \vec{J} + \frac{1}{c^2}\nabla \cdot \frac{\partial \vec{E}}{\partial t}$$

Since divergence of a curl of vector is zero, we have

$$\nabla \cdot \vec{J} + \epsilon_0 \frac{\partial}{\partial t}(\nabla \cdot \vec{E}) = 0$$

Using the equation of continuity, we have

$$-\frac{\partial \rho}{\partial t} + \epsilon_0 \frac{\partial}{\partial t}(\nabla \cdot \vec{E}) = 0 \qquad \text{or} \qquad \frac{\partial}{\partial t}(\epsilon_0 \nabla \cdot \vec{E} - \rho) = 0$$

This equation is valid for arbitrary time, and therefore we have

$$\epsilon_0 \nabla \cdot \vec{E} - \rho = 0 \qquad \text{or} \qquad \nabla \cdot \vec{E} = \frac{\rho}{\epsilon_0}$$

Example 3: Using Maxwell's equation

$$\nabla \cdot \vec{E} = \frac{\rho}{\epsilon_0} \qquad \text{and} \qquad \nabla \times \vec{B} = \mu_0 \vec{J} + \frac{1}{c^2}\frac{\partial \vec{E}}{\partial t}$$

prove (i) Coulomb's law and (ii) equation of continuity.

Solution: (i) Integrating the relation $\nabla \cdot \vec{E} = \rho/\epsilon_0$ over volume of a sphere of radius r, we have

$$\int_V (\nabla \cdot \vec{E})\mathrm{d}V = \int_V \frac{\rho}{\epsilon_0}\,\mathrm{d}V = \frac{q}{\epsilon_0}$$

where q is total charge. Using Gauss' divergence theorem on left side, we have

$$\int_S \vec{E} \cdot \mathrm{d}\vec{S} = \frac{q}{\epsilon_0} \qquad E(4\pi r^2) = \frac{q}{\epsilon_0} \qquad E = \frac{q}{4\pi\epsilon_0\, r^2}$$

The force F on a test charge q_0 is

$$F = q_0 E = \frac{1}{4\pi\epsilon_0}\frac{qq_0}{r^2}$$

It is Coulomb's law.

(ii) Taking divergence of equation

$$\nabla \times \vec{B} = \mu_0 \vec{J} + \frac{1}{c^2}\frac{\partial \vec{E}}{\partial t}$$

we have

$$\nabla \cdot (\nabla \times \vec{B}) = \mu_0\, \nabla \cdot \vec{J} + \frac{1}{c^2}\,\nabla \cdot \frac{\partial \vec{E}}{\partial t}$$

Since divergence of a curl of vector is zero, we have

$$0 = \mu_0\, \nabla \cdot \vec{J} + \frac{1}{c^2}\frac{\partial}{\partial t}(\nabla \cdot \vec{E})$$

We know $\nabla \cdot \vec{E} = \rho/\epsilon_0$. Therefore,

$$\nabla \cdot \vec{J} + \frac{\partial \rho}{\partial t} = 0$$

This is equation of continuity.

4. Energy in electromagnetic fields: Poynting vector

Energy may be transported through space with the help of electromagnetic waves. Let the material inside a surface be isotropic, homogeneous, and characterized by permeability μ, permittivity ϵ and conductivity σ. Let us consider a volume V bounded by a closed surface S. Maxwell's third and fourth relations are

$$\nabla \times \vec{E} = -\frac{\partial \vec{B}}{\partial t} \qquad \text{and} \qquad \nabla \times \vec{B} = \mu \vec{J} + \mu\epsilon \frac{\partial \vec{E}}{\partial t}$$

Taking scalar product of both sides of above equations with $\vec{B}$ and $\vec{E}$, respectively, and subtracting, we get

$$\vec{E} \cdot (\nabla \times \vec{B}) - \vec{B} \cdot (\nabla \times \vec{E}) = \mu \vec{J} \cdot \vec{E} + \left[\mu\epsilon\, \vec{E} \cdot \frac{\partial \vec{E}}{\partial t} + \vec{B} \cdot \frac{\partial \vec{B}}{\partial t}\right]$$

Using $\nabla \cdot (\vec{E} \times \vec{B}) = \vec{B} \cdot (\nabla \times \vec{E}) - \vec{E} \cdot (\nabla \times \vec{B})$, we have

$$-\nabla \cdot (\vec{E} \times \vec{B}) = \mu\vec{J} \cdot \vec{E} + \left[\mu\epsilon\, \vec{E} \cdot \frac{\partial \vec{E}}{\partial t} + \vec{B} \cdot \frac{\partial \vec{B}}{\partial t}\right]$$

$$\vec{J} \cdot \vec{E} + \left[\frac{\epsilon}{2}\frac{\partial E^2}{\partial t} + \frac{1}{2\mu}\frac{\partial B^2}{\partial t}\right] + \nabla \cdot (\vec{E} \times \vec{B}) = 0$$

$$\vec{J} \cdot \vec{E} + \frac{\partial}{\partial t}\left[\frac{\epsilon}{2}\, E^2 + \frac{1}{2\mu}\, B^2\right] + \nabla \cdot (\vec{E} \times \vec{B}) = 0$$

Integrating over the volume V bounded by surface S, we get

$$\int_V \vec{J} \cdot \vec{E}\, \mathrm{d}V + \int_V \frac{\partial}{\partial t}\left[\frac{\epsilon}{2}\, E^2 + \frac{1}{2\mu}\, B^2\right]\mathrm{d}V + \int_V \nabla \cdot (\vec{E} \times \vec{B})\mathrm{d}V = 0$$

Applying Gauss' divergence theorem on last term on left, we have

$$\int_V \vec{J} \cdot \vec{E}\, \mathrm{d}V + \int_V \frac{\partial}{\partial t}\left[\frac{\epsilon}{2}\, E^2 + \frac{1}{2\mu}\, B^2\right]\mathrm{d}V = -\int_S (\vec{E} \times \vec{B}) \cdot \mathrm{d}\vec{S} \tag{8.2}$$

Now, we interpret first term on left side of this equation.

(i) Interpretation of first term $\int_V \vec{J} \cdot \vec{E}\, \mathrm{d}V$

The current distribution represented by the vector $\vec{J}$ can be considered as made up of various charges q_j moving with velocity $\vec{v}_j$ so that

$$\begin{aligned}\int_V \vec{J} \cdot \vec{E}\, \mathrm{d}V &= \int_V (\vec{J}\, \mathrm{d}V) \cdot \vec{E} \\ &= \int_V \vec{J}\, (\mathrm{d}\vec{S} \cdot \mathrm{d}\vec{l}) \cdot \vec{E} = \int_V (\vec{J} \cdot \mathrm{d}\vec{S})(\mathrm{d}\vec{l} \cdot \vec{E}) = \int_V I(\mathrm{d}\vec{l} \cdot \vec{E}) \\ &= \int_V \frac{\mathrm{d}q}{\mathrm{d}t}\, (\mathrm{d}\vec{l} \cdot \vec{E}) = \int_V \mathrm{d}q\, (\vec{v} \cdot \vec{E}) = \sum_j q_j\, (\vec{v}_j \cdot \vec{E}_j)\end{aligned} \tag{8.3}$$

where $\vec{E}_j$ denotes the electric field at the position of charge q_j. We know that Lorentz force $\vec{F}$ experienced by a point charge q is

$$\vec{F} = q(\vec{E} + \vec{v} \times \vec{B})$$

Then the work done by this force in displacing the charge by an infinitesimal distance $\mathrm{d}\vec{\ell}$ is

$$\begin{aligned}\mathrm{d}W = \vec{F} \cdot \mathrm{d}\vec{\ell} &= q\vec{E} \cdot \mathrm{d}\vec{\ell} + q(\vec{v} \times \vec{B}) \cdot \mathrm{d}\vec{\ell} \\ &= q\vec{E} \cdot \mathrm{d}\vec{\ell} + q\, \mathrm{d}\vec{\ell} \cdot \left(\frac{\mathrm{d}\vec{\ell}}{\mathrm{d}t} \times \vec{B}\right)\end{aligned}$$

$$= q\vec{E} \ . \ \mathrm{d}\vec{\ell} + q\left(\mathrm{d}\vec{\ell} \times \frac{\mathrm{d}\vec{\ell}}{\mathrm{d}t}\right) \ . \ \vec{B} = q\vec{E} \ . \ \mathrm{d}\vec{\ell}$$

Thus, the rate of doing work is

$$\frac{\mathrm{d}W}{\mathrm{d}t} = q\vec{E} \ . \ \frac{\mathrm{d}\vec{\ell}}{\mathrm{d}t} = q\vec{E} \ . \ \vec{v}$$

Considering all charges, we can write

$$\frac{\mathrm{d}W}{\mathrm{d}t} = \sum_j q_j(\vec{E}_j \ . \ \vec{v}_j) \tag{8.4}$$

The right side of equations (8.3) and (8.4) are the same. Therefore,

$$\int_V \vec{J} \ . \ \vec{E} \ \mathrm{d}V = \frac{\mathrm{d}W}{\mathrm{d}t} \tag{8.5}$$

Thus, the term $\int_V \vec{J} \ . \ \vec{E} \ \mathrm{d}V$ represents the rate at which the work is done by the field on the charges.

The first and second terms on right side represent the time rate of change of electromagnetic energy stored in the electric and magnetic fields, respectively, within the volume V. Thus, the left side of equation (8.2) represents the sum of:

(i) the rate at which work is done by the field on the charge, *i.e.*, the power spent by the fields due to motion of charges (equation 8.5), and

(ii) the time rate of change of electromagnetic energy stored in electric and magnetic fields with in the volume V.

Considering (i) and (ii), we can conclude that the right side

$$\int_S (\vec{E} \times \vec{B}) \ . \ \mathrm{d}\vec{S}$$

of equation (8.2) must represent energy flowing per second into the volume V through surface S or the power flowing out of volume V through the surface S. We denote

$$\vec{E} \times \vec{B} = \vec{P}$$

where $\vec{P}$ is known as Poynting vector and is defined as:

The amount of field energy passing through unit area of surface in unit time which is normal to the direction of energy flow.

Its unit is watt/m^2

4.1 Poynting theorem: Law of conservation of energy of electromagnetic field

The energy equation can be written as

$$\vec{J} \,.\, \vec{E} + \frac{1}{2}\frac{\partial}{\partial t}\Big(\epsilon E^2 + \frac{B^2}{\mu}\Big) + \nabla \,.\, (\vec{E} \times \vec{B}) = 0$$

$$-\vec{J} \,.\, \vec{E} = \frac{\partial}{\partial t}\Big(\frac{\epsilon}{2}\, E^2 + \frac{1}{2\mu}\, B^2\Big) + \nabla \,.\, (\vec{E} \times \vec{B})$$

$$= \frac{\partial U}{\partial t} + \nabla \,.\, \vec{P}$$

where

$$U = \frac{1}{2}\,\epsilon E^2 + \frac{1}{2\mu}\, B^2$$

represents energy density of electromagnetic field.

For nonconducting situation, we have $\vec{J} = 0$ and therefore

$$\frac{\partial U}{\partial t} + \nabla \,.\, \vec{P} = 0$$

which resembles with the equation of continuity, which expresses the law of conservation of charge. By analogy, this equation can be termed expressing the law of conservation of energy of electromagnetic fields.

5. Maxwell's field equations in terms of vector and scalar potentials

The Maxwell's field equations are

$$\left.\begin{array}{ll} \text{(i)}\ \nabla \,.\, \vec{E} = \rho/\epsilon_0 & \text{(ii)}\ \nabla \,.\, \vec{B} = 0 \\ \text{(iii)}\ \nabla \times \vec{E} + \frac{\partial \vec{B}}{\partial t} = 0 & \text{(iv)}\ \nabla \times \vec{B} - \frac{1}{c^2}\frac{\partial \vec{E}}{\partial t} = \mu_0 \vec{J} \end{array}\right\} \quad (8.6)$$

where $c = 1/\sqrt{\mu_0\epsilon_0}$. Equations (8.6) (i) and (iv) form inhomogeneous pair while (8.6) (ii) and (iii) form homogeneous pair. For vector potential $\vec{A}$ and scalar potential ϕ, we consider

$$\vec{E} = -\nabla\phi - \frac{\partial \vec{A}}{\partial t} \qquad \text{and} \qquad \vec{B} = \nabla \times \vec{A}$$

homogeneous pair of field equations is satisfied. We now consider inhomogeneous pair of field equations. From Maxwell's field equation

$$\nabla \times \vec{B} - \frac{1}{c^2}\,\frac{\partial \vec{E}}{\partial t} = \mu_0 \vec{J}$$

$$\nabla \times (\nabla \times \vec{A}) + \frac{1}{c^2}\frac{\partial \nabla \phi}{\partial t} + \frac{1}{c^2}\frac{\partial^2 \vec{A}}{\partial t^2} = \mu_0 \vec{J}$$

$$\nabla(\nabla \,.\, \vec{A}) - \nabla^2 \vec{A} + \nabla\Big(\frac{1}{c^2}\frac{\partial \phi}{\partial t}\Big) + \frac{1}{c^2}\frac{\partial^2 \vec{A}}{\partial t^2} = \mu_0 \vec{J}$$

$$\nabla^2 \vec{A} - \frac{1}{c^2}\frac{\partial^2 \vec{A}}{\partial t^2} - \nabla\Big[(\nabla \,.\, \vec{A}) + \frac{1}{c^2}\frac{\partial \phi}{\partial t}\Big] = -\mu_0 \vec{J}$$

Using Lorentz condition $\nabla \,.\, \vec{A} + (1/c^2)\, \partial\phi/\partial t = 0$, we get

$$\nabla^2 \vec{A} - \frac{1}{c^2}\frac{\partial^2 \vec{A}}{\partial t^2} = -\mu_0 \vec{J} \qquad \text{or} \qquad \Box \vec{A} = -\mu_0 \vec{J} \tag{8.7}$$

where

$$\Box \equiv \nabla^2 - \frac{1}{c^2}\frac{\partial^2}{\partial t^2} = \frac{\partial^2}{\partial x^2} + \frac{\partial^2}{\partial y^2} + \frac{\partial^2}{\partial y^2} + \frac{\partial^2}{\partial (ict)^2}$$

$$= \frac{\partial^2}{\partial (x^1)^2} + \frac{\partial^2}{\partial (x^2)^2} + \frac{\partial^2}{\partial (x^3)^2} + \frac{\partial^2}{\partial (x^4)^2} = \frac{\partial^2}{\partial (x^i)^2}$$

From Gauss' law, we have

$$\nabla \,.\, \vec{E} = \frac{\rho}{\epsilon_0}$$

Using the expression for $\vec{E}$, we have

$$\nabla \,.\, \Big(-\frac{\partial \vec{A}}{\partial t} - \text{grad } \phi\Big) = \frac{\rho}{\epsilon_0} \qquad\qquad -\frac{\partial(\nabla \,.\, \vec{A})}{\partial t} - \nabla^2\phi = \frac{\rho}{\epsilon_0}$$

Using Lorentz condition $\nabla \,.\, \vec{A} + (1/c^2)\, \partial\phi/\partial t = 0$, we get

$$\nabla^2 \phi - \frac{1}{c^2}\frac{\partial^2 \phi}{\partial t^2} = -\frac{\rho}{\epsilon_0} \qquad \text{or} \qquad \Box \phi = -\frac{\rho}{\epsilon_0} \tag{8.8}$$

Maxwell's field equations (8.6) are equivalent to two equations (8.7) and (8.8) expressed in terms of vector potential A and scalar potential ϕ.

5.1 Covariance of electrodynamics in terms of inhomogeneous field equations

Maxwell's four field equations are reducible to two inhomogeneous field equations either in terms of potential functions ($\vec{A}$, ϕ) or in terms of field vectors ($\vec{E}$, $\vec{B}$) together with the respective sources ($\vec{J}$, ρ). Let us consider the following form to investigate the covariance of electrodynamics

$$\Box\, \vec{A} = -\mu_0 \vec{J} \qquad \text{and} \qquad \Box\, \phi = -\frac{\rho}{\epsilon_0}$$

where the quantities have usual meaning.

If A^i and J^i represent **four vectors** of potential and charge, respectively, these two relations reduce to the following form

$$\Box\, A^i = -\mu_0\, J^i \tag{8.9}$$

where

$$A^i = (\vec{A},\ i\phi/c) \qquad \text{and} \qquad J^i = (\vec{J},\ ic\rho)$$

The Lorentz transformations are

$$\left.\begin{aligned} A'^1 &= \gamma\left(A^1 - \tfrac{v}{c^2}\,\phi\right) \\ A'^2 &= A^2 \\ A'^3 &= A^3 \\ \phi' &= \gamma(\phi - v\,A^1) \end{aligned}\right\} \qquad \left.\begin{aligned} J'^1 &= \gamma(J^1 - v\,\rho) \\ J'^2 &= J^2 \\ J'^3 &= J^3 \\ \rho' &= \gamma\left(\phi - \tfrac{v}{c^2}\,J^1\right) \end{aligned}\right\} \tag{8.10}$$

The transformations given by (8.10) have already been established. The covariance of relation (8.9) demands that it retains its form even on transforming it to frame of reference S′. That is

$$\Box' A'^i = -\mu_0\, J'^i \tag{8.11}$$

where

$$\Box' \equiv \nabla'^2 - \frac{1}{c^2}\frac{\partial^2}{\partial t'^2} = \frac{\partial^2}{\partial (x'^i)^2}$$

We know $\Box = \Box'$. **Let us use component relations to establish the relation (8.11).**

$$\Box' A'^1 = \Box\, A'^1 = -\mu_0\, J'^1$$

Using relation (8.10), we get

$$\Box\gamma\left(A^1 - \frac{v}{c^2}\,\phi\right) = -\mu_0\,\gamma(J^1 - v\,\rho)$$

$$\Box A^1 - \frac{v}{c^2}\,\Box\phi = -\mu_0\,(J^1 - v\,\rho)$$

Substituting $\Box\,\phi = -\rho/\epsilon_0$ and $c^2 = 1/\mu_0\epsilon_0$, we get

$$\Box A^1 = -\mu_0\, J^1 \tag{8.12}$$

In the similar manner, we have

$$\Box' A'^2 = \Box\, A'^2 = -\mu_0\, J'^2$$

But, $A'^2 = A^2$ and $J'^2 = J^2$. Therefore,

$$\Box\, A^2 = -\mu_0\, J^2 \tag{8.13}$$

Similarly, for third component, we have

$$\Box' A'^3 = \Box\ A'^3 = -\mu_0\ J'^3$$

But, $A'^3 = A^3$ and $J'^3 = J^3$. Therefore,

$$\Box\ A^3 = -\mu_0\ J^3 \tag{8.14}$$

For the fourth component, we have

$$\Box' A'^4 = \Box\ A'^4 = -\mu_0\ J'^4$$

Substituting $A'^4 = i\phi'/c$ and $J'^4 = ic\rho'$ and using (8.10) , we get

$$\Box\ \frac{i\phi'}{c} = -\mu_0\ (ic\rho') \qquad\qquad \Box\ \phi' = -\mu_0\ c^2 \rho'$$

Now, transformations for ϕ' and ρ', we have

$$\Box\gamma[\phi - vA^1] = -\mu_0\ c^2\gamma\Big[\rho - \frac{v}{c^2}J^1\Big]$$

Using $\Box\phi = -\rho/\epsilon_0$ and $c^2 = 1/\mu_0\epsilon_0$, we have

$$\Box A^1 = -\mu_0 J^1$$

It also shows that the fourth imaginary component arises due to relativistic motion along the first component, *i.e.*, say, x-axis.

6. Constant electromagnetic field

By a constant electromagnetic field we mean a field which does not depend on the time. Clearly the potentials of a constant field can be chosen so that they are functions only of the coordinates and not of the time. A constant magnetic field is equal, as before, to $\vec{B} = \text{curl}\vec{A}$. A constant electric field is equal to

$$\vec{E} = -\text{grad}\phi \tag{8.15}$$

Thus, a constant electric field is determined only by the scalar potential and a constant magnetic field only by the vector potential. However, we can add to the scalar potential, without changing the fields, only an arbitrary constant (not depending on either the coordinates or the time). Further, we can add to the vector potential,without changing the magnetic field, a gradient of an arbitrary function of the coordinates.

We now determine the energy of a charge in a constant electromagnetic field. When the field is constant, then the Lagrangian for the charge also does not depend explicitly on the time. As we know, in this case the energy is conserved and coincides with the Hamiltonian. Thus, we have

$$\mathcal{H} = \frac{m_0 c^2}{\sqrt{1 - v^2/c^2}} + e\ \phi \tag{8.16}$$

Thus, the presence of the field adds to the energy of the particle the term $e\phi$, the potential energy of the charge in the field. We note the important fact that the energy depends only on the scalar and not on the vector potential. This means that the magnetic field does not affect the energy of the charge. Only the electric field can change the energy of the particle. This is related to the fact that the magnetic field does no work on the charge.

Example 4: Show that a scalar potential expressing constant electric field is not unique, and a vector potential expressing constant magnetic field is not unique.

Solution: When we add an arbitrary constant c, which neither depends on the coordinates nor on the time, to a scalar potential ϕ, the corresponding constant electric field is

$$\vec{B} = -\text{grad}(\phi + c) = -\text{grad}\phi$$

It shows that the scalar potential is not unique for a constant electric field. Further, when we add a gradient of an arbitrary function f to the vector potential $\vec{A}$, the constant magnetic field is

$$\vec{B} = \text{curl}(\vec{A} + \text{ grad} f) = \text{curl}\vec{A}$$

It shows that the vector potential is not unique for a constant magnetic field.

Example 5: Show that the scalar potential ϕ and vector potential $\vec{A}$ can be expressed as $\phi = -\vec{E} \cdot \vec{r}$ and $\vec{A} = \frac{1}{2}\, \vec{B} \times \vec{r}$, where $\vec{E}$, $\vec{B}$, $\vec{r}$, are constant uniform electric field, uniform magnetic field and position vector, respectively.

Solution: The constant (time-independent) electric field is expressed as $\vec{E} = -$ grad ϕ. We have

$$\nabla(\vec{E} \cdot \vec{r}) = (\vec{E} \cdot \nabla)\vec{r} + (\vec{r} \cdot \nabla)\vec{E} + \vec{r} \times \text{ curl } \vec{E} + \vec{E} \times \text{ curl } \vec{r}$$

The electric field is uniform, (*i.e.*, it does not vary with coordinates) and curl $\vec{r} = 0$. Then, we have

$$\nabla(\vec{E} \cdot \vec{r}) = (\vec{E} \cdot \nabla)\vec{r} = \left(E_x \frac{\partial}{\partial x} + E_y \frac{\partial}{\partial y} + E_z \frac{\partial}{\partial z}\right)(x\ \hat{x} + y\ \hat{y} + z\ \hat{z})$$

$$= E_x\ \hat{x} + E_y\ \hat{y} + E_z\ \hat{z} = \vec{E}$$

Comparing the expressions, $\vec{E} = -$ grad ϕ and $\vec{E} = \nabla(\vec{E} \cdot \vec{r})$, we get

$$\phi = -\vec{E} \cdot \vec{r} \tag{8.17}$$

We have

$$\text{curl}(\vec{B} \times \vec{r}) = \vec{B} \text{ div } \vec{r} + \vec{r} \text{ div } \vec{B} - (\vec{B} \cdot \nabla)\vec{r} - (\vec{r} \cdot \nabla)\vec{B}$$

The magnetic field is uniform, (*i.e.*, it does not vary in space). Then, we have

$$\text{curl } (\vec{B} \times \vec{r}) = \vec{B} \text{ div } \vec{r} - (\vec{B} \cdot \nabla)\vec{r}$$

$$= \vec{B}\Big(\frac{\partial}{\partial x}\ \hat{x} + \frac{\partial}{\partial y}\ \hat{y} + \frac{\partial}{\partial z}\ \hat{z}\Big)\ .\ (x\ \hat{x} + y\ \hat{y} + z\ \hat{z})$$

$$-\Big(B_x\ \frac{\partial}{\partial x} + B_y\ \frac{\partial}{\partial y} + B_z\ \frac{\partial}{\partial z}\Big)(x\ \hat{x} + y\ \hat{y} + z\ \hat{z})$$

$$= 3\vec{B} - (B_x\ \hat{x} + B_y\ \hat{y} + B_z\ \hat{z}) = 2\vec{B}$$

Comparing the expressions, $\vec{B} =$ curl $\vec{A}$ and $\vec{B} = \frac{1}{2}$ curl $(\vec{B} \times \vec{r})$, we get

$$\vec{A} = \frac{1}{2}\ (\vec{B} \times \vec{r}) \tag{8.18}$$

Review Questions

1. Show that Ampere's law $\nabla.\ \vec{B} = \mu_0 \vec{J}$ valid for magnetostatics, cannot be accommodated for time-dependent field. What modification was made to this equation. Discuss that the modification becomes zero for magnetostatics.
2. State and derive equation of continuity.
3. What is displacement current? Discuss about its significance.
4. Write Maxwell's equations in differential form, and derive them.
5. Write Maxwell's equations in differential form, and transform them in the integral form. What these equations signify?
6. Write notes on the following:

   (i) Displacement vector and displacement current density
   (ii) Maxwell's equations in differential form
   (iii) Maxwell's equations in integral form
   (iv) Poynting theorem

Multiple-Choice Questions

1. The number of Maxwell's field equations is

   A. 3 B. 4 C. 6 D. 8

   Ans. B

2. The charge density ρ is expressed as

   A. $\nabla\ .\ \vec{E}$ B. $\nabla\ .\ \vec{D}$ C. $\nabla \times \vec{E}$ D. $\nabla \times \vec{D}$

   Ans. B

3. Law of conservation of energy of electromagnetic field is known as

A. Stokes' theorem B. Gauss' theorem

C. Poynting theorem D. Green's theorem

Ans. C

4. Which is not Maxwell's equation?

A. $\nabla \cdot \vec{E} = \frac{\rho}{\epsilon_0}$ B. $\nabla \cdot \vec{B} = 0$ C. $\nabla \times \vec{E} = -\frac{\partial \vec{B}}{\partial t}$ D. $\nabla \times \vec{B} = \mu_0 \vec{J}$

Ans. D

5. For electromagnetic field, the Poynting theorem is statement of the law of conservation of

A. energy B. mass C. charge D. momentum

Ans. A

6. For nonconducting situation, for Poynting vector $\vec{P}$ and electromagnetic field energy density U, the law of conservation of energy of electromagnetic fields is expressed as

A. $\frac{\partial U}{\partial t} - \nabla \cdot \vec{P} = 0$ B. $\frac{\partial U}{\partial t} + \nabla \cdot \vec{P} = 0$

C. $\frac{\partial \vec{P}}{\partial t} + \nabla U = 0$ D. $\frac{\partial \vec{P}}{\partial t} - \nabla U = 0$

Ans. B

Chapter 9

Dispersion and Scattering

When the refractive index n of a medium varies with frequency of radiation propagating through it, the medium is said to be dispersive and the variation of refractive index n of the medium with wavelength λ (*i.e.*, frequency) gives the phenomenon of **dispersion** and is represented by $\mathrm{d}n/\mathrm{d}\lambda$. It has been generally found that the refractive index of medium decreases with the increase of wavelength. It is known as **normal dispersion**.

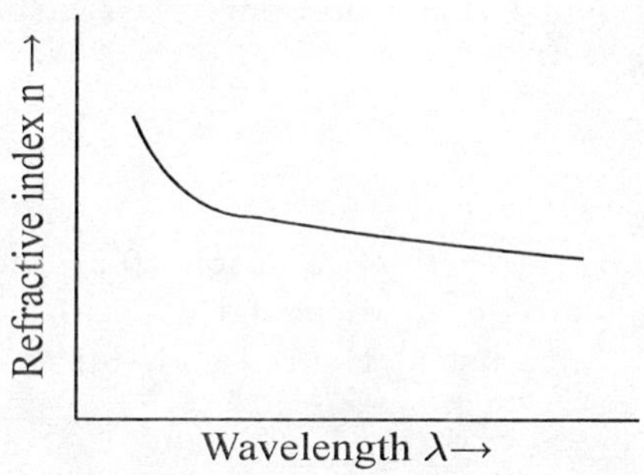

Figure 1: Typical variation of refractive index n with wavelength λ.

But, in some small wavelength ranges, there is often increase of refractive index with the increase of wavelength, due to an increased absorption of radiation propagating through the medium. It is known as **anomalous dispersion**.

Normal dispersion has following characteristics:

(i) The refractive index decreases with the increase of wavelength.

(ii) The slope of the curve (*i.e.*, $\mathrm{d}n/\mathrm{d}\lambda$) is greater at shorter wavelengths. Hence dispersion increases with the decrease of wavelength.

For example, in the visible region of electromagnetic spectrum, the dispersion of violet colour is more than that of the red colour. When white light incident on a surface of prism, it disperses into various colours (wavelengths), in the sequence of violet, indigo, blue, green, yellow, orange and red (VIBGYOR) (Figure 2). The characteristics of normal dispersion may be explained with the help of *Cauchy's empirical equation*, expressed as

$$n^2 = A + \frac{B}{\lambda^2} + \frac{C}{\lambda^4} + \dots \tag{9.1}$$

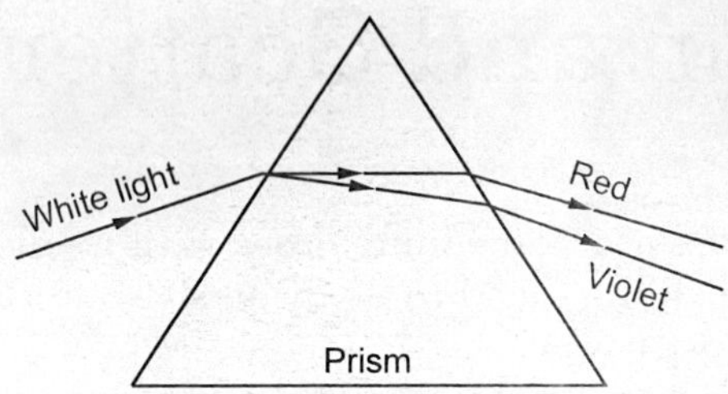

Figure 2: On dispersion in a prism, a white light disperses into various colours, in the sequence of VIBGYOR.

where A, B, C are constants, which depend on medium and their value decreases rapidly in magnitude in sequence. For some purposes, without loosing much accuracy, equation (9.1) may be written as

$$n^2 = A + \frac{B}{\lambda^2} \tag{9.2}$$

Here, A and B both are positive. This equation (9.2) shows that the refractive index n decreases with the increase of wavelength λ. So, for the visible light, n is maximum at the violet end and least at the red end. Further,

$$2n\,\frac{\mathrm{d}n}{\mathrm{d}\lambda} = -\frac{2B}{\lambda^3}$$

Therefore,

$$\frac{\mathrm{d}n}{\mathrm{d}\lambda} = -\frac{B}{\lambda^3}\left[A + \frac{B}{\lambda^2}\right]^{-1/2} = -\frac{AB}{\lambda^3} \tag{9.3}$$

where higher order terms are neglected. Equation (9.3) shows that dispersion varies inversely as the cube of wavelength. The negative sign indicates that the slope of dispersion curve is negative and dispersion (magnitude of slope) decreases as λ increases.

When the wavelength range is not too large, then normal dispersion may be expressed with the help of Hartmann's empirical formula:

$$n = A + \frac{B}{\lambda - C}$$

where A, B and C are constants to be determined from observations.

1. Dispersion relation for gases

Based on the electromagnetic theory, a dispersion relation for gaseous medium has been derived by Lorentz as the following. The derivation is based on the following assumptions:

(i) There is no appreciable interaction between the constituent particles of medium.

(ii) In each particle, nucleus is regarded very heavy as compared to the electrons. Hence the reduced mass of nucleus and electron is equal to the mass of electron.

(iii) The electric field of electromagnetic wave propagation through the medium induces electric dipole in the constituent particles due to relative displacement of electrons and nucleus of a particle.

(iv) These electric dipoles execute forced vibrations under the oscillatory electric field of radiation.

(v) The electrons are bound to the nucleus by a linear restoring force which is proportional to the displacement $\vec{r}$ from the equilibrium position.

(vi) The electrons experience a damping or frictional force, which is proportional to the velocity of electron.

(vii) Over an atom or a molecule, $\vec{E}$ is constant in space *i.e.*, it varies only with time:

$$\vec{E} = \vec{E}_0 \, \mathrm{e}^{-i\,\omega t}$$

The electric force on the electron of mass m and charge e is $e\vec{E}$ and the magnetic force can be neglected on account of its motion. In addition to the electric force, the electron is acted upon by two mechanical forces:

(a) restoring force proportional to its displacement $\vec{r}$ from its equilibrium position, *i.e.*, equal to $-k\,\vec{r}$, where k is a positive constant, depending on the structure and properties of particles.

(b) damping frictional force against the direction of motion of electron, which is proportional to its velocity, *i.e.*, equal to $-\ell\, \mathrm{d}\vec{r}/\mathrm{d}t$, where l is a positive constant of proportionality.

Hence, according to Newton's second law of motion, the equation of motion is

$$m\,\frac{\mathrm{d}^2\vec{r}}{\mathrm{d}t^2} = e\vec{E} - k\vec{r} - \ell\,\frac{\mathrm{d}\vec{r}}{\mathrm{d}t} \qquad \text{or} \qquad \frac{\mathrm{d}^2\vec{r}}{\mathrm{d}t^2} + \gamma\,\frac{\mathrm{d}\vec{r}}{\mathrm{d}t} + \omega_0^2\,\vec{r} = \frac{e}{m}\,\vec{E} \tag{9.4}$$

where

$$\gamma = \frac{\ell}{m} \qquad \text{and} \qquad \omega_0^2 = \frac{k}{m}$$

ω_0 is natural frequency of oscillation. Equation (9.4) is second order linear inhomogeneous equation. Its solution has two parts: (i) complimentary function and (ii) particular integral.

(i) Complimentary function is the solution of equation

$$\frac{\mathrm{d}^2\vec{r}}{\mathrm{d}t^2} + \gamma\,\frac{\mathrm{d}\vec{r}}{\mathrm{d}t} + \omega_0^2\,\vec{r} = 0$$

Let the solution of this equation be

$$\vec{r} = \vec{A}\,\mathrm{e}^{\alpha t}$$

where $\vec{A}$ and α are unknown real constants. Then,

$$\frac{\mathrm{d}\vec{r}}{\mathrm{d}t} = \alpha\vec{A}\,\mathrm{e}^{\alpha t} \qquad \text{and} \qquad \frac{\mathrm{d}^2\vec{r}}{\mathrm{d}t^2} = \alpha^2\vec{A}\,\mathrm{e}^{\alpha t}$$

Therefore,

$$\alpha^2\vec{A}\,\mathrm{e}^{\alpha t} + \gamma\alpha\vec{A}\,\mathrm{e}^{\alpha t} + \omega_0^2\,\vec{A}\,\mathrm{e}^{\alpha t} = 0 \qquad \text{or} \qquad \alpha^2 + \gamma\alpha + \omega_0^2 = 0$$

This quadratic equation has solutions:

$$\alpha = \frac{-\gamma \pm \sqrt{\gamma^2 - 4\omega_0^2}}{2} = -\frac{\gamma}{2} \pm \sqrt{\frac{\gamma^2}{4} - \omega_0^2}$$

When the damping is small, $(\gamma^2/4) << \omega_0^2$, we have

$$\alpha = -\frac{\gamma}{2} \pm i\sqrt{\omega_0^2 - \frac{\gamma^2}{4}} = -\frac{\gamma}{2} \pm i\beta$$

where

$$\beta = \sqrt{\omega_0^2 - \frac{\gamma^2}{4}}$$

So, we have

$$\vec{r} = \vec{A}_1\,\exp\Big[-\frac{\gamma}{2} + i\beta\Big]t + \vec{A}_2\,\exp\Big[-\frac{\gamma}{2} - i\beta\Big]t$$

$$= \exp\Big(-\frac{\gamma t}{2}\Big)[\vec{A}_1\,\exp(i\beta t) + \vec{A}_2\,\exp(-i\beta t)]$$

(ii) Particular integral is obtained by solving the equation

$$\frac{\mathrm{d}^2\vec{r}}{\mathrm{d}t^2} + \gamma\,\frac{\mathrm{d}\vec{r}}{\mathrm{d}t} + \omega_0^2\,\vec{r} = \frac{e}{m}\,\vec{E}_0\,\mathrm{e}^{-i\omega t}$$

Let the solution of this equation be

$$\vec{r} = \vec{B}\,\mathrm{e}^{-i\omega t}$$

where $\vec{B}$ is constant. Then,

$$\frac{\mathrm{d}\vec{r}}{\mathrm{d}t} = -i\omega\,\vec{B}\,\mathrm{e}^{-i\omega t} \qquad \text{and} \qquad \frac{\mathrm{d}^2\vec{r}}{\mathrm{d}t^2} = -\omega^2\vec{B}\,\mathrm{e}^{-i\omega t}$$

Therefore,

$$-\omega^2\ \vec{B}\ \mathrm{e}^{-i\omega t} - i\gamma\ \omega\ \vec{B}\ \mathrm{e}^{-i\omega t} + \omega_0^2\ \vec{B}\ \mathrm{e}^{-i\omega t} = \frac{e}{m}\ \vec{E}_0\ \mathrm{e}^{-i\omega t}$$

$$\vec{B}(-\omega^2 - i\gamma\ \omega + \omega_0^2) = \frac{e}{m}\ \vec{E}_0 \qquad \text{or} \qquad \vec{B} = \frac{e\vec{E}_0}{m(\omega_0^2 - i\gamma\ \omega - \omega^2)}$$

So, we have

$$\vec{r} = \frac{e\vec{E}_0}{m(\omega_0^2 - i\gamma\ \omega - \omega^2)}\ \mathrm{e}^{-i\omega t}$$

Now, general solution of equation (9.4) is

$$\vec{r} = \exp\Big(-\frac{\gamma t}{2}\Big)[\vec{A}_1\ \exp(i\beta t) + \vec{A}_2\ \exp(-i\beta t)] + \frac{e\vec{E}_0}{m(\omega_0^2 - i\gamma\ \omega - \omega^2)}\ \mathrm{e}^{-i\omega t}$$

In this equation the first term represents free damped vibrations which soon die out due to the term $\exp(-\gamma t/2)$. So, the first term is neglected. Now, the solution of equation (9.4) is

$$\vec{r} = \frac{e\vec{E}_0}{m(\omega_0^2 - \omega^2 - i\gamma\ \omega)}\ \mathrm{e}^{-i\omega t} = \frac{e\vec{E}}{m(\omega_0^2 - \omega^2 - i\gamma\ \omega)} \tag{9.5}$$

Suppose, N is the number of dispersion electrons per unit volume. The polarization $\vec{P}$ is equal to the dipole moment per unit volume:

$$\vec{P} = Ne\ \vec{r} = Ne\ \frac{e\vec{E}}{m(\omega_0^2 - \omega^2 - i\gamma\ \omega)} = \frac{Ne^2\vec{E}/m}{(\omega_0^2 - \omega^2) - i\gamma\ \omega} \tag{9.6}$$

The imaginary term in the denominator means that $\vec{P}$ and $\vec{E}$ are not in phase. The phase difference is $\tan^{-1}[-\gamma\ \omega/(\omega^2 - \omega_0^2)]$.

In the derivation of equation (9.6), it is assumed that all the oscillators have the same natural frequency ω_0. This shows that the medium absorbs only one spectral line. To take into account the possibility of existence of other absorption lines, we assume that there are N molecules per unit volume and in each molecule, there are f_1 oscillators whose constants are ω_1 and γ_1; f_2 oscillators whose constants are ω_2 and γ_2; and so on.

The generalization of equation (9.6) can be written as

$$\vec{P} = N \sum_k f_k\ (e\ \vec{r}_k)$$

where

$$\vec{r}_k = \frac{e\ \vec{E}/m}{\omega_k^2 - \omega^2 - i\gamma_k\ \omega} \tag{9.7}$$

so that

$$\vec{P} = \frac{Ne^2\vec{E}}{m} \sum_k \frac{f_k}{\omega_k^2 - \omega^2 - i\ \gamma_k\ \omega} \tag{9.8}$$

The electrical polarizability α is given by

$$\alpha = \frac{\vec{P}}{\vec{E}} = \frac{Ne^2}{m} \sum_k \frac{f_k}{\omega_k^2 - \omega^2 - i\,\gamma_k\,\omega} \tag{9.9}$$

The dielectric constant ϵ_r is

$$\begin{aligned} \epsilon_r &= 1 + \frac{1}{\epsilon_0}\frac{\vec{P}}{\vec{E}} \\ &= 1 + \frac{1}{\epsilon_0}\frac{Ne^2}{m} \sum_k \frac{f_k}{\omega_k^2 - \omega^2 - i\,\gamma_k\,\omega} \end{aligned} \tag{9.10}$$

Equation (9.10) shows that the dielectric constant is complex quantity. Thus, the refractive index, which is equal to the square root of dielectric constant, also is a complex. Hence,

$$n^{*2} = 1 + \frac{1}{4\pi\epsilon_0}\frac{4\pi Ne^2}{m} \sum_k \frac{f_k}{\omega_k^2 - \omega^2 - i\,\gamma_k\,\omega} \tag{9.11}$$

where n^* denotes the complex refractive index. This relation is known as dispersion formula. The significance of the complex refractive index is that besides the dispersion, there is absorption of energy in the medium. The refractive index has become frequency dependent. Now we consider the following situations:

- Normal dispersion
- Anomalous dispersion

1.1 Normal dispersion

In the region remote from the natural frequencies of oscillators, *i.e.*, the absorption frequencies of the medium, the term $\gamma_k\,\omega$ in the denominator of equation (9.11) is so small that it can be neglected in comparison to $(\omega_k^2 - \omega^2)$. Equation (9.11) now reduces to

$$n^2 = 1 + \frac{1}{4\pi\epsilon_0}\frac{4\pi Ne^2}{m} \sum_k \frac{f_k}{\omega_k^2 - \omega^2} \tag{9.12}$$

This expression shows that the refractive index n is real and increases with the increase of frequency ω of incident wave. This is the situation of normal dispersion. We know that $\omega = 2\pi f = 2\pi c/\lambda$. Substituting the value of ω in equation (9.12), we have

$$\begin{aligned} n^2 &= 1 + \frac{1}{4\pi\epsilon_0}\frac{4\pi Ne^2}{m} \sum_k \frac{f_k}{(4\pi^2c^2/\lambda_k^2) - (4\pi^2c^2/\lambda^2)} \\ &= 1 + \frac{Ne^2}{m\epsilon_0(4\pi^2c^2)} \sum_k \frac{f_k\lambda^2\lambda_k^2}{\lambda^2 - \lambda_k^2} \end{aligned}$$

$$= 1 + \sum_k \frac{A_k \lambda^2}{\lambda^2 - \lambda_k^2} \tag{9.13}$$

where

$$A_k = \frac{Ne^2 f_k \lambda_k^2}{m\epsilon_0 4\pi^2 c^2}$$

Formula (9.13) is known as **Sellmeier's equation**.

For the situation, $\lambda >> \lambda_k$, equation (9.13) can be written as

$$n^2 = 1 + \sum_k A_k \left(1 - \frac{\lambda_k^2}{\lambda^2}\right)^{-1}$$

$$= 1 + \sum_k A_k \left(1 + \frac{\lambda_k^2}{\lambda^2} + \frac{\lambda_k^4}{\lambda^4} + \ldots\right)$$

$$= 1 + \sum_k A_k + \sum_k A_k \frac{\lambda_k^2}{\lambda^2} + \sum_k A_k \frac{\lambda_k^4}{\lambda^4} + \ldots$$

$$= A + \frac{B}{\lambda^2} + \frac{C}{\lambda^4} + \ldots \tag{9.14}$$

where

$$A = 1 + \sum_k A_k; \qquad B = \sum_k A_k \lambda_k^2; \qquad C = \sum_k A_k \lambda_k^4$$

Equation (9.14) is the **Cauchy's dispersion formula**. The coefficient A is the **coefficient of refraction** while the coefficient B is the **coefficient of dispersion**. From equation (9.14), it is clear that

$$2n \frac{\mathrm{d}n}{\mathrm{d}\lambda} = -\frac{2B}{\lambda^3} - \frac{4C}{\lambda^5} + \ldots$$

$$\frac{\mathrm{d}n}{\mathrm{d}\lambda} = -\frac{1}{n\lambda^3}\left(B + \frac{2C}{\lambda^2} + \ldots\right)$$

$$\frac{\mathrm{d}n}{\mathrm{d}\lambda} = -\frac{1}{\lambda^3}\left(B + \frac{2C}{\lambda^2} + \ldots\right)\left(A + \frac{B}{\lambda^2} + \ldots\right)^{-1/2}$$

$$\frac{\mathrm{d}n}{\mathrm{d}\lambda} = -\frac{BA}{\lambda^3} = -\frac{k}{\lambda^3}$$

This is expression for dispersive power. The expression shows that the dispersive power varies inversely as the cube of wavelength. The minus sign signifies that the slope of the dispersive power is negative.

1.2 Anomalous dispersion

As long as the impressed frequencies do not include a natural frequency of electrons, the dispersion is always normal. But, when the impressed frequencies contain the natural frequency from the so many frequencies of the electron, the normal course of dispersion is modified.

$$n^{*2} = 1 + \frac{1}{4\pi\epsilon_0} \frac{4\pi N e^2}{m} \sum_k \frac{f_k}{\omega_k^2 - \omega^2 - i\,\gamma_k\,\omega} \tag{9.15}$$

$$n^* = \left[1 + \frac{1}{4\pi\epsilon_0} \frac{4\pi N e^2}{m} \sum_k \frac{f_k}{\omega_k^2 - \omega^2 - i\,\gamma_k\,\omega}\right]^{1/2}$$

$$= 1 + \frac{1}{4\pi\epsilon_0} \frac{2\pi N e^2}{m} \sum_k \frac{f_k}{\omega_k^2 - \omega^2 - i\,\gamma_k\,\omega}$$

Multiplying numerator and denominator of second term on right side by $(\omega_k^2 - \omega^2 + i\gamma_k\omega)$, we get

$$n^* = 1 + \frac{N e^2}{2m\epsilon_0} \sum_k \frac{f_k(\omega_k^2 - \omega^2 + i\,\gamma_k\,\omega)}{(\omega_k^2 - \omega^2)^2 + \gamma_k^2\,\omega^2}$$

Expressing the complex refractive index as $n^* = n' + i\,n''$, where n' is real part and n'' is imaginary part, we have

$$n' = 1 + \frac{N e^2}{2m\epsilon_0} \sum_k \frac{f_k(\omega_k^2 - \omega^2)}{(\omega_k^2 - \omega^2)^2 + \gamma_k^2\,\omega^2}$$

and

$$n'' = \frac{N e^2 \omega}{2m\epsilon_0} \sum_k \frac{f_k\,\gamma_k}{(\omega_k^2 - \omega^2)^2 + \gamma_k^2\,\omega^2}$$

Let us consider only one value ω_1, γ_1 and $f_k = 1$ (the situation of one resonance), so that

$$n' = 1 + \frac{N e^2}{2m\epsilon_0} \frac{(\omega_1^2 - \omega^2)}{(\omega_1^2 - \omega^2)^2 + \gamma_1^2\,\omega^2}$$

and

$$n'' = \frac{N e^2 \omega}{2m\epsilon_0} \frac{\gamma_1}{(\omega_1^2 - \omega^2)^2 + \gamma_1^2\,\omega^2}$$

The n'' has the largest value when $\omega = \omega_1$. This is the resonant frequency. On the two sides of the resonance, the distribution is symmetrical. At $\omega = \omega_1$, we have $n' = 1$. For maximum and minimum of n', the requirement

$$\frac{\partial n'}{\partial \omega} = 0$$

provides

$$[(\omega_1^2 - \omega^2)^2 + \gamma_1^2\,\omega^2](-2\omega) - (\omega_1^2 - \omega^2)^2[2(\omega_1^2 - \omega^2)^2(-2\omega) + \gamma_1^2(2\omega)] = 0$$

so that

$$\omega = [\omega_1^2 \pm \omega_1\gamma_1]^{1/2}$$

$$= \omega_1\left[1 \pm \frac{\gamma_1}{\omega_1}\right]^{1/2} = \omega_1 \pm \frac{\gamma_1}{2}$$

Therefore, n' is the minimum at $\omega = \omega_1 + \gamma_1/2$ and is the maximum at $\omega = \omega_1 - \gamma_1/2$. Figure 3 shows the variation of real and imaginary parts of n^* plotted as a function of frequency ω.

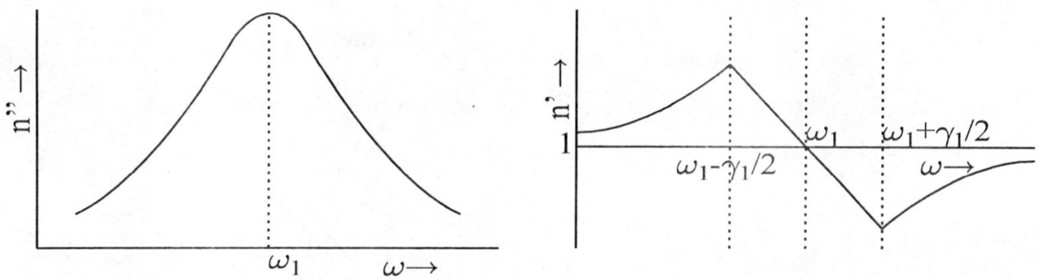

Figure 3: Variation of real and imaginary parts of n^* versus frequency ω.

It is clear from the figure that at low frequencies, n' is slightly greater than 1 and it increases as ω increases. It reaches a maximum for $\omega = \omega_1 - \gamma/2$ and then decreases rapidly until $\omega = \omega_1 + \gamma/2$. When $\omega = \omega_1$, $n' = 1$. After that, the curve again increases and approaches 1 asymptotically. The imaginary part corresponds to the absorption of electromagnetic waves propagating through the gas. It is also clear from the figure that n'' is maximum when $\omega = \omega_1$, when n' is 1. It has a width at half maximum approximately equal to γ. Thus, in the region where n' changes rapidly, the gas is relatively highly absorbing.

2. Scattering

When an electromagnetic radiation passes through a medium (substance), it is absorbed by the electrons of the medium. These electrons are set into vibrations. Now, either its energy is lost by inelastic collisions of molecules, containing the energy, with the others or by radiating energy because of its vibrational property. Thus, the electrons absorb energy from the incident electromagnetic radiations and re-emit into space. This process is known as the **scattering** of electromagnetic radiation by the system (Figure 4). The scattered rays are in all directions, and the scattering is characterized in terms of following parameters.

2.1 Differential scattering cross section

While dealing with the scattering, one of the physical quantities of interest is the distribution of scattered energy as a function of scattering angle. This information is expressed in terms of a quantity σ known as the **scattering cross section**.

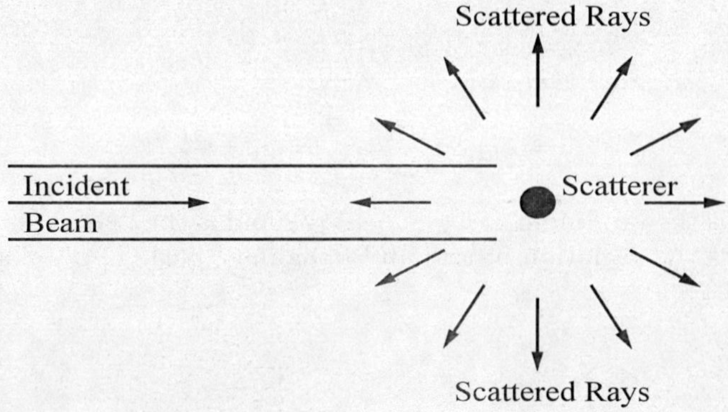

Figure 4: Scattering of electromagnetic radiation.

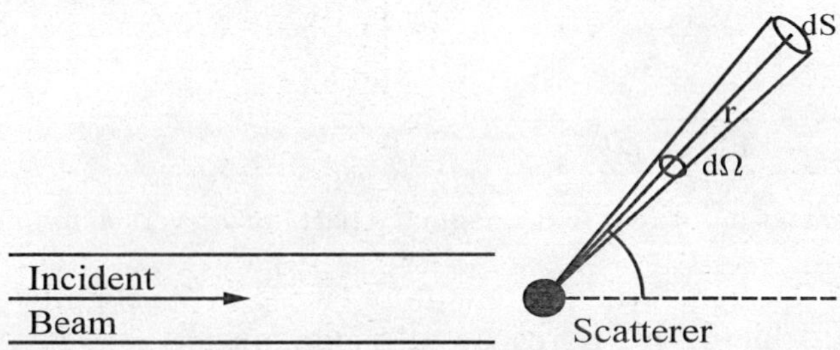

Figure 5: The beam incident on a scatterer.

The differential scattering cross section defined as the average energy per unit solid angle per unit time scattered in the direction of angles θ and ϕ divided by the average energy per unit area per unit time in a direction normal to the incident wave:

$$\frac{d\sigma}{d\Omega} = \frac{\text{Average energy per unit solid angle per unit time}}{\text{Average energy per unit area per unit time}}$$

Let the solid angle $d\Omega$ subtend an area dS (Figure 5). The energy scattered per unit time is

$$dP_{\text{scatt}} = W_{\text{scatt}}\, dS \tag{9.16}$$

where W_{scatt} is the intensity of scattered energy per unit time per unit solid angle. Thus, we have

$$\frac{dP_{\text{scatt}}}{d\Omega} = \frac{W_{\text{scatt}}\, dS}{d\Omega} = W_{\text{scatt}}\, r^2 \tag{9.17}$$

because $d\Omega = dS/r^2$. Hence, the differential scattering cross section is given by

$$\frac{d\sigma}{d\Omega} = \frac{dP_{\text{scatt}}}{d\Omega}\,\frac{1}{P_{\text{incident}}} = \frac{W_{\text{scatt}}}{P_{\text{incident}}}\, r^2 \tag{9.18}$$

where P_{incident} is incident flux density. Equation (9.18) shows that the differential scattering cross section has the dimensions of area.

2.2 Total scattering cross section

The total scattering cross section is defined as the total average per unit time scattered in all directions divided by the intensity of the incoming plane wave. From equation (9.18), we have

$$d\sigma = \frac{W_{\text{scatt}}}{P_{\text{incident}}}\, r^2\, d\Omega = \frac{W_{\text{scatt}}}{P_{\text{incident}}}\, dS \tag{9.19}$$

On integration, we get scattering cross section

$$\sigma = \int \frac{W_{\text{scatt}}}{P_{\text{incident}}}\, dS \tag{9.20}$$

3. Scattering by a free electron (Thomson scattering)

Consider an electron of mass m and charge e in the path of a plane monochromatic wave. Let the incident electromagnetic wave be moving along z-axis. The motion of electron is given by

$$\vec{F} = e\,\vec{E}$$

Considering $\vec{E} = \vec{E}_0\, e^{i(kz-\omega t)}$ and following Newton's second law of motion, the equation of motion is

$$m\,\frac{d^2x}{dt^2} = e\,E_0\, e^{i(kz-\omega t)} \qquad \text{or} \qquad \frac{d^2x}{dt^2} = \frac{e\,E_0}{m}\, e^{i(kz-\omega t)}$$

The solution of this equation is

$$x = -\frac{e\,E_0}{m\,\omega^2}\, e^{i(kz-\omega t)} \tag{9.21}$$

We know that an oscillating charge behaves like an electric dipole with electric dipole moment p:

$$p = ex = -\frac{e^2\,E_0}{m\,\omega^2}\, e^{i(kz-\omega t)} = -|p_0|\, e^{i(kz-\omega t)} \tag{9.22}$$

where

$$|p_0| = \frac{e^2\,E_0}{m\omega^2} \tag{9.23}$$

The oscillating dipole radiates energy and the average energy radiated per second per unit area is given by

$$W_{\text{scatt}} = \frac{1}{4\pi\epsilon_0}\,\frac{\omega^4|p_0|^2}{8\pi c^3 r^2}\,\sin^2\alpha$$

where α is the angle between the direction of observation and the axis of the electric dipole, which is along the electric vector of incident radiation. Using the expression for $|p_0|$, we get

$$W_{\text{scatt}} = \frac{1}{4\pi\epsilon_0}\,\frac{\omega^4 e^4 E_0^2}{8\pi c^3 r^2 m^2 \omega^4}\,\sin^2\alpha$$

$$= \frac{1}{4\pi\epsilon_0} \frac{e^4 E_0^2}{8\pi c^3 r^2 m^2} \sin^2\alpha \tag{9.24}$$

For monochromatic plane waves propagating in z direction,

$$\vec{S} = \frac{1}{\mu_0}\left[E_0\cos(kz-\omega t)\,\hat{x} \times \frac{E_0}{c}\cos(kz-\omega t)\,\hat{y}\right] = c\,\epsilon_0\,E_0^2\cos^2(kz-\omega t)\,\hat{z}$$

Thus,

$$P_{\text{incident}} = \frac{1}{2}\,c\,\epsilon_0\,E_0^2$$

The differential scattering cross section is given by

$$\frac{d\sigma}{d\Omega} = \frac{W_{\text{scatt}}}{P_{\text{incident}}}\,r^2$$

$$= \left(\frac{1}{4\pi\epsilon_0}\frac{e^4 E_0^2}{8\pi c^3 r^2 m^2}\sin^2\alpha\right)\left(\frac{2}{c\,\epsilon_0\,E_0^2}\right) r^2 = \left(\frac{e^2}{4\pi\epsilon_0 mc^2}\right)^2 \sin^2\alpha$$

Let $r_0 = e^2/(4\pi\epsilon_0 mc^2)$ is classical radius of electron, then

$$\frac{d\sigma}{d\Omega} = r_0^2\,\sin^2\alpha$$

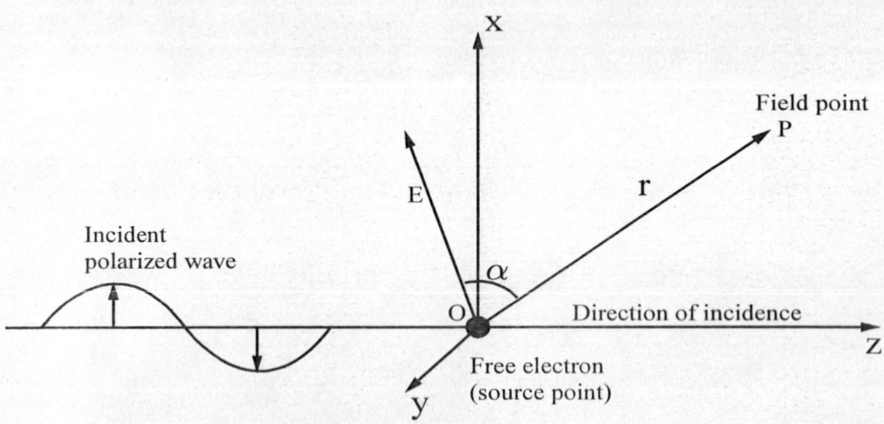

Figure 6: Scattering by free electron.

Figure 6 shows the free electron as the source, placed at the origin of coordinate system. The field point is P where the radiation field is E_{rad}. The incident polarized wave is with electric vector $\vec{E}$. Angle θ is the scattering angle and ϕ is the polarizing angle. In order to obtain relationship between angle θ, ϕ and α, we draw a plane PMN perpendicular to the plane containing $\vec{E}$, then

$$\text{ON} = r\cos\alpha$$

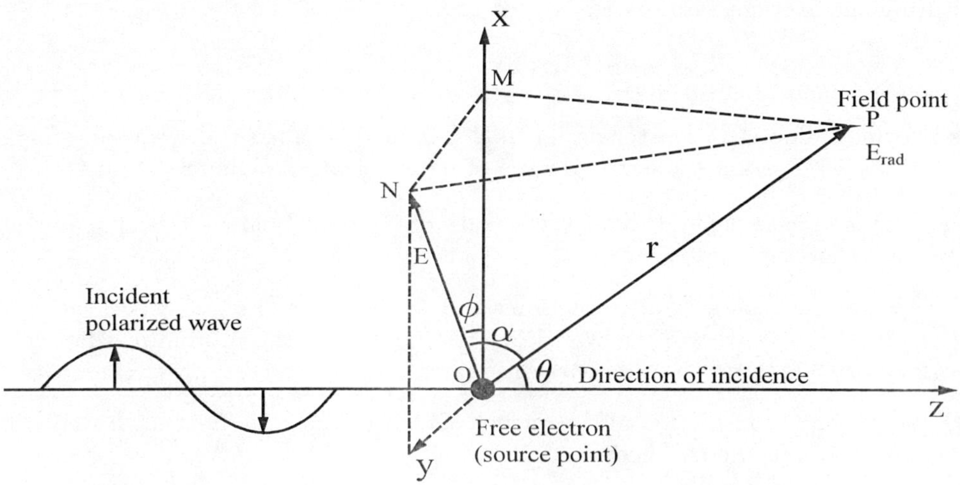

Figure 7: Scattering by free electron.

From Figure 7

$$\text{OM} = r\cos(90^\circ - \theta) = r\sin\theta$$

and

$$\text{ON} = \text{OM}\cos\phi = r\sin\theta\cos\phi$$

Comparing the two expressions for ON, we get

$$\cos\alpha = \sin\theta\ \cos\phi \qquad \textit{or} \qquad \cos^2\alpha = \sin^2\theta\ \cos^2\phi$$

Now,

$$\sin^2\alpha = 1 - \cos^2\alpha = 1 - \sin^2\theta\cos^2\phi$$

$$= 1 - \cos^2\phi(1 - \cos^2\theta)$$

For a plane polarized wave with polarization along x-axis, we have $\phi = 0$ (Figure 7). Therefore,

$$\sin^2\alpha = \cos^2\theta$$

Hence,

$$\frac{\mathrm{d}\sigma}{\mathrm{d}\Omega} = r_0^2\ \cos^2\theta$$

If the primary wave is unpolarized, *i.e.*, randomly polarized, we must average over ϕ, and thus, we have

$$\overline{\sin^2\alpha} = 1 - \overline{\cos^2\phi}\ (1 - \cos^2\theta) = 1 - \frac{1}{2}\ (1 - \cos^2\theta) = \frac{1}{2}\ (1 + \cos^2\theta)$$

Substituting the average value of $\sin^2 \alpha$, we get

$$\frac{d\sigma}{d\Omega} = \frac{r_0^2}{2} (1 + \cos^2 \theta) \tag{9.25}$$

This is known as Thomson scattering formula. Angle θ is called the scattering angle while $(1 + \cos^2 \theta)/2$ is known as the degree of depolarization. It shows that:

(i) the scattering is independent of the nature of incident beam *i.e.*, ω (frequency and wavelength)

(ii) the scattering occurs in all directions and is the maximum when $\theta = 0$ or π (*i.e.*, in the forward and backward directions), and is the minimum when $\theta = \pi/2$ or $3\pi/2$ (*i.e.*, in the side way directions).

(iii) the scattering is a function of the nature of charged particle because it depends on the radius of the charged particle.

The total scattering cross section is given by

$$\sigma = \int \frac{d\sigma}{d\Omega}\, d\Omega = \int \frac{r_0^2}{2} (1 + \cos^2 \theta)\, d\Omega$$

$$= \frac{r_0^2}{2} \int_0^{\pi} \int_0^{2\pi} (1 + \cos^2 \theta)\, \sin\theta\, d\theta\, d\phi = \frac{8\pi}{3}\, r_0^2$$

If the incident radiation is plane polarized, then

$$\sigma = \int \frac{d\sigma}{d\Omega}\, d\Omega = \int r_0^2 \cos^2 \theta\, d\Omega$$

$$= r_0^2 \int_0^{\pi} \int_0^{2\pi} \cos^2 \theta\, \sin\theta\, d\theta\, d\phi = \frac{4\pi}{3}\, r_0^2$$

The dimension of σ is the same as that of r_0^2. That is, the dimension of σ is L^2, *i.e.*, area. From equation (9.25), we get the following information:

(i) If $(d\sigma/d\Omega)/r_0^2$ is plotted against the scattering angle θ, it is symmetrical about the line $\theta = \pi/2$.

(ii) Much of the scattered energy is concentrated in the forward direction [equation (9.24)]. This shows that Thomson scattering is for small scattering angles.

It may be noted that classically the scattering by electrons and particles is to be treated as a case of re-emission. It means that the incident energy is first absorbed, then by virtue of its excited state substance radiates. The electron is treated free in such scattering. Of course, to assume that the electrons are free in case of low energy photons is not justified.

In case of Compton scattering in which the energy of striking photon is extremely high (x-rays), the electron may be treated practically free as it is knocked from the parent atom or molecule very easily.

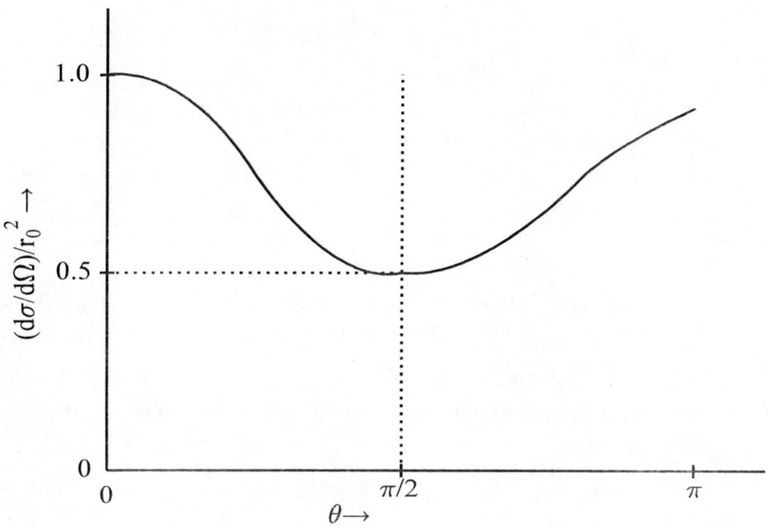

Figure 8: Variation of $(\mathrm{d}\sigma/\mathrm{d}\Omega)/r_0^2$ versus θ.

4. Rayleigh scattering (Scattering by a bound electron)

Consider the case that an electron system is subjected to a monochromatic electromagnetic radiation. The vibrations of electrons are of forced type. Though in the beginning, an electron first vibrates with its natural frequency, on which the external electric field is superimposed. Ultimately it starts and continues to vibrate with the frequency of the applied field. In this case, the vibrations of bound electrons under the action of incident electromagnetic radiation is important to discuss the scattering.

In order to set up a differential equation for the forced vibration of the bound electron, let us consider the following influencing factors:

(i) The applied periodic electromagnetic force may be represented by $e\, E_0\, \mathrm{e}^{i(kz-\omega t)}$, where e is the charge of electron.

(ii) The restoring force which is proportional to the instantaneous value of displacement and is given by $-k\, x$. The electron oscillates with natural frequency $\omega_0 = \sqrt{k/m}$.

(iii) The self force of the electron which is equivalent to the functional force in case of mechanical motion, is directly proportional to the instantaneous velocity of the particle. Let it be represented by $-\ell\, \mathrm{d}x/\mathrm{d}t$, where, ℓ is constant.

For the net force acting on a bound electron, the equation of motion can be expressed as

$$m\, \frac{\mathrm{d}^2 x}{\mathrm{d}t^2} = eE_0\, \mathrm{e}^{i(kz-\omega t)} - k\, x - \ell\, \frac{\mathrm{d}x}{\mathrm{d}t}$$

$$\frac{d^2x}{dt^2} + \frac{\ell}{m}\frac{dx}{dt} + \frac{k}{m}\,x = \frac{e}{m}\,E_0\,e^{i(kz-\omega t)}$$

$$\frac{d^2x}{dt^2} + \gamma\,\frac{dx}{dt} + \omega_0^2\,x = \frac{e}{m}\,E_0\,e^{i(kz-\omega t)} \tag{9.26}$$

where

$$\gamma = \frac{\ell}{m} \qquad \text{and} \qquad \omega_0^2 = \frac{k}{m}$$

Here, ω_0 is natural frequency of electron. Equation (9.26) is second order linear inhomogeneous differential equation. Its solution consists of two parts: (i) complimentary function and (ii) particular integral.

(i) The **complimentary function** is the solution of the differential equation

$$\frac{d^2x}{dt^2} + \gamma\,\frac{dx}{dt} + \omega_0^2\,x = 0 \tag{9.27}$$

Let the solution of equation (9.27) be of the form

$$x = A\,e^{\alpha t} \tag{9.28}$$

where A and α are unknown real constants. Then,

$$\frac{dx}{dt} = \alpha\,A\,e^{\alpha t} \qquad \text{and} \qquad \frac{d^2x}{dt^2} = \alpha^2 A\,e^{\alpha t}$$

Therefore,

$$\alpha^2 A\,e^{\alpha t} + \gamma\alpha A\,e^{\alpha t} + \omega_0^2\,\vec{A}\,e^{\alpha t} = 0 \qquad \text{or} \qquad \alpha^2 + \gamma\alpha + \omega_0^2 = 0$$

This quadratic equation has solutions:

$$\alpha = \frac{-\gamma \pm \sqrt{\gamma^2 - 4\omega_0^2}}{2} = -\frac{\gamma}{2} \pm \sqrt{\frac{\gamma^2}{4} - \omega_0^2}$$

When the damping is small, $(\gamma^2/4) << \omega_0^2$, we have

$$\alpha = -\frac{\gamma}{2} \pm i\beta$$

where

$$\beta = \sqrt{\omega_0^2 - \frac{\gamma^2}{4}}$$

Thus, we have

$$x = A_1\,\exp\Big[\Big\{-\frac{\gamma}{2} + i\beta\Big\}t\Big] + A_2\,\exp\Big[\Big\{-\frac{\gamma}{2} - i\beta\Big\}t\Big]$$

or

$$x = e^{-\gamma t/2}\Big(A_1\,\exp[i\beta t] + A_2\,\exp[-i\beta t]\Big)$$

(ii) The **particular integral** can be obtained by solving the equation

$$\frac{d^2x}{dt^2} + \gamma \frac{dx}{dt} + \omega_0^2\, x = \frac{e}{m} E_0\, e^{i(kz-\omega t)} \tag{9.29}$$

Let the solution of equation (9.29) be of the form

$$x = B\, e^{i(kz-\omega t)} \tag{9.30}$$

where B is constant. Then,

$$\frac{dx}{dt} = -i\omega\, B\, e^{i(kz-\omega t)} \qquad \text{and} \qquad \frac{d^2x}{dt^2} = -\omega^2 B\, e^{i(kz-\omega t)}$$

Therefore,

$$-\omega^2\, B\, e^{i(kz-\omega t)} - i\gamma\, \omega\, B\, e^{i(kz-\omega t)} + \omega_0^2\, B\, e^{i(kz-\omega t)} = \frac{e}{m} E_0\, e^{i(kz-\omega t)}$$

$$B(-\omega^2 - i\gamma\,\omega + \omega_0^2) = \frac{e}{m} E_0 \qquad \text{or} \qquad B = \frac{eE_0}{m(\omega_0^2 - i\gamma\,\omega - \omega^2)}$$

Thus, the solution becomes

$$x = \left[\frac{e\, E_0}{m(\omega_0^2 - \omega^2 - i\gamma\omega)}\right] \exp[i(kz - \omega t)] \tag{9.31}$$

The value of x can be written as

$$x = \frac{eE_0}{m}\left[\frac{(\omega_0^2 - \omega^2) + i\gamma\omega}{(\omega_0^2 - \omega^2)^2 + \gamma^2\omega^2}\right] \exp[i(kz - \omega t)]$$

$$= \frac{eE_0\, \exp[i(kz - \omega t)]}{m[(\omega_0^2 - \omega^2)^2 + \gamma^2\omega^2]^{1/2}}\left[\frac{\omega_0^2 - \omega^2}{[(\omega_0^2 - \omega^2)^2 + \gamma^2\omega^2]^{1/2}} + \frac{i\,\gamma\omega}{[(\omega_0^2 - \omega^2)^2 + \gamma^2\omega^2]^{1/2}}\right]$$

$$= \frac{eE_0\, \exp[i(kz - \omega t)]}{m[(\omega_0^2 - \omega^2)^2 + \gamma^2\omega^2]^{1/2}}\, (\cos\delta + i\sin\delta) \tag{9.32}$$

where

$$\cos\delta = \frac{\omega_0^2 - \omega^2}{[(\omega_0^2 - \omega^2)^2 + \gamma^2\omega^2]^{1/2}} \qquad \sin\delta = \frac{\gamma\omega}{[(\omega_0^2 - \omega^2)^2 + \gamma^2\omega^2]^{1/2}}$$

$$\tan\delta = \frac{\gamma\omega}{\omega_0^2 - \omega^2} \qquad \delta = \tan^{-1}\left[\frac{\gamma\omega}{\omega_0^2 - \omega^2}\right]$$

Now, equation (9.32) can be written as

$$x = \frac{eE_0\, \exp[i(kz - \omega t + \delta)]}{m[(\omega_0^2 - \omega^2)^2 + \gamma^2\omega^2]^{1/2}} \tag{9.33}$$

So, the general solution of equation (9.26) is given by the sum of complimentary function and particular integral. That is

$$x = e^{-\gamma t/2}(A_1 \exp[i\beta t] + A_2 \exp[-i\beta t]) + \frac{eE_0 \exp[i(kz - \omega t + \delta)]}{m[(\omega_0^2 - \omega^2)^2 + \gamma^2\omega^2]^{1/2}}$$

The first term on right side represents the free damped vibration which dies soon due to factors $\exp(-\gamma t/2)$. Thus, we have

$$x = \frac{eE_0 \exp[i(kz - \omega t + \delta)]}{m[(\omega_0^2 - \omega^2)^2 + \gamma^2\omega^2]^{1/2}}$$

The oscillating charge is equivalent to an induced electric dipole moment p. Thus, we have

$$p = ex = \frac{e^2 E_0 \exp[i(kz - \omega t + \delta)]}{m[(\omega_0^2 - \omega^2)^2 + \gamma^2\omega^2]^{1/2}} = |p_0| \exp[i(kz - \omega t + \delta)]$$

where

$$|p_0| = \frac{e^2 E_0}{m[(\omega_0^2 - \omega^2)^2 + \gamma^2\omega^2]^{1/2}}$$

This oscillating dipole radiates energy and the average energy radiated per second per unit area is

$$W_{\text{scatt}} = \frac{1}{4\pi\epsilon_0} \frac{\omega^4 |p_0|^2}{8\pi c^3 r^2} \sin^2\alpha$$

where α is the angle between the direction of observation and the axis of electric dipole, which is along the electric vector of incident radiation (Figure 9). Using the expression for $|p_0|$, we get

$$W_{\text{scatt}} = \frac{1}{4\pi\epsilon_0} \frac{\omega^4}{8\pi c^3 r^2} \frac{e^4 E_0^2 \sin^2\alpha}{m^2[(\omega_0^2 - \omega^2)^2 + \gamma^2\omega^2]} \tag{9.34}$$

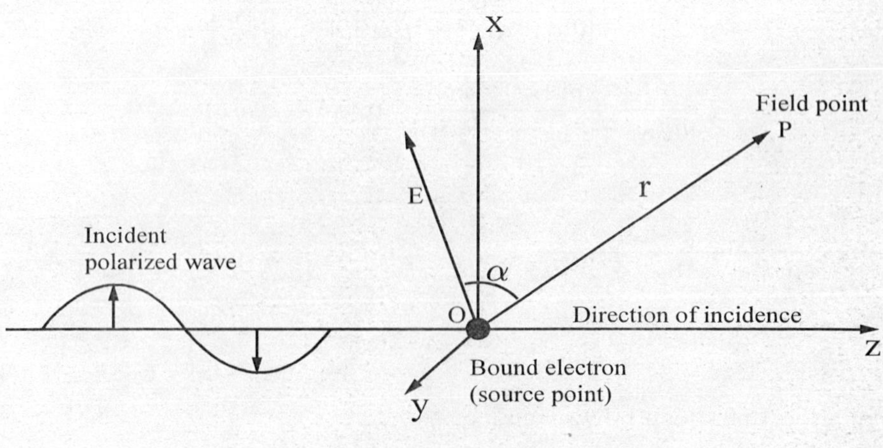

Figure 9: Scattering by bound electron.

For monochromatic plane waves propagating in the z direction,

$$\vec{S} = \frac{1}{\mu_0}\left[E_0 \cos(kz - \omega t)\ \hat{x} \times \frac{E_0}{c}\ \cos(kz - \omega t)\ \hat{y}\right] = c\ \epsilon_0\ E_0^2 \cos^2(kz - \omega t)\ \hat{z}$$

Thus,

$$P_{\text{incident}} = \frac{1}{2}\ c\ \epsilon_0\ E_0^2$$

The differential scattering cross section is given by

$$\begin{aligned}\frac{d\sigma}{d\Omega} &= \frac{W_{\text{scatt}}}{P_{\text{incident}}}\ r^2 \\ &= \left(\frac{1}{4\pi\epsilon_0}\ \frac{\omega^4}{8\pi c^3 r^2}\ \frac{e^4 E_0^2 \sin^2\alpha}{m^2[(\omega_0^2 - \omega^2)^2 + \gamma^2\omega^2]}\right)\left(\frac{2}{c\epsilon_0 E_0^2}\right)\ r^2 \\ &= \left(\frac{e^2}{4\pi\epsilon_0 mc^2}\right)^2\ \frac{\omega^4 \sin^2\alpha}{(\omega_0^2 - \omega^2)^2 + \gamma^2\omega^2} \\ &= r_0^2\ \frac{\omega^4 \sin^2\alpha}{(\omega_0^2 - \omega^2)^2 + \gamma^2\omega^2}\end{aligned}$$

where $r_0 = e^2/(4\pi\epsilon_0 mc^2)$ is known as the classical radius of electron. From Figure 10, we have (as in the last section)

$$\sin^2\alpha = 1 - \cos^2\phi(1 - \cos^2\theta)$$

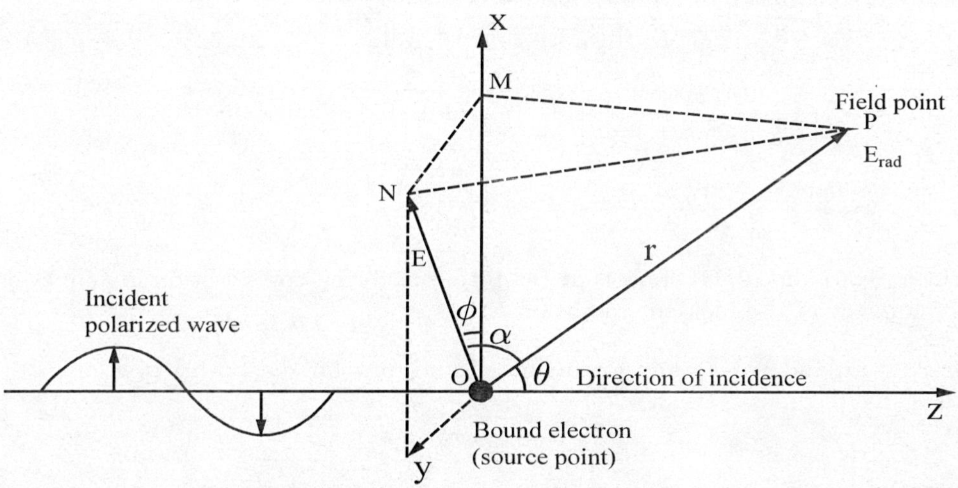

Figure 10: Scattering by bound electron.

For a plane polarized wave with polarization along x-axis, we have $\phi = 0$. Therefore

$$\sin^2\alpha = \cos^2\theta$$

Hence,

$$\frac{d\sigma}{d\Omega} = r_0^2\ \frac{\omega^4 \cos^2\theta}{(\omega_0^2 - \omega^2)^2 + \gamma^2\omega^2} \tag{9.35}$$

The total scattering cross section is

$$\sigma = \int \frac{d\sigma}{d\Omega}\, d\Omega = \int \frac{r_0^2\, \omega^4 \cos^2\theta}{(\omega_0^2 - \omega^2)^2 + \gamma^2\omega^2}\, d\Omega$$

$$= \frac{r_0^2\, \omega^4}{(\omega_0^2 - \omega^2)^2 + \gamma^2\omega^2} \int_0^{\pi} \int_0^{2\pi} \cos^2\theta\ \sin\theta\ d\theta\ d\phi$$

$$= \frac{4\pi}{3}\, \frac{r_0^2\, \omega^4}{(\omega_0^2 - \omega^2)^2 + \gamma^2\omega^2} \tag{9.36}$$

If the primary wave is unpolarized, *i.e.*, randomly polarized, we must average over ϕ, and thus, we have

$$\overline{\sin^2\alpha} = 1 - \overline{\cos^2\phi}(1 - \cos^2\theta) = 1 - \frac{1}{2}\,(1 - \cos^2\theta) = \frac{1}{2}\,(1 + \cos^2\theta)$$

Hence,

$$\frac{d\sigma}{d\Omega} = r_0^2\, \frac{\omega^4(1 + \cos^2\theta)}{2[(\omega_0^2 - \omega^2)^2 + \gamma^2\omega^2]} \tag{9.37}$$

The total scattering cross section is

$$\sigma = \int \frac{d\sigma}{d\Omega}\, d\Omega = \int \frac{r_0^2\, \omega^4(1 + \cos^2\theta)}{2[(\omega_0^2 - \omega^2)^2 + \gamma^2\omega^2]}\, d\Omega$$

$$= \frac{r_0^2\, \omega^4}{2[(\omega_0^2 - \omega^2)^2 + \gamma^2\omega^2]} \int_0^{\pi} \int_0^{2\pi} (1 + \cos^2\theta)\ \sin\theta\ d\theta\ d\phi$$

$$= \frac{8\pi}{3}\, \frac{r_0^2\, \omega^4}{(\omega_0^2 - \omega^2)^2 + \gamma^2\omega^2} \tag{9.38}$$

Equations (9.36) and (9.38) show that the total scattering cross section is a function of the frequency of the incident radiation. Further,

(i) If $\omega >> \omega_0$ and $\gamma \to 0$, and for polarized and unpolarized both waves

$$\sigma = \frac{8\pi}{3}\, r_0^2$$

i.e., **Thomson scattering** occurs and the charge behaves as if it was free.

(ii) If $\omega << \omega_0$, then for polarized and unpolarized both waves

$$\sigma = \frac{8\pi}{3}\, r_0^2 \left(\frac{\omega}{\omega_0}\right)^4 \tag{9.39}$$

This scattering is known as the **Rayleigh scattering**. The equation (9.39) shows that the amount of scattered light is proportional to $1/\lambda^4$, where λ is the wavelength of incident radiation.

(iii) If $\omega = \omega_0$, then for polarized and unpolarized waves, respectively.

$$\sigma = \frac{4\pi}{3} r_0^2 \left(\frac{\omega_0}{\gamma}\right)^2 \qquad \text{and} \qquad \sigma = \frac{8\pi}{3} r_0^2 \left(\frac{\omega_0}{\gamma}\right)^2 \tag{9.40}$$

This scattering is known as the **resonant scattering**. In this case, the total scattering cross section is maximum and the natural frequency of the electron as an oscillator becomes equal to the frequency of incident radiation.

On the basis of Rayleigh scattering, we can explain the following facts:

(a) Sky appears blue

(b) Red colour of sunset and sunrise.

(c) Use of red light for danger signals

(a) Sky appears blue:

According to Rayleigh scattering

$$\sigma \propto \frac{1}{\lambda^4}$$

i.e., shorter wavelength is scattered more strongly. The light coming from the sun is scattered by air molecules and as we know that shorter visible wavelengths correspond to blue light, hence it is scattered strongly. We see this scattered light which is blue, and therefore, the sky appears blue.

(b) Red colour of sunset and sunrise:

At the sunset or sunrise, we see the red colour. This is because the red colour has the longest wavelength and its scattering is the least. The blue light is almost scattered from the direct light coming to our eyes from the Sun leaving only the red colour to be seen.

(c) Use of red light for danger signals:

Although our eye is most sensitive to yellow and green colours even then red light is used for danger signals. We have seen that the red light has the longest wavelength in the visible region and therefore, its scattering is the least. This shows that red light can travel longer distances in atmosphere in comparison to the other colours and can be seen from larger distances. Due to this fact, the red light is used for danger signals.

4.1 Application of Rayleigh scattering in radio astronomy

We are aware that out of electromagnetic spectrum (γ-rays, x-rays, ultraviolet radiation, visible light, infrared radiation, microwaves, radio waves), the Earth's atmosphere is transparent in two wide bands, called **optical window** and **radio window**. Remaining part of electromagnetic spectrum is not allowed to transmit through the Earth's atmosphere. The wavelength in optical window ranges from

3000 Å to 10000 Å whereas in the radio window, it ranges from few mm to few decameters. Thus, the ratio of average wavelength in radio window, λ_r to that in the optical window λ_o is

$$\frac{\lambda_r}{\lambda_o} \approx 10^5$$

Following Rayleigh scattering, the ratio of scattering in the radio window, σ_r to that in the optical window σ_o is

$$\frac{\sigma_r}{\sigma_o} \approx 10^{-20}$$

showing that the scattering in radio window is extremely small as compared to that in the optical window. Consequently, the radiations in radio window can travel very long distance as compared to those in the optical window. So, for the study of very distant cosmic objects, the radio window is extremely helpful as compared to optical window. The radiations lying in the radio window are produced by molecules existing in the cosmic objects through rotational transitions.

Review Questions

1. Prove that the two frequencies at which the refractive index is maximum and minimum are separated by a distance γ.

2. What are normal and anomalous dispersions? Discuss about the dispersion relation for gases.

3. Derive expression for Thomson scattering.

4. Derive expression for scattering by a free electron.

5. Derive expression for Rayleigh scattering.

6. Derive expression for scattering by a bound electron.

7. Write notes on the following

   (i) Normal dispersion
   (ii) Anomalous dispersion
   (iii) Scattering
   (iv) Differential scattering cross section and total scattering cross section
   (v) Colour of sky
   (vi) Colour of rising and setting sun
   (vii) Cauchy's empirical equation for dispersion
   (viii) Hartmann's empirical equation for dispersion
   (ix) Application of radio window in Radio Astronomy

Multiple-Choice Questions

1. For normal dispersion,

   A. refractive index of medium decreases with the increase of wavelength.

   B. refractive index of medium increases with the increase of wavelength.

   C. refractive index of medium does not depend on the wavelength.

   Ans. A

2. White light is allowed to pass through a prism of glass. Dispersion of which colour is the largest?

   A. Red B. Yellow C. Green D. Violet

   Ans. D

3. White light is allowed to pass through a prism of glass. Refractive index for which colour is the largest?

   A. Red B. Yellow C. Green D. Violet

   Ans. D

4. Which scattering is for free electrons?

   A. Thomson B. Rayleigh C. Maxwell

   Ans. A

5. Which scattering is for bound electrons?

   A. Thomson B. Rayleigh C. Maxwell

   Ans. B

6. For Thomson scattering, total cross section varies with frequency ω of electromagnetic radiation as

   A. ω^0 B. ω^2 C. ω^3 D. ω^4

   Ans. A

7. For Rayleigh scattering, total cross section varies with frequency ω of electromagnetic radiation as

   A: ω^0 B. ω^2 C. ω^3 D. ω^4

   Ans. D

8. For resonant scattering, total cross section varies with frequency ω of electromagnetic radiation as

   A. ω^0 B. ω^2 C. ω^3 D. ω^4

   Ans. B

9. Because of which scattering, the colour of sky appears blue?

   A. Thomson B. Rayleigh C. Maxwell D. Compton

   Ans. B

10. Because of which scattering, the colour of rising Sun appears red?

    A. Thomson B. Rayleigh C. Maxwell D. Compton

    Ans. B

11. Because of which scattering, the colour of setting Sun appears red?

    A. Thomson B. Rayleigh C. Maxwell D. Compton

    Ans. B

12. For scattering of radiation by a bound electron, the scattering cross section is inversely proportional to which power of wavelength?

    A. 1 B. 2 C. 3 D. 4

    Ans. D

13. For which electromagnetic radiation, the refractive index is the largest?

    A. γ-rays B. x-rays C. Visible light D. Radio waves

    Ans. A

Chapter 10

Electromagnetic Waves

1. Waves in one dimension

A wave in a continuous medium is a kind of disturbance that propagates with a fixed shape at constant velocity. When the medium has absorption, the intensity of wave decreases during its propagation. Frequencies travel at different speeds. When the propagation of wave is in two or three dimensions, its amplitude decreases. We have to remember that the standing waves do not propagate at all. A wave can be produced through various methods. For example, a wave may be produced by shaking one end of a taut string.

Let us try to represent a wave mathematically. Figure 1 shows an example of wave propagating from left towards right at two different times; one representing at initial time, say, $t = 0$, and the other at some later time t. In absence of absorption and dispersion, for one-dimensional wave, each point on the wave-form simply shifts to the right by an amount vt, where v represents the velocity of wave.

Let z-axis is taken the direction of propagation of one-dimensional wave and at a time t, its form is represented by a function $f(z,t)$. Let the initial shape of the wave be $g(z) = f(z,0)$. In time t, each point on the shape shifts a distance vt. We can represent this phenomenon as the following:

$$f(z,t) = f(z - vt, 0) = g(z - vt)$$

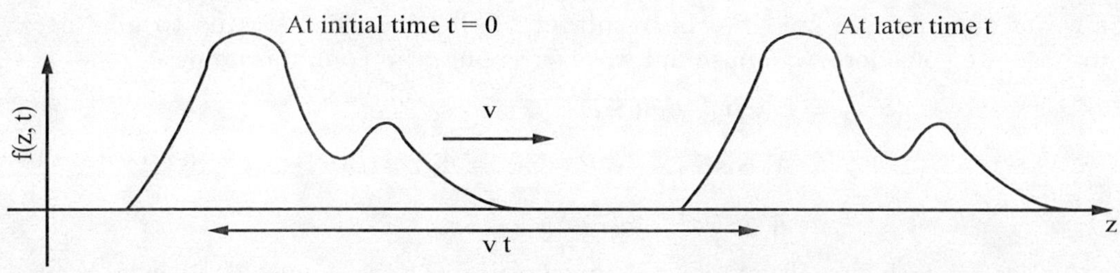

Figure 1: One dimensional wave.

Some examples of functions showing such relation are:

$$f_1(z,t) = A\ \mathrm{e}^{-b(z-vt)^2}, \qquad f_2(z,t) = A\sin[b(z-vt)]$$

$$f_3(z,t) = \frac{A}{b(z-vt)^2+1}$$

where A and b are constants. These functions however show different shapes. Some examples of functions not showing such relation are:

$$f_4(z,t) = A\ \mathrm{e}^{-b(2z^2+vt)}, \qquad f_5(z,t) = A\sin(bz)\cos(bvt)^3$$

Small disturbances on a wave satisfy the relation

$$\frac{\partial^2 f}{\partial z^2} = \frac{1}{v^2}\frac{\partial^2 f}{\partial t^2}$$

where v is the speed of propagation of wave. This relation is known as the **wave equation**. Out of several possible wave forms, the sinusoidal one is

$$f(z,t) = A\cos[k(z-vt)+\delta] \tag{10.1}$$

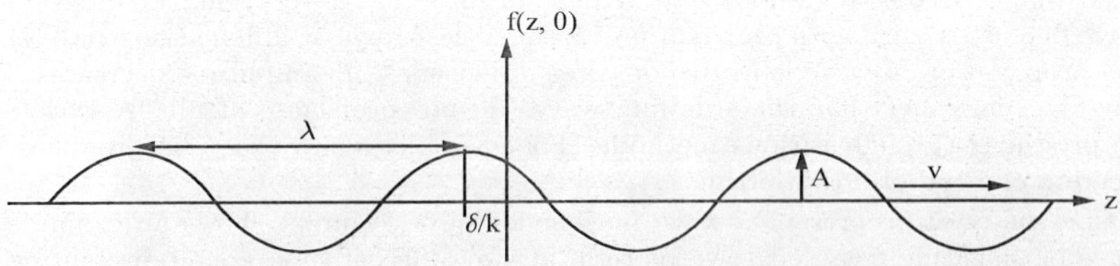

Figure 2: Typical variation of $f(z,0)$ versus z.

where A is amplitude, δ the phase constant. It is common to write sinusoidal waves (equation 10.1) in terms of ω, rather than v:

$$f(z,t) = A\cos(kz-\omega t+\delta) \tag{10.2}$$

where $\omega = kv$, k is wave number, $\lambda = 2\pi/k$ is wavelength, $T = 2\pi/\omega$ is time-period.

1.1 Superposition

Two or more waves may arrive simultaneously at a common point and the disturbance at that point would be the resultant of the disturbances due to all waves. Suppose, we consider two sinusoidal waves arriving at a common point:

$$\begin{aligned} f_r(z,t) &= f_1(z,t) + f_2(z,t) \\ &= A_1\cos(k_1 z - \omega_1 t - \delta_1) + A_2\cos(k_2 z - \omega_2 t - \delta_2) \\ &= A\cos(kz - \omega t - \delta) \end{aligned}$$

Notice that the resultant of the superposition of two sinusoidal waves is also a sinusoidal wave. However, the values of the amplitude, wave number, angular frequency, and phase constant, in general, change.

1.2 Boundary conditions: Reflection and Transmission

A wave may be propagating in a string or through a continuous medium. Let us now consider a situation that the wave arrives at the end of the string or end of the medium. In fact, the further course of action would depend on the situation at the end. That is, it would be subject to the condition at the boundary. Suppose, there may be another medium, which is different, so that the wave velocities v_1 and v_2 in the two mediums are different.

Let us say, for convenience, that the interface of two mediums occurs at $z = 0$. Let the incident wave

$$f_I(z,t) = A_I \cos(k_1 z - \omega t - \delta_I), \qquad (z < 0), \tag{10.3}$$

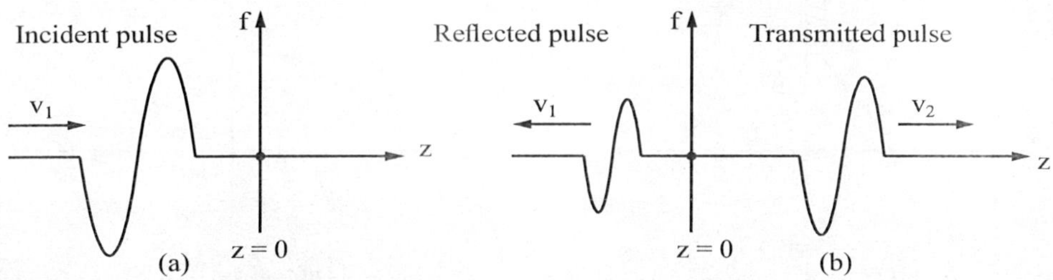

Figure 3: Reflection and transmission of wave.

is coming in from the left. It gives rise to a reflected wave

$$f_R(z,t) = A_R \cos(-k_1 z - \omega t - \delta_R), \qquad (z < 0), \tag{10.4}$$

traveling back along string 1 (medium 1) (hence the minus sign in front of k_1). In addition, there may be a transmitted wave

$$f_T(z,t) = A_T \cos(k_2 z - \omega t - \delta_T), \qquad (z > 0), \tag{10.5}$$

which continues through the string 2 (medium 2).

All the three, the incident wave $f_I(z,t)$, reflected wave $f_R(z,t)$, and transmitted wave $f_T(z,t)$, are oscillating at the same frequency ω. Since the wave velocities are different in the two strings (mediums), their wavelengths and wave numbers are also different:

$$\frac{v_1}{v_2} = \frac{\omega/k_1}{\omega/k_2} = \frac{k_2}{k_1}$$

Further, we have

$$\frac{k_2}{k_1} = \frac{2\pi/\lambda_2}{2\pi/\lambda_1} = \frac{\lambda_1}{\lambda_2}$$

Therefore, we get

$$\frac{v_1}{v_2} = \frac{k_2}{k_1} = \frac{\lambda_1}{\lambda_2} \tag{10.6}$$

For a sinusoidal incident wave, the net disturbance of the string (medium) is:

$$f(z,t) = \begin{cases} A_I \cos(k_1 z - \omega t - \delta_I) + A_R \cos(-k_1 z - \omega t - \delta_R), & (z < 0) \\ A_T \cos(k_2 z - \omega t - \delta_T), & (z > 0) \end{cases} \tag{10.7}$$

At the interface ($z = 0$), the displacement just slightly to the left ($z = 0^-$) must equal the displacement slightly to the right ($z = 0^+$), otherwise there would be a break between the two mediums. Mathematically, $f(z,t)$ is continuous at $z = 0$:

$$f(0,^- t) = f(0,^+ t) \tag{10.8}$$

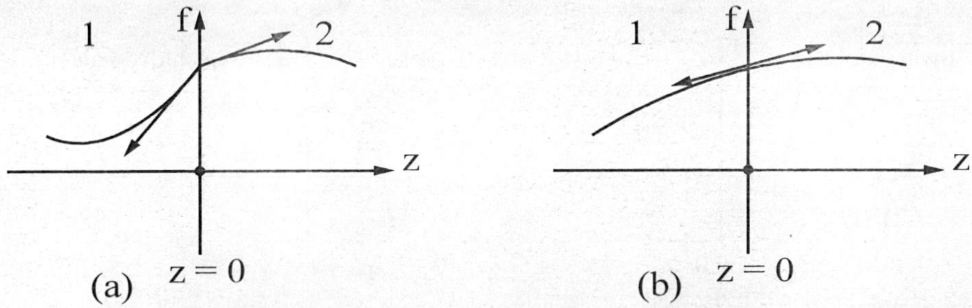

Figure 4: Boundary conditions at the interface.

Further, for smooth transit between the two strings (mediums), the derivative of f at the joint must also be continuous:

$$\left.\frac{\partial f}{\partial z}\right|_{0^-} = \left.\frac{\partial f}{\partial z}\right|_{0^+} \tag{10.9}$$

These two equations (10.8) and (10.9) are known as the boundary conditions. Applying the boundary condition (10.8) to equation (10.7), we get

$$\Big[A_I \cos(k_1 z - \omega t - \delta_I) + A_R \cos(-k_1 z - \omega t - \delta_R)\Big]_{0^-} = \Big[A_T \cos(k_2 z - \omega t - \delta_T)\Big]_{0^+}$$

or

$$A_I \cos(-\omega t - \delta_I) + A_R \cos(-\omega t - \delta_R) = A_T \cos(-\omega t - \delta_T)$$

The only way that this equation is satisfied for all values of time t is that $\delta_I = \delta_R = \delta_T$. Then, we have

$$A_I + A_R = A_T \tag{10.10}$$

Applying the boundary condition (10.9) to equation (10.7), we get

$$\Big[k_1 A_I \sin(k_1 z - \omega t - \delta_I) - k_1 A_R \sin(-k_1 z - \omega t - \delta_R)\Big]_{0^-} = \Big[k_2 A_T \sin(k_2 z - \omega t - \delta_T)\Big]_{0^+}$$

With $\delta_I = \delta_R = \delta_T$, we get

$$k_1(A_I - A_R) = k_2 A_T \tag{10.11}$$

From equations (10.10) and (10.11), we have

$$A_R = \left(\frac{k_1 - k_2}{k_1 + k_2}\right) A_I \qquad A_T = \left(\frac{2k_1}{k_1 + k_2}\right) A_I \tag{10.12}$$

Or, in terms of the velocities (using equation 10.6):

$$A_R = \left(\frac{v_2 - v_1}{v_2 + v_1}\right) A_I \qquad A_T = \left(\frac{2v_2}{v_2 + v_1}\right) A_I \tag{10.13}$$

It may be kept in mind that the difference between v_1 and v_2 is not very large, but significant. For the situation, $v_2 > v_1$, we have $\lambda_2 > \lambda_1$, A_T is a bit larger than A_I, and A_R is very small as compared to A_I, but in the same direction (Figure 5); so the incident and reflected waves are in the same phase.

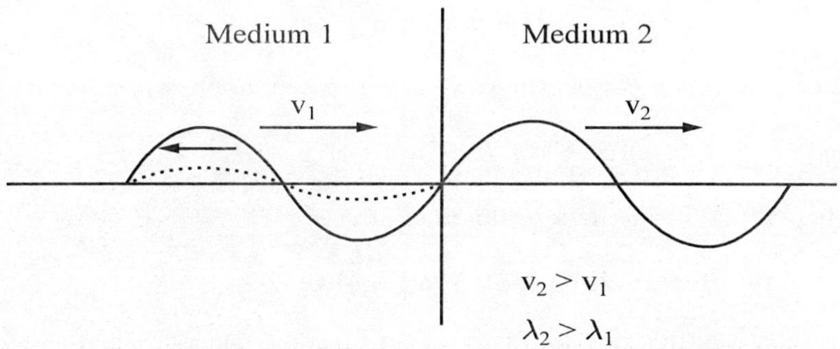

Figure 5: Reflected wave is shown by the dotted curve.

For the situation, $v_2 < v_1$, we have $\lambda_2 < \lambda_1$, A_T is a bit smaller than A_I, and A_R is very small as compared to A_I (in magnitude), but in the opposite direction (Figure 6), which is equivalent to the introduction of phase difference of 180°.

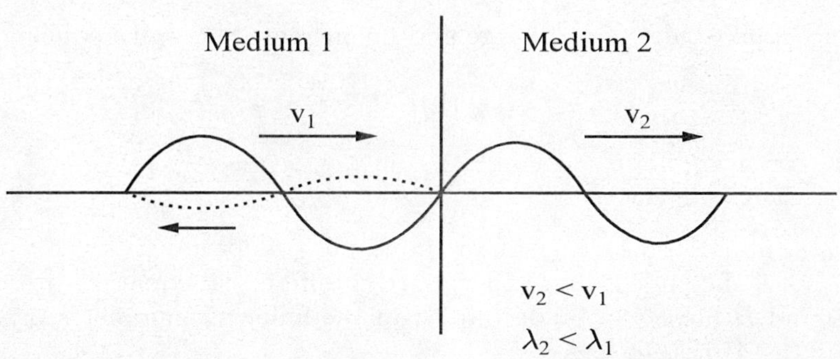

Figure 6: Reflected wave is shown by the dotted curve.

1.3 Polarization

The electromagnetic radiations are transverse, *i.e.*, the electric and magnetic fields lie in the xy-plane, whose normal is the direction of propagation of wave (z-axis). When the electric vector is only in one direction, the wave is said to be polarized. For the electric vector along x-axis, the wave is expressed, as shown in Figure 7 (a):

$$f_x(z,t) = A\ \cos(kz - \omega t - \delta)\ \hat{x}$$

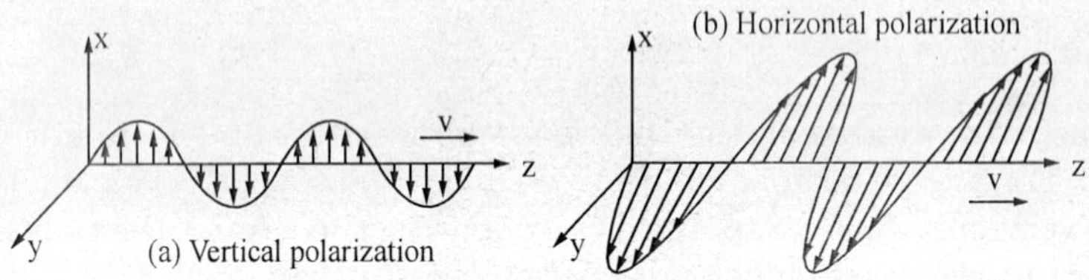

Figure 7: Polarization of wave.

or for the electric vector along y-axis, the wave is expressed, as shown in Figure 7 (b):

$$f_y(z,t) = A\ \cos(kz - \omega t - \delta)\ \hat{y}$$

or along any other direction in the xy plane:

$$f(z,t) = A\ \cos(kz - \omega t - \delta)\ \hat{n}$$

The polarization vector $\hat{n}$ defines the plane of vibration of electric vector. Because the waves are transverse, $\hat{n}$ is perpendicular to the direction of propagation:

$$\hat{n}\ .\ \hat{z} = 0$$

2. Electromagnetic waves in vacuum

In regions of space where there is no charge and no current, Maxwell's equations are

$$\left.\begin{array}{ll} \text{(i)}\ \nabla\ .\ \vec{E} = 0 & \text{(ii)}\ \nabla \times \vec{E} = -\frac{\partial \vec{B}}{\partial t} \\ \text{(iii)}\ \nabla\ .\ \vec{B} = 0 & \text{(iv)}\ \nabla \times \vec{B} = \frac{1}{c^2}\frac{\partial \vec{E}}{\partial t} \end{array}\right\} \quad (10.14)$$

where $c^2 = 1/\mu_0\epsilon_0$; μ_0 is permeability of vacuum and ϵ_0 is permittivity of vacuum. They constitute a set of coupled, first-order, partial differential equations for $\vec{E}$ and $\vec{B}$ fields. The $\vec{E}$ and $\vec{B}$ fields can be decoupled in the following manner. Apply the curl to equation (10.14) (ii), to have

$$\nabla \times (\nabla \times \vec{E}) = -\nabla \times \left[\frac{\partial \vec{B}}{\partial t}\right] \qquad \nabla(\nabla\ .\ \vec{E}) - \nabla^2\vec{E} = -\frac{\partial}{\partial t}(\nabla \times \vec{B})$$

Using equations. 10.14 (i) and (iv), we get

$$-\nabla^2 \vec{E} = -\frac{\partial}{\partial t}\left[\frac{1}{c^2}\frac{\partial \vec{E}}{\partial t}\right] \qquad \frac{1}{c^2}\frac{\partial^2 \vec{E}}{\partial t^2} - \nabla^2 \vec{E} = 0 \tag{10.15}$$

It gives that the electric field satisfies the wave equation. Now, taking curl of equation 10.14 (iv), we have

$$\nabla \times (\nabla \times \vec{B}) = \nabla \times \left[\frac{1}{c^2}\frac{\partial \vec{E}}{\partial t}\right] \qquad \nabla(\nabla.\vec{B}) - \nabla^2 \vec{B} = \frac{1}{c^2}\frac{\partial}{\partial t}(\nabla \times \vec{E})$$

Using equations. (10.14) (ii) and (iii), we get

$$-\nabla^2 \vec{B} = -\frac{\partial}{\partial t}\left[\frac{1}{c^2}\frac{\partial \vec{B}}{\partial t}\right] \qquad \frac{1}{c^2}\frac{\partial^2 \vec{B}}{\partial t^2} - \nabla^2 \vec{B} = 0 \tag{10.16}$$

It gives that the magnetic field satisfies the wave equation. This way, the electric and magnetic fields are decoupled from each other. However, now we have second order differential equations. So Maxwell's equations imply that empty space supports the propagation of electromagnetic waves, traveling at a speed

$$c = \frac{1}{\sqrt{\epsilon_0 \mu_0}}$$

It says that light is an electromagnetic wave.

2.1 Monochromatic plane waves

Let us consider sinusoidal waves of frequency ω. Since different frequencies in the visible range correspond to different colours, such waves are called monochromatic. Suppose, the waves are traveling in z direction and have no x or y dependence; these are called plane waves, because the fields are uniform over every plane perpendicular to the direction of propagation (Figure 8). We are interested, then, in fields of the form

$$\vec{E}(z,t) = \vec{E}_0 \cos(kz - \omega t) \qquad \text{and} \qquad \vec{B}(z,t) = \vec{B}_0 \cos(kz - \omega t)$$

where $\vec{E}_0$ and $\vec{B}_0$ are the amplitudes, and $\omega = ck$. Expressing the amplitudes in terms of components, we have

$$\vec{E}(z,t) = [E_{0x}\,\hat{x} + E_{0y}\,\hat{y} + E_{0z}\,\hat{z}]\cos(kz - \omega t) \tag{10.17}$$

and

$$\vec{B}(z,t) = [B_{0x}\,\hat{x} + B_{0y}\,\hat{y} + B_{0z}\,\hat{z}]\cos(kz - \omega t) \tag{10.18}$$

Now, the wave equations for $\vec{E}$ and $\vec{B}$ (equations. 10.15 and 10.16) have been derived from Maxwell's equations. Maxwell's equations impose extra constraints on $\vec{E}_0$ and $\vec{B}_0$. Using equation (10.17) in $\nabla \,.\, \vec{E} = 0$, we have

$$\left(\frac{\partial}{\partial x}\,\hat{x} + \frac{\partial}{\partial y}\,\hat{y} + \frac{\partial}{\partial z}\,\hat{z}\right) . \,[E_{0x}\,\hat{x} + E_{0y}\,\hat{y} + E_{0z}\,\hat{z}]\cos(kz - \omega t) = 0$$

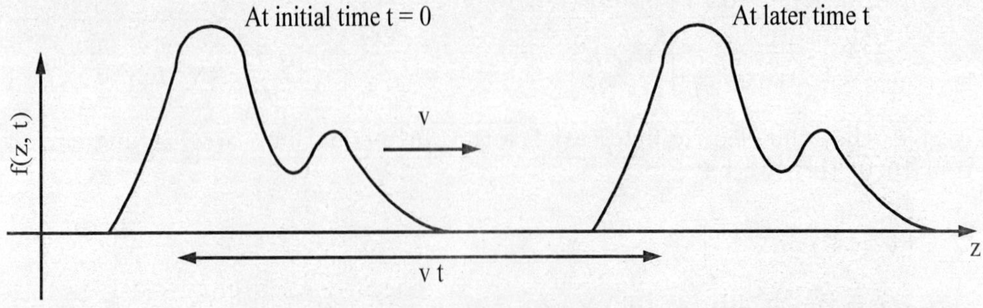

Figure 8: Plane wave propagating in the z direction.

$$-E_{0z}\ \sin(kz-\omega t)\ k=0$$

giving

$$E_{0z}=0$$

Thus, the electric field does not have component along z-axis.

Similarly, using equation (10.18) in $\nabla\ .\ \vec{B}=0$, we have

$$B_{0z}=0$$

Hence, the magnetic field does not have component along z-axis. Therefore, we have

$$\vec{E}(z,t)=[E_{0x}\ \hat{x}+E_{0y}\ \hat{y}]\cos(kz-\omega t) \tag{10.19}$$

and

$$\vec{B}(z,t)=[B_{0x}\ \hat{x}+B_{0y}\ \hat{y}]\cos(kz-\omega t) \tag{10.20}$$

This implies that the electromagnetic waves are transverse, as the electric and magnetic fields lie in the xy-plane, and are perpendicular to the direction of propagation of wave. The Faraday's law,

$$\nabla\times\vec{E}=-\frac{\partial\vec{B}}{\partial t}$$

can be written as

$$\frac{\partial E_z}{\partial y}-\frac{\partial E_y}{\partial z}=-\frac{\partial B_x}{\partial t} \tag{10.21}$$

$$\frac{\partial E_x}{\partial z}-\frac{\partial E_z}{\partial x}=-\frac{\partial B_y}{\partial t} \tag{10.22}$$

$$\frac{\partial E_y}{\partial x}-\frac{\partial E_x}{\partial y}=-\frac{\partial B_z}{\partial t} \tag{10.23}$$

Using equation (10.19) and (10.20) in (10.21), we get

$$E_{0y} \sin(kz - \omega t)\ k = B_{0x} \sin(kz - \omega t)\ (-\omega) \qquad \text{or} \qquad -kE_{0y} = \omega B_{0x}$$

Using equation (10.19) and (10.20) in (10.22), we get

$$-E_{0x} \sin(kz - \omega t)\ k = B_{0y} \sin(kz - \omega t)\ (-\omega) \qquad \text{or} \qquad kE_{0x} = \omega B_{0y}$$

These two relations can be combined as

$$k(\hat{z} \times \vec{E}_0) = \omega\ \vec{B}_0 \qquad \text{or} \qquad \vec{B}_0 = \frac{k}{\omega}\ (\hat{z} \times \vec{E}_0) \tag{10.24}$$

Evidently, $\vec{E}$ and $\vec{B}$ are in phase and mutually perpendicular to each other; their amplitudes are related by

$$|\vec{B}_0| = \frac{k}{\omega}\ |\vec{E}_0| = \frac{1}{c}\ |\vec{E}_0| \tag{10.25}$$

It shows that the amplitude of electric field is c times that of the magnetic field. The Maxwell's equation, $\nabla \times \vec{B} = (\partial \vec{E}/\partial t)/c^2$, does not yield an independent condition; it simply reproduces equation (10.24).

Example 1: Derive equation (10.24) using Maxwell's equation,

$$\nabla \times \vec{B} = \frac{1}{c^2} \frac{\partial \vec{E}}{\partial t}$$

If the electric field $\vec{E}$ points in the x direction, then the magnetic field $\vec{B}$ points in the y direction:

$$\vec{E}(z,t) = \vec{E}_0 \cos(kz - \omega t + \delta)\ \hat{x}, \qquad \vec{B}(z,t) = \vec{B}_0 \cos(kz - \omega t + \delta)\ \hat{y}$$

There is nothing special about the z direction, the above discussion can be easily generalized for a monochromatic plane waves traveling in an arbitrary direction. The notation is facilitated by the introduction of the propagation (or wave) vector, $\vec{k}$, pointing in the direction of propagation, having magnitude k. The scalar product $\vec{k}\ .\ \vec{r}$ is the appropriate generalization of kz, so

$$\left.\begin{aligned} \vec{E}(\vec{r},t) &= \vec{E}_0\ \cos(\vec{k}\ .\ \vec{r} - \omega t)\ \hat{n} \\ \vec{B}(\vec{r},t) &= \tfrac{1}{c}\ \vec{E}_0\ \cos(\vec{k}\ .\ \vec{r} - \omega t)\ (\hat{k} \times \hat{n}) = \tfrac{1}{c}\ (\hat{k} \times \vec{E}) \end{aligned}\right\}$$

where $\hat{n}$ is the polarization vector. Because $\vec{E}$ is transverse,

$$\hat{n}\ .\ \hat{k} = 0$$

2.2 Energy and momentum in electromagnetic waves

The energy per unit volume in electromagnetic fields is

$$u = \frac{1}{2}\left(\epsilon_0\ E^2 + \frac{1}{\mu_0}\ B^2\right)$$

In the case of a monochromatic plane wave (equation 10.25)

$$B^2 = \frac{1}{c^2}\ E^2 = \epsilon_0 \mu_0\ E^2 \qquad \text{so that} \qquad \epsilon_0\ E^2 = \frac{1}{\mu_0}\ B^2$$

It shows that the **contributions of electric and magnetic fields are equal.** Thus, we have

$$u = \epsilon_0\ E^2 = \epsilon_0\ E_0^2 \cos^2(kz - \omega t)$$

This is the amount of energy carried by a plane wave during its propagation. The energy flux density (energy per unit area, per unit time) transported by the fields is given by the Poynting vector:

$$\vec{S} = \frac{1}{\mu_0}\ (\vec{E} \times \vec{B})$$

For monochromatic plane waves propagating in the z direction,

$$\vec{S} = \frac{1}{\mu_0}\left[E_0 \cos(kz - \omega t)\ \hat{x} \times \frac{E_0}{c}\ \cos(kz - \omega t)\ \hat{y}\right]$$

$$= c\ \epsilon_0\ E_0^2 \cos^2(kz - \omega t)\ \hat{z}$$

Thus, $\vec{S}$ is the energy density (u) times the velocity of waves ($c\hat{k}$). Electromagnetic fields not only carry energy, they also carry momentum. In fact, the momentum density stored in the fields is

$$\vec{g} = \frac{1}{c^2}\ \vec{S}$$

For monochromatic plane waves, then,

$$\vec{g} = \frac{1}{c}\ \epsilon_0\ E_0^2 \cos^2(kz - \omega t)\ \hat{z} = \frac{1}{c}\ u\ \hat{z}$$

In the case of light, the wavelength is so short($\sim 5 \times 10^{-7}$ m), and the period so brief ($\sim 10^{-15}$ s), that any macroscopic measurement encompasses many cycles. Typically, therefore, we are not interested in the fluctuating cosine - squared term in the energy and momentum densities; all we want is the average value. Now, the average of cosine-squared over a complete cycle is 1/2,[1] so

$$\langle u \rangle = \frac{1}{2}\ \epsilon_0\ E_0^2 \qquad \langle \vec{S} \rangle = \frac{1}{2}\ c\ \epsilon_0\ E_0^2\ \hat{z} \qquad \langle \vec{g} \rangle = \frac{1}{2c}\ \epsilon_0\ E_0^2\ \hat{z}$$

---

[1] $\left\langle \cos^2\theta \right\rangle = \frac{1}{2\pi} \int_0^{2\pi} \cos^2\theta\ d\theta$

The average power per unit area transported by an electromagnetic wave is called the **intensity**:

$$I \equiv \langle \vec{S} \rangle = \frac{1}{2}\, c\, \epsilon_0\, E_0^2$$

When light falls (at normal incidence) on a perfect absorber, it delivers its momentum to the surface. In a time Δt, the momentum transfer is $\Delta p = \langle \vec{g} \rangle\, A\, c\, \Delta t$, so the **radiation pressure** (average force per unit area) is

$$P = \frac{1}{A}\frac{\Delta p}{\Delta t} = \frac{1}{2}\, \epsilon_0\, E_0^2 = \frac{I}{c}$$

On a perfect reflector the pressure is twice as great, because the momentum switches direction, instead of simply being absorbed.

$$P = \begin{cases} I/c & \text{For perfect absorber} \\ 2I/c & \text{For perfect reflector} \end{cases}$$

2.3 Boundary conditions

In general, the fields $\vec{E}$, $\vec{B}$, $\vec{D}$, and $\vec{H}$ will be discontinuous at a boundary between two different media, or at a surface that carries a charge density σ or a current density K. In the case of linear media, the boundary conditions are:

$$\left.\begin{array}{ll} \text{(i) } \epsilon_1 E_1^{\perp} - \epsilon_2 E_2^{\perp} = \sigma_f & \text{(ii) } E_1^{\parallel} - E_2^{\parallel} = 0 \\ \text{(iii) } B_1^{\perp} - B_2^{\perp} = 0 & \text{(iv) } \frac{1}{\mu_1}\, B_1^{\parallel} - \frac{1}{\mu_2}\, B_2^{\parallel} = \vec{K}_f \times \hat{n} \end{array}\right\}$$

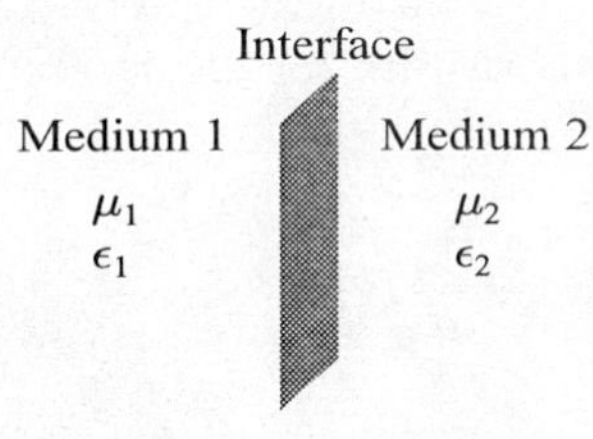

Figure 9: Interface between two mediums 1 and 2.

where $\hat{n}$ is a unit vector perpendicular to the interface (pointing from 2 toward 1), and σ_f and $\vec{K}_f$ are surface free-charge and free-current densities. The prefixes parallel ($\parallel$) and perpendicular ($\perp$) denotes the parallel and perpendicular to the interface between the two media. In particular, if there is no free charge or free current at the interface, then the boundary conditions are:

$$\left.\begin{array}{ll} \text{(i) } \epsilon_1 E_1^{\perp} - \epsilon_2 E_2^{\perp} = 0 & \text{(ii) } E_1^{\parallel} - E_2^{\parallel} = 0 \\ \text{(iii) } B_1^{\perp} - B_2^{\perp} = 0 & \text{(iv) } \frac{1}{\mu_1}\, B_1^{\parallel} - \frac{1}{\mu_2}\, B_2^{\parallel} = 0 \end{array}\right\}$$

As we can find that these equations are the basis for the theory of reflection and refraction.

3. Electromagnetic waves in linear medium

Inside a matter, but in the regions where there is no free charge or no free current, and Maxwell's equations are

$$\left.\begin{array}{ll}\text{(i) } \nabla \cdot \vec{E} = 0 & \text{(ii) } \nabla \times \vec{E} = -\frac{\partial \vec{B}}{\partial t} \\ \text{(iii) } \nabla \cdot \vec{B} = 0 & \text{(iv) } \nabla \times \vec{B} = \mu\epsilon \frac{\partial \vec{E}}{\partial t}\end{array}\right\}$$

Here, the medium is linear and the permeability μ and permittivity ϵ of the medium do not vary from point to point. These equations differ from those for vacuum only in the replacement of μ_0 by μ and ϵ_0 by ϵ. Evidently electromagnetic waves propagate through a linear homogeneous medium at a speed

$$v = \frac{1}{\sqrt{\mu\epsilon}} = \frac{c}{n} \tag{10.26}$$

where

$$n = \frac{c}{v} = c\sqrt{\mu\,\epsilon} = \sqrt{\frac{\mu\,\epsilon}{\mu_0\epsilon_0}}$$

is the index of refraction of the medium (substance). For most materials, μ is very close to μ_0 (that is, the permeability does not depend much on the medium), so

$$n \approx \sqrt{\epsilon_r}$$

where $\epsilon_r = \epsilon/\epsilon_0$ is the relative permittivity (dielectric constant) of medium. Since ϵ_r is almost always greater than 1, light travels more slowly through matter - a well known fact from optics. All the previous results for electromagnetic waves through vacuum are valid here with the simple transcription $\epsilon_0 \to \epsilon$, $\mu_0 \to \mu$, and $c \to v$. The energy density is

$$u = \frac{1}{2}\left(\epsilon E^2 + \frac{1}{\mu}\,B^2\right)$$

and the Poynting vector is

$$\vec{S} = \frac{1}{\mu}\,(\vec{E} \times \vec{B})$$

For monochromatic plane waves, the frequency and wave number are related by $\omega = kv$, the amplitude of $\vec{B}$ is $1/v$ times the amplitude of $\vec{E}$:

$$|\vec{B}_0| = \frac{1}{v}\,|\vec{E}_0|$$

The intensity is

$$I = \frac{1}{2}\,\epsilon\, v\, E_0^2 \tag{10.27}$$

As we can have different mediums, so we would like to see what happens when a wave propagates from one transparent medium into another, for example, from air to water, or from glass to plastic. In the case of linear media, the boundary conditions are:

$$\left.\begin{array}{ll}\text{(i) } \epsilon_1 E_1^{\perp} - \epsilon_2 E_2^{\perp} = \sigma_f & \text{(ii) } E_1^{\|} - E_2^{\|} = 0 \\ \text{(iii) } B_1^{\perp} - B_2^{\perp} = 0 & \text{(iv) } \frac{1}{\mu_1} B_1^{\|} - \frac{1}{\mu_2} B_2^{\|} = \vec{K}_f \times \hat{n}\end{array}\right\}$$

where $\hat{n}$ is a unit vector perpendicular to the interface (pointing from 2 toward 1), and σ_f and $\vec{K}_f$ are surface free-charge and free-current densities. We expect to get a reflected wave and a transmitted wave. The details depend on the exact nature of the electrodynamic boundary conditions, which are as the following.

$$\left.\begin{array}{ll}\text{(i) } \epsilon_1 E_1^{\perp} = \epsilon_2 E_2^{\perp} & \text{(ii) } E_1^{\|} = E_2^{\|} \\ \text{(iii) } B_1^{\perp} = B_2^{\perp} & \text{(iv) } \frac{1}{\mu_1} B_1^{\|} = \frac{1}{\mu_2} B_2^{\|}\end{array}\right\} \qquad (10.28)$$

These equations relate the electric and magnetic fields just to the left and just to the right of the interface between two linear media.

3.1 Reflection and transmission at normal incidence

Suppose the xy plane forms the boundary between two linear mediums. A plane wave of frequency ω, traveling in the z direction and polarized in the x direction, approaches the interface from the left, as shown in Figure 10.

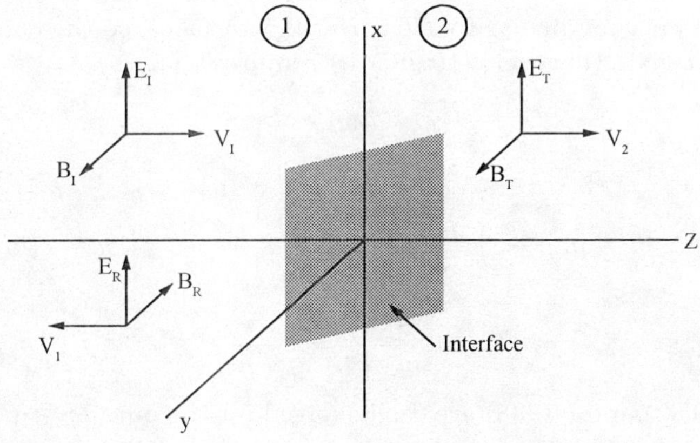

Figure 10: xy plane is boundary between two linear mediums.

The plane polarized incident wave is

$$\left.\begin{array}{l}\vec{E}_I(z,t) = \vec{E}_{0I} \cos(k_1 z - \omega t)\ \hat{x} \\ \vec{B}_I(z,t) = \frac{1}{v_1} \vec{E}_{0I} \cos(k_1 z - \omega t)\ \hat{y}\end{array}\right\} \qquad (10.29)$$

It gives rise to a reflected wave

$$\left.\begin{aligned} \vec{E}_R(z,t) &= \vec{E}_{0R}\ \cos(-k_1 z - \omega t)\ \hat{x} \\ \vec{B}_R(z,t) &= -\tfrac{1}{v_1}\ \vec{E}_{0R}\ \cos(-k_1 z - \omega t)\ \hat{y} \end{aligned}\right\} \tag{10.30}$$

which travels back to the left in medium (1) with velocity v_1, and a transmitted wave

$$\left.\begin{aligned} \vec{E}_T(z,t) &= \vec{E}_{0T}\ \cos(k_2 z - \omega t)\ \hat{x} \\ \vec{B}_T(z,t) &= \tfrac{1}{v_2}\ \vec{E}_{0T}\ \cos(k_2 z - \omega t)\ \hat{y} \end{aligned}\right\} \tag{10.31}$$

which continues on to the right in medium (2). Note the minus sign in $\vec{B}_R$, if we prefer, by the fact that the Poynting vector aims in the direction of propagation. For a sinusoidal incident wave, the net disturbance for the electric field is:

$$\begin{cases} \vec{E}_{0I}\ \cos(k_1 z - \omega t)\ \hat{x} + \vec{E}_{0R}\ \cos(-k_1 z - \omega t)\ \hat{x}, & (z < 0) \\ \vec{E}_{0T}\ \cos(k_2 z - \omega t)\ \hat{x}, & (z > 0) \end{cases}$$

and the magnetic field is:

$$\begin{cases} \tfrac{1}{v_1}\ \vec{E}_{0I}\ \cos(k_1 z - \omega t)\ \hat{y} - \tfrac{1}{v_1}\ \vec{E}_{0R}\ \cos(-k_1 z - \omega t)\ \hat{y}, & (z < 0) \\ \tfrac{1}{v_2}\ \vec{E}_{0T}\ \cos(k_2 z - \omega t)\ \hat{y}, & (z > 0) \end{cases}$$

At $z = 0$, the combined fields on the left must join the fields on the right, in accordance with the boundary conditions (equation 10.28). In this case there are no components perpendicular to the surface (interface), so the conditions (10.28) (i) and (iii) are trivial. However, (10.28) (ii) requires that

$$\vec{E}_{0I} + \vec{E}_{0R} = \vec{E}_{0T} \tag{10.32}$$

while (10.28) (iv) says

$$\frac{1}{\mu_1}\Big(\frac{1}{v_1}\ \vec{E}_{0I} - \frac{1}{v_1}\ \vec{E}_{0R}\Big) = \frac{1}{\mu_2}\Big(\frac{1}{v_2}\ \vec{E}_{0T}\Big) \qquad \text{or} \qquad \vec{E}_{0I} - \vec{E}_{0R} = \beta \vec{E}_{0T} \tag{10.33}$$

where

$$\beta = \frac{\mu_1\ v_1}{\mu_2\ v_2} = \frac{\mu_1\ n_2}{\mu_2\ n_1} \tag{10.34}$$

Equations (10.32) and (10.33) are easily solved for the outgoing amplitudes, in terms of the incident amplitude:

$$\vec{E}_{0R} = \Big(\frac{1-\beta}{1+\beta}\Big)\ \vec{E}_{0I} \qquad \text{and} \qquad \vec{E}_{0T} = \Big(\frac{2}{1+\beta}\Big)\ \vec{E}_{0I} \tag{10.35}$$

Indeed, if the permeabilities μ, are close to their values in vacuum (as, remember, they are for most media), then $\beta = v_1/v_2$, and we have

$$\vec{E}_{0R} = \Big(\frac{v_2 - v_1}{v_2 + v_1}\Big)\ \vec{E}_{0I} \qquad \text{and} \qquad \vec{E}_{0T} = \Big(\frac{2v_2}{v_2 + v_1}\Big)\ \vec{E}_{0I}$$

The reflected wave is in phase if $v_2 > v_1$ and out of phase if $v_2 < v_1$

$$\vec{E}_{0R} = -\left(\frac{v_1 - v_2}{v_2 + v_1}\right) \vec{E}_{0I} \qquad \text{and} \qquad \vec{E}_{0T} = \left(\frac{2v_2}{v_2 + v_1}\right) \vec{E}_{0I}$$

The negative sign is equivalent to a phase difference of 180°. The real amplitudes are related by

$$\vec{E}_{0R} = \left|\frac{v_2 - v_1}{v_2 + v_1}\right| \vec{E}_{0I} \qquad \text{and} \qquad \vec{E}_{0T} = \left|\frac{2v_2}{v_2 + v_1}\right| \vec{E}_{0I}$$

or, in terms of the indices of refraction,

$$\vec{E}_{0R} = \left|\frac{n_1 - n_2}{n_1 + n_2}\right| \vec{E}_{0I} \qquad \text{and} \qquad \vec{E}_{0T} = \left|\frac{2n_1}{n_1 + n_2}\right| \vec{E}_{0I}$$

Let us calculate reflected and transmitted fraction of the incident energy. According to equation (10.27), the intensity (average power per unit area) is

$$I = \frac{1}{2}\,\epsilon v E_0^2$$

If $\mu_1 = \mu_2 = \mu_0$, then the ratio of the reflected intensity to the incident intensity is

$$R \equiv \frac{I_R}{I_I} = \left(\frac{E_{0R}}{E_{0I}}\right)^2 = \left(\frac{n_1 - n_2}{n_1 + n_2}\right)^2 \tag{10.36}$$

whereas the ratio of the transmitted intensity to the incident intensity is

$$T \equiv \frac{I_T}{I_I} = \frac{\epsilon_2 v_2}{\epsilon_1 v_1}\left(\frac{E_{0T}}{E_{0I}}\right)^2 = \frac{4n_1 n_2}{(n_1 + n_2)^2} \tag{10.37}$$

R is called the reflection coefficient and T the transmission coefficient; they measure the fraction of the incident energy that is reflected and transmitted, respectively. Notice that

$$R + T = 1 \tag{10.38}$$

as conservation of energy, of course, requires. For instance, when light passes from air ($n_1 = 1$) into glass ($n_2 = 1.5$), R = 0.05 and T = 0.95. No surprise: most of the light is transmitted.

3.2 Reflection and transmission at oblique incidence

We have discussed about the reflection and transmission at normal incidence. Now, we take the case that incoming wave hits the interface at an arbitrary angle θ_I as shown in Figure 11. Though, the normal incidence is just a special case of oblique incidence, with $\theta_I = 0$, but due to its importance, it is treated separately. Suppose, a monochromatic plane wave

$$\vec{E}_I(\vec{r}, t) = \vec{E}_{0I}\,\cos(\vec{k}_I\,.\,\vec{r} - \omega t) \qquad \text{and} \qquad \vec{B}_I(\vec{r}, t) = \frac{1}{v_1}\,(\hat{k}_I \times \vec{E}_I)$$

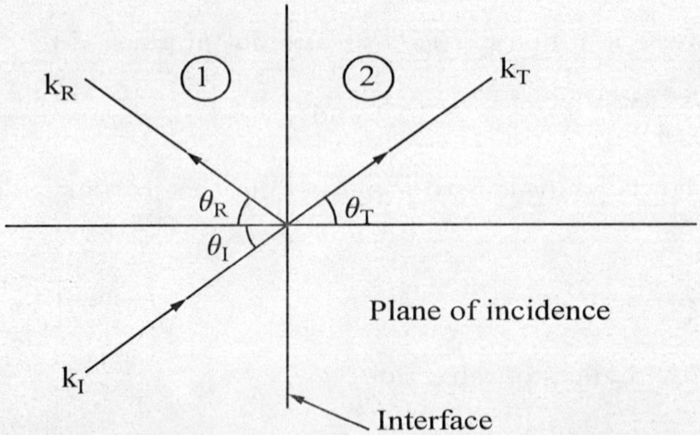

Figure 11: Oblique incidence at the interface.

approaches from the left, giving rise to a reflected wave,

$$\vec{E}_R(\vec{r},t) = \vec{E}_{0R}\ \cos(\vec{k}_R\ .\ \vec{r} - \omega t) \qquad \text{and} \qquad \vec{B}_R(\vec{r},t) = \frac{1}{v_1}\ (\hat{k}_R \times \vec{E}_R)$$

and a transmitted wave

$$\vec{E}_T(\vec{r},t) = \vec{E}_{0T}\ \cos(\vec{k}_T\ .\ \vec{r} - \omega t) \qquad \text{and} \qquad \vec{B}_T(\vec{r},t) = \frac{1}{v_2}\ (\hat{k}_T \times \vec{E}_T)$$

All three waves have the same frequency ω. Therefore, the three wave numbers are related as:

$$k_I v_1 = k_R v_1 = k_T v_2 = \omega \qquad \text{or} \qquad k_I = k_R = \frac{v_2}{v_1}\ k_T = \frac{n_1}{n_2}\ k_T \tag{10.39}$$

For a sinusoidal incident wave, the net disturbance for electric field is:

$$\begin{cases} \vec{E}_{0I}\ \cos(\vec{k}_I\ .\ \vec{r} - \omega t) + \vec{E}_{0R}\ \cos(\vec{k}_R\ .\ \vec{r} - \omega t), & (z < 0) \\ \vec{E}_{0T}\ \cos(\vec{k}_T\ .\ \vec{r} - \omega t), & (z > 0) \end{cases}$$

and the net disturbance for magnetic field is:

$$\begin{cases} \frac{1}{v_1}\left[\hat{k}_I \times \vec{E}_{0I}\ \cos(\vec{k}_I\ .\ \vec{r} - \omega t)\right] + \frac{1}{v_1}\left[\hat{k}_R \times \vec{E}_{0R}\ \cos(\vec{k}_R\ .\ \vec{r} - \omega t)\right], & (z < 0) \\ \frac{1}{v_2}\left[\hat{k}_T \times \vec{E}_{0T}\ \cos(\vec{k}_T\ .\ \vec{r} - \omega t)\right], & (z > 0) \end{cases}$$

The combined fields in medium (1) must now be joined to the fields in medium (2), using the boundary conditions:

$$\left.\begin{array}{ll} \text{(i)}\ \epsilon_1 E_1^{\perp} = \epsilon_2 E_2^{\perp} & (ii)\ E_1^{\parallel} = E_2^{\parallel} \\ \text{(iii)}\ B_1^{\perp} = B_2^{\perp} & (iv)\ \frac{1}{\mu_1}\ B_1^{\parallel} = \frac{1}{\mu_2}\ B_2^{\parallel} \end{array}\right\} \tag{10.40}$$

Because the boundary conditions must hold at all points on the plane, and for all times, these cosine factors must be equal (when $z = 0$). Otherwise, a slight change in x, say, would destroy the equality. Of course, the time factors are already equal, as the incidence, reflection and transmission are occurring simultaneously. As for the spatial terms, evidently

$$\vec{k}_I \cdot \vec{r} = \vec{k}_R \cdot \vec{r} = \vec{k}_T \cdot \vec{r} \qquad \text{when } z = 0 \tag{10.41}$$

or, more explicitly,

$$x\, k_{Ix} + y\, k_{Iy} = x\, k_{Rx} + y\, k_{Ry} = x\, k_{Tx} + y\, k_{Ty} \tag{10.42}$$

for all x and all y. But, equation (10.42) can only hold if the components are separately equal. Putting $x = 0$, we get

$$k_{Iy} = k_{Ry} = k_{Ty} \tag{10.43}$$

Putting $y = 0$, we get

$$k_{Ix} = k_{Rx} = k_{Tx} \tag{10.44}$$

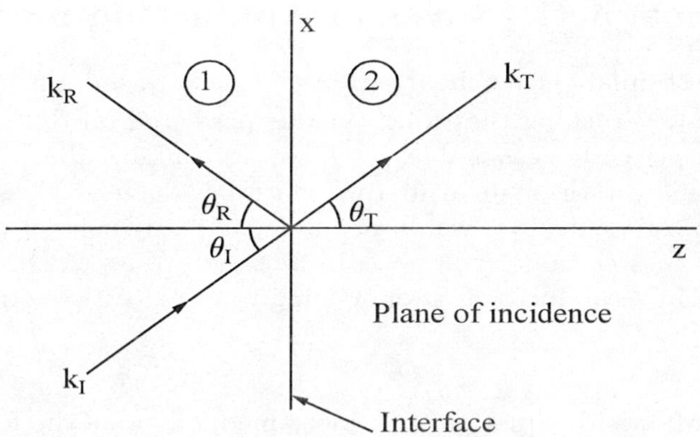

Figure 12: Oblique incidence.

We may as well orient our axes so that $\vec{k}_I$ lies in the xz plane (i.e. $k_{Iy} = 0$); according to equation (10.43), we have $k_{Ry} = 0$ (*i.e.*, $\vec{k}_R$ lies in the xz plane) and $k_{Ty} = 0$ (*i.e.*, $\vec{k}_T$ lies in the xz plane). Conclusion:

First Law: The incident, reflected, and transmitted wave vectors form a plane (called the **plane of incidence**), which also includes the normal to the surface (here, the z axis).

Meanwhile, equation (10.44) implies that

$$k_I \sin\theta_I = k_R \sin\theta_R = k_T \sin\theta_T \tag{10.45}$$

where θ_I is the angle of incidence, θ_R is the angle of reflection, and θ_T is the angle of transmission (more commonly known as the angle of refraction), all of them measured with respect to the normal (Figure 12).

Second Law: In view of equation (10.39), we have $k_I = k_R$, and therefore,

$$\sin\theta_I = \sin\theta_R \qquad \theta_I = \theta_R \tag{10.46}$$

This is the **law of reflection**, stating that the angle of incidence is equal to the angle of reflection.

Third Law: In view of equation (10.39), we have $k_I = k_T(n_1/n_2)$, and therefore,

$$\frac{\sin\theta_I}{\sin\theta_T} = \frac{k_T}{k_I} = \frac{n_2}{n_1} \tag{10.47}$$

This is the **law of refraction - Snell's law**.

These are the three fundamental laws of geometrical optics. Any other waves (water waves, for instance, or sound waves) can be expected to obey the same optical laws when they pass from one medium into another.

4. Electromagnetic waves in conducting medium

Earlier we have stipulated that the free charge density ρ_f and the free current density $\vec{J}_f$ are zero, and everything that followed was predicted on that assumption. Such a restriction is perfectly reasonable when we are talking about wave propagation through a vacuum or through insulating materials such as glass or (pure) water. But, in the case of conductors we do not independently control the flow of charge, and in general $\vec{J}_f$ is certainly not zero. In fact, according to Ohm's law, the (free) current density in a conductor is proportional to the electric field:

$$\vec{J}_f = \sigma\,\vec{E} \tag{10.48}$$

With this, Maxwell's equations for linear media assume the form

$$\left.\begin{array}{ll} \text{(i)}\ \nabla\cdot\vec{E} = \rho_f/\epsilon & \text{(ii)}\ \nabla\times\vec{E} = -\frac{\partial\vec{B}}{\partial t} \\ \text{(iii)}\ \nabla\cdot\vec{B} = 0 & \text{(iv)}\ \nabla\times\vec{B} = \mu\sigma\vec{E} + \mu\epsilon\,\frac{\partial\vec{E}}{\partial t} \end{array}\right\} \tag{10.49}$$

Now, the continuity equation for free charge,

$$\nabla\cdot\vec{J}_f = -\frac{\partial\rho_f}{\partial t} \tag{10.50}$$

Using equation (10.48) in (10.50), we have

$$\frac{\partial\rho_f}{\partial t} = -\nabla\cdot\vec{J}_f = -\sigma(\nabla\cdot\vec{E}) \tag{10.51}$$

Using equation (10.49) (i) in (10.51), we have

$$\frac{\partial\rho_f}{\partial t} = -\frac{\sigma}{\epsilon}\,\rho_f$$

For a homogeneous linear medium, it follows that

$$\rho_f(t) = \mathrm{e}^{-(\sigma/\epsilon)t}\ \rho_f(0) \tag{10.52}$$

Thus, any initial free charge $\rho_f(0)$ dissipates in a characteristic time $\tau \equiv \epsilon/\sigma$. This reflects the familiar fact that if we put some free charge on a conductor, it flows out to the edges. The time constant τ affords a measure of how good a conductor is. For a perfect conductor, $\sigma = \infty$ and $\tau = 0$; for a good conductor, τ is much less than the other relevant times in the problem. We wait for any accumulated free charge to disappear. From then on $\rho_f = 0$, and Maxwell's equations are

$$\left.\begin{array}{ll}\text{(i) } \nabla\,.\,\vec{E} = 0 & \text{(ii) } \nabla\times\vec{E} = -\frac{\partial\vec{B}}{\partial t} \\ \text{(iii) } \nabla\,.\,\vec{B} = 0 & \text{(iv) } \nabla\times\vec{B} = \mu\epsilon\,\frac{\partial\vec{E}}{\partial t} + \mu\sigma\vec{E}\end{array}\right\} \tag{10.53}$$

These differ from the corresponding equations for nonconducting media only in the last term in (iv) - which is absent, obviously, when $\sigma = 0$. Apply the curl to equation (10.53) (ii), to have

$$\nabla\times(\nabla\times\vec{E}) = -\nabla\times\left[\frac{\partial\vec{B}}{\partial t}\right] \qquad \nabla(\nabla\,.\,\vec{E}) - \nabla^2\vec{E} = -\frac{\partial}{\partial t}(\nabla\times\vec{B})$$

Using equations 10.53 (i) and (iv), we get

$$-\nabla^2\vec{E} = -\frac{\partial}{\partial t}\left[\mu\epsilon\,\frac{\partial\vec{E}}{\partial t}\right] - \mu\sigma\frac{\partial\vec{E}}{\partial t} \qquad \nabla^2\vec{E} = \mu\epsilon\,\frac{\partial^2\vec{E}}{\partial t^2} + \mu\sigma\,\frac{\partial\vec{E}}{\partial t} \tag{10.54}$$

Apply the curl to equation (10.53) (iv), to have

$$\nabla\times(\nabla\times\vec{B}) = \nabla\times\left[\mu\epsilon\,\frac{\partial\vec{E}}{\partial t}\right] + \mu\sigma\nabla\times\vec{E}$$

$$\nabla(\nabla.\vec{B}) - \nabla^2\vec{B} = \mu\epsilon\,\frac{\partial}{\partial t}(\nabla\times\vec{E}) + \mu\sigma\nabla\times\vec{E}$$

Using equations (10.53) (ii) and (iii), we get

$$-\nabla^2\vec{B} = -\mu\epsilon\,\frac{\partial^2\vec{B}}{\partial t^2} - \mu\sigma\,\frac{\partial\vec{B}}{\partial t} \qquad \nabla^2\vec{B} = \mu\epsilon\,\frac{\partial^2\vec{B}}{\partial t^2} + \mu\sigma\,\frac{\partial\vec{B}}{\partial t} \tag{10.55}$$

These equations (10.54) and (10.55) still admit plane-wave solutions,

$$\vec{E}(z,t) = \vec{E}_0\ \mathrm{e}^{i(k^*z-\omega t)}, \qquad \vec{B}(z,t) = \vec{B}_0\ \mathrm{e}^{i(k^*z-\omega t)} \tag{10.56}$$

but this time the wave number k^* is complex:

$$(k^*)^2 = \mu\epsilon\,\omega^2 + i\mu\sigma\,\omega \tag{10.57}$$

as one can easily check by plugging equation (10.56) into equations (10.54) and (10.55).

Example 2: Using equation (10.56) in equations (10.54) and (10.55), verify equation (10.57).

Taking $k^* = k' + ik''$, where $i = \sqrt{-1}$ and

$$k' \equiv \omega\sqrt{\frac{\epsilon\mu}{2}}\left[\sqrt{1+\left(\frac{\sigma}{\epsilon\omega}\right)^2}+1\right]^{1/2} \qquad k'' \equiv \omega\sqrt{\frac{\epsilon\mu}{2}}\left[\sqrt{1+\left(\frac{\sigma}{\epsilon\omega}\right)^2}-1\right]^{1/2} \tag{10.58}$$

Example 3: Using equations (10.58), verify $k^{*2} = k'^2 - k''^2 + 2ik'k''$.

The imaginary part of k^* results in an attenuation of the wave (decreasing amplitude with increasing z):

$$\vec{E}(z,t) = \vec{E}_0 \ \mathrm{e}^{-k''z} \ \mathrm{e}^{i(k'z-\omega t)}, \tag{10.59}$$

$$\vec{B}(z,t) = \vec{B}_0 \ \mathrm{e}^{-k''z} \ \mathrm{e}^{i(k'z-\omega t)} \tag{10.60}$$

or

$$\vec{E}(z,t) = [E_{0x} \ \hat{x} + E_{0y} \ \hat{y} + E_{0z} \ \hat{z}] \ \mathrm{e}^{-k''z} \ \mathrm{e}^{i(k'z-\omega t)} \tag{10.61}$$

$$\vec{B}(z,t) = [B_{0x} \ \hat{x} + B_{0y} \ \hat{y} + B_{0z} \ \hat{z}] \ \mathrm{e}^{-k''z} \ \mathrm{e}^{i(k'z-\omega t)} \tag{10.62}$$

The distance it takes to reduce the amplitude by a factor of $1/e$ (about a third) is called the **skin depth**:

$$d \equiv \frac{1}{k''}$$

It is a measure of how far the wave penetrates into the conductor. Meanwhile, the real part of k^* determines the wavelength, the propagation speed, and the index of refraction, in the usual way:

$$\lambda = \frac{2\pi}{k'}, \qquad v = \frac{\omega}{k'}, \qquad n = \frac{ck'}{\omega} \tag{10.63}$$

The attenuated plane waves (equations 10.59 and 10.61) satisfy the modified wave equations (10.54) and (10.55) for any $\vec{E}_0$ and $\vec{B}_0$. But, Maxwell's equations (10.53) impose further constraints, which serve to determine the relative amplitudes, phases, and polarizations of $\vec{E}$ and $\vec{B}$. Equation (10.53) (i), $\nabla \, . \, \vec{E} = 0$, gives

$$\left(\frac{\partial}{\partial x} \ \hat{x} + \frac{\partial}{\partial y} \ \hat{y} + \frac{\partial}{\partial z} \ \hat{z}\right) . \ [E_{0x} \ \hat{x} + E_{0y} \ \hat{y} + E_{0z} \ \hat{z}] \ \mathrm{e}^{i(k^*z-\omega t)} = 0$$

$$E_{0z} \ ik^* \ \mathrm{e}^{i(k^*z-\omega t)} = 0 \qquad \text{giving} \qquad E_{0z} = 0$$

Similarly, equation (10.53) (iii), $\nabla \, . \, \vec{B} = 0$, gives

$$B_{0z} = 0$$

That is, the electric and magnetic fields do not have z-component, and therefore, the electromagnetic waves are transverse, *i.e.*, the electric and magnetic fields are perpendicular to the direction of propagation.

We may orient our axes so that $\vec{E}$ is polarized along the x direction:

$$\vec{E}(z,t) = E_{0x}\ \mathrm{e}^{i(k^* z - \omega t)}\ \hat{x} \tag{10.64}$$

Equation (10.53) (ii) can be written as

$$\frac{\partial E_z}{\partial y} - \frac{\partial E_y}{\partial z} = -\frac{\partial B_x}{\partial t} \tag{10.65}$$

$$\frac{\partial E_x}{\partial z} - \frac{\partial E_z}{\partial x} = -\frac{\partial B_y}{\partial t} \tag{10.66}$$

$$\frac{\partial E_y}{\partial x} - \frac{\partial E_x}{\partial y} = -\frac{\partial B_z}{\partial t} \tag{10.67}$$

In view of equation (10.64), this set of three equations reduces to

$$\frac{\partial E_x}{\partial z} = -\frac{\partial B_y}{\partial t} \tag{10.68}$$

Using equation (10.64) in (10.68), we have

$$\frac{\partial B_y}{\partial t} = -E_{0x}\ \mathrm{e}^{i(k^* z - \omega t)}\ ik^* \tag{10.69}$$

Integration of equation (10.69) with respect to t gives

$$B_y = \frac{k^*}{\omega}\ E_{0x}\ \mathrm{e}^{i(k^* z - \omega t)}$$

so that

$$\vec{B}(z,t) = \frac{k^*}{\omega}\ E_{0x}\ \mathrm{e}^{i(k^* z - \omega t)}\ \hat{y} \tag{10.70}$$

Equations (10.64) and (10.70) show that the electric and magnetic fields are perpendicular to each other. Further, both the electric and magnetic fields are perpendicular to the direction of propagation of the wave. Equation (10.53) (iv) can be written as

$$\frac{\partial B_z}{\partial y} - \frac{\partial B_y}{\partial z} = \mu\epsilon\ \frac{\partial E_x}{\partial t} + \mu\sigma E_x \tag{10.71}$$

$$\frac{\partial B_x}{\partial z} - \frac{\partial B_z}{\partial x} = \mu\epsilon\ \frac{\partial E_y}{\partial t} + \mu\sigma E_y \tag{10.72}$$

$$\frac{\partial B_y}{\partial x} - \frac{\partial B_x}{\partial y} = \mu\epsilon\ \frac{\partial E_z}{\partial t} + \mu\sigma E_z \tag{10.73}$$

In view of equation (10.64) and the fact that z-components of electric and magnetic fields are absent, this set of three equations reduces to

$$-\frac{\partial B_y}{\partial z} = \mu\epsilon\,\frac{\partial E_x}{\partial t} + \mu\sigma E_x \tag{10.74}$$

Using equation (10.64) in (10.74), we have

$$\frac{\partial B_y}{\partial z} = i(\mu\epsilon\omega + i\mu\sigma)E_{0x}\ \mathrm{e}^{i(k^* z-\omega t)} \tag{10.75}$$

Integration of equation (10.75) with respect to z gives

$$B_y = \frac{(\mu\epsilon\omega + i\mu\sigma)}{k^*}\ E_{0x}\ \mathrm{e}^{i(k^* z-\omega t)}$$

so that

$$\vec{B}(z,t) = \frac{(\mu\epsilon\omega + i\mu\sigma)}{k^*}\ E_{0x}\ \mathrm{e}^{i(k^* z-\omega t)}\ \hat{y} \tag{10.76}$$

Equations (10.64) and (10.76) show that the electric and magnetic fields are perpendicular to each other. Further, both the electric and magnetic fields are perpendicular to the direction of propagation of the wave.

4.1 Reflection at a conducting surface

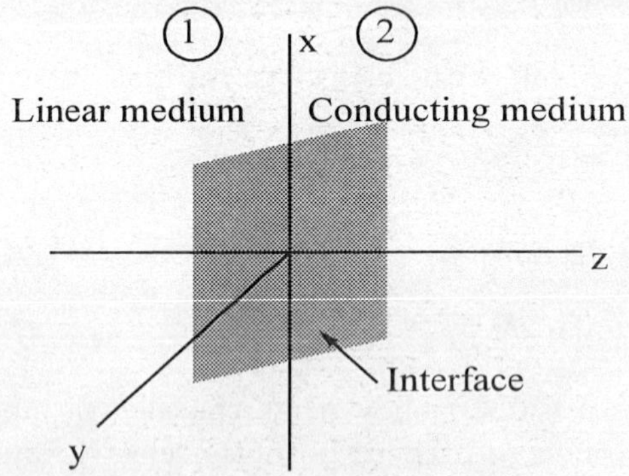

Figure 13: Reflection at a conducting surface

The boundary conditions we have used to analyze reflection and refraction at an interface between two dielectrics do not hold in the presence of free charges and currents. Instead, we have more general relations:

$$\left.\begin{array}{ll} \text{(i)}\ \epsilon_1 E_1^{\perp} - \epsilon_2 E_2^{\perp} = \sigma_f & \text{(iii)}\ E_1^{\parallel} - E_2^{\parallel} = 0 \\ \text{(ii)}\ B_1^{\perp} - B_2^{\perp} = 0 & \text{(iv)}\ \frac{1}{\mu_1}\,B_1^{\parallel} - \frac{1}{\mu_2}\,B_2^{\parallel} = \vec{K}_f \times \hat{n} \end{array}\right\} \tag{10.77}$$

where σ_f (not to be confused with conductivity) is the free surface charge, $\vec{K}_f$ is the free surface current, and $\hat{n}$ is a unit vector perpendicular to the surface, pointing from medium (2) into medium (1).

For ohmic conductors ($\vec{J}_f = \sigma\vec{E}$) there can be no free surface current, since this would require an infinite electric field at the boundary. Suppose now that the xy plane forms the boundary between a nonconducting linear medium (1) and a conductor (2). A monochromatic plane wave, traveling in the z direction and polarized in the x direction, approaches from the left, as in Figure 13:

$$\vec{E}_I(z,t) = \vec{E}_{0I}\, e^{i(k_1 z-\omega t)}\,\hat{x} \qquad \text{and} \qquad \vec{B}_I(z,t) = \frac{1}{v_1}\vec{E}_{0I}\, e^{i(k_1 z-\omega t)}\,\hat{y}$$

This incident wave gives rise to a reflected wave,

$$\vec{E}_R(z,t) = \vec{E}_{0R}\, e^{i(-k_1 z-\omega t)}\,\hat{x} \qquad \text{and} \qquad \vec{B}_R(z,t) = -\frac{1}{v_1}\vec{E}_{0R}\, e^{i(-k_1 z-\omega t)}\,\hat{y}$$

propagating back to the left in medium (1), and a transmitted wave

$$\vec{E}_T(z,t) = \vec{E}_{0T}\, e^{i(k_2^* z-\omega t)}\,\hat{x} \qquad \text{and} \qquad \vec{B}_T(z,t) = \frac{k_2^*}{\omega}\,\vec{E}_{0T}\, e^{i(k_2^* z-\omega t)}\,\hat{y}$$

which is attenuated as it penetrates into the conductor. The net disturbance for the electric field is:

$$\begin{cases} \vec{E}_{0I}\, e^{i(k_1 z-\omega t)}\,\hat{x} + \vec{E}_{0R}\, e^{i(-k_1 z-\omega t)}\,\hat{x}, & (z<0) \\ \vec{E}_{0T}\, e^{i(k_2^* z-\omega t)}\,\hat{x}, & (z>0) \end{cases}$$

and for the magnetic field is:

$$\begin{cases} \frac{1}{v_1}\vec{E}_{0I}\, e^{i(k_1 z-\omega t)}\,\hat{y} - \frac{1}{v_1}\vec{E}_{0R}\, e^{i(-k_1 z-\omega t)}\,\hat{y}, & (z<0) \\ \frac{k_2^*}{\omega}\,\vec{E}_{0T}\, e^{i(k_2^* z-\omega t)}\,\hat{y}, & (z>0) \end{cases}$$

At $z = 0$, the combined wave in medium (1) must join the wave in medium (2), pursuant to the boundary conditions (equation 10.77). Since $E^{\perp} = 0$ on both sides, boundary condition (i) yields $\sigma_f = 0$. Since $B^{\perp} = 0$, (ii) is automatically satisfied. Meanwhile, (iii) gives

$$\vec{E}_{0I} + \vec{E}_{0R} = \vec{E}_{0T} \tag{10.78}$$

and (iv) (with $\vec{K}_f = 0$) says

$$\frac{1}{\mu_1 v_1}\left(\vec{E}_{0I} - \vec{E}_{0R}\right) - \frac{k_2^*}{\mu_2\omega}\,\vec{E}_{0T} = 0 \qquad \text{or} \qquad \vec{E}_{0I} - \vec{E}_{0R} = \beta^*\vec{E}_{0T} \tag{10.79}$$

where

$$\beta^* = \frac{\mu_1 v_1}{\mu_2\omega}\,k_2^* \tag{10.80}$$

It follows that

$$\vec{E}_{0R} = \left(\frac{1-\beta^*}{1+\beta^*}\right)\vec{E}_{0I} \qquad \vec{E}_{0T} = \left(\frac{2}{1+\beta^*}\right)\vec{E}_{0I} \tag{10.81}$$

These results are formally identical to the ones that apply at the boundary between nonconductors, but the resemblance is deceptive since β^* is now a complex number. For a perfect conductor ($\sigma = \infty$), $k_2 = \infty$, so β^* is infinite, and

$$\vec{E}_{0R} = -E_{0I} \qquad \text{and} \qquad \vec{E}_{0T} = 0 \tag{10.82}$$

In this case the wave is totally reflected, with a 180° phase shift. That is why excellent conductors make good mirrors. In practice, we paint a thin coating of silver onto the back of a pane of glass - the glass has nothing to do with the reflection; it is just there to support the silver and to keep it from tarnishing. Since the skin depth in silver at optical frequencies is less than 100 Å, we do not need a very thick layer.

Review Questions

1. Describe about the propagation of one-dimensional wave.
2. Show that on reflection at the denser medium, a phase difference of 180° is introduced.
3. Using Maxwell's equations, discuss the propagation of electromagnetic waves in vacuum. Show that the electric and magnetic vectors, and the direction of propagation constitute a orthogonal system, and the electric and magnetic vectors are in phase with each other.
4. Using Maxwell's equations, discuss the propagation of electromagnetic waves in a linear medium. Show that the electric and magnetic vectors, and the direction of propagation constitute a orthogonal system, and the electric and magnetic vectors are in phase with each other.
5. For propagation of electromagnetic waves in a linear medium, discuss the reflection and transmission at normal incidence, and at oblique incidence.
6. For oblique incidence of electromagnetic waves in linear mediums, discuss the laws of reflection and of refraction.
7. Using Maxwell's equations, discuss the propagation of electromagnetic waves in a conducting medium. Show that the electric and magnetic vectors, and the direction of propagation constitute a orthogonal system,
8. Write notes on the following

   (i) One-dimensional wave
   (ii) Superposition of waves
   (iii) Laws of reflection
   (iv) Laws of refraction

Multiple-Choice Questions

1. How many degrees phase is introduced when an electromagnetic wave is reflected back at a rarer medium?

   A. 0 B. 60 C. 90 D. 180

   Ans. A

2. How many degrees phase is introduced when an electromagnetic wave is reflected back at a denser medium?

   A. 0 B. 60 C. 90 D. 180

   Ans. D

3. How much is relative permittivity of free space?

   A. 0 B. 1 C. 1.33 D. infinite

   Ans. B

4. Which of the following statements is correct for vacuum?

   A. Only electric field satisfies the wave equation.

   B. Only magnetic field satisfies the wave equation.

   C. Both electric and magnetic fields satisfy the wave equation.

   D. Neither electric field nor magnetic field satisfies the wave equation.

   Ans. C

5. For a plane polarized electromagnetic wave, how much is the angle between the direction of vibration of electric field and the direction of propagation of wave?

   A. 0 B. 60 C. 90 D. 180

   Ans. C

6. For a plane polarized electromagnetic wave, how much is the angle between the direction of vibration of magnetic field and the direction of propagation of wave?

   A. 0 B. 60 C. 90 D. 180

   Ans. C

7. For an unpolarized electromagnetic wave, how much is the angle between the direction of vibration of electric field and the direction of propagation of wave?

   A. 0 B. 60 C. 90 D. 180

   Ans. C

8. For an unpolarized electromagnetic wave, how much is the angle between the direction of vibration of magnetic field and the direction of propagation of wave?

   A. 0 B. 60 C. 90 D. 180

   Ans. C

9. In an unpolarized electromagnetic wave, how much is the phase difference between the electric and magnetic fields?

A. 0 B. 60 C. 90 D. 180

Ans. A

10. In an electromagnetic wave, the amplitude of electric field is how many times that of the magnetic field?

A. 1 B. c C. $1/c$ D. 2

Ans. B

11. In an electromagnetic wave, the contribution of electric field to energy is how many times that of the magnetic field?

A. 1 B. c C. $1/c$ D. 2

Ans. A

12. In an electromagnetic wave propagating along the z-axis, the electric vector is vibrating along x-axis. Along which axis are the vibrations of corresponding magnetic field?

A. x B. y C. z

Ans. B

13. For linear mediums, which parameter(s) remain nearly equal to that of the vacuum?

A. permeability B. permittivity

C. permeability and permittivity both

D. neither permeability nor permittivity

Ans. A

14. In an electromagnetic wave propagating in a linear medium with velocity v, the amplitude of electric field is how many times that of the magnetic field?

A. 1 B. v C. $1/v$ D. 2

Ans. B

15. Electromagnetic radiation propagating from one linear medium to another linear medium is partly reflected back to the first medium and is partly transmitted to the second medium. The angle of incidence is equal to

A. angle of reflection B. angle of refraction

C. both the angle of reflection and angle of refraction

D. neither the angle of reflection nor the angle of refraction

Ans. A

16. Electromagnetic radiation propagating from one linear medium to another linear medium is partly reflected back to the first medium and is partly transmitted to the second medium. The angle of refraction is equal to

    A. angle of reflection   B. angle of incidence

    C. both the the angle of reflection and angle of incidence

    D. neither the angle of reflection nor the angle of incidence

    Ans. D

17. Electromagnetic radiation propagating from one linear medium to another linear medium is partly reflected back to the first medium and is partly transmitted to the second medium. The angle of reflection is equal to

    A. angle of refraction   B. angle of incidence

    C. both the the angle of refraction and angle of incidence

    D. neither the angle of refraction nor the angle of incidence

    Ans. B

18. Electromagnetic radiation propagating from one linear medium to another linear medium is partly reflected back to the first medium and is partly transmitted to the second medium. The amplitude of reflected wave is 0.6 times that of the incident wave. What is the ratio of amplitude of transmitted wave to that of incident wave?

    A. 0.2   B. 0.4   C. 0.6   D. 0.8

    Ans. D

19. Electromagnetic radiation propagating from one linear medium to another linear medium is partly reflected back to the first medium and is partly transmitted to the second medium. The amplitude of reflected wave is 0.8 times that of the incident wave. What is the ratio of amplitude of transmitted wave to that of incident wave?

    A. 0.2   B. 0.4   C. 0.6   D. 0.8

    Ans. C

20. Electromagnetic radiation propagating from one linear medium to another linear medium is partly reflected back to the first medium and is partly transmitted to the second medium. The reflection coefficient of wave is 0.6. What is the transmission coefficient?

    A. 0.2   B. 0.4   C. 0.6   D. 0.8

    Ans. B

Chapter 11

Waveguides and Antennas

1. Basic concepts of waveguides

A waveguide is very long pipe (tube) made of a conductor of high conductivity and filled with a dielectric of low loss. In most of the waveguides, the cross section is either rectangular or circular. In the present discussion, we consider an arbitrary shape of the cross section of waveguide.

So far, we have discussed about the plane waves moving through a region which is unlimited (unbound). Now, we want to discuss about the propagation of wave through a bound region. When an electromagnetic wave is introduced in a bound region (a tube), it undergoes multiple reflections from the walls of the tube and propagates along the axis of the tube (z-axis here), as shown in Figure 1.

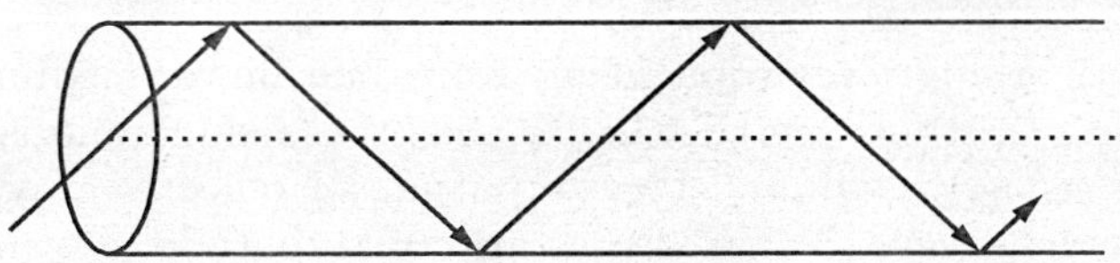

Figure 1: A waveguide with axis along z-axis.

The dielectric medium in the waveguide does not have free charge and free current, and therefore, the Maxwell's equations are

$$\left.\begin{array}{ll} \text{(i) } \nabla \cdot \vec{E} = 0 & \text{(ii) } \nabla \cdot \vec{B} = 0 \\ \text{(iii) } \nabla \times \vec{E} = -\frac{\partial \vec{B}}{\partial t} & \text{(iv) } \nabla \times \vec{B} = \epsilon\mu \, \frac{\partial \vec{E}}{\partial t} \end{array}\right\} \tag{11.1}$$

where μ and ϵ are, respectively, the permeability and permittivity of dielectric medium; $\vec{B} = \mu\vec{H}$, and $\vec{D} = \epsilon\vec{E}$. The fields $\vec{E}$ and $\vec{B}$ are functions of the position coordinates α_1, α_2, z, and time coordinate t. For a rectangular waveguide, α_1 and α_2 are x and y coordinates, respectively, and for a cylindrical waveguide, α_1 and α_2 are ρ and ϕ coordinates, respectively.

The coordinates α_1 and α_2 obviously lie in a plane (which is cross section of the waveguide) whose normal is parallel to the axis of the waveguide (z-axis). A plane wave propagating along $+z$ direction is expressed as

$$e^{i(kz-\omega t)} \tag{11.2}$$

.

Thus, the time-dependence of $\vec{E}$ and $\vec{B}$ fields may be described by $e^{-i\omega t}$ and therefore Maxwell's equation (11.1) (iii) gives

$$\nabla \times \vec{E}(\alpha_1, \alpha_2, z, t) = -\frac{\partial}{\partial t}\vec{B}(\alpha_1, \alpha_2, z, t)$$

$$\nabla \times [\vec{E}(\alpha_1, \alpha_2, z)\ e^{-i\omega t}] = -\frac{\partial}{\partial t}[\vec{B}(\alpha_1, \alpha_2, z)\ e^{-i\omega t}]$$

$$\nabla \times \vec{E}(\alpha_1, \alpha_2, z) = i\omega\ \vec{B}(\alpha_1, \alpha_2, z) \tag{11.3}$$

Similarly, from equation (11.1) (iv), (i) and (ii), we get, respectively,

$$\nabla \times \vec{B}(\alpha_1, \alpha_2, z) = -i\omega\epsilon\mu\ \vec{E}(\alpha_1, \alpha_2, z) \tag{11.4}$$

$$\nabla\ .\ \vec{E}(\alpha_1, \alpha_2, z) = 0 \tag{11.5}$$

$$\nabla\ .\ \vec{B}(\alpha_1, \alpha_2, z) = 0 \tag{11.6}$$

In equations (11.3), (11.4), (11.5) and (11.6), the field $\vec{E}$ and $\vec{B}$ are functions of space coordinates (α_1, α_2, z) and for simplicity, we drop writing the coordinates each time. Taking the curl of (11.3), we get

$$\nabla \times \left[\nabla \times \vec{E}\right] = i\omega\ \nabla \times \vec{B}$$

$$\nabla(\nabla\ .\ \vec{E}) - \nabla^2\vec{E} = \omega^2\epsilon\mu\ \vec{E}$$

Here, we have used equation (11.4) and the fact that gradient of a divergence of vector is zero, and we have

$$\nabla^2\vec{E} + \omega^2\epsilon\mu\ \vec{E} = 0 \tag{11.7}$$

Similarly, we get

$$\nabla^2\vec{B} + \omega^2\epsilon\mu\ \vec{B} = 0 \tag{11.8}$$

Now,

$$\nabla^2 \equiv \frac{\partial^2}{\partial x^2} + \frac{\partial^2}{\partial y^2} + \frac{\partial^2}{\partial z^2} \tag{11.9}$$

in Cartesian coordinates (rectangular waveguide) and

$$\nabla^2 \equiv \frac{1}{\rho}\frac{\partial}{\partial \rho}\left(\rho\ \frac{\partial}{\partial \rho}\right) + \frac{1}{\rho^2}\frac{\partial^2}{\partial \phi^2} + \frac{\partial^2}{\partial z^2} \tag{11.10}$$

in cylindrical coordinates (circular waveguide). These two expressions (11.9) and (11.10) can be written as

$$\nabla^2 \equiv \nabla_t^2 + \frac{\partial^2}{\partial z^2} \tag{11.11}$$

where transverse component ∇_t^2 is

$$\nabla_t^2 = \frac{\partial^2}{\partial x^2} + \frac{\partial^2}{\partial y^2} \tag{11.12}$$

in Cartesian coordinates (rectangular waveguide), and

$$\nabla_t^2 = \frac{1}{\rho}\frac{\partial}{\partial \rho}\left(\rho\,\frac{\partial}{\partial \rho}\right) + \frac{1}{\rho^2}\frac{\partial^2}{\partial \phi^2} \tag{11.13}$$

in cylindrical coordinates (circular waveguide). Considering expression (11.2) for plane wave moving along $+z$ direction, it is again useful to separate out the spacial variation of $\vec{E}(\alpha_1, \alpha_2, z)$ and $\vec{B}(\alpha_1, \alpha_2, z)$ in the z-direction by considering

$$\vec{E}(\alpha_1, \alpha_2, z) = \vec{E}_1(\alpha_1, \alpha_2)\ \mathrm{e}^{ikz}$$

and

$$\vec{B}(\alpha_1, \alpha_2, z) = \vec{B}_1(\alpha_1, \alpha_2)\ \mathrm{e}^{ikz}$$

where k is guide propagation constant. Applying this to equation (11.7), and using equation (11.11) we get

$$\left[\nabla_t^2 + \frac{\partial^2}{\partial z^2}\right]\vec{E}_1(\alpha_1, \alpha_2)\ \mathrm{e}^{ikz} + \omega^2\epsilon\mu\ \vec{E}_1(\alpha_1, \alpha_2)\ \mathrm{e}^{ikz} = 0$$

$$\nabla_t^2\vec{E}_1(\alpha_1, \alpha_2) + (\omega^2\epsilon\mu - k^2)\vec{E}_1(\alpha_1, \alpha_2) = 0 \tag{11.14}$$

Similarly, we get

$$\nabla_t^2\vec{B}_1(\alpha_1, \alpha_2) + (\omega^2\epsilon\mu - k^2)\vec{B}_1(\alpha_1, \alpha_2) = 0 \tag{11.15}$$

It is useful to separate the field $\vec{E}_1(\alpha_1, \alpha_2)$ and $\vec{B}_1(\alpha_1, \alpha_2)$ into components parallel and perpendicular to the direction of propagation (z-axis). Thus,

$$\vec{E}_1(\alpha_1, \alpha_2) = \vec{E}_t(\alpha_1, \alpha_2) + \vec{E}_z(\alpha_1, \alpha_2) \tag{11.16}$$

and

$$\vec{B}_1(\alpha_1, \alpha_2) = \vec{B}_t(\alpha_1, \alpha_2) + \vec{B}_z(\alpha_1, \alpha_2) \tag{11.17}$$

If $\hat{u}$ denotes the unit vector in the direction of z-axis, the parallel components are given by

$$\vec{E}_z(\alpha_1, \alpha_2) = [\hat{u}\ .\ \vec{E}_1(\alpha_1, \alpha_2)]\hat{u} \tag{11.18}$$

and

$$\vec{B}_z(\alpha_1, \alpha_2) = [\hat{u}\ .\ \vec{B}_1(\alpha_1, \alpha_2)]\hat{u} \tag{11.19}$$

and the transverse components are given by

$$\vec{E}_t(\alpha_1, \alpha_2) = [\hat{u} \times \vec{E}_1(\alpha_1, \alpha_2)] \times \hat{u} \tag{11.20}$$

and

$$\vec{B}_t(\alpha_1, \alpha_2) = [\hat{u} \times \vec{B}_1(\alpha_1, \alpha_2)] \times \hat{u} \tag{11.21}$$

Suppose in Cartesian coordinates, we write

$$\vec{E}(\alpha_1, \alpha_2, z) = [E_x(x, y)\ \hat{x} + E_y(x, y)\ \hat{y} + E_z(x, y)\ \hat{z}]\ \mathrm{e}^{ikz}$$

and

$$\vec{B}(\alpha_1, \alpha_2, z) = [B_x(x, y)\ \hat{x} + B_y(x, y)\ \hat{y} + B_z(x, y)\ \hat{z}]\ \mathrm{e}^{ikz}$$

then equation (11.3) gives

$$\frac{\partial E_z}{\partial y} - ikE_y = i\omega B_x \tag{11.22}$$

$$ikE_x - \frac{\partial E_z}{\partial x} = i\omega B_y \tag{11.23}$$

$$\frac{\partial E_y}{\partial x} - \frac{\partial E_x}{\partial y} = i\omega B_z \tag{11.24}$$

and equation (11.4) gives

$$\frac{\partial B_z}{\partial y} - ikB_y = -i\omega\epsilon\mu E_x \tag{11.25}$$

$$ikB_x - \frac{\partial B_z}{\partial x} = -i\omega\epsilon\mu E_y \tag{11.26}$$

$$\frac{\partial B_y}{\partial x} - \frac{\partial B_x}{\partial y} = -i\omega\epsilon\mu E_z \tag{11.27}$$

Eliminating E_y from equations (11.22) and (11.26), we get

$$B_x = \frac{1}{(\omega^2\epsilon\mu - k^2)}\left[ik\ \frac{\partial B_z}{\partial x} - i\omega\epsilon\mu\frac{\partial E_z}{\partial y}\right] \tag{11.28}$$

Eliminating E_x from equations (11.23) and (11.25), we get

$$B_y = \frac{1}{(\omega^2\epsilon\mu - k^2)}\left[ik\ \frac{\partial B_z}{\partial y} + i\omega\epsilon\mu\frac{\partial E_z}{\partial x}\right] \tag{11.29}$$

Using B_y from equation (11.29) in equation (11.23), we get

$$E_x = \frac{1}{(\omega^2\epsilon\mu - k^2)}\left[ik\ \frac{\partial E_z}{\partial x} + i\omega\frac{\partial B_z}{\partial y}\right] \tag{11.30}$$

Using B_x from equation (11.28) in equation (11.22), we get

$$E_y = \frac{1}{(\omega^2\epsilon\mu - k^2)}\Big[ik\,\frac{\partial E_z}{\partial y} - i\omega\frac{\partial B_z}{\partial x}\Big] \tag{11.31}$$

It shows that x, y components (*i.e.*, transverse component) can be expressed in terms of z component (*i.e.*, axial component). We have transverse electric (TE) mode and transverse magnetic (TM) mode.

In the TM mode, $B_z = 0$ and $E_z \neq 0$ and

$$\vec{E}_t = E_x\,\hat{x} + E_y\,\hat{y} = \frac{ik}{(\omega^2\epsilon\mu - k^2)}\Big[\frac{\partial E_z}{\partial x}\,\hat{x} + \frac{\partial E_z}{\partial y}\,\hat{y}\Big] = \frac{ik}{(\omega^2\epsilon\mu - k^2)}\nabla_t E_z \tag{11.32}$$

where

$$\nabla_t \equiv \hat{x}\,\frac{\partial}{\partial x} + \hat{y}\,\frac{\partial}{\partial y}$$

and

$$\begin{aligned}\vec{B}_t = B_x\,\hat{x} + B_y\,\hat{y} &= \frac{i\omega\epsilon\mu}{(\omega^2\epsilon\mu - k^2)}\Big[-\frac{\partial E_z}{\partial y}\,\hat{x} + \frac{\partial E_z}{\partial x}\,\hat{y}\Big]\\ &= \frac{i\omega\epsilon\mu\,\hat{z}}{(\omega^2\epsilon\mu - k^2)} \times \Big[\frac{\partial E_z}{\partial x}\,\hat{x} + \frac{\partial E_z}{\partial y}\,\hat{y}\Big] = \frac{i\omega\epsilon\mu\,\hat{z}}{(\omega^2\epsilon\mu - k^2)} \times \nabla_t E_z\end{aligned} \tag{11.33}$$

In the TE mode, $E_z = 0$ and $B_z \neq 0$ and

$$\begin{aligned}\vec{E}_t = E_x\,\hat{x} + E_y\,\hat{y} &= \frac{i\omega}{(\omega^2\epsilon\mu - k^2)}\Big[\frac{\partial B_z}{\partial y}\,\hat{x} - \frac{\partial B_z}{\partial x}\,\hat{y}\Big]\\ &= -\frac{i\omega\,\hat{z}}{(\omega^2\epsilon\mu - k^2)} \times \Big[\frac{\partial B_z}{\partial y}\,\hat{y} + \frac{\partial B_z}{\partial x}\,\hat{x}\Big] = -\frac{i\omega\,\hat{z}}{(\omega^2\epsilon\mu - k^2)} \times \nabla_t B_z\end{aligned} \tag{11.34}$$

and

$$\vec{B}_t = B_x\,\hat{x} + B_y\,\hat{y} = \frac{ik}{(\omega^2\epsilon\mu - k^2)}\Big[\frac{\partial B_z}{\partial x}\,\hat{x} + \frac{\partial B_z}{\partial y}\,\hat{y}\Big] = \frac{ik}{(\omega^2\epsilon\mu - k^2)}\nabla_t B_z \tag{11.35}$$

Since B_z and E_z do not exist simultaneously, equations (11.32) and (11.34) can be combined to

$$\vec{E}_t = \frac{1}{(\omega^2\epsilon\mu - k^2)}\Big[ik\nabla_t E_z(\alpha_1, \alpha_2) - i\omega\hat{u} \times \nabla_t B_z(\alpha_1, \alpha_2)\Big] \tag{11.36}$$

and equations (11.33) and (11.35) can be combined to

$$\vec{B}_t = \frac{1}{(\omega^2\epsilon\mu - k^2)}\Big[i\omega\epsilon\mu\;\hat{u} \times \nabla_t E_z(\alpha_1, \alpha_2) + ik\nabla_t B_z(\alpha_1, \alpha_2)\Big] \tag{11.37}$$

Since equations (11.36) and (11.37) are in terms of operator ∇_t, they can be generalized and would also be applicable for circular waveguides with

$$\nabla_t \equiv \hat{\rho}\,\frac{\partial}{\partial \rho} + \hat{\phi}\,\frac{1}{\rho}\,\frac{\partial}{\partial \phi}$$

Equations (11.36) and (11.37) show that the transverse components can be calculated from the axial components. Therefore, it is sufficient to determine the axial components $\vec{E}_z(\alpha_1, \alpha_2)$ and $\vec{B}_z(\alpha_1, \alpha_2)$ as approximate solutions of equations

$$\nabla_t^2\vec{E}_z(\alpha_1, \alpha_2) + (\omega^2\epsilon\mu - k^2)\vec{E}_z(\alpha_1, \alpha_2) = 0 \tag{11.38}$$

and

$$\nabla_t^2\vec{B}_z(\alpha_1, \alpha_2) + (\omega^2\epsilon\mu - k^2)\vec{B}_z(\alpha_1, \alpha_2) = 0 \tag{11.39}$$

1.1 Boundary conditions

At the perfectly conducting bounding surface of **waveguide**, the fields $\vec{E}(\alpha_1, \alpha_2, z, t)$ and $\vec{B}(\alpha_1, \alpha_2, z, t)$ satisfy the boundary conditions:

(i) Tangent component of electric field is continuous, *i.e.*,

$$|\hat{n} \times \vec{E}(\alpha_1, \alpha_2, z, t)|_s = 0 \tag{11.40}$$

(ii) Normal component of magnetic field is continuous, *i.e.*,

$$|\hat{n} \,.\, \vec{B}(\alpha_1, \alpha_2, z, t)|_s = 0 \tag{11.41}$$

where $\hat{n}$ is a unit vector normal to the surface. The boundary condition (11.40) is equivalent to

$$|\vec{E}_z(\alpha_1, \alpha_2)|_s = 0 \tag{11.42}$$

and the boundary condition (11.41) is equivalent to

$$\left|\frac{\partial}{\partial n}\vec{B}_z(\alpha_1, \alpha_2)\right|_s = 0 \tag{11.43}$$

where $\partial/\partial n$ is the normal derivative at a point on the surface.

Because the boundary conditions on $\vec{E}_z(\alpha_1, \alpha_2)$ and $\vec{B}_z(\alpha_1, \alpha_2)$ are different, they cannot be satisfied simultaneously. Consequently, the electromagnetic fields may be divided into distinct categories, based on the boundary conditions: either $\vec{E}_z(\alpha_1, \alpha_2)$ is present or $\vec{B}_z(\alpha_1, \alpha_2)$ is present or both are absent.

(i) Transverse electric (TE) mode:

When $\vec{B}_z(\alpha_1, \alpha_2)$ is present and $\vec{E}_z(\alpha_1, \alpha_2)$ is absent everywhere, such waves are called transverse electric waves (TE mode). The boundary condition is

$$\left|\frac{\partial}{\partial n}\vec{B}_z(\alpha_1, \alpha_2)\right|_s = 0 \tag{11.44}$$

The transverse components are given by

$$\vec{B}_t(\alpha_1, \alpha_2) = \frac{ik}{(\omega^2\epsilon\mu - k^2)} \nabla_t \vec{B}_z(\alpha_1, \alpha_2) \tag{11.45}$$

and

$$\vec{E}_t(\alpha_1, \alpha_2) = -\frac{i\omega}{(\omega^2\epsilon\mu - k^2)} \hat{u} \times \nabla_t \vec{B}_z(\alpha_1, \alpha_2) \tag{11.46}$$

(ii) Transverse magnetic (TM) mode:

When $\vec{E}_z(\alpha_1, \alpha_2)$ is present and $\vec{B}_z(\alpha_1, \alpha_2)$ is absent everywhere, such waves are called transverse magnetic waves (TM mode). The boundary condition is

$$|\vec{E}_z(\alpha_1, \alpha_2)|_s = 0 \tag{11.47}$$

The transverse components are given by

$$\vec{B}_t(\alpha_1, \alpha_2) = \frac{i\omega\mu\epsilon}{(\omega^2\epsilon\mu - k^2)} \hat{u} \times \nabla_t \vec{E}_z(\alpha_1, \alpha_2) \tag{11.48}$$

and

$$\vec{E}_t(\alpha_1, \alpha_2) = \frac{ik}{(\omega^2\epsilon\mu - k^2)} \nabla_t \vec{E}_z(\alpha_1, \alpha_2) \tag{11.49}$$

(iii) Transverse electromagnetic (TEM) mode:

When both $\vec{E}_z(\alpha_1, \alpha_2)$ and $\vec{B}_z(\alpha_1, \alpha_2)$ are absent everywhere, such waves are called transverse electromagnetic waves (TEM mode). It can be found that in case of hollow conducting pipes, either TE or TM waves can propagate, but TEM waves do not propagate. In case of waveguide, the only problem is to solve equations (11.38) and (11.39) for different systems by applying the boundary conditions (11.42) and (11.43).

Example 1: Using Maxwell's equations, derive relations:

$$\vec{E}_t = \frac{1}{(\omega^2\epsilon\mu - k^2)} \Big[ik\nabla_t E_z(\alpha_1, \alpha_2) - i\omega\hat{u} \times \nabla_t B_z(\alpha_1, \alpha_2)\Big]$$

and

$$\vec{B}_t = \frac{1}{(\omega^2\epsilon\mu - k^2)} \Big[i\omega\epsilon\mu\ \hat{u} \times \nabla_t E_z(\alpha_1, \alpha_2) + ik\nabla_t B_z(\alpha_1, \alpha_2)\Big]$$

in cylindrical coordinates (ρ, ϕ, z).

Solution: Suppose in cylindrical coordinates, we write

$$\vec{E}(\alpha_1, \alpha_2, z) = [E_\rho(\rho, \phi)\ \hat{\rho} + E_\phi(\rho, \phi))\ \hat{\phi} + E_z(\rho, \phi)\ \hat{z}]\ e^{ikz}$$

and

$$\vec{B}(\alpha_1, \alpha_2, z) = [B_\rho(\rho, \phi)\ \hat{\rho} + B_\phi(\rho, \phi))\ \hat{\phi} + B_z(\rho, \phi)\ \hat{z}]\ e^{ikz}$$

then equation (11.3) gives

$$\frac{1}{\rho}\frac{\partial E_z}{\partial \phi} - ikE_\phi = i\omega B_\rho \tag{11.50}$$

$$ikE_\rho - \frac{\partial E_z}{\partial \rho} = i\omega B_\phi \tag{11.51}$$

$$\frac{\partial(\rho E_\phi)}{\partial \rho} - \frac{\partial E_\rho}{\partial \phi} = i\omega\rho B_z \tag{11.52}$$

and equation (11.4) gives

$$\frac{1}{\rho}\frac{\partial B_z}{\partial \phi} - ikB_\phi = -i\omega\epsilon\mu E_\rho \tag{11.53}$$

$$ikB_\rho - \frac{\partial B_z}{\partial \rho} = -i\omega\epsilon\mu E_\phi \tag{11.54}$$

$$\frac{\partial(\rho B_\phi)}{\partial \rho} - \frac{\partial B_\rho}{\partial \phi} = -i\omega\epsilon\mu\rho E_z \tag{11.55}$$

Eliminating E_ϕ from equations (11.50) and (11.54), we get

$$B_\rho = \frac{1}{(\omega^2\epsilon\mu - k^2)}\left[ik\,\frac{\partial B_z}{\partial \rho} - \frac{i\omega\epsilon\mu}{\rho}\,\frac{\partial E_z}{\partial \phi}\right] \tag{11.56}$$

Eliminating E_ρ from equations (11.51) and (11.53), we get

$$B_\phi = \frac{1}{(\omega^2\epsilon\mu - k^2)}\left[\frac{ik}{\rho}\,\frac{\partial B_z}{\partial \phi} + i\omega\epsilon\mu\,\frac{\partial E_z}{\partial \rho}\right] \tag{11.57}$$

Using B_ϕ from equation (11.57) in equation (11.51), we get

$$E_\rho = \frac{1}{(\omega^2\epsilon\mu - k^2)}\left[\frac{i\omega}{\rho}\,\frac{\partial B_z}{\partial \phi} + ik\,\frac{\partial E_z}{\partial \rho}\right] \tag{11.58}$$

Using B_ρ from equation (11.56) in equation (11.50), we get

$$E_\phi = \frac{1}{(\omega^2\epsilon\mu - k^2)}\left[\frac{ik}{\rho}\,\frac{\partial E_z}{\partial \phi} - i\omega\,\frac{\partial B_z}{\partial \rho}\right] \tag{11.59}$$

It shows that ρ, ϕ components (*i.e.*, transverse component) can be expressed in terms of z component (*i.e.*, axial component). We have transverse electric (TE) mode and transverse magnetic (TM) mode.

In the TM mode, $B_z = 0$ and $E_z \neq 0$ and

$$\vec{E}_t = E_\rho\,\hat{\rho} + E_\phi\,\hat{\phi} = \frac{ik}{(\omega^2\epsilon\mu - k^2)}\left[\frac{\partial E_z}{\partial \rho}\,\hat{\rho} + \frac{1}{\rho}\,\frac{\partial E_z}{\partial \phi}\,\hat{\phi}\right] = \frac{ik}{(\omega^2\epsilon\mu - k^2)}\nabla_t E_z \tag{11.60}$$

where

$$\nabla_t \equiv \hat{\rho}\,\frac{\partial}{\partial \rho} + \hat{\phi}\,\frac{1}{\rho}\,\frac{\partial}{\partial \phi}$$

and

$$\vec{B}_t = B_\rho\,\hat{\rho} + B_\phi\,\hat{\phi} = \frac{1}{(\omega^2\epsilon\mu - k^2)}\left[-\frac{i\omega\epsilon\mu}{\rho}\,\frac{\partial E_z}{\partial \phi}\,\hat{\rho} + i\omega\epsilon\mu\,\frac{\partial E_z}{\partial \rho}\,\hat{\phi}\right]$$

$$= \frac{i\omega\epsilon\mu\,\hat{z}\times}{(\omega^2\epsilon\mu - k^2)}\left[\frac{1}{\rho}\,\frac{\partial E_z}{\partial \phi}\,\hat{\phi} + \frac{\partial E_z}{\partial \rho}\,\hat{\rho}\right] = \frac{i\omega\epsilon\mu}{(\omega^2\epsilon\mu - k^2)}\,\hat{z}\times\nabla_t E_z \tag{11.61}$$

In the TE mode, $E_z = 0$ and $B_z \neq 0$ and

$$\vec{E}_t = E_\rho\,\hat{\rho} + E_\phi\,\hat{\phi} = \frac{1}{(\omega^2\epsilon\mu - k^2)}\left[\frac{i\omega}{\rho}\,\frac{\partial B_z}{\partial \phi}\,\hat{\rho} - i\omega\,\frac{\partial B_z}{\partial \rho}\,\hat{\phi}\right]$$

$$= -\frac{i\omega\ \hat{z}\times}{(\omega^2\epsilon\mu - k^2)}\left[\frac{1}{\rho}\frac{\partial B_z}{\partial\phi}\ \hat{\phi} + \frac{\partial B_z}{\partial\rho}\ \hat{\rho}\right] = -\frac{i\omega}{(\omega^2\epsilon\mu - k^2)}\ \hat{z}\times\nabla_t B_z \tag{11.62}$$

and

$$\vec{B}_t = B_\rho\ \hat{\rho} + B_\phi\ \hat{\phi} = \frac{1}{(\omega^2\epsilon\mu - k^2)}\left[ik\ \frac{\partial B_z}{\partial\rho}\ \hat{\rho} + \frac{ik}{\rho}\ \frac{\partial B_z}{\partial\phi}\ \hat{\phi}\right]$$

$$= \frac{ik}{(\omega^2\epsilon\mu - k^2)}\nabla_t B_z \tag{11.63}$$

Since B_z and E_z do not exist simultaneously, equations (11.60) and (11.62) can be combined to

$$\vec{E}_t = \frac{1}{(\omega^2\epsilon\mu - k^2)}\left[ik\nabla_t E_z(\alpha_1,\alpha_2) - i\omega\hat{u}\times\nabla_t B_z(\alpha_1,\alpha_2)\right] \tag{11.64}$$

and equations (11.61) and (11.63) can be combined to

$$\vec{B}_t = \frac{1}{(\omega^2\epsilon\mu - k^2)}\left[i\omega\epsilon\mu\ \hat{u}\times\nabla_t E_z(\alpha_1,\alpha_2) + ik\nabla_t B_z(\alpha_1,\alpha_2)\right] \tag{11.65}$$

2. Rectangular waveguide

When the cross section of metal tube (waveguide) is rectangular, it is called the rectangular waveguide, as shown in Figure 2. It has cross sectional lengths a and b, and z-axis is parallel to axis of the tube. The conductivity of the metal is assumed to be infinite, and the interior is filled with a loss-free dielectric. For the rectangular waveguide, we have TE and TM modes of propagation.

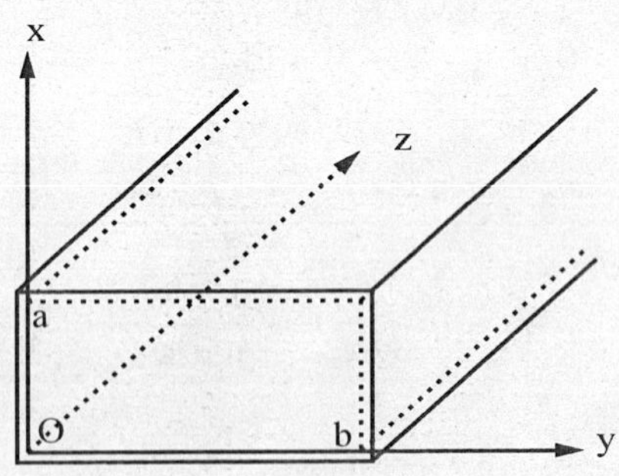

Figure 2: Rectangular waveguide having dimensions a and b of cross section.

2.1 TE wave in rectangular waveguide

In this case $E_z(x,y) = 0$ everywhere and $B_z(x,y)$ satisfies the wave equation

$$\nabla_t^2 B_z(x,y) + k'^2 B_z(x,y) = 0 \tag{11.66}$$

where $k'^2 = (\omega^2\epsilon\mu - k^2)$. Thus,

$$\Big(\frac{\partial^2}{\partial x^2} + \frac{\partial^2}{\partial y^2}\Big)B_z(x,y) + k'^2 B_z(x,y) = 0 \tag{11.67}$$

The boundary conditions for $B_z(x,y)$ are

$$\left.\begin{aligned} &\tfrac{\partial}{\partial x}B_z(x,y) = 0 \qquad \text{at} \quad x = 0 \quad \text{and} \quad x = a \\ &\tfrac{\partial}{\partial y}B_z(x,y) = 0 \qquad \text{at} \quad y = 0 \quad \text{and} \quad y = b \end{aligned}\right\} \tag{11.68}$$

Equation (11.67) can be written as

$$\frac{\partial^2}{\partial x^2}B_z(x,y) + \frac{\partial^2}{\partial y^2}B_z(x,y) + k'^2 B_z(x,y) = 0 \tag{11.69}$$

This is second order, linear partial differential equation and can be solved using the method of separation of variables. Let $B_z(x,y) = X(x)Y(y)$, where X is a function of x only and Y is a function of y only. Thus, we get

$$Y\frac{\partial^2 X}{\partial x^2} + X\frac{\partial^2 Y}{\partial y^2} + k'^2 XY = 0 \qquad \text{or} \qquad \frac{1}{X}\frac{\partial^2 X}{\partial x^2} + k'^2 = -\frac{1}{Y}\frac{\partial^2 Y}{\partial y^2}$$

The left side is a function of x only and the right side is a function of y only, and both are equal to each other. This is possible only when they are separately equal to a constant, say, A^2. Hence,

$$\frac{1}{X}\frac{\partial^2 X}{\partial x^2} + k'^2 = -\frac{1}{Y}\frac{\partial^2 Y}{\partial y^2} = A^2$$

Thus, we should solve two equations

$$\frac{1}{X}\frac{\mathrm{d}^2 X}{\mathrm{d}x^2} + k'^2 = A^2 \qquad\qquad -\frac{1}{Y}\frac{\mathrm{d}^2 Y}{\mathrm{d}y^2} = A^2$$

or

$$\frac{\mathrm{d}^2 X}{\mathrm{d}x^2} + B^2 X = 0 \qquad\qquad \frac{\mathrm{d}^2 Y}{\mathrm{d}y^2} + A^2 Y = 0$$

where $B^2 = k'^2 - A^2$. Solutions of these second order differential equations are

$$X = C_1 \cos Bx + C_2 \sin Bx \qquad \text{and} \qquad Y = C_3 \cos Ay + C_4 \sin Ay$$

where C_1, C_2, C_3 and C_4 are arbitrary constants. For non-trivial solutions, both C_1, C_2 cannot be zero simultaneously; and both C_3, C_4 cannot be zero simultaneously. The boundary conditions (11.68) are:

$$\frac{\mathrm{d}X}{\mathrm{d}x} = 0 \quad \text{at } x = 0 \text{ and } x = a \qquad \text{and} \qquad \frac{\mathrm{d}Y}{\mathrm{d}y} = 0 \quad \text{at } y = 0 \text{ and } y = b$$

We have

$$\frac{dX}{dx} = -C_1 B \ \sin Bx + C_2 B \ \cos Bx$$

When $x = 0$, we have $C_2 B = 0$, giving $C_2 = 0$.

When $x = a$, we have $-C_1 B \ \sin Ba = 0$.

Since $C_1 \neq 0$, we have $\sin Ba = 0$, giving $Ba = \ell\pi$; $B = \ell\pi/a$.

where ℓ is an integer. Therefore, we have

$$X = C_1 \ \cos(\ell\pi/a)x$$

We have

$$\frac{dY}{dy} = -C_3 A \ \sin Ay + C_4 A \ \cos Ay$$

When $y = 0$, we have $C_4 A = 0$, giving $C_4 = 0$.

When $y = b$, we have $-C_3 A \ \sin Ab = 0$.

Since $C_3 \neq 0$, we have $\sin Ab = 0$, giving $Ab = m\pi$; $A = m\pi/b$.

where m is an integer. Therefore, we have

$$Y = C_3 \ \cos(m\pi/b)y$$

Finally, the solution becomes

$$B_z(x, y) = X \ Y = C_1 C_3 \ \cos\frac{\ell\pi x}{a} \ \cos\frac{m\pi y}{b}$$

or

$$B_z(x, y) = B_0 \ \cos\frac{\ell\pi x}{a} \ \cos\frac{m\pi y}{b}$$

Hence, ℓ and m are two integer parameters and a set of given values of ℓ and m defines the mode of a **waveguide**, which is called the (ℓ, m) **waveguide** mode for TE wave designated as $\text{TE}_{\ell m}$ mode. $\ell = m = 0$, gives $B_z(x, y) = B_0$; it is static mode. For the propagating mode, one of the ℓ and m must be non-zero.

Cut-off frequency and cut-off wavelength

A wave guide only carries or propagates signals below a certain wavelength, known as the cut-off wavelength. Above the waveguide cut-off wavelength, the waveguide does not allow the propagation of signal. We have

$$k'^2 = B^2 + A^2 = \left(\frac{\ell\pi}{a}\right)^2 + \left(\frac{m\pi}{b}\right)^2 \tag{11.70}$$

where the indices ℓ and m specify the mode. Using $k' = 2\pi/\lambda$, the cut-off wavelength λ_c is given by

$$\left(\frac{1}{\lambda_c}\right)_{\ell m} = \frac{1}{2}\left(\frac{\ell^2}{a^2} + \frac{m^2}{b^2}\right)^{1/2} \tag{11.71}$$

using $\omega = 2\pi c/\lambda$, the cut-off frequency ω_c is

$$(\omega_c)_{\ell m} = \pi c \left(\frac{\ell^2}{a^2} + \frac{m^2}{b^2}\right)^{1/2} \tag{11.72}$$

The modes corresponding to ℓ and m are designated as $TE_{\ell m}$ mode. The case $\ell = m = 0$ gives a static field, and does not represent a wave propagation. So, TE_{00} mode does not exist. If $a < b$, the lowest cut off frequency results for $\ell = 0$ and $m = 1$, that is,

$$(\omega_c)_{01} = \frac{\pi c}{b} \quad \text{or} \quad k' = \frac{\pi}{b} \tag{11.73}$$

The TE_{01} mode is called the **principal** or **dominant mode**.

Phase velocity and group velocity

We have

$$k'^2 = \mu\epsilon\omega^2 - k^2$$

For free space $\mu_0\epsilon_0 = 1/c^2$ and therefore,

$$k'^2 = \frac{\omega^2}{c^2} - k^2$$

Using $\omega/c = k_0 = 2\pi/\lambda_0$ and $k' = 2\pi/\lambda_c$, we have

$$k = \sqrt{k_0^2 - k'^2} = k_0\sqrt{1 - (\lambda_0/\lambda_c)^2}$$

The phase velocity is

$$v_p = \frac{\omega}{k} = \frac{ck_0}{k_0\sqrt{1 - (\lambda_0/\lambda_c)^2}} = \frac{c}{\sqrt{1 - (\lambda_0/\lambda_c)^2}}$$

which is greater than c. However, the energy carried by the wave travels with the group velocity. We have $\omega^2 = c^2(k^2 + k'^2)$ and thus,

$$\omega\,\frac{\partial\omega}{\partial k} = c^2 k$$

So, the group velocity v_g is

$$v_g = \frac{\partial\omega}{\partial k} = \frac{c^2 k}{\omega} = \frac{ck}{k_0} = c\sqrt{1 - (\lambda_0/\lambda_c)^2}$$

which is smaller than c.

2.2 Transverse components of electric field and magnetic field

The transverse components of electric field and magnetic field intensity may now be calculated in the following manner:

$$\vec{E}_t(x,y) = \frac{-i\omega}{(\omega^2\epsilon\mu - k^2)}\,[\hat{u}\times\nabla_t\vec{B}_z(x,y)]$$

$$= \frac{-i\omega}{(\omega^2\epsilon\mu - k^2)}\,\hat{z}\times\left(\hat{x}\,\frac{\partial}{\partial x}+\hat{y}\,\frac{\partial}{\partial y}\right)B_0\,\cos\frac{\ell\pi x}{a}\,\cos\frac{m\pi y}{b}$$

$$= \frac{-i\omega B_0}{(\omega^2\epsilon\mu - k^2)}\left(\hat{y}\,\frac{\partial}{\partial x}-\hat{x}\,\frac{\partial}{\partial y}\right)\cos\frac{\ell\pi x}{a}\,\cos\frac{m\pi y}{b}$$

$$= \frac{-i\omega B_0}{(\omega^2\epsilon\mu - k^2)}\left(\hat{x}\,\frac{m\pi}{b}\cos\frac{\ell\pi x}{a}\,\sin\frac{m\pi y}{b}-\hat{y}\,\frac{\ell\pi}{a}\sin\frac{\ell\pi x}{a}\,\cos\frac{m\pi y}{b}\right)$$

From this equation, the components $\vec{E}_x(x,y)$ and $\vec{E}_y(x,y)$ can be calculated.

Similarly,

$$\vec{B}_t(x,y) = \frac{ik}{(\omega^2\epsilon\mu - k^2)}\,\nabla_t\vec{B}_z(x,y)$$

$$= \frac{ik}{(\omega^2\epsilon\mu - k^2)}\left(\hat{x}\,\frac{\partial}{\partial x}+\hat{y}\,\frac{\partial}{\partial y}\right)B_0\,\cos\frac{\ell\pi x}{a}\,\cos\frac{m\pi y}{b}$$

$$= \frac{ikB_0}{(\omega^2\epsilon\mu - k^2)}\left(-\hat{x}\,\frac{\ell\pi}{a}\sin\frac{\ell\pi x}{a}\,\cos\frac{m\pi y}{b}-\hat{y}\,\frac{m\pi}{b}\cos\frac{\ell\pi x}{a}\,\sin\frac{m\pi y}{b}\right)$$

From this equation, the components $\vec{B}_x(x,y)$ and $\vec{B}_y(x,y)$ can be calculated.

2.3 TM wave in rectangular waveguide

In this case $B_z(x,y) = 0$ everywhere and $E_z(x,y)$ satisfies the wave equation

$$\nabla_t^2 E_z(x,y) + k'^2 E_z(x,y) = 0$$

where $k'^2 = (\omega^2\epsilon\mu - k^2)$. Thus,

$$\left(\frac{\partial^2}{\partial x^2}+\frac{\partial^2}{\partial y^2}\right)E_z(x,y) + k'^2 E_z(x,y) = 0 \tag{11.74}$$

The boundary conditions for $E_z(x,y)$ are

$$\left.\begin{aligned} E_z(x,y) &= 0 \quad \text{at} \quad x=0 \quad \text{and} \quad x=a \\ E_z(x,y) &= 0 \quad \text{at} \quad y=0 \quad \text{and} \quad y=b \end{aligned}\right\} \tag{11.75}$$

This case can also be solved using the technique of separation of variables, as discussed for the case of TE mode, and we have

$$E_z(x,y) = E_0 \sin\frac{\ell\pi x}{a} \sin\frac{m\pi y}{b} \tag{11.76}$$

where ℓ and m are integers. If $\ell = 0$ or $m = 0$, we have $E_z(x,y) = 0$. For the propagating mode, both ℓ and m must be non-zero. We have

$$k'^2 = \left(\frac{\ell\pi}{a}\right)^2 + \left(\frac{m\pi}{b}\right)^2$$

k' is always positive and would have minimum allowed value, and corresponding maximum allowed value of wavelength. The **cut off wavelength** λ_c is

$$\left(\frac{1}{\lambda_c}\right)_{\ell m} = \frac{1}{2}\left(\frac{\ell^2}{a^2} + \frac{m^2}{b^2}\right)^{1/2} \tag{11.77}$$

and **cut off frequency** ω_c is

$$(\omega_c)_{\ell m} = \pi c\left(\frac{\ell^2}{a^2} + \frac{m^2}{b^2}\right)^{1/2} \tag{11.78}$$

This shows that the TE and TM modes of a rectangular waveguide have the same set of cut off frequencies. However, in this case also, $\ell = 0$ or $m = 0$ represent a trivial solution, since this gives $E_z(x,y) = 0$ and hence, all the components of $\vec{E}$ are zero. Of course, the mode $\ell = 1$ and $m = 0$ was dominant in the TE mode. In the TM mode, the lowest mode has $\ell = m = 1$ and may be represented by TM_{11}.

The cut off frequency of the lowest mode is given by

$$(\omega_c)_{11} = \pi c\left(\frac{1}{a^2} + \frac{1}{b^2}\right)^{1/2} = \frac{\pi c}{a}\left(1 + \frac{a^2}{b^2}\right)^{1/2} \tag{11.79}$$

When $a > b$, the cut off frequency of lowest TM mode is greater than that of the lowest TE mode by a factor $(1 + a^2/b^2)^{1/2}$.

Example 2: Show that in a rectangular **waveguide** TE_{00} mode does not exist.

Solution: For TE_{00} mode, we have $\ell = 0$ and $m = 0$. On substitution of $\ell = 0$ and $m = 0$, we have

$$B_z(x,y) = B_0, \qquad B_x(x,y) = 0, \qquad B_y(x,y) = 0$$

$$E_x(x,y) = 0, \qquad E_y(x,y) = 0, \qquad E_z(x,y) = 0$$

This shows that $\vec{E}$ does not exist. Hence, TE_{00} mode does not exist.

Example 3: For the dominant mode, find out the guide wavelength when free space wavelength is 10 cm for a rectangular **waveguide** whose width is 6.0 cm? What is the cut off wavelength for the guide?

Solution: The cut off frequency

$$(\lambda_c)_{\ell m} = \frac{2}{[(\ell^2/a^2) + (m^2/b^2)]^{1/2}}$$

Thus,

$$(\lambda_c)_{10} = \frac{2}{[(1^2/6^2) + (0^2/b^2)]^{1/2}} = 12 \text{ cm}$$

We know that

$$\frac{1}{\lambda_g^2} = \frac{1}{\lambda_f^2} - \frac{1}{\lambda_c^2}$$

Thus,

$$\frac{1}{\lambda_g^2} = \frac{1}{10^2} - \frac{1}{12^2} \qquad \lambda_g = 18 \text{ cm}$$

Example 4: What is width of a rectangular wave guide so that the energy of electromagnetic radiations having space wavelength of 3 cm travels down the guide at 95% of the speed of light?

Solution: The group velocity is given by

$$v_g = c\sqrt{1 - (\lambda_0/\lambda_c)^2}$$

where cut off wavelength

$$(\lambda_c)_{\ell m} = \frac{2}{[(\ell^2/a^2) + (m^2/b^2)]^{1/2}}$$

Thus,

$$(\lambda_c)_{10} = \frac{2}{[(1^2/a^2) + (0^2/b^2)]^{1/2}} = 2a$$

We are given $v_g = 0.95c$ and $\lambda_0 = 3$ cm. On substitution of values, we have

$$0.95c = c\sqrt{1 - (3/2a)^2} \qquad a = 4.8 \text{ cm}$$

Example 5: A rectangular wave guide has dimensions of 2.5 cm and 5 cm. Determine the guide wavelength, phase velocity and phase constant at the wavelength of 4.5 cm for the dominant mode.

Solution: Since $a < b$, we have $\ell = 0$ and $m = 1$. For the dominant mode, we have

$$(\lambda_c)_{01} = \frac{2}{[(0^2/a^2) + (1^2/b^2)]^{1/2}} = 2b \qquad (\lambda_c)_{01} = 2 \times 5 = 10 \text{ cm}$$

Wavelength within the guide is given by

$$\lambda_g = \frac{\lambda_0}{[1 - (\lambda_0/\lambda_c)^2]^{1/2}} = \frac{4.5}{[1 - (4.5/10)^2]^{1/2}} = 5 \text{ cm}$$

The corresponding propagation constant is

$$k_g = \frac{2\pi}{\lambda_g} = \frac{2 \times 3.14}{5} = 1.256 \text{ cm}^{-1}$$

The phase velocity v_p is given by

$$v_p = \frac{c}{[1 - (\lambda_0/\lambda_c)^2]^{1/2}} = \frac{3 \times 10^{10}}{[1 - (4.5/10)^2]^{1/2}} = 3.3 \times 10^{10} \text{ cm/sec}$$

Example 6: A hollow rectangular waveguide has $a = 6$ cm and $b = 4$ cm. The frequency of the impressed signal is 3 GHz. Compute the TE_{10} mode:

(i) cut off wavelength
(ii) guide wavelength λ_g
(iii) phase velocity v_p
(iv) group velocity v_g

Solution: (i) Cut off wavelength

$$(\lambda_c)_{\ell m} = \frac{2}{[(\ell^2/a^2) + (m^2/b^2)]^{1/2}}$$

$$(\lambda_c)_{10} = \frac{2}{[(1^2/6^2) + (0^2/4^2)]^{1/2}} = 12 \text{ cm}$$

(ii) Guide wavelength

$$\lambda_g = \frac{\lambda_0}{[1 - (\lambda_0/\lambda_c)^2]^{1/2}}$$

where

$$\lambda_0 = \frac{c}{f} = \frac{3 \times 10^{10}}{3 \times 10^9} = 10 \text{ cm}$$

Therefore,

$$\lambda_g = \frac{10}{[1 - (10/12)^2]^{1/2}} = 18.2 \text{ cm}$$

The corresponding propagation constant is

$$k_g = \frac{2\pi}{\lambda_g} = \frac{2 \times 3.14}{18.2} = 0.345 \text{ cm}^{-1}$$

(iii) The phase velocity v_p is

$$v_p = \frac{c}{[1 - (\lambda_0/\lambda_c)^2]^{1/2}} = \frac{3 \times 10^{10}}{[1 - (10/12)^2]^{1/2}} = 5.5 \times 10^{10} \text{ cm/sec}$$

(iv) The group velocity v_g is given by

$$v_g = c[1 - (\lambda_0/\lambda_c)^2]^{1/2} = 3 \times 10^{10}[1 - (10/12)^2]^{1/2} = 1.65 \times 10^{10} \text{ cm/sec}$$

Example 7: Find (a) what transmission modes are available at an operating frequency of 3 GHz in a hollow rectangular waveguide of inner dimensions 3.44 × 7.22

cm, and (b) the corresponding values of phase constant, phase velocity and group velocity.

Solution: For this problem

$$\lambda_0 = \frac{c}{f} = \frac{3 \times 10^{10}}{3 \times 10^9} = 10 \text{ cm}$$

We know the cut off wavelength

$$(\lambda_c)_{\ell m} = \frac{2}{[(\ell^2/a^2) + (m^2/b^2)]^{1/2}}$$

(a) A mode can be transmitted provided its corresponding cut off wavelength λ_c is greater than λ_0, *i.e.* $\lambda_c > 10$ cm.

Now, we consider the following cases,

(i) For TE_{00} mode, $\ell = 0$, $m = 0$, there will be no transmission.

(ii) For TE_{10} mode, $\ell = 1$, $m = 0$, $(\lambda_c)_{10} = 2a = 6.88$ cm
Since $(\lambda_c)_{10} < \lambda_0$, TE_{10} mode does not exist.

(iii) For TE_{01} mode, $\ell = 0$, $m = 1$, $(\lambda_c)_{01} = 2b = 14.44$ cm
Since $(\lambda_c)_{01} > \lambda_0$, TE_{01} mode can propagate.

It can be verified that higher TE modes do not propagate as in every case $(\lambda_c)_{\ell m} < \lambda_0$. Similarly, TM modes also do not propagate because $(\lambda_c)_{\ell m} < \lambda_0$ (lowest TM mode is TM_{11}). So, only TE_{01} mode propagates.

(b) Guide wavelength

$$\lambda_g = \frac{\lambda_0}{[1 - (\lambda_0/\lambda_c)^2]^{1/2}} = \frac{10}{[1 - (10/14.44)^2]^{1/2}} = 14.3 \text{ cm}$$

The corresponding propagation constant is

$$k_g = \frac{2\pi}{\lambda_g} = \frac{2 \times 3.14}{14.3} = 0.45 \text{ cm}^{-1}$$

The phase velocity v_p is

$$v_p = \frac{c}{[1 - (\lambda_0/\lambda_c)^2]^{1/2}} = \frac{3 \times 10^{10}}{[1 - (10/14.44)^2]^{1/2}} = 4.3 \times 10^{10} \text{ cm/sec}$$

The group velocity v_g is

$$v_g = c[1 - (\lambda_0/\lambda_c)^2]^{1/2} = 3 \times 10^{10}[1 - (10/14.44)^2]^{1/2} = 2.1 \times 10^{10} \text{ cm/sec}$$

3. Circular waveguide

When the cross section of metal tube is circular, it is called the circular waveguide, as shown in Figure 3. It has cross sectional radius a, and z-axis is parallel to axis of the tube. The conductivity of the metal is assumed to be infinite, and the interior is filled with a loss-free dielectric. For convenience, here, we use cylindrical coordinates (ρ, ϕ, z).

Figure 3: Circular waveguide.

3.1 TE waves in circular waveguide

In this case $E_z(\rho,\phi) = 0$ everywhere and $B_z(\rho,\phi)$ satisfies the wave equation

$$\nabla_t^2 B_z(\rho,\phi) + k'^2 B_z(\rho,\phi) = 0$$

$$\left[\frac{1}{\rho}\frac{\partial}{\partial\rho}\left(\rho\,\frac{\partial}{\partial\rho}\right) + \frac{1}{\rho^2}\frac{\partial^2}{\partial\phi^2}\right] B_z(\rho,\phi) + k'^2 B_z(\rho,\phi) = 0 \tag{11.80}$$

The boundary condition for $B_z(\rho,\phi)$ is

$$\frac{\partial}{\partial n}B_z(\rho,\phi) = \frac{\partial}{\partial\rho}B_z(\rho,\phi) = 0 \qquad \text{at} \quad \rho = a \quad \text{for all values of } \phi \tag{11.81}$$

Let $B_z(\rho,\phi) = X(\rho)\,Y(\phi)$, where X is a function of ρ only and Y is a function of ϕ only. Thus, we get

$$\frac{Y}{\rho}\frac{\partial}{\partial\rho}\left(\rho\,\frac{\partial X}{\partial\rho}\right) + \frac{X}{\rho^2}\frac{\partial^2 Y}{\partial\phi^2} + k'^2 XY = 0$$

On multiplying by ρ^2/XY and rearranging, we get

$$\frac{\rho}{X}\left[\frac{\partial X}{\partial\rho} + \rho\,\frac{\partial^2 X}{\partial\rho^2}\right] + k'^2\rho^2 = -\frac{1}{Y}\frac{\partial^2 Y}{\partial\phi^2} \tag{11.82}$$

The left side of (11.82) is a function of ρ only and the right side is a function of ϕ only, and both are equal to each other. This is possible only when they are separately equal to a constant, say, ℓ^2. Thus, we have

$$\frac{\rho}{X}\left[\frac{\partial X}{\partial\rho} + \rho\,\frac{\partial^2 X}{\partial\rho^2}\right] + k'^2\rho^2 = \ell^2 \qquad\qquad -\frac{1}{Y}\frac{\partial^2 Y}{\partial\phi^2} = \ell^2$$

$$\rho^2\,\frac{\partial^2 X}{\partial\rho^2} + \rho\,\frac{\partial X}{\partial\rho} + (k'^2\rho^2 - \ell^2)X = 0 \qquad\qquad \frac{\partial^2 Y}{\partial\phi^2} + \ell^2 Y = 0$$

$$\frac{\partial^2 X}{\partial\rho^2} + \frac{1}{\rho}\frac{\partial X}{\partial\rho} + \left(k'^2 - \frac{\ell^2}{\rho^2}\right)X = 0 \qquad\qquad \frac{\partial^2 Y}{\partial\phi^2} + \ell^2 Y = 0 \tag{11.83}$$

The solution of second of (11.83) is

$$Y = A_\ell \cos \ell\phi + B_\ell \sin \ell\phi$$

The first of (11.83) can be rearranged as

$$\frac{1}{k'^2}\frac{\partial^2 X}{\partial \rho^2} + \frac{1}{k'^2 \rho}\frac{\partial X}{\partial \rho} + \left(1 - \frac{\ell^2}{k'^2 \rho^2}\right) X = 0$$

$$\frac{\partial^2 X}{\partial (k'\rho)^2} + \frac{1}{k'\rho}\frac{\partial X}{\partial (k'\rho)} + \left(1 - \frac{l^2}{(k'\rho)^2}\right) X = 0 \tag{11.84}$$

Equation (11.84) is similar to the following Bessel equation

$$\frac{\partial^2 y}{\partial x^2} + \frac{1}{x}\frac{\partial y}{\partial x} + \left(1 - \frac{n^2}{x^2}\right) y = 0 \tag{11.85}$$

Solution of equation (11.85) is in form of Bessel's function $J_n(x)$. The solution of equation (11.85) is

$$X = J_\ell(k'\rho) \tag{11.86}$$

Thus, the solution of differential equation (11.80) becomes

$$B_z(\rho, \phi) = X\ Y = J_\ell(k'\rho)[A_\ell \cos \ell\phi + B_\ell \sin \ell\phi] \tag{11.87}$$

Applying the boundary condition

$$\frac{\partial}{\partial \rho} B_z(\rho, \phi) = \frac{\partial}{\partial \rho} J_\ell(k'\rho) = 0$$

at $\rho = a$ for all values of ϕ, we get

$$J'_\ell(k'a) = 0 \tag{11.88}$$

For a given value of ℓ, equation (11.88) has infinite number of solutions for $k'a$. If m-th solution is $s_{\ell m}$, then $k' = s_{\ell m}/a$. Now, solution (11.87) becomes

$$B_z(\rho, \phi) = J_\ell(s_{\ell m}\rho/a)[A_\ell \cos \ell\phi + B_\ell \sin \ell\phi] \tag{11.89}$$

3.2 Transverse components of electric field and magnetic field

The transverse components of electric field and magnetic field intensity may now be calculates as follows:

$$\vec{E}_t(\rho, \phi) = \frac{-i\omega}{(\omega^2 \epsilon\mu - k^2)} [\hat{u} \times \nabla_t \vec{B}_z(\rho, \phi)]$$

$$= \frac{-i\omega}{(\omega^2 \epsilon\mu - k^2)} [\hat{u} \times \left(\hat{\rho}\frac{\partial}{\partial \rho} + \hat{\phi}\frac{1}{\rho}\frac{\partial}{\partial \phi}\right) \vec{B}_z(\rho, \phi)]$$

$$= \frac{-i\omega}{(\omega^2 \epsilon\mu - k^2)} \left(\hat{\phi}\frac{\partial}{\partial \rho} - \hat{\rho}\frac{1}{\rho}\frac{\partial}{\partial \phi}\right) J_\ell(s_{\ell m}\rho/a)[A_\ell \cos \ell\phi + B_\ell \sin \ell\phi]$$

$$= \frac{-i\omega}{(\omega^2\epsilon\mu - k^2)} \Big[\hat{\phi}\ \frac{s_{\ell m}}{a} J'_\ell(s_{\ell m}\rho/a)[A_\ell \cos \ell\phi + B_\ell \sin \ell\phi]$$

$$-\hat{\rho}\ \frac{\ell}{\rho}\ J_\ell(s_{\ell m}\rho/a)[-A_\ell \sin \ell\phi + B_\ell \cos \ell\phi]\Big]$$

From this equation, the components $\vec{E}_\rho(\rho, \phi)$ and $\vec{E}_\phi(\rho, \phi)$ can be calculated.

Similarly,

$$\vec{B}_t(\rho, \phi) = \frac{ik}{(\omega^2\epsilon\mu - k^2)}\ \nabla_t \vec{B}_z(\rho, \phi)$$

$$= \frac{ik}{(\omega^2\epsilon\mu - k^2)} \Big(\hat{\rho}\ \frac{\partial}{\partial\rho} + \hat{\phi}\ \frac{1}{\rho}\ \frac{\partial}{\partial\phi}\Big) \vec{B}_z(\rho, \phi)$$

$$= \frac{ik}{(\omega^2\epsilon\mu - k^2)} \Big(\hat{\rho}\ \frac{\partial}{\partial\rho} + \hat{\phi}\ \frac{1}{\rho}\ \frac{\partial}{\partial\phi}\Big) J_\ell(s_{\ell m}\rho/a)[A_\ell \cos \ell\phi + B_\ell \sin \ell\phi]$$

$$= \frac{ik}{(\omega^2\epsilon\mu - k^2)} \Big(\hat{\rho}\ \frac{s_{\ell m}}{a}\ J'_\ell(s_{\ell m}\rho/a)[A_\ell \cos \ell\phi + B_\ell \sin \ell\phi]$$

$$+\hat{\phi}\ \frac{\ell}{\rho}\ J_\ell(s_{\ell m}\rho/a)[-A_\ell \sin \ell\phi + B_\ell \cos \ell\phi]\Big)$$

From this equation, the components $\vec{B}_\rho(\rho, \phi)$ and $\vec{B}_\phi(\rho, \phi)$ can be calculated.

Cut off wavelength

$$\lambda_c = \frac{2\pi}{k'} = \frac{2\pi a}{s_{\ell m}}$$

If λ_f denotes free space wavelength, wave guide wavelength λ_g is expressed as

$$\lambda_g = \frac{\lambda_f}{\sqrt{1 - (\lambda_f/\lambda_c)^2}} = \frac{\lambda_f}{\sqrt{1 - (\lambda_f s_{\ell m}/2\pi a)^2}}$$

Impedance offered is

$$\text{Impedance} = \sqrt{\frac{\mu}{\epsilon}}\ \frac{\lambda_g}{\lambda_f} = \sqrt{\frac{\mu}{\epsilon}}\ \frac{1}{\sqrt{1 - (\lambda_f/\lambda_c)^2}}$$

3.3 TM waves in circular waveguide

In this case $B_z(\rho, \phi) = 0$ everywhere and $E_z(\rho, \phi)$ satisfies the wave equation

$$\nabla_t^2 E_z(\rho, \phi) + k'^2 E_z(\rho, \phi) = 0$$

or

$$\Big[\frac{1}{\rho}\frac{\partial}{\partial\rho}\Big(\rho\ \frac{\partial}{\partial\rho}\Big) + \frac{1}{\rho^2}\frac{\partial^2}{\partial\phi^2}\Big] E_z(\rho, \phi) + k'^2 E_z(\rho, \phi) = 0 \tag{11.90}$$

The boundary condition for $E_z(\rho, \phi)$ is

$$E_z(\rho, \phi) = 0 \qquad \text{at} \quad \rho = a \quad \text{for all values of} \quad \phi \tag{11.91}$$

The solution of (11.90) is

$$E_z(\rho, \phi) = J_\ell(k'\rho)[A_\ell \cos \ell\phi + B_\ell \sin \ell\phi] \tag{11.92}$$

Applying the boundary condition (11.91), we have

$$E_z(a, \phi) = J_\ell(k'a)[A_\ell \cos \ell\phi + B_\ell \sin \ell\phi] = 0 \qquad \text{or} \qquad J_\ell(k'a) = 0 \tag{11.93}$$

For a given value of ℓ, equation (11.93) has infinite number of solutions for $k'a$. If mth solution is $s_{\ell m}$, then $k' = s_{\ell m}/a$. Now, solution (11.92) becomes

$$E_z(\rho, \phi) = J_\ell(s_{\ell m}\rho/a)[A_\ell \cos \ell\phi + B_\ell \sin \ell\phi] \tag{11.94}$$

3.4 Transverse components of electric field and magnetic field

The transverse components of electric field and magnetic field intensity may now be calculated as follows:

$$\begin{aligned}
\vec{B}_t(\rho, \phi) &= \frac{i\omega\mu\epsilon}{(\omega^2\epsilon\mu - k^2)} \left[\hat{u} \times \nabla_t \vec{E}_z(\rho, \phi)\right] \\
&= \frac{i\omega\mu\epsilon}{(\omega^2\epsilon\mu - k^2)} \left[\hat{u} \times \left(\hat{\rho}\, \frac{\partial}{\partial\rho} + \hat{\phi}\, \frac{1}{\rho}\, \frac{\partial}{\partial\phi}\right) \vec{E}_z(\rho, \phi)\right] \\
&= \frac{i\omega\mu\epsilon}{(\omega^2\epsilon\mu - k^2)} \left(\hat{\phi}\, \frac{\partial}{\partial\rho} - \hat{\rho}\, \frac{1}{\rho}\, \frac{\partial}{\partial\phi}\right) J_\ell(s_{\ell m}\rho/a)[A_\ell \cos \ell\phi + B_\ell \sin \ell\phi] \\
&= \frac{i\omega\mu\epsilon}{(\omega^2\epsilon\mu - k^2)} \left[\hat{\phi}\, \frac{s_{\ell m}}{a}\, J'_\ell(s_{\ell m}\rho/a)[A_\ell \cos \ell\phi + B_\ell \sin \ell\phi] \right. \\
&\qquad \left. -\hat{\rho}\, \frac{\ell}{\rho}\, J_\ell(s_{\ell m}\rho/a)[-A_\ell \sin \ell\phi + B_\ell \cos \ell\phi]\right]
\end{aligned}$$

From this equation, the components $\vec{B}_\rho(\rho, \phi)$ and $\vec{B}_\phi(\rho, \phi)$ can be calculated.

Similarly,

$$\begin{aligned}
\vec{E}_t(\rho, \phi) &= \frac{ik}{(\omega^2\epsilon\mu - k^2)} \nabla_t \vec{E}_z(\rho, \phi) \\
&= \frac{ik}{(\omega^2\epsilon\mu - k^2)} \left(\hat{\rho}\, \frac{\partial}{\partial\rho} + \hat{\phi}\, \frac{1}{\rho}\, \frac{\partial}{\partial\phi}\right) \vec{E}_z(\rho, \phi) \\
&= \frac{ik}{(\omega^2\epsilon\mu - k^2)} \left(\hat{\rho}\, \frac{\partial}{\partial\rho} + \hat{\phi}\, \frac{1}{\rho}\, \frac{\partial}{\partial\phi}\right) J_\ell(s_{\ell m}\rho/a)[A_\ell \cos \ell\phi + B_\ell \sin \ell\phi]
\end{aligned}$$

$$= \frac{ik}{(\omega^2\epsilon\mu - k^2)}\left(\hat{\rho}\,\frac{s_{\ell m}}{a}\,J'_\ell(s_{\ell m}\rho/a)[A_\ell\cos\ell\phi + B_\ell\sin\ell\phi]\right.$$

$$\left. +\,\hat{\phi}\,\frac{\ell}{\rho}\,J_\ell(s_{\ell m}\rho/a)[-A_\ell\sin\ell\phi + B_\ell\cos\ell\phi]\right)$$

From this equation, the components $\vec{E}_\rho(\rho,\phi)$ and $\vec{E}_\phi(\rho,\phi)$ can be calculated.

Cut off wavelength

$$\lambda_c = \frac{2\pi}{k'} = \frac{2\pi a}{s_{\ell m}}$$

If λ_f denotes free space wavelength, wave guide wavelength λ_g is expressed as

$$\lambda_g = \frac{\lambda_f}{\sqrt{1-(\lambda_f/\lambda_c)^2}} = \frac{\lambda_f}{\sqrt{1-(\lambda_f s_{\ell m}/2\pi a)^2}}$$

Impedance offered is

$$\text{Impedance} = \sqrt{\frac{\mu}{\epsilon}}\,\frac{\lambda_g}{\lambda_f} = \sqrt{\frac{\mu}{\epsilon}}\,\frac{1}{\sqrt{1-(\lambda_f/\lambda_c)^2}}$$

4. Transmission line (circuit representation)

For radio-frequency engineering, a transmission line is a specialized cable designed to conduct AC current of radio frequency. For the currents with enough high frequency, their wave nature must be considered. Transmission lines are used for purposes such as connecting radio transmitters and radio receivers with their antennas (they are then called the feed lines or feeders), distributing cable television signals, etc.

Ordinary electrical cables suffice to carry low frequency AC current, such as mains power, whose frequency is 100 to 120 Hz, and audio signals. However, they cannot be used to carry currents in the radio frequency range, above about 30 kHz, because the energy tends to radiate off the cable as radio waves, causing power losses. Radio frequency currents also tend to reflect from discontinuities in the cable such as connectors and joints, and travel back down the cable toward the source. These reflections act as bottlenecks and prevent the signal power from reaching the destination.

Transmission lines have specialized construction, and impedance matching, to carry electromagnetic signals with minimal reflections and power losses. The distinguishing feature of most transmission lines is that they have uniform cross sectional dimensions along their length, giving them a uniform impedance, called the characteristic impedance, to prevent reflections. The power losses in transmission lines increase at microwave frequencies and above, and waveguides are used instead of transmission lines. The wave guides are very long pipes to confine and guide the propagation of electromagnetic waves.

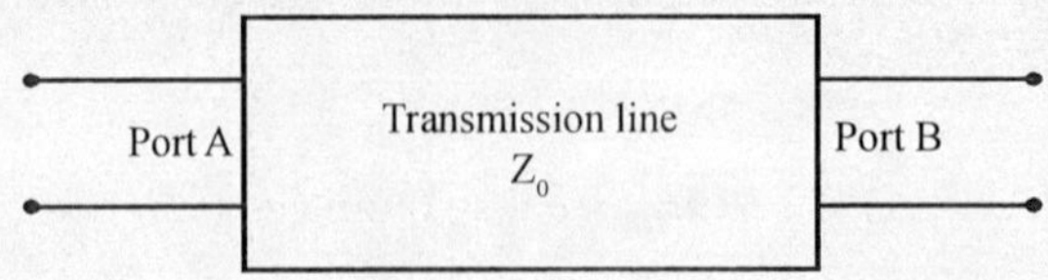

Figure 4: Two-port network of transmission line.

4.1 Four-terminal model

For analysis of an electrical transmission line, it can be modeled as a two-port network (also called a quadrupole), as shown in Figure 4.

As a simple case, the network is assumed to be linear (i.e. the complex voltage across either port is proportional to the complex current flowing into it when there are no reflections), and the two ports are considered to be interchangeable. If the transmission line is uniform along its length, then its behaviour is mainly described by a single parameter called the characteristic impedance, denoted by Z_0:

$$Z_0 = \frac{\text{complex voltage of a given wave}}{\text{complex current of the same wave}}$$

This is the ratio of complex voltage of a given wave to the complex current of the same wave at any point on the line. Typical values of Z_0 are 50 or 75 ohms for a coaxial cable, about 100 ohms for a twisted pair of wires, and about 300 ohms for a common type of untwisted pair used in radio transmission. When sending power down a transmission line, it is usually desirable that as much power as possible is absorbed by the load and as little as possible is reflected back to the source. This can be ensured by making the load impedance equal to Z_0, in which case the transmission line is said to be matched.

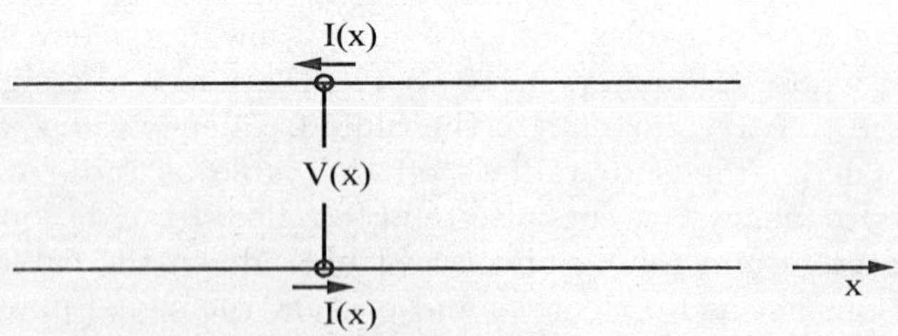

Figure 5: A transmission line is shown as two wires. At a distance x into the line, there is current $I(x)$ flowing through each wire, and there is a voltage $V(x)$ between the wires. If the current and voltage come from a single wave (with no reflection), then $V(x)/I(x) = Z_0$, where Z_0 is line characteristic impedance of the line.

Some of the power that is fed into a transmission line is lost because of its resistance. This effect is called ohmic or resistive loss. At high frequencies of transmission, another effect called dielectric loss becomes important, adding to the losses caused by resistance. Dielectric loss is caused when the insulating material inside the transmission line absorbs energy from the alternating electric field and converts it to heat.

The transmission line is modeled with a resistance (R) and inductance (L) in series with a capacitance (C) and conductance (G) in parallel, as shown in Figure 6. The resistance and conductance contribute to the loss in a transmission line. The total loss of power in a transmission line is often specified in decibels per meter (dB/m), which usually depends on the frequency of the signal. The intensity level β is defined as

$$\beta(\text{dB}) = 10 \log_{10}\left(\frac{I}{I_0}\right)$$

where I is intensity in W/m^2, I_0 the reference intensity. When $I = I_0$, we have $\beta = 0$. The unit called decibel (dB) is used to indicate that this ratio is multiplied by 10. Each factor of 10 in intensity corresponds to 10 dB. For two sources A and B, we have:

I_B/I_A	$\beta_B - \beta_A$
1	0
10	10
100	20
1000	30
10000	40

High-frequency transmission lines can be defined as those designed to carry electromagnetic waves whose wavelengths are shorter than or comparable to the length of the line.

4.2 Telegrapher's equations

The telegrapher's equations (or just telegraph equations) are a pair of linear differential equations which describe the voltage (V) and current (I) on an electrical transmission line with distance and time. They were developed by Oliver Heaviside who created the transmission line model, and are based on Maxwell's equations.

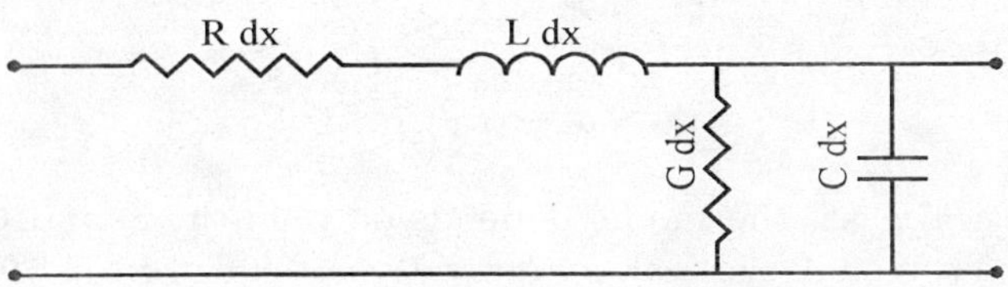

Figure 6: Transmission line element.

- The distributed resistance R of the conductors is represented by a series resistor (expressed in ohms per unit length).
- The distributed inductance L (due to the magnetic field around the wires, self-inductance, etc.) is represented by a series inductor (in henries per unit length).

- The capacitance C between the two conductors is represented by a shunt capacitor between the signal wire and the return wire (in farads per unit length).
- The conductance G of the dielectric material separating the two conductors is represented by a shunt resistor between the signal wire and the return wire (in siemens per unit length).

The model consists of an infinite series of the elements shown in Figure 6, and the values of the components are specified per unit length. R, L, C, and G may also be functions of frequency. The line voltage $V(x)$ and the $I(x)$ can be expressed in the frequency domain as

$$\frac{\partial V(x)}{\partial x} = -(R + i\,\omega\,L)I(x) \tag{11.95}$$

$$\frac{\partial I(x)}{\partial x} = -(G + i\,\omega\,C)V(x) \tag{11.96}$$

Here, $i = \sqrt{-1}$. The characteristic impedance Z_0 is

$$Z_0 = \sqrt{\frac{(R + i\omega L)}{(G + i\omega C)}}$$

Special case as a loss-less line

The transmission line is considered as a loss-less structure, when the resistance R and conductance G are negligibly small, then the equations are:

$$\frac{\partial V(x)}{\partial x} = -i\,\omega\,L\,I(x) \tag{11.97}$$

$$\frac{\partial I(x)}{\partial x} = -i\,\omega\,C\,V(x) \tag{11.98}$$

and the characteristic impedance Z_0 is

$$Z_0 = \sqrt{\frac{L}{C}}$$

In this hypothetical case, the model depends only on the L and C elements which greatly simplifies the analysis. Taking derivative of equation (11.97), we have

$$\frac{\partial^2 V(x)}{\partial x^2} = -i\,\omega\,L\,\frac{\partial I(x)}{\partial x} \tag{11.99}$$

Using equation (11.98) in (11.99), we get

$$\frac{\partial^2 V(x)}{\partial x^2} + \omega^2 LC\,V(x) = 0 \tag{11.100}$$

Similarly, we get

$$\frac{\partial^2 I(x)}{\partial x^2} + \omega^2 LC\ I(x) = 0 \tag{11.101}$$

Equation (11.100) is second order linear, homogeneous differential equation, having solution:

$$V(x) = V_{(+)}\ \mathrm{e}^{-i\omega\sqrt{LC}\ x} + V_{(-)}\ \mathrm{e}^{+i\omega\sqrt{LC}\ x} \tag{11.102}$$

where $V_{(+)}$ and $V_{(-)}$ are constants. Similarly, solution of equation (11.101) is:

$$I(x) = \sqrt{\frac{C}{L}}\ V_{(+)}\ \mathrm{e}^{-i\omega\sqrt{LC}\ x} - \sqrt{\frac{C}{L}}\ V_{(-)}\ \mathrm{e}^{+i\omega\sqrt{LC}\ x}$$

or

$$I(x) = \frac{1}{Z_0}[V_{(+)}\ \mathrm{e}^{-i\omega\sqrt{LC}\ x} - V_{(-)}\ \mathrm{e}^{+i\omega\sqrt{LC}\ x}] \tag{11.103}$$

These equations (11.102) and (11.103) are wave equations having solution as have plane waves with equal propagation in forward and reverse directions. The physical significance of this is that electromagnetic waves propagate down transmission lines and in general, there is a reflected component that interferes with the original signal. These equations are fundamental to transmission line theory.

4.3 General case of a line with losses

In the general case the loss terms, R and G, both are present. Differentiation of equation (11.95) gives

$$\frac{\partial^2 V(x)}{\partial x^2} = -(R + i\ \omega\ L)\frac{\partial I(x)}{\partial x} \tag{11.104}$$

Using equation (11.96) in (11.104), we get

$$\frac{\partial^2 V(x)}{\partial x^2} = \gamma^2\ V(x) \tag{11.105}$$

where the propagation constant γ given by

$$\gamma = \sqrt{(R + i\omega L)(G + i\omega C)}$$

is complex quantity. Similarly, we get

$$\frac{\partial^2 I(x)}{\partial x^2} = \gamma^2\ I(x) \tag{11.106}$$

These equations (11.105) and (11.106) are fundamental to transmission line theory. They are also wave equations, and have solutions similar to the special case, but these are a mixture of sines and cosines with exponential decay factors.

The solution for $V(x)$ is:

$$V(x) = V_{(+)}\ \mathrm{e}^{-\gamma x} + V_{(-)}\ \mathrm{e}^{+\gamma x}$$

where $V_{(+)}$ and $V_{(-)}$ are constants, which can be determined from boundary conditions. Suppose, $\gamma = \alpha + i\,\beta$, where α and β are, respectively, real and imaginary parts of γ. Then, we have

$$V(x) = V_{(+)}\,\mathrm{e}^{-\alpha x}\,\mathrm{e}^{-i\beta x} + V_{(-)}\,\mathrm{e}^{+\alpha x}\,\mathrm{e}^{+i\beta x}$$

Similarly, the solution for $I(x)$ is:

$$I(x) = \frac{1}{Z_0}\left[V_{(+)}\,\mathrm{e}^{-\alpha x}\,\mathrm{e}^{-i\beta x} - V_{(-)}\,\mathrm{e}^{+\alpha x}\,\mathrm{e}^{+i\beta x}\right]$$

The real and imaginary parts of γ can be computed as the following.

$$\gamma = \sqrt{(R + i\omega L)(G + i\omega C)} = \alpha + i\,\beta$$

so that,

$$\alpha^2 - \beta^2 = RG - \omega^2 LC \equiv a \qquad\qquad 2\alpha\beta = \omega(CR + LG) \equiv b$$

and

$$\alpha = \pm\sqrt{\frac{1}{2}\left(a + \sqrt{a^2 + b^2}\right)} \qquad\qquad \beta = \frac{\pm b}{\sqrt{2\left(a + \sqrt{a^2 + b^2}\right)}}$$

with the plus or minus signs chosen opposite to the direction of the wave's motion through the conducting medium. The value of a is generally negative and b is always positive.

5. Antenna

An antenna is a device to transmit and/or receive electromagnetic radio waves. It is an interface between radio waves propagating through the space and electric currents moving in metal conductors. An antenna must be tuned (matched) to the same frequency band as the radio system to which it is connected, otherwise reception and/or transmission is impaired. Characteristics of an antenna are:

- Bandwidth
- Radiation pattern
- Directivity
- Far field
- Gain
- Impedance

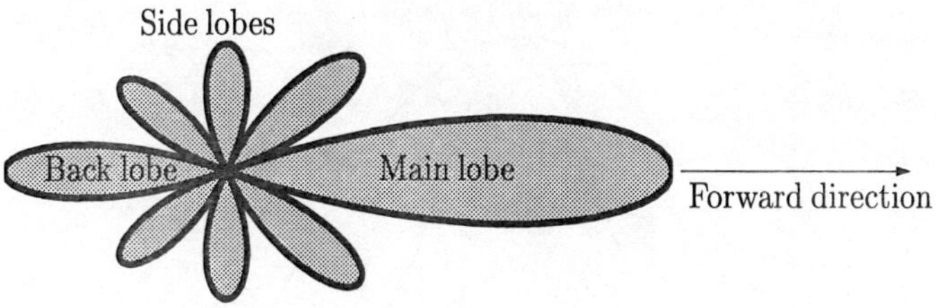

Figure 7: Radiation pattern.

Bandwidth is a fundamental parameter of antenna. It describes the range of frequencies over which the antenna can properly radiate or receive energy. The bandwidth of a signal is the difference between the signal's high and low frequencies. For instance, a signal transmitting between 40 and 50 MHz has a bandwidth of 10 MHz. This means that the energy of the signal is contained between 40 and 50 MHz (and the energy in any other frequency range is negligible).

Radiation pattern defines the variation of the power radiated by an antenna as a function of the direction away from the antenna. The main beam is the region around the direction of maximum radiation. The side-lobes are smaller beams that are away from the main beam. These side-lobes are usually radiation in undesired directions which can never be eliminated completely. The side-lobes occur at roughly from 45° to 135°.

Far field is the region far from the antenna, where the radiation pattern does not change shape with distance (R) from the antenna. If the maximum linear dimension of an antenna is D and the wavelength is λ, then the following three conditions must be for the far field region:

$$R > \frac{2D^2}{\lambda} \qquad\qquad R >> D \qquad\qquad R >> \lambda$$

Directivity is another fundamental parameter of antenna. It is a measure of the directionality of antenna's radiation pattern. An omnidirectional antenna, in theory, has zero directionality *i.e.,* it has no preferential direction. Whereas an antenna that sends or receives signals focused in one direction would have higher directivity. The intensity of radiations emitted or received by the antenna depend on the direction. The radiation pattern (or antenna pattern) refers to the directional (angular) dependence of the strength of the radio waves emitted from the antenna.

Suppose, $U(\theta, \phi)$ denotes the intensity of radiation emitted in the direction (θ, ϕ) (in spherical polar coordinate system). Here, θ is known as polar angle and ϕ is azimuthal angle. For two cases:

$$\text{(i)}\;\; U(\theta, \phi) = \sin\theta \qquad\qquad \text{(ii)}\;\; U(\theta, \phi) = (\sin\theta)^5$$

the radiation patterns are plotted in Figure 8. The radiation pattern for the antenna (i) is less directional than that for the antenna (ii); hence we expect the directivity of (ii) to be lower.

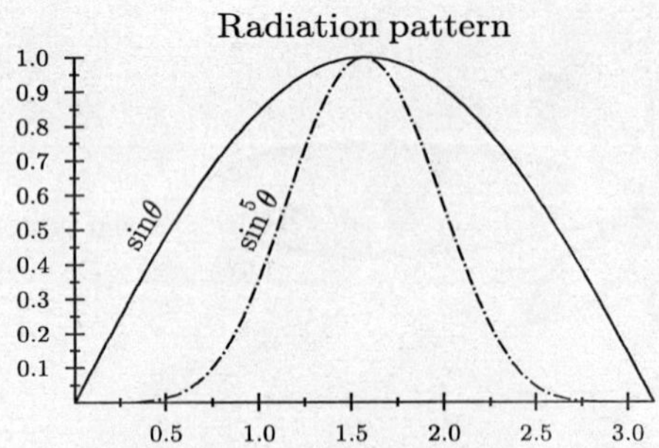

Figure 8: Radiation patterns.

Total output (radiated) power can be obtained by integrating the radiation intensity over all solid angles $d\Omega = \sin\theta \; d\theta \; d\phi$ (directions):

$$P_0 = \int_{\phi=0}^{2\pi}\int_{\theta=0}^{\pi} U(\theta,\phi)\; d\Omega = \int_{\phi=0}^{2\pi}\int_{\theta=0}^{\pi} U(\theta,\phi)\; \sin\theta \; d\theta \; d\phi$$

Solid angle of sphere is

$$\int_{\phi=0}^{2\pi}\int_{\theta=0}^{\pi} \sin\theta \; d\theta \; d\phi = 4\pi \text{ steradian}$$

Thus, the mean radiation intensity $\overline{U}$ is given by

$$\overline{U} = \frac{P_0}{4\pi} = \frac{\eta \; P_{in}}{4\pi}$$

where η is efficiency of antenna, defined as the total radiated power P_0 divided by the input power P_{in} at the feed-point:

$$\eta = \frac{P_0}{P_{in}}$$

The directive gain or directivity $D(\theta, \phi)$ of an antenna in a given direction is the ratio of its radiation intensity $U(\theta, \phi)$ in that direction to its mean radiation intensity $\overline{U}$. That is

$$D(\theta,\phi) = \frac{U(\theta,\phi)}{\overline{U}}$$

An isotropic antenna, meaning one with the same radiation intensity in all directions, therefore has directivity, D = 1, in all directions.

Antenna gain indicates the strength of a signal that an antenna sends or receives in a particular direction when compared to an ideal antenna in hypothetically the same situation. For a television antenna, where you know the signals originating direction, one would like to have an antenna with high gain. There is a relationship between **gain** G and **directivity** D as the following.

$$G = \eta \; D$$

Antenna impedance relates the voltage to the current at the input to the antenna. This is an extremely important parameter. Let us say an antenna has an impedance of 50 ohms. This means that if a sinusoidal voltage is applied at the antenna terminals with an amplitude of 1 volt, then the current has an amplitude of $1/50 = 0.02$ amp. Since the impedance is a real number, the voltage is in phase with the current. Suppose, the impedance is given by a complex number, say $Z = 50 + i\ 50$ ohms. Then $|Z| = \sqrt{(50)^2 + (50)^2} = 70.71$ and phase angle is

$$\phi = \tan^{-1}\left(\frac{50}{50}\right) = 45° = \frac{\pi}{4} \text{ rad}$$

This means the phase of the current lags the voltage by 45°. That is, the current waveform is delayed relative to the voltage waveform. To spell it out, if the voltage (with frequency f) at the antenna terminals is given by

$$V(t) = \cos(2\pi f t)$$

The electric current is then equal to:

$$I(t) = \frac{1}{70.71} \cos\left(2\pi f t - \frac{\pi}{4}\right)$$

The current lags behind the voltage.

6. Half-wave dipole antenna

A **dipole** is the prototype antenna, on the basis of which various classes of antennas are designed. A basic dipole antenna consists of two conductors (usually metal rods or wires) arranged symmetrically, with one side of the balanced feed-line from the transmitter or receiver attached to each. The most common type, the half-wave dipole, consists of two resonant elements just under a quarter wavelength long. A half-wave dipole antenna consists of two quarter-wavelength conductors placed end to end for a total length of approximately $\ell = \lambda/2$.

The current distribution is that of a standing wave, approximately sinusoidal along the length of the dipole, with a node at each end and an anti-node (peak current) at the center (feed point):

$$I(z) = I_0 \cos \omega t \ \cos kz$$

where $k = 2\pi/\lambda$ and z runs from $-\ell/2$ to $\ell/2$.
This produces a radiation pattern whose electric field is given by

$$E_\theta = \frac{\eta I_0}{2\pi r} \frac{cos(\pi \cos\theta/2)}{\sin\theta} \sin(\omega t - kr)$$

The directional factor $cos(\pi \cos\theta/2)/\sin\theta$ is barely different from $\sin\theta$ applying to the short dipole, resulting in a very similar radiation pattern as noted above. Thin linear conductors of length ℓ are in fact resonant at any integer multiple of a half-wavelength:

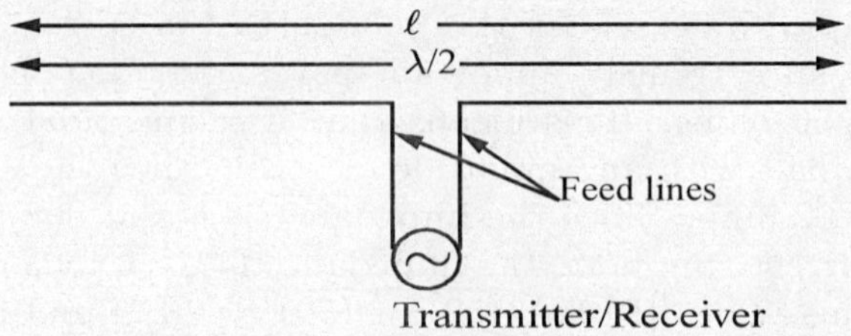

Figure 9: Half-wave dipole antenna.

$$\ell = n \, \frac{\lambda}{2}$$

where λ is the wavelength and n is an integer. For a center-fed dipole, however, there is a great dissimilarity between n being odd or being even. Dipoles which are an odd number of half-wavelengths in length have reasonably low driving point impedances (which are purely resistive at that resonant frequency).

However ones which are an even number of half-wavelengths in length, that is, an integer number of wavelengths in length, have a huge driving point impedance (albeit purely resistive at that resonant frequency).

7. Quarter-wave mono-pole antenna

A mono-pole antenna is a class of radio antenna consisting of a straight rod-shaped conductor, often mounted perpendicular over some type of conductive surface, called a ground plane.

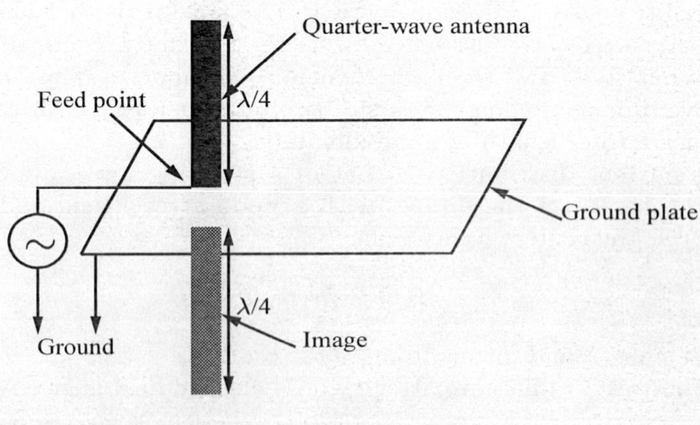

Figure 10: Quarter-wave mono-pole antenna.

The driving signal from the transmitter is applied, or for receiving antennas the output signal to the receiver is taken, between the lower end of the mono-pole and the ground plane. One side of the antenna feed-line is attached to the lower end of the mono-pole, and the other side is attached to the ground plane, which is often the Earth.

This is in contrast with a dipole antenna which consists of two identical rod conductors, with the signal from the transmitter applied between the two halves of the antenna.

The mono-pole is often used as a resonant antenna; the rod functions as an open resonator for radio waves, oscillating with standing waves of voltage and current along its length. Therefore the length of the antenna is determined by the wavelength of the radio waves, used there. The most common form is the quarter-wave mono-pole, in which the antenna is approximately one quarter of the wavelength of the radio waves.

8. Short-dipole antenna

The short-dipole antenna is the simplest of all antennas. It is simply an open-circuited wire, fed at its center as shown in Figure 11.

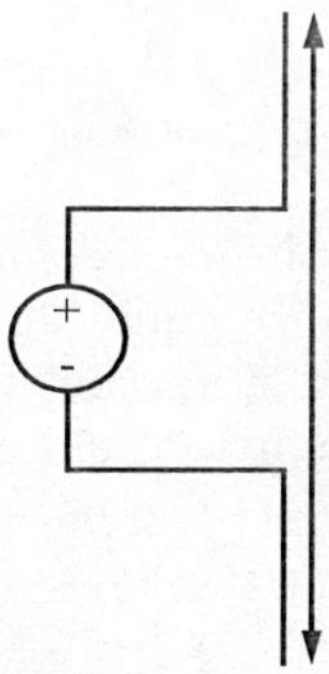

Figure 11: Short dipole antenna of length L.

The words *short* or *small* in antenna engineering always imply *relative to wavelength.* Therefore, the absolute size of the short-dipole antenna does not matter, only the size of the wire relative to the wavelength is needed. Typically, a dipole is short if its length is less than a tenth of the wavelength:

$$L < \frac{\lambda}{10}$$

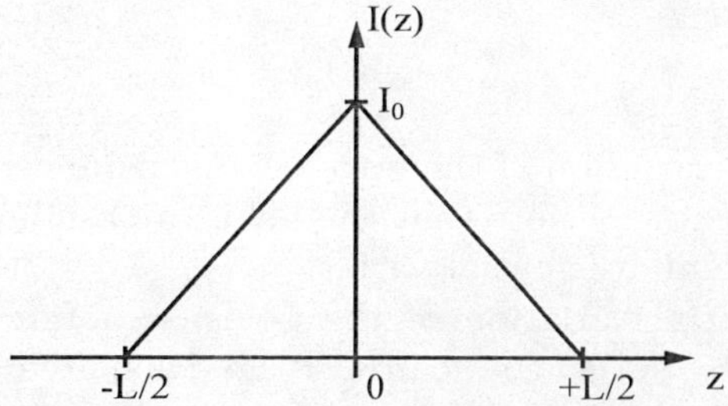

Figure 12: Current distribution along a short dipole antenna.

If the short dipole antenna is oriented along the z-axis with the center of the dipole at $z = 0$, then the current distribution on a thin, short dipole is given by:

$$I(z) = I_0\left(1 - \frac{2|z|}{L}\right)$$

This current distribution is plotted in Figure 12; the highest current is I_0 at $z = 0$. The feed-point is usually at the center of the dipole as shown in Figure 12. Note that this is the amplitude of the current distribution; it is oscillating in time sinusoidally at frequency f. The fields radiated from the short dipole antenna in the far field (in spherical polar coordinates r, θ, ϕ) are:

$$E_\theta = \frac{i\eta k I_0 L\, \mathrm{e}^{-ikr}}{8\pi r} \sin\theta \tag{11.107}$$

$$H_\phi = \frac{ikI_0 L\, \mathrm{e}^{-ikr}}{8\pi r} \sin\theta \tag{11.108}$$

where $i = \sqrt{-1}$, $k = 2\pi/\lambda$ is wave number, η is efficiency, and $E_r = E_\phi = H_r = H_\theta = 0$.

In case of AC source with frequency ω, the I_0 is replaced by $2I_h \exp(i\omega t)$, where I_h is the amplitude of AC source. The above equations (11.107) and (11.108) can be understood somewhat intuitively. First, note that in the far-field, only the E_θ electric field in the theta direction and magnetic field H_ϕ in the azimuth direction fields are nonzero. Further, these fields are orthogonal and in-phase.

Next, the fields are perpendicular to the direction of propagation, which is always in the $\hat{r}$ direction of propagation direction (away from the antenna). Also, the ratio of the E_θ to the H_ϕ is given by η (characteristic or intrinsic impedance of free space). This indicates that in the far-field region the fields are propagating like a plane-wave.

Second, the fields die off as $1/r$, which indicates the power falls off as

$$P(r) \propto \frac{1}{r^2}$$

Third, the fields are proportional to L, indicated a longer dipole radiates more power. This is true as long as increasing the length does not cause the short dipole assumption to become invalid. Also, the fields are proportional to the current amplitude I_0, which should make sense (more current, more power). The exponential term:

$$\mathrm{e}^{-ikr}$$

describes the phase-variation of the wave versus distance. The parameter k is known as the wave-number. Also, note that the fields are oscillating in time at a frequency f in addition to the above spatial variation.

Finally, the spatial variation of the fields as a function of direction from the antenna are given by $\sin\theta$. One can use the calculated fields, to find the radiated flux (power per unit area) at any point as the magnitude of the real part of the Poynting vector, given by $\vec{E} \times \vec{H}^*/2$. With $\vec{E}$ and $\vec{H}$ being at right angles and in phase, there is no imaginary part and is simply equal to $E_\theta H_\phi^*/2$ with the phase factors (the exponentials) canceling out, and we have:

$$\begin{aligned} S &= \frac{1}{2}\ E_\theta\ H_\phi^* \\ &= \frac{1}{2}\Big(\frac{i\eta k I_0 L\ \mathrm{e}^{-ikr}}{8\pi r}\ \sin\theta\Big)\Big(\frac{-ikI_0 L\ \mathrm{e}^{ikr}}{8\pi r}\ \sin\theta\Big) \\ &= \frac{1}{2}\ \frac{\eta I_0^2 L^2 k^2}{(8\pi r)^2}\ \sin^2\theta = \frac{\eta I_0^2}{32}\Big(\frac{L}{\lambda}\Big)^2\ \frac{1}{r^2}\ \sin^2\theta \end{aligned}$$

Using the above expression for the radiation in the far field for a given feed-point current, we can integrate over all solid angle to obtain the total radiated power.

$$P_{\text{total}} = \int_0^{2\pi}\int_0^{\pi} S\ (4\pi r^2)\ \frac{\mathrm{d}\Omega}{4\pi}$$

where $\mathrm{d}\Omega = \sin\theta\ \mathrm{d}\theta\ \mathrm{d}\phi$ is solid angle element. Thus, we have

$$\begin{aligned} P_{\text{total}} &= \int_0^{2\pi}\int_0^{\pi}\Big[\frac{\eta I_0^2}{32}\Big(\frac{L}{\lambda}\Big)^2\ \frac{1}{r^2}\ \sin^2\theta\Big](4\pi r^2)\ \frac{\sin\theta\ \mathrm{d}\theta\ \mathrm{d}\phi}{4\pi} \\ &= \frac{\eta I_0^2}{32}\Big(\frac{L}{\lambda}\Big)^2\int_0^{2\pi}\int_0^{\pi}\sin^3\theta\ \mathrm{d}\theta\ \mathrm{d}\phi = \frac{\pi}{12}\ \eta\ I_0^2\Big(\frac{L}{\lambda}\Big)^2 \end{aligned}$$

From this, one may find the radiation resistance, equal to the resistive (real) part of the feed-point impedance, neglecting a component due to ohmic losses. The factor 1/2 averaging of $\cos^2\theta$ for sinusoidal waves. Putting

$$P_{\text{total}} = \frac{1}{2}\ I_0^2\ R_{\text{rad}}$$

we get

$$R_{\text{rad}} = \frac{2}{I_0^2}\ P_{\text{total}} = \frac{2}{I_0^2}\ \frac{\pi}{12}\ \eta\ I_0^2\Big(\frac{L}{\lambda}\Big)^2 = \frac{\pi}{6}\ \eta\Big(\frac{L}{\lambda}\Big)^2$$

For antenna of length: $\lambda/50 < L < \lambda/10$, we have

$$R_{\text{rad}} \approx 20\pi^2\Big(\frac{L}{\lambda}\Big)^2$$

Skin effect

Skin effect is the tendency of an alternating electric current (AC) to become distributed within a conductor, such that the current density is the largest near the surface of the conductor and decreases exponentially with greater depths in the conductor. The electric current flows mainly at the *skin* of the conductor, between the outer surface and a level called the skin depth. Skin depth depends on the frequency of the alternating current; as frequency increases, current flow moves to the surface,

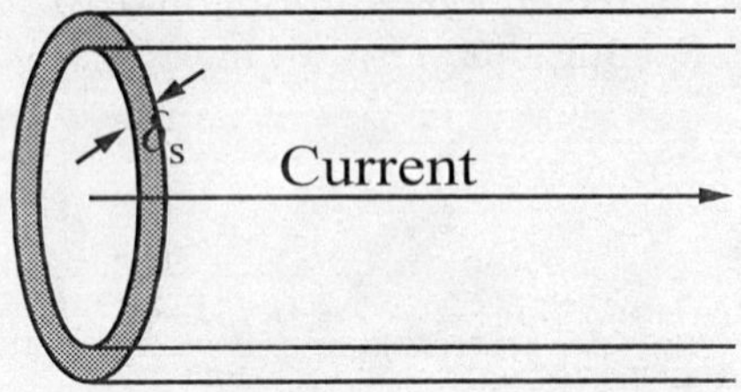

Figure 13: AC current carrying circular wire showing the skin depth.

resulting in less skin depth. Skin effect reduces the effective cross-section of the conductor and thus increases its effective resistance.

Skin effect is caused by opposing eddy currents induced by the changing magnetic field resulting from the alternating current. At 60 Hz in copper, the skin depth is about 8.5 mm. At high frequencies the skin depth becomes much smaller. The general formula for the skin depth is

$$\delta_s = \sqrt{\frac{2\rho}{\omega\mu}}\sqrt{\sqrt{1+(\rho\omega\epsilon)^2}+\rho\omega\epsilon}$$

where ρ is resistivity of the conductor, ω the angular frequency of current, μ the permeability of the conductor, ϵ the permittivity of the conductor, At frequencies much below $1/\rho\epsilon$ the quantity inside the large radical is close to unity and the formula is more usually given as:

$$\delta_s = \sqrt{\frac{2\rho}{\omega\mu}} = \sqrt{\frac{\rho}{\pi f \mu}}$$

Loss resistance

In order to calculate, the loss resistance, let us start with the equation for the DC resistance (R) of a material with resistivity (ρ), length L and cross sectional area A:

$$R = \frac{\rho\, L}{A}$$

Let us calculate the power dissipated as a result of the finite-conductivity (or the resistivity) of a material with electric current flowing through it, then we can use equation

$$P = IV = I^2 R$$

In this equation, R is the total resistance, and I is the current that flows through the material. For the case of the short dipole antenna, we have a current that drops off linearly with distance:

$$I(z) = I_0\Big(1 - \frac{2\,|z|}{L}\Big)$$

For this current, the power dissipated is

$$P_{\text{diss}} = \frac{1}{2} \int_{-L/2}^{+L/2} I^2 \, \mathrm{d}R \tag{11.109}$$

In the above equation, dR is the resistance of a small section of the antenna. We can approximate the current along a small section as constant, and so we can rewrite dR as:

$$\mathrm{d}R = \frac{\rho \, \mathrm{d}z}{A} = \frac{\rho \, \mathrm{d}z}{2\pi a \delta_s} \tag{11.110}$$

In equation (11.110), we have written the cross sectional area as the skin depth δ_s times the circumference of the wire $2\pi a$, where a denotes the radius of the wire. We can rewrite the electric current distribution of I(x) as the following:

$$I(z) = \begin{cases} I_0 \, \frac{2z}{L} & 0 \le z \le L/2 \\ 2I_0 \left[1 - \frac{z}{L}\right] & L/2 \le z \le L \end{cases} \tag{11.111}$$

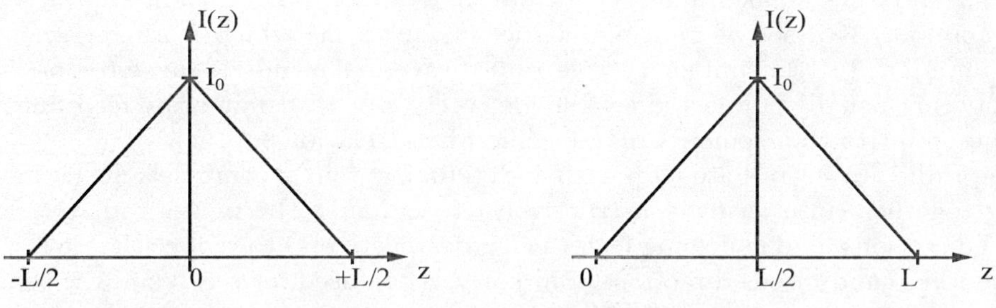

Figure 14: Distribution of current.

The power dissipation equation then becomes:

$$\begin{aligned} P_{\text{diss}} &= \frac{1}{2} \int_0^L I^2(z) \, \mathrm{d}R \\ &= \frac{1}{2} \int_0^{L/2} I^2(z) \, \mathrm{d}R + \frac{1}{2} \int_{L/2}^{L} I^2(z) \, \mathrm{d}R \\ &= \int_0^{L/2} I^2(z) \, \mathrm{d}R \end{aligned} \tag{11.112}$$

Example 8: Show that

$$\int_0^{L/2} I^2(z) \, \mathrm{d}z = \int_{L/2}^{L} I^2(z) \, \mathrm{d}z$$

Solution: We can substitute Equation (11.110) and $I(z)$ to get:

$$P_{\text{diss}} = \int_0^{L/2} \left(\frac{I_0 2z}{L}\right)^2 \frac{\rho \, \mathrm{d}z}{2\pi a \delta_s} = \left(\frac{I_0 2}{L}\right)^2 \frac{\rho}{2\pi a \delta_s} \int_0^{L/2} z^2 \, \mathrm{d}z = \frac{\rho L I_0^2}{12\pi a \delta_s} \quad (11.113)$$

On substituting the expression for the skin depth, we get

$$P_{\text{diss}} = \frac{L I_0^2}{12\pi a} \sqrt{\rho \pi f \mu} = \frac{L I_0^2}{12\pi a} \sqrt{\frac{\pi f \mu}{\sigma}} \quad (11.114)$$

Therefore, the loss-resistance is

$$R_{\text{loss}} = \frac{2}{I_0^2} P_{\text{diss}} = \frac{L}{6\pi a} \sqrt{\frac{\pi f \mu}{\sigma}}$$

σ represents the conductivity of the dipole (usually very high, if made of metal).

9. Antenna array

The main purpose of antenna array is to have very strong intensity in the desired direction and very weak intensity in undesired directions. An antenna array (or array antenna) is a set of multiple connected antennas which work together as a single antenna, to transmit or receive radio waves. An individual antenna (called elements) are usually connected to a single receiver or transmitter by feed-lines that feed the power to the elements in a specific phase relationship.

The radio waves radiated by each individual antenna combine and superpose, adding together (interfering constructively) to enhance the power radiated in the desired directions, and canceling (interfering destructively) to reduce the power radiated in other (undesired) directions. Similarly, when used for receiving, the separate radio frequency currents from the individual antennas combine in the receiver with the correct phase relationship to enhance signals received from the desired directions and cancel signals from undesired directions.

Figure 15: Antenna array. Courtesy: Google.

An antenna array can achieve higher gain (directivity), that is a narrower beam of radio waves, than could be achieved by a single element. In general, the larger the number of individual antenna elements used, the higher the gain and the narrower the beam.

The radiation pattern of such an antenna consists of a strong beam in one direction, the main lobe, plus a series of weaker beams at different angles called side-lobes, usually representing residual radiation in unwanted directions. The larger the width of the antenna and the greater the number of component antenna elements, the narrower the main lobe, and the higher the gain which can be achieved, and smaller side-lobes.

10. Radar

Radar is a detection system that uses radio waves to determine the range, angle, or velocity of an object. It can be used to detect aircrafts, ships, spacecrafts, guided missiles, motor vehicles, and to gather weather information, etc. A radar system consists of a transmitter producing electromagnetic waves in the radio or microwaves domain, a transmitting antenna, a receiving antenna (often the same antenna is used for transmitting and receiving) and a receiver and processor to determine properties of the object(s).

Figure 16: Radar. Courtesy: Google.

Radio waves (pulsed or continuous) from the transmitter are reflected back from the object and they return to the receiver, giving information about the object's location and speed. Radar was developed secretly for military use by several nations in the period before and during the World War II.

The term RADAR was coined in 1940 by the United States Navy as an acronym for **RAdio Detection And Ranging**. The term radar has since entered English and other languages as a common noun, losing all capitalization.

The modern uses of radar are highly diverse, including air and terrestrial traffic control, radar astronomy, air-defense systems, antimissile systems, marine radars to locate landmarks and other ships, aircraft anti-collision systems, ocean surveillance systems, outer space surveillance and rendezvous systems, meteorological precipitation monitoring, altimetry and flight control systems, guided missile target locating systems, self-driving cars, and ground-penetrating radar for geological observations. High tech radar systems are associated with digital signal processing, machine learning and are capable of extracting useful information from very high noise levels.

Review Questions

1. Discuss basic concepts of waveguides.
2. Describe the theory of a rectangular waveguide.
3. Describe the theory of a circular waveguide.
4. Telegrapher's equations for a transmission line and discuss their solutions.
5. Discuss the circuit representation of transmission line.
6. What is an antenna? Describe about the half-wave dipole antenna.
7. What is an antenna? Describe about the quarter-wave mono-pole antenna.
8. What is an antenna? Describe about the short-dipole antenna.
9. Write notes on the following
   (i) TE mode in waveguide
   (ii) TM mode in waveguide
   (iii) Telegrapher's equations
   (iv) Antenna
   (v) Skin effect
   (vi) Radiation pattern
   (vii) Band width
   (viii) Directivity
   (ix) Antenna array
   (x) Radar

Multiple-Choice Questions

1. A waveguide is
   A. made of high conducting material filled with dielectric.
   B. made of high conducting material filled with low conductivity material
   C. made of dielectric filled with high conducting material.
   D. made of dielectric filled with low conducting material.

   Ans. A

2. Which is not the shape of the cross section of waveguide?
   A. circle B. square C. triangle

   Ans. C

3. In Transverse Electric (TE) mode, z-component (along the direction of propagation of wave) of what is always absent?

   A. Electric field B. Magnetic field

   C. Electric and magnetic fields both

   D. Neither electric field nor magnetic field

   Ans. A

4. In Transverse Magnetic (TM) mode, z-component (along the direction of propagation of wave) of what is always absent?

   A. Electric field B. Magnetic field

   C. Electric and magnetic fields both

   D. Neither electric field nor magnetic field

   Ans. B

5. In Transverse Electromagnetic (TEM) mode, z-component (along the direction of propagation of wave) of what is always absent?

   A. Electric field B. Magnetic field

   C. Electric and magnetic fields both

   D. Neither electric field nor magnetic field

   Ans. C

6. The indices (ℓ, m) correspond to the length a and width b of a rectangular waveguide. When $a < b$, which Transverse Electric (TE) mode of the waveguide does not represent a wave propagation?

   A. 0 0 B. 0 1 C. 1 0 D. 1 1

   Ans. A

7. The indices (ℓ, m) correspond to the length a and width b of a rectangular waveguide. When $a < b$, which of the Transverse Electric (TE) mode of the waveguide is the principal mode.

   A. 0 0 B. 0 1 C. 1 0 D. 1 1

   Ans. B

8. The indices (ℓ, m) correspond to the length a and width b of a rectangular waveguide. When $a < b$, which of the Transverse Magnetic (TM) mode of the waveguide is the lowest mode.

   A. 0 0 B. 0 1 C. 1 0 D. 1 1

   Ans. D

9. In a rectangular waveguide, in the lowest mode, frequency in which mode is large?

   A. TE B. TM C. depend on dimensions of waveguide.

   Ans. B

10. Intensity of source A is 1000 times larger than that of the source B. How much is the difference in their intensity levels in dB.

    A. 3 B. 20 C. 30 D. 40

    Ans. C

11. In the radiation pattern of an antenna, the main lobe (in degrees) occurs, generally, between

    A. -20 and 20 B. -45 and 45 C. -90 and 90 D. -100 and 100

    Ans. B

12. In the radiation pattern of an antenna, the side lobes (in degrees) occur on both sides, generally, between

    A. 20 and 90 B. 45 and 135 C. 45 and 90 D. 30 and 100

    Ans. B

13. For a half-wave dipole antenna, the ratio of the size of antenna to the wavelength of radiation is

    A. 0.25 B. 0.5 C. 2.0 D. 4.0

    Ans. B

14. For a quarter-wave mono-pole antenna, the ratio of the size of antenna to the wavelength of radiation is

    A. 0.25 B. 0.5 C. 2.0 D. 4.0

    Ans. A

15. For a short-dipole antenna, the ratio of the size of antenna to the wavelength of radiation is

    A. < 0.1 B. 0.25 C. 0.5 D. > 1.0

    Ans. A

16. If impedance of a circuit is real, which of the following statements is correct?

    A. current and voltage are in phase

    B. current leads to the voltage

    C. current lags behind the voltage

    Ans. A

17. If impedance of a circuit is complex, which of the following statements is correct?

    A. current and voltage are in phase

    B. current and voltage are not in phase

    Ans. B

18. If impedance of a circuit is $Z = 50+i\,50$ ohm, which of the following statements is correct?

    A. current and voltage are in phase

    B. current leads to the voltage

    C. current lags behind the voltage

    Ans. C

19. If impedance of a circuit is $Z = 50-i\,50$ ohm, which of the following statements is correct?

    A. current and voltage are in phase

    B. current leads to the voltage

    C. current lags behind the voltage

    Ans. B

Chapter 12

Principle of Relativity

After several attempts, when the result of Michelson-Morley experiment could not be explained, in 1905, Einstein came with two bold postulates which gave birth to a new type of physics, known as the special theory of relativity. It is based on two postulates:

1. All the physical laws are the same in all the inertial (non-accelerated or non-rotating) reference frames.

2. The speed of light in the free space (vacuum) does not depend on the the motion of the source and/or of the observer. That is, the speed of light in vacuum is the same in all inertial reference frames.

This constancy of the speed of light in all inertial reference frames makes the electromagnetic waves to behave differently from the mechanical waves which require a material medium for their propagation. The second postulate of Einstein could help in explaining the result of the Michelson - Morley experiment.

1. Galilean transformations

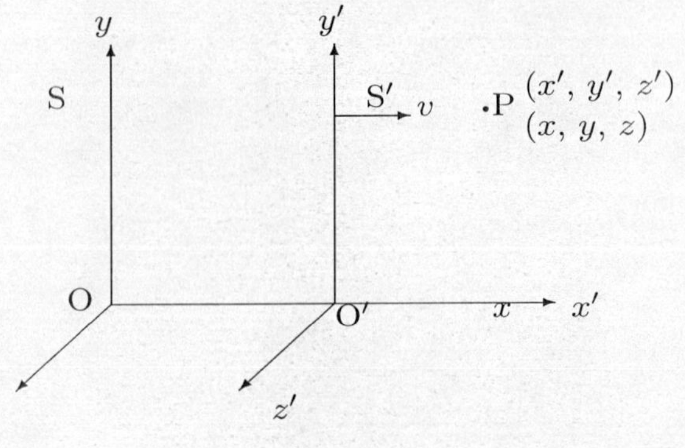

Figure 1: Inertial reference frame S′ is moving with velocity v along x-axis with to another inertial reference frame S.

Let us consider two inertial reference frames S and S′, as shown in figure 1. For convenience, suppose S′ is moving with a constant linear velocity v along the x-axis with respect to S. Suppose, at the time $t = 0$ both S and S′ coincided with each other. At time t, the distance OO′ is vt. Suppose, at time t, the coordinates of P measured in S and S′ are (x, y, z) and (x', y', z'), respectively. Then, we have

$$x' = x - vt \qquad y' = y \qquad z' = z \qquad t' = t \tag{12.1}$$

In classical treatment, the time is taken as absolute, *i.e.*, the time does not depend on the reference frame. It is equivalent to say that the time and space are independent of each other. That is, there is no coupling between the space and time. (Note that in equation (12.1) x' depends on t, but t' does not depend on x.) The inverse transformations are

$$x = x' + vt \qquad y = y' \qquad z = z' \qquad t = t' \tag{12.2}$$

The equations (12.1) and (12.2) are known as the Galilean transformations for the coordinates. Remember that here we have assumed the motion of S′ along the x-axis.

According to the Galilean transformations, when two events are simultaneous in one inertial reference frame, they are simultaneous in all the inertial reference frames. It was because the **speed of light was assumed to be infinite.**

2. Occurrence of two events

Another important consequence of the constancy of the speed of light is regarding the concept of the simultaneity. The speed of light was measured as 3×10^8 m/s in the free space. According to the second postulate of Einstein, two simultaneous events occurring in two different places for an observer in S is not simultaneous for another observer in S′.

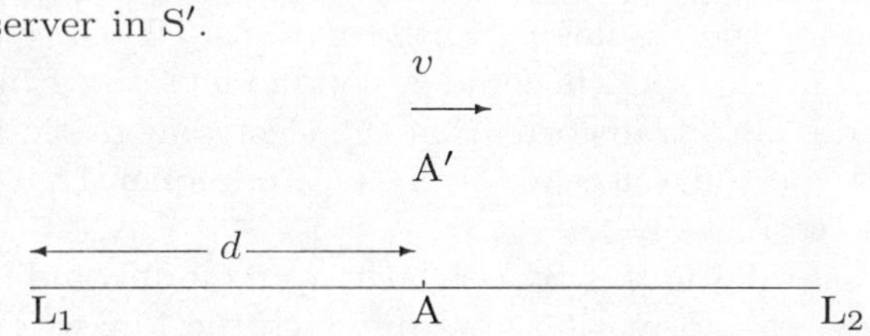

Figure 2: Relative nature of the simultaneity. The observer A′ is moving relative to the observer A with velocity v along the x-axis.

In order to illustrate this, let us consider two observers A and A′ in the inertial reference frames S and S′, respectively. Obviously the observer A′ is moving with a velocity v relative to A in the direction of x-axis. Suppose, there are two flash lamps L_1 and L_2, equidistant from A (Figure 2). The observer A can fire the lamps L_1 and L_2 by some remote control device.

Suppose, these lamps are fired simultaneously by A when A′ was overhead of A. Since the lamps are at equal distance from A, and A is not moving, the lights from both the lamps reach to A at the same instant, so that the firing is simultaneous for him. Let us see what is the case with the observer A′. As the light takes finite time to reach A and during this time A′ would move towards L_2. The signal from L_2 would travel a shorter distance for reaching to A′ whereas the signal from L_1 would travel a longer distance for reaching to A′. Thus, A′ would see that the lamp L_2 was fired earlier than L_1. Thus, for the observer A′, the events are not simultaneous.

Let, $L_1A = L_2A = d$. Then the time taken by light from L_1 and from L_2, to reach at A is $t = d/c$. During this time t, the observer A′ would move a distance $vt = vd/c$ towards the lamp L_2. Thus, the time taken by light from L_2 to reach A′ is

$$t'_2 = \frac{1}{c}\left(d - \frac{vd}{c}\right) = \frac{d}{c}\left(1 - \frac{v}{c}\right)$$

and the time t'_1 taken by light from L_1 to reach to A′ is

$$t'_1 = \frac{d}{c}\left(1 + \frac{v}{c}\right)$$

Here, we have assumed that the speed of light does not depend on the motion of the source and/or the observer (Einstein's second postulate). Thus, the time difference is

$$\Delta t' = t'_1 - t'_2 = \frac{2vd}{c^2}$$

It shows that the flashing of the two lamps L_1 and L_2 are simultaneous for the observer A whereas it is not simultaneous for the observer A′. This shows that the simultaneity is not absolute. That is, the simultaneity is relative. Hence, the second postulate of Einstein is not consistent with the Galilean transformations, which tells that the simultaneity is absolute, as the time is absolute.

The second postulate of Einstein says that the time is not absolute and it intermixes the space and time coordinates and therefore it is no more correct to take $t = t'$, which has been true for the Galilean transformations. Thus, some new law for the transformation was needed which could accommodate the Einstein's second postulate. Further, under that transformation the electromagnetic laws should remain invariant in all the inertial reference frames. Consequently, the Galilean transformations are not the correct one.

For finding out such transformations, scientists attempted, but could not succeed. Lorentz in 1904 derived new transformation laws in which the Maxwell's equations were invariant from one inertial reference frame to another. But, it was based on a number of ad hoc assumptions. Based on his postulates, Einstein in 1905 rederived Lorentz transformations, as discussed in the following section.

Example 1. A rod AB is placed parallel to the x-axis and is observed from the inertial reference frames S and S′. Show that a nonlinear relationship between x and x' given by $x' = ax^2$ does not preserve the homogeneity of the space.

Solution: Let us consider a rod AB of length 3 m in the reference frame S is placed at $x_1 = 1$ m and $x_2 = 4$ m. When these coordinates are observed in S′, we have $x'_1 = a(1)^2$ and $x'_2 = a(4)^2$, so that the length of the rod observed in S′ is $x'_2 - x'_1 = 15a$ m.

Suppose, the rod is shifted in S so that the new positions of A and B are at $x_1 = 2$ m and $x_2 = 5$ m. When these coordinates are observed in S′, we have $x'_1 = a(2)^2$ and $x'_2 = a(5)^2$, so that the length of the rod observed in S′ is $x'_2 - x'_1 = 21a$ m.

Thus, for a mere shifting of the rod in S, its length observed in S′ changes from $15a$ m to $21a$ m. This clearly shows that the space is inhomogeneous.

Example 2. Show that for a homogeneous medium, the transformations between two reference frames must be linear.

Solution: Here, we prove that when the transformation is linear, the medium is homogeneous. Suppose the transformation is given by $x' = ax$. Let us consider a rod AB of length 3 m in the reference frame S is placed at $x_1 = 1$ m and $x_2 = 4$ m. When these coordinates are observed in S′, we have $x'_1 = a(1)$ and $x'_2 = a(4)$, so that the length of the rod observed in S′ is $x'_2 - x'_1 = 3a$ m.

Suppose, the rod is shifted in S to that the new positions of A and B are at $x_1 = 2$ m and $x_2 = 5$ m. When these coordinates are observed in S′, we have $x'_1 = a(2)$ and $x'_2 = a(5)$, so that the length of the rod observed in S′ is $x'_2 - x'_1 = 3a$ m.

Thus, for a mere shifting of the rod in S, its length observed in S′ does not change. This clearly shows that the space is homogeneous.

3. Lorentz transformations

Let us consider two inertial reference frames S and S′ moving with uniform velocities (Figure 3). Suppose, the two reference frames coincide at the starting time. Now, we consider a four-dimensional space in which a point P represents an event occurring at $(x,\ y,\ z,\ t)$ when observed in S and at $(x',\ y',\ z',\ t')$ when observed in S′.

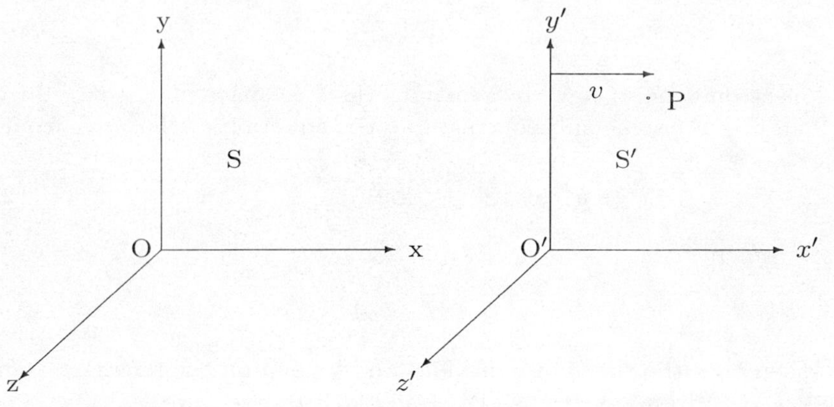

Figure 3: An inertial reference frame S′ is moving with velocity $\vec{v}$ relative to another inertial reference frame S along their common x-axis.

Here, the aim is to find out some transformations between the two coordinate systems, which satisfy the Einstein's second postulate that the speed of light does not depend on the reference frame. Moreover, the transformation should be linear so that the homogeneity of the space is maintained.

Since S′ is moving parallel to x-axis, we have $y = y'$ and $z = z'$. However, in the new transformation x and t are intermixed. Let us take x' and t' as linear combinations of x and t, and write

$$x' = lx + mt \tag{12.3}$$

and

$$t' = nx + pt \tag{12.4}$$

Now, we need to find out the values of the coefficients l, m, n, p. Figure 3 shows that the origin O′ of reference frame S′ satisfies

$$x' = 0 \tag{12.5}$$

However, in the reference frame S, the motion of O′ is described as

$$x = vt \qquad \text{or} \qquad x - vt = 0 \tag{12.6}$$

Equations (12.5) and (12.6) both describe the motion of the same point O′, the two equations must have a linear relationship between them. Thus, we write

$$x' = q(x - vt) \tag{12.7}$$

Comparison of equations (12.3) and (12.7) shows that

$$l = q \qquad \text{and} \qquad m = -qv$$

Thus, we have

$$x' = l(x - vt) \tag{12.8}$$

Let a spherical electromagnetic wave be sent into the free space at $t = t' = 0$ from the origins of S and S′. At some subsequent time, the equation of the wave front in S is

$$x^2 + y^2 + z^2 - c^2t^2 = 0 \tag{12.9}$$

and of the wave front in S′ is

$$x'^2 + y'^2 + z'^2 - c^2t'^2 = 0 \tag{12.10}$$

Here, we have taken that the speed of light does not depend on the reference frame. Since $y = y'$ and $z = z'$, from equations (12.9) and (12.10), we have

$$x'^2 - c^2t'^2 = x^2 - c^2t^2 \tag{12.11}$$

Using equations (12.4) and (12.8) in (12.11), we get

$$l^2(x^2 - 2vxt + v^2t^2) - c^2(n^2x^2 + 2npxt + p^2t^2) = x^2 - c^2t^2$$

Equating the coefficients of x^2, xt and t^2 on the two sides of this equation, we get

$$\ell^2 - c^2n^2 = 1 \tag{12.12}$$

$$v\ell^2 + c^2np = 0 \tag{12.13}$$

$$v^2\ell^2 - c^2p^2 = -c^2 \tag{12.14}$$

Solution of these equations (12.12), (12.13) and (12.14) is a bit tricky task. Equations (12.3) and (12.4) show that when there is no coupling between space and time, we have $m = n = 0$ and l may be equal to p, so that the transformation of space and of time has similar kind of variation. Using $\ell = p$ in equation (12.14), we get

$$\ell = p = \frac{1}{\sqrt{1-\beta^2}} \tag{12.15}$$

Using equation (12.15) in (12.12) or in (12.13), we get

$$n = -\frac{v}{c^2\sqrt{1-\beta^2}} \tag{12.16}$$

where $\beta = v/c$. Using equation (12.15) in (12.8), we have

$$x' = \frac{x - vt}{\sqrt{1-\beta^2}}$$

Using equations (12.15) and (12.16) in (12.4), we get

$$t' = \frac{1}{\sqrt{1-\beta^2}}\left[t - \frac{\beta x}{c}\right]$$

Thus, finally, the Lorentz transformations from S to S$'$ are

$$x' = \frac{x - vt}{\sqrt{1-\beta^2}}; \qquad y' = y; \qquad z' = z; \qquad t' = \frac{t - \beta x/c}{\sqrt{1-\beta^2}} \tag{12.17}$$

The inverse Lorentz transformations from S$'$ to S are

$$x = \frac{x' + vt'}{\sqrt{1-\beta^2}}; \qquad y = y'; \qquad z = z'; \qquad t = \frac{t' + \beta x'/c}{\sqrt{1-\beta^2}} \tag{12.18}$$

It can be easily noted that when $v << c$, the Lorentz transformations reduce to the Galilean transformations and we have $t = t'$.

3.1 Inverse Lorentz transformations

Though we have written inverse transformations above, but the derivation of them is as the following. Since S$'$ is moving parallel to x-axis, we have $y' = y$ and $z' = z$. However, in the new transformation x and t are intermixed. Let us take x and t as linear combinations of x' and t', and write

$$x = \ell x' + mt' \tag{12.19}$$

and

$$t = nx' + pt' \tag{12.20}$$

Now, we need to find out the values of the coefficients l, m, n, p. Figure 3 shows that the origin O of reference frame S satisfies

$$x = 0 \tag{12.21}$$

However, in the reference frame S$'$, the motion of O is described as

$$x' = -vt' \qquad \text{or} \qquad x' + vt' = 0 \tag{12.22}$$

Equations (12.21) and (12.22) both describe the motion of the same point O, the two equations must have a linear relationship between them. Thus, we write

$$x = q(x' + vt') \tag{12.23}$$

Comparison of equations (12.19) and (12.23) shows that

$$\ell = q \qquad \text{and} \qquad m = qv$$

Thus,

$$x = \ell(x' + vt') \tag{12.24}$$

Let a spherical electromagnetic wave be sent into the free space at $t = t' = 0$ from the origins of S and S$'$. At some subsequent time, the equation of the wave front in S is

$$x^2 + y^2 + z^2 - c^2t^2 = 0 \tag{12.25}$$

and of the wave front in S$'$ is

$$x'^2 + y'^2 + z'^2 - c^2t'^2 = 0 \tag{12.26}$$

Here, we have taken that the speed of light does not depend on the reference frame. Since $y' = y$ and $z' = z$, from equations (12.25) and (12.10), we have

$$x^2 - c^2t^2 = x'^2 - c^2t'^2 \tag{12.27}$$

Using equations (12.20) and (12.24) in (12.27), we get

$$l^2(x'^2 + 2vx't' + v^2t'^2) - c^2(n^2x'^2 + 2npx't' + p^2t'^2) = x'^2 - c^2t'^2$$

Equating the coefficients of x'^2, $x't'$ and t'^2 on the two sides of this equation, we get

$$\ell^2 - c^2n^2 = 1 \tag{12.28}$$

$$v\ell^2 - c^2np = 0 \tag{12.29}$$

$$v^2\ell^2 - c^2p^2 = -c^2 \tag{12.30}$$

Solution of these equations (12.28), (12.29) and (12.30) is a bit tricky task. Equations (12.19) and (12.20) show that when there is no coupling between space and time, we have $m = n = 0$ and ℓ may be equal to p, so that the transformation of space and of time has similar kind of variation. Using $\ell = p$ in equation (12.30), we get

$$\ell = p = \frac{1}{\sqrt{1-\beta^2}} \tag{12.31}$$

Using equation (12.31) in (12.28) or in (12.29), we get

$$n = \frac{v}{c^2\sqrt{1-\beta^2}} \tag{12.32}$$

Using equation (12.31) in (12.24), we have

$$x = \frac{x' + vt'}{\sqrt{1-\beta^2}}$$

Using equations (12.31) and (12.32) in (12.20), we get

$$t = \frac{1}{\sqrt{1-\beta^2}}\left[t' + \frac{\beta x'}{c}\right]$$

Thus, finally, the Lorentz transformations from S′ to S are

$$x = \frac{x' + vt'}{\sqrt{1-\beta^2}}; \qquad y = y'; \qquad z = z'; \qquad t = \frac{t' + \beta x'/c}{\sqrt{1-\beta^2}} \tag{12.33}$$

It can be easily noted that when $v << c$, the Lorentz transformations reduce to the Galilean transformations and we have $t = t'$.

Example 3. Show that $x^2 + y^2 + z^2 - c^2t^2 = x'^2 + y'^2 + z'^2 - c^2t'^2$ under Lorentz transformations.

Solution: We can start from any set, Equations (12.17) or (12.18), of Lorentz transformations. Using

$$x' = \frac{x - vt}{\sqrt{1-\beta^2}}; \qquad y' = y; \qquad z' = z; \qquad t' = \frac{t - \beta x/c}{\sqrt{1-\beta^2}}$$

we have

$$x'^2 + y'^2 + z'^2 - c^2t'^2 = \frac{(x - vt)^2}{1-\beta^2} + y^2 + z^2 - \frac{c^2(t - \beta x/c)^2}{1-\beta^2}$$

$$= \frac{(x^2 - 2xvt + v^2t^2) - (c^2t^2 - 2xvt + \beta^2 x^2)}{1-\beta^2} + y^2 + z^2$$

$$= \frac{x^2(1-\beta^2) - c^2t^2(1 - v^2/c^2)}{1-\beta^2} + y^2 + z^2 = x^2 + y^2 + z^2 - c^2t^2$$

Example 4. An event occurs at the position (50, 20, 30) and time $t = 1 \times 10^{-4}$ s in an inertial reference frame S. Consider another inertial reference frame S′ which is moving with velocity $v = 2.4 \times 10^8$ m/s relative to that of S along the direction of x-axis. Calculate the position and time of the event when measured in S′: (i) under Galilean transformations and (ii) under Lorentz transformations.

Solution: In the reference frame S, we have

$$x = 50 \qquad y = 20 \qquad z = 30 \qquad t = 1 \times 10^{-4}$$

(i) The Galilean transformations are

$$x' = x - vt \qquad y' = y \qquad z' = z \qquad t' = t$$

Thus, we have

$$x' = 50 - 2.4 \times 10^8 \times 1 \times 10^{-4} = -23950$$

$$y' = 20 \qquad z' = 30 \qquad t' = 1 \times 10^{-4}$$

(ii) The Lorentz transformations are

$$x' = \frac{x - vt}{\sqrt{1-\beta^2}}; \qquad y' = y; \qquad z' = z; \qquad t' = \frac{t - \beta x/c}{\sqrt{1-\beta^2}}$$

We have $\beta = 2.4 \times 10^8 / 3 \times 10^8 = 0.8$. Using the values, we have

$$x' = \frac{x - vt}{\sqrt{1-\beta^2}} = \frac{50 - 2.4 \times 10^8 \times 1 \times 10^{-4}}{\sqrt{1 - 0.64}} = -9916.7$$

$$t' = \frac{t - \beta x/c}{\sqrt{1-\beta^2}} = \frac{1 \times 10^{-4} - 0.8 \times 50/3 \times 10^8}{\sqrt{1 - 0.64}} = 1.66 \times 10^{-4}$$

Thus, we have

$$x' = -9916.7 \qquad y' = 20 \qquad z' = 30 \qquad t' = 1.66 \times 10^{-4}$$

4. Four-vectors

As in relativistic dynamics, time and space are coupled, let us consider a four-dimensional space with coordinates (x, y, z, ict). In four-dimensional space, the components are denoted by x^i, where the superscript i can take the values 1, 2, 3, 4, so that

$$x^1 = x \qquad x^2 = y \qquad x^3 = z \qquad x^4 = ict$$

The square of the length of radius four-vector is

$$\begin{aligned}(\text{length of radius four-vector})^2 &= (x^1)^2 + (x^2)^2 + (x^3)^2 + (x^4)^2 \\ &= x^2 + y^2 + z^2 - c^2t^2\end{aligned}$$

Line element ds is expressed as

$$\begin{aligned}\mathrm{d}s^2 &= (\mathrm{d}x^1)^2 + (\mathrm{d}x^2)^2 + (\mathrm{d}x^3)^2 + (\mathrm{d}x^4)^2 \\ &= (\mathrm{d}x)^2 + (\mathrm{d}y)^2 + (\mathrm{d}z)^2 + (ic\mathrm{d}t)^2 \\ &= \mathrm{d}x^2 + \mathrm{d}y^2 + \mathrm{d}z^2 - c^2\mathrm{d}t^2\end{aligned}$$

The Lorentz transformations are

$$x' = \gamma(x - vt); \qquad y' = y; \qquad z' = z \qquad t' = \gamma(t - \beta x/c)$$

where $\beta = v/c$ and $\gamma = 1/\sqrt{1-\beta^2}$. Thus, we have

$$x' = \gamma(x - \beta ct); \qquad y' = y; \qquad z' = z; \qquad ict' = \gamma(ict - i\beta x)$$

or

$$x'^1 = \gamma(x^1 + i\beta x^4); \qquad x'^2 = x^2; \qquad x'^3 = x^3; \qquad x'^4 = \gamma(x^4 - i\beta x^1)$$

It can be expressed as

$$\begin{bmatrix} x'^1 \\ x'^2 \\ x'^3 \\ x'^4 \end{bmatrix} = \begin{bmatrix} \gamma & 0 & 0 & i\gamma\beta \\ 0 & 1 & 0 & 0 \\ 0 & 0 & 1 & 0 \\ -i\gamma\beta & 0 & 0 & \gamma \end{bmatrix} \begin{bmatrix} x^1 \\ x^2 \\ x^3 \\ x^4 \end{bmatrix}$$

The inverse Lorentz transformations are

$$x^1 = \gamma(x'^1 - i\beta x'^4); \qquad x^2 = x'^2; \qquad x^3 = x'^3 \qquad x^4 = \gamma(x'^4 + i\beta x'^1)$$

It can be expressed as

$$\begin{bmatrix} x^1 \\ x^2 \\ x^3 \\ x^4 \end{bmatrix} = \begin{bmatrix} \gamma & 0 & 0 & -i\gamma\beta \\ 0 & 1 & 0 & 0 \\ 0 & 0 & 1 & 0 \\ i\gamma\beta & 0 & 0 & \gamma \end{bmatrix} \begin{bmatrix} x'^1 \\ x'^2 \\ x'^3 \\ x'^4 \end{bmatrix}$$

Example 5: For coordinates, the four-vector x^i, show that $x^i x^i$ is invariant under Lorentz transformation.

Solution: The coordinates four-vector follow Lorentz transformation

$$x'^1 = \gamma(x^1 + i\beta x^4); \qquad x'^2 = x^2; \qquad x'^3 = x^3; \qquad x'^4 = \gamma(x^4 - i\beta x^1)$$

We want to show that $x^i x^i = x'^i x'^i$. That is,

$$x^1x^1 + x^2x^2 + x^3x^3 + x^4x^4 = x'^1x'^1 + x'^2x'^2 + x'^3x'^3 + x'^4x'^4$$

or

$$(x^1)^2 + (x^2)^2 + (x^3)^2 + (x^4)^2 = (x'^1)^2 + (x'^2)^2 + (x'^3)^2 + (x'^4)^2$$

Using Lorentz transformation, we have

$$(x'^1)^2 + (x'^2)^2 + (x'^3)^2 + (x'^4)^2 = \gamma^2(x^1 + i\beta x^4)^2 + (x^2)^2 + (x^3)^2 + \gamma^2(x^4 - i\beta x^1)^2$$

$$= \gamma^2[(x^1)^2 + 2i\beta x^1x^4 - \beta^2(x^4)^2] + (x^2)^2 + (x^3)^2 + \gamma^2[(x^4)^2 - 2i\beta x^1x^4 - \beta^2(x^1)^2]$$

$$= \gamma^2[(x^1)^2(1-\beta^2) + (x^4)^2(1-\beta^2)] + (x^2)^2 + (x^3)^2 = (x^1)^2 + (x^2)^2 + (x^3)^2 + (x^4)^2$$

5. Four-vector of physical quantities

In general, A^i (with components A^1, A^2, A^3, A^4) is a four-dimensional vector (or four-vector) whose components follow the Lorentz transformations. That is, we have Lorentz transformations

$$A'^1 = \gamma(A^1 + i\beta A^4); \qquad A'^2 = A^2; \qquad A'^3 = A^3; \qquad A'^4 = \gamma(A^4 - i\beta A^1)$$

and inverse Lorentz transformations

$$A^1 = \gamma(A'^1 - i\beta A'^4); \qquad A^2 = A'^2; \qquad A^3 = A'^3; \qquad A^4 = \gamma(A'^4 + i\beta A'^1)$$

The component A^4 is known as the time component, and A^1, A^2, A^3 the space components of the four-vector. Often we write $A^i = (\vec{A}, A^4)$

5.1 Invariance of A^iA^i

A^iA^i is invariant under Lorentz transformations. That is

$$A^1A^1 + A^2A^2 + A^3A^3 + A^4A^4 = A'^1A'^1 + A'^2A'^2 + A'^3A'^3 + A'^4A'^4$$

or

$$(A^1)^2 + (A^2)^2 + (A^3)^2 + (A^4)^2 = (A'^1)^2 + (A'^2)^2 + (A'^3)^2 + (A'^4)^2$$

Proof

We have

$$(A^1)^2+(A^2)^2+(A^3)^2+(A^4)^2=\gamma^2(A'^1-i\beta A'^4)^2+(A'^2)^2+(A'^3)^2$$

$$+\gamma^2(A'^4+i\beta A'^1)^2=\gamma^2[(A'^1)^2-2i\beta A'^1A'^4-\beta^2(A'^4)^2]+(A'^2)^2+(A'^3)^2$$

$$+\gamma^2[(A'^4)^2+2i\beta A'^1A'^4-\beta^2(A'^1)^2]=\gamma^2[(A'^1)^2(1-\beta^2)+(A'^4)^2(1-\beta^2)]$$

$$+(A'^2)^2+(A'^3)^2=(A'^1)^2+(A'^2)^2+(A'^3)^2+(A'^4)^2$$

Notes:

1. For a physical quantity A^i which can be expressed as four-vector, A^iA^i is invariant under Lorentz transformation.
2. For the norm, the squares of the space component and time component are added. That is, $A^iA^i=[(A^1)^2+(A^2)^2+(A^3)^2]+(A^4)^2=\vec{A}^2+(A^4)^2$.

5.2 Four-velocity

We can define velocity four vector u^i as the derivative

$$u^i=\frac{\mathrm{d}x^i}{\mathrm{d}\tau} \tag{12.34}$$

where x^i is a position four vector and τ is a scalar invariant of four dimensional space. It is the operator of four-dimensional space. It is called proper time and is defined by $\mathrm{d}\tau=\mathrm{d}t\sqrt{1-\beta^2}$. The spatial and temporal components of u^i may be written by writing spatial and temporal components of x^i as $\vec{r}$ and x^4. Thus, we get

$$u^i=\gamma\left[\frac{\mathrm{d}\vec{r}}{\mathrm{d}t},\ \frac{\mathrm{d}x^4}{\mathrm{d}t}\right]$$

Writing $\mathrm{d}\vec{r}/\mathrm{d}t=\vec{v}$ and $\mathrm{d}x^4/\mathrm{d}t=ic$, we get

$$u^i=\gamma[\vec{v},\quad ic] \tag{12.35}$$

where $\gamma=1/\sqrt{1-\beta^2}$ and $\beta=u/c$. The norm of this vector gives

$$u^iu^i=u^1u^1+u^2u^2+u^3u^3+u^4u^4=\gamma^2[u^2-c^2]=-c^2 \tag{12.36}$$

5.3 Momentum four-vector

Now, knowing four-velocity u^i and that the product of mass with velocity gives momentum, we define momentum four-vector by the following relation

$$p^i = m_0 i\ u^i \tag{12.37}$$

where m_0 is the rest-mass of the particle and u^i the four-velocity. The norm of this on using equation (12.37) gives

$$p^i p^i = m_0^2\ u^i u^i = m_0^2 \gamma^2 (\vec{v} + ic)\ .\ (\vec{v} + ic) = -m_0^2 c^2 \tag{12.38}$$

We now write its spatial and temporal components by keeping equation (12.35) in mind

$$p^i = \gamma(m_0 \vec{v},\ i m_0 c)$$

As $m = m_0/\sqrt{1-\beta^2}$ and $\vec{p} = m\vec{v}$ and $mc^2 = E$, we write

$$p^i = (m\vec{v},\ imc) = (\vec{p},\ \ iE/c) \tag{12.39}$$

in which $\vec{p}$ is the usual three-momentum and E is the total energy associated with particle. This relation gives

$$p^i p^i = (\vec{p})^2 + \left(\frac{iE}{c}\right)^2 = p^2 - \frac{E^2}{c^2} \tag{12.40}$$

Using equation (12.38) in above equation, we get

$$-m_0^2 c^2 = p^2 - \frac{E^2}{c^2} \qquad \text{or} \qquad E^2 = p^2 c^2 + m_0^2 c^4 \tag{12.41}$$

This energy expressed in terms of momentum if written in terms of Hamiltonian function H, becomes

$$H = c\sqrt{(p^2 + m_0^2 c^2)} = m_0 c^2 \left(1 + \frac{p^2}{m_0^2 c^2}\right)^{1/2}$$

In case the particle velocity $\vec{u}$ is small compared to c, then $m_0 c >> p$ so that above relation becomes

$$H = m_0 c^2 + \frac{p^2}{2m_0} \tag{12.42}$$

This is modified form of classical Hamiltonian with added term of rest energy $m_0 c^2$. In the case of photon $m_0 \to 0$ and $u \to c$, then equation (12.41) gives

$$p = E/c$$

This substitution in equation (12.39) gives

$$p^i = (\vec{p},\ ip) \qquad \text{or} \qquad p^i = (\vec{p},\ iE/c)$$

It may be noted that the operator i on p makes it to belong to time part. Taking norm of it, we get

$$p^i p^i = (\vec{p})^2 + (ip)^2 = p^2 - p^2 = 0$$

In this case, p^i is a null four-vector. The transformed components of p^i may be written as

$$\left.\begin{aligned} p'^1 &= \gamma[p^1 + i\beta\, p^4] = \gamma\left[p^1 + i\beta\, \tfrac{iE}{c}\right] = \gamma\left[p^1 - \tfrac{v}{c^2}E\right] \\ p'^2 &= p^2 \\ p'^3 &= p^3 \\ p'^4 &= \gamma[p^4 - i\beta\, p^1] \qquad \tfrac{iE'}{c} = \gamma\left[\tfrac{iE}{c} - \tfrac{iv}{c}\, p^1\right] \qquad E' = \gamma[E - vp^1] \end{aligned}\right\} \quad (12.43)$$

where v is relative velocity of frame S$'$ with respect to the frame S.

Invariance of $(E^2 - p^2c^2)$

Using transformations given in equation (12.43), the invariance of $(E^2 - p^2c^2)$ under Lorentz transformations can be shown as below. For invariance, we should prove

$$E'^2 - p'^2c^2 = E^2 - p^2c^2$$

Using equation (12.43), we have

$$\begin{aligned} E'^2 - p'^2c^2 &= E'^2 - c^2[(p'^1)^2 + (p'^2)^2 + (p'^3)^2] \\ &= \gamma^2[E - vp^1]^2 - c^2\left[\gamma^2(p^1 - \frac{v}{c^2}E)^2 + (p^2)^2 + (p^3)^2\right] \\ &= \gamma^2\left([E - vp^1]^2 - c^2\left[(p^1 - \frac{v}{c^2}E)^2\right]\right) - c^2[(p^2)^2 + (p^3)^2] \\ &= E^2 - c^2[(p^1)^2 + (p^2)^2 + (p^3)^2] = E^2 - p^2c^2 \end{aligned}$$

The invariance of $(E^2 - p^2c^2)$ establishes that energy and momentum are conserved in inertial frames.

5.4 Four-current

Four-current is expressed as

$$J^i = (\vec{J},\ ic\rho)$$

Lorentz transformations are

$$\left.\begin{aligned} J'^1 &= \gamma(J^1 + i\beta J^4) = \gamma(J^1 - v\rho) \\ J'^2 &= J^2 \\ J'^3 &= J^3 \\ J'^4 &= \gamma(J^4 - i\beta J^1) \qquad ic\rho' = \gamma(ic\rho - i\beta J^1) \qquad \rho' = \gamma\left(\rho - \tfrac{v}{c^2}\, J^1\right) \end{aligned}\right\} \quad (12.44)$$

Inverse Lorentz transformations are

$$\left.\begin{aligned} J^1 &= \gamma(J'^1 - i\beta J'^4) = \gamma(J'^1 + v\rho') \\ J^2 &= J'^2 \\ J^3 &= J'^3 \\ J^4 &= \gamma(J'^4 + i\beta J'^1) \qquad ic\rho = \gamma(ic\rho' + i\beta J'^1) \qquad \rho = \gamma\left(\rho' + \tfrac{v}{c^2}\, J'^1\right) \end{aligned}\right\}$$

Invariance of $(J^2 - \rho^2 c^2)$

Using transformations given in equation (12.44), the invariance of $(E^2 - p^2c^2)$ under Lorentz transformations can be shown as below. For invariance, we should prove

$$J'^2 - \rho'^2 c^2 = J^2 - \rho^2 c^2$$

Using equation (12.44), we have

$$\begin{aligned} J'^2 - \rho'^2 c^2 &= (J'^1)^2 + (J'^2)^2 + (J'^3)^2 - \rho'^2 c^2 \\ &= \gamma^2 [J^1 - v\rho]^2 + (J^2)^2 + (J^3)^2 - c^2\gamma^2 \left[\rho - \frac{v}{c^2} J^1\right]^2 \\ &= \gamma^2[(J^1)^2 + v^2\rho^2 - 2vJ^1\rho] + (J^2)^2 + (J^3)^2 - c^2\gamma^2\left[\rho^2 + \frac{v^2}{c^4}(J^1)^2 - 2\frac{v}{c^2}\rho J^1\right] \\ &= (J^1)^2 + (J^2)^2 + (J^3)^2 - \rho^2 c^2 = J^2 - \rho^2 c^2 \end{aligned}$$

5.5 Electromagnetic four-potential

Four-electromagnetic potential is expressed as

$$A^i = (\vec{A},\ i\phi/c)$$

Lorentz transformations are

$$\left.\begin{aligned} A'^1 &= \gamma(A^1 + i\beta A^4) = \gamma\left(A^1 - \tfrac{v}{c^2}\,\phi\right) \\ A'^2 &= A^2 \\ A'^3 &= A^3 \\ A'^4 &= \gamma(A^4 - i\beta A^1) \qquad i\phi'/c = \gamma(i\phi/c - i\beta A^1) \qquad \phi' = \gamma(\phi - v\,A^1) \end{aligned}\right\} \quad (12.45)$$

Inverse Lorentz transformations are

$$\left.\begin{aligned} A^1 &= \gamma(A'^1 - i\beta A'^4) = \gamma\left(A'^1 + \tfrac{v}{c^2}\,\phi'\right) \\ A^2 &= A'^2 \\ A^3 &= A'^3 \\ A^4 &= \gamma(A'^4 + i\beta A'^1) \qquad i\phi/c = \gamma(i\phi'/c + i\beta A'^1) \qquad \phi = \gamma(\phi' + v\,A'^1) \end{aligned}\right\}$$

Invariance of $(\phi^2 - A^2c^2)$

Using transformations given in equation (12.45), the invariance of $(\phi^2 - A^2c^2)$ under Lorentz transformations can be shown as below. For invariance, we should prove

$$\phi'^2 - A'^2c^2 = \phi^2 - A^2c^2$$

Using equation (12.43), we have

$$\begin{aligned}
\rho'^2 - A'^2c^2 &= \phi'^2 - c^2[(A'^1)^2 + (A'^2)^2 + (A'^3)^2] \\
&= \gamma^2[\phi - vA^1]^2 - c^2\left[\gamma^2(A^1 - \frac{v}{c^2}\phi)^2 + (A^2)^2 + (A^3)^2\right] \\
&= \gamma^2\left([\phi - vA^1]^2 - c^2\left[(A^1 - \frac{v}{c^2}\phi)^2\right]\right) - c^2[(A^2)^2 + (A^3)^2] \\
&= \phi^2 - c^2[(A^1)^2 + (A^2)^2 + (A^3)^2] = \phi^2 - A^2c^2
\end{aligned}$$

5.6 Invariant form of continuity equation

The basis of electrodynamics is the conservation of charge which is precisely expressed by the equation of continuity

$$\nabla \, . \, \vec{J} + \frac{\partial \rho}{\partial t} = 0$$

where $\vec{J}$ is current density and ρ the charge density. It can be written as

$$\frac{\partial J_x}{\partial x} + \frac{\partial J_y}{\partial y} + \frac{\partial J_z}{\partial z} + \frac{\partial (ic\rho)}{\partial (ict)} = 0$$

For four current $J^i = (\vec{J}, \ ic\rho)$, we have

$$\frac{\partial J^1}{\partial x^1} + \frac{\partial J^2}{\partial x^2} + \frac{\partial J^3}{\partial x^3} + \frac{\partial J^4}{\partial x^4} = 0 \qquad \text{or} \qquad \frac{\partial J^i}{\partial x^i} = 0$$

5.7 Invariant form of Lorentz condition

The field equation corresponding to the equation of continuity is **Lorentz condition** which interrelate scalar potential function ϕ and vector potential function $\vec{A}$. The relation for free space is written as

$$\nabla \, . \, \vec{A} + \frac{1}{c^2}\frac{\partial \phi}{\partial t} = 0$$

$\vec{A}$ and ϕ are known as electromagnetic or electrodynamic potentials. The equation shows the possibility of four-vector representation and thereby covariant transformation. The Lorentz condition can be expressed as

$$\frac{\partial A_x}{\partial x} + \frac{\partial A_y}{\partial y} + \frac{\partial A_z}{\partial z} + \frac{\partial (i\phi/c)}{\partial (ict)} = 0$$

For four potential $A^i = (\vec{A}, \ i\phi/c)$, we have

$$\frac{\partial A^1}{\partial x^1} + \frac{\partial A^2}{\partial x^2} + \frac{\partial A^3}{\partial x^3} + \frac{\partial A^4}{\partial x^4} = 0 \qquad \text{or} \qquad \frac{\partial A^i}{\partial x^i} = 0$$

5.8 Lorentz variance of three operators

The three operators are defined as

$$\nabla \equiv \frac{\partial}{\partial x}\,\hat{x} + \frac{\partial}{\partial y}\,\hat{y} + \frac{\partial}{\partial z}\,\hat{z}$$

$$\nabla^2 \equiv \frac{\partial^2}{\partial x^2} + \frac{\partial^2}{\partial y^2} + \frac{\partial^2}{\partial z^2}$$

$$\Box \equiv \nabla^2 - \frac{1}{c^2}\frac{\partial^2}{\partial t^2} = \frac{\partial^2}{\partial x^2} + \frac{\partial^2}{\partial y^2} + \frac{\partial^2}{\partial z^2} - \frac{1}{c^2}\frac{\partial^2}{\partial t^2}$$

It may be seen that three-dimensional operators and functions do vary when subjected to Lorentz transformation, while four-dimensional operators and functions exhibit invariance. Since operators ∇, ∇^2, and $\Box$ are common in formulating electrodynamics, we establish that ∇, ∇^2 are variants and $\Box$ is invariant under Lorentz transformation. That is, four-dimensional operators and functions show relativistic invariance, and one should formulate electrodynamics in terms of four-vectors of current and potential. We intend to show that $\nabla^2 \neq \nabla'^2$ and $\nabla \neq \nabla'$, but $\Box = \Box'$. The Lorentz transformations are

$$x' = \gamma(x - vt); \qquad y' = y; \qquad z' = z; \qquad t' = \gamma(t - \beta x/c) \tag{12.46}$$

where $\gamma = 1/\sqrt{1-\beta^2}$ and $\beta = v/c$ The inverse Lorentz transformations are

$$x = \gamma(x' + vt'); \qquad y = y'; \qquad z = z'; \qquad t = \gamma(t' + \beta x'/c) \tag{12.47}$$

Functional relations given by these transformations may be written as

$$x' = f_1(x,t) \qquad y' = f_2(y) \qquad z' = f_3(z) \qquad t' = f_4(x,t)$$

Therefore, we write

$$\frac{\partial}{\partial x'} = \frac{\partial}{\partial x}\frac{\partial x}{\partial x'} + \frac{\partial}{\partial t}\frac{\partial t}{\partial x'} \qquad \frac{\partial}{\partial y'} = \frac{\partial}{\partial y}\frac{\partial y}{\partial y'} \tag{12.48}$$

$$\frac{\partial}{\partial z'} = \frac{\partial}{\partial z}\frac{\partial z}{\partial z'} \qquad \frac{\partial}{\partial t'} = \frac{\partial}{\partial x}\frac{\partial x}{\partial t'} + \frac{\partial}{\partial t}\frac{\partial t}{\partial t'} \tag{12.49}$$

Using inverse transformations, we get

$$\frac{\partial x}{\partial x'} = \gamma \qquad \frac{\partial x}{\partial t'} = \gamma v \qquad \frac{\partial y}{\partial y'} = 1 \qquad \frac{\partial z}{\partial z'} = 1 \qquad \frac{\partial t}{\partial x'} = \frac{\gamma v}{c^2} \qquad \frac{\partial t}{\partial t'} = \gamma$$

Substitution of these relations in (12.48) and (12.49)

$$\frac{\partial}{\partial x'} = \gamma\left[\frac{\partial}{\partial x} + \frac{v}{c^2}\frac{\partial}{\partial t}\right] \qquad \frac{\partial}{\partial y'} = \frac{\partial}{\partial y} \tag{12.50}$$

$$\frac{\partial}{\partial z'} = \frac{\partial}{\partial z} \qquad \frac{\partial}{\partial t'} = \gamma\left[v\frac{\partial}{\partial x} + \frac{\partial}{\partial t}\right] \tag{12.51}$$

Substitution of (12.50) and (12.51) gives

$$\nabla'^2 = \frac{\partial^2}{\partial x'^2} + \frac{\partial^2}{\partial y'^2} + \frac{\partial^2}{\partial z'^2} = \left(\frac{\partial}{\partial x'}\right)\left(\frac{\partial}{\partial x'}\right) + \left(\frac{\partial}{\partial y'}\right)\left(\frac{\partial}{\partial y'}\right) + \left(\frac{\partial}{\partial z'}\right)\left(\frac{\partial}{\partial z'}\right)$$

$$= \gamma^2\left[\frac{\partial}{\partial x} + \frac{v}{c^2}\frac{\partial}{\partial t}\right]^2 + \frac{\partial^2}{\partial y^2} + \frac{\partial^2}{\partial z^2} \neq \nabla^2$$

We have

$$\nabla' = \frac{\partial}{\partial x'}\,\hat{x} + \frac{\partial}{\partial y'}\,\hat{y} + \frac{\partial}{\partial z'}\,\hat{z} = \gamma\left[\frac{\partial}{\partial x} + \frac{v}{c^2}\frac{\partial}{\partial t}\right]\hat{x} + \frac{\partial}{\partial y}\,\hat{y} + \frac{\partial}{\partial z}\,\hat{z} \neq \nabla$$

Further,

$$\Box' = \nabla'^2 - \frac{1}{c^2}\frac{\partial^2}{\partial t'^2} = \frac{\partial^2}{\partial x'^2} + \frac{\partial^2}{\partial y'^2} + \frac{\partial^2}{\partial z'^2} - \frac{1}{c^2}\frac{\partial^2}{\partial t'^2}$$

$$= \left(\frac{\partial}{\partial x'}\right)\left(\frac{\partial}{\partial x'}\right) + \left(\frac{\partial}{\partial y'}\right)\left(\frac{\partial}{\partial y'}\right) + \left(\frac{\partial}{\partial z'}\right)\left(\frac{\partial}{\partial z'}\right) - \frac{1}{c^2}\left(\frac{\partial}{\partial t'}\right)\left(\frac{\partial}{\partial t'}\right)$$

$$= \gamma^2\left[\frac{\partial}{\partial x} + \frac{v}{c^2}\frac{\partial}{\partial t}\right]^2 + \frac{\partial^2}{\partial y^2} + \frac{\partial^2}{\partial z^2} - \frac{\gamma^2}{c^2}\left[v\frac{\partial}{\partial x} + \frac{\partial}{\partial t}\right]^2$$

$$= \gamma^2\left[\frac{\partial^2}{\partial x^2} + \frac{2v}{c^2}\frac{\partial^2}{\partial t\partial x} + \frac{v^2}{c^4}\frac{\partial^2}{\partial t^2}\right] + \frac{\partial^2}{\partial y^2} + \frac{\partial^2}{\partial z^2} - \frac{\gamma^2}{c^2}\left[v^2\frac{\partial^2}{\partial x^2} + 2v\frac{\partial^2}{\partial t\partial x} + \frac{\partial^2}{\partial t^2}\right]$$

$$= \frac{\partial^2}{\partial x^2} + \frac{\partial^2}{\partial y^2} + \frac{\partial^2}{\partial z^2} - \frac{1}{c^2}\frac{\partial^2}{\partial t^2} = \Box$$

It establishes invariance of $\Box$ under Lorentz transformation.

5.9 Lorentz invariance of $\partial A^i/\partial x^i$

We want to show that

$$\frac{\partial A'^i}{\partial x'^i} = \frac{\partial A^i}{\partial x^i}$$

or

$$\frac{\partial A'^1}{\partial x'} + \frac{\partial A'^2}{\partial y'} + \frac{\partial A'^3}{\partial z'} + \frac{1}{ic}\frac{\partial A'^4}{\partial t'} = \frac{\partial A^1}{\partial x} + \frac{\partial A^2}{\partial y} + \frac{\partial A^3}{\partial z} + \frac{1}{ic}\frac{\partial A^4}{\partial t}$$

For Lorentz transformations

$$x' = \gamma(x - vt); \qquad y' = y; \qquad z' = z; \qquad t' = \gamma(t - \beta x/c)$$

the functional relations may be written as

$$x' = f_1(x,t) \qquad y' = f_2(y) \qquad z' = f_3(z) \qquad t' = f_4(x,t)$$

Therefore, we write

$$\frac{\partial}{\partial x'} = \frac{\partial}{\partial x}\frac{\partial x}{\partial x'} + \frac{\partial}{\partial t}\frac{\partial t}{\partial x'} \qquad \frac{\partial}{\partial y'} = \frac{\partial}{\partial y}\frac{\partial y}{\partial y'}$$

$$\frac{\partial}{\partial z'} = \frac{\partial}{\partial z}\frac{\partial z}{\partial z'} \qquad \frac{\partial}{\partial t'} = \frac{\partial}{\partial x}\frac{\partial x}{\partial t'} + \frac{\partial}{\partial t}\frac{\partial t}{\partial t'}$$

Using inverse transformations,

$$x = \gamma(x' + vt'); \qquad y = y'; \qquad z = z'; \qquad t = \gamma(t' + \beta x'/c)$$

we get

$$\frac{\partial x}{\partial x'} = \gamma \qquad \frac{\partial x}{\partial t'} = \gamma v \qquad \frac{\partial y}{\partial y'} = 1 \qquad \frac{\partial z}{\partial z'} = 1 \qquad \frac{\partial t}{\partial x'} = \frac{\gamma v}{c^2} \qquad \frac{\partial t}{\partial t'} = \gamma$$

Now,

$$\begin{aligned}\frac{\partial A'^1}{\partial x'} &= \gamma\frac{\partial A'^1}{\partial x} + \frac{\gamma v}{c^2}\frac{\partial A'^1}{\partial t}\\ &= \gamma^2\frac{\partial(A^1 + i\beta A^4)}{\partial x} + \frac{\gamma^2 v}{c^2}\frac{\partial(A^1 + i\beta A^4)}{\partial t}\\ &= \gamma^2\frac{\partial A^1}{\partial x} + i\beta\gamma^2\frac{\partial A^4}{\partial x} + \frac{\gamma^2 v}{c^2}\frac{\partial A^1}{\partial t} + \frac{i\beta\gamma^2 v}{c^2}\frac{\partial A^4}{\partial t}\end{aligned} \tag{12.52}$$

$$\frac{\partial A'^2}{\partial y'} = \frac{\partial A^2}{\partial y} \tag{12.53}$$

$$\frac{\partial A'^3}{\partial z'} = \frac{\partial A^3}{\partial z} \tag{12.54}$$

$$\begin{aligned}\frac{1}{ic}\frac{\partial A'^4}{\partial t'} &= \frac{\gamma}{ic}\frac{\partial A'^4}{\partial t} + \frac{\gamma v}{ic}\frac{\partial A'^4}{\partial x}\\ &= \frac{\gamma^2}{ic}\frac{\partial(A^4 - i\beta A^1)}{\partial t} + \frac{\gamma^2 v}{ic}\frac{\partial(A^4 - i\beta A^1)}{\partial x}\\ &= \frac{\gamma^2}{ic}\frac{\partial A^4}{\partial t} - \frac{\beta\gamma^2}{c}\frac{\partial A^1}{\partial t} + \frac{\gamma^2 v}{ic}\frac{\partial A^4}{\partial x} - \frac{\beta\gamma^2 v}{c}\frac{\partial A^1}{\partial x}\end{aligned} \tag{12.55}$$

On addition of equations (12.52), (12.53), (12.54), (12.55), we get

$$\begin{aligned}\frac{\partial A'^1}{\partial x'} + \frac{\partial A'^2}{\partial y'} + \frac{\partial A'^3}{\partial z'} + \frac{1}{ic}\frac{\partial A'^4}{\partial t'} &= \gamma^2\frac{\partial A^1}{\partial x} + i\beta\gamma^2\frac{\partial A^4}{\partial x} + \frac{\gamma^2 v}{c^2}\frac{\partial A^1}{\partial t}\\ &+ \frac{i\beta\gamma^2 v}{c^2}\frac{\partial A^4}{\partial t} + \frac{\partial A^2}{\partial y} + \frac{\partial A^3}{\partial z} + \frac{\gamma^2}{ic}\frac{\partial A^4}{\partial t} - \frac{\beta\gamma^2}{c}\frac{\partial A^1}{\partial t} + \frac{\gamma^2 v}{ic}\frac{\partial A^4}{\partial x} - \frac{\beta\gamma^2 v}{c}\frac{\partial A^1}{\partial x}\\ &= \gamma^2\frac{\partial A^1}{\partial x}\left[1 - \frac{\beta v}{c}\right] + \frac{\partial A^2}{\partial y} + \frac{\partial A^3}{\partial z} + \frac{\gamma^2}{ic}\frac{\partial A^4}{\partial t}\left[1 - \frac{\beta v}{c}\right]\\ &\quad + \frac{\partial A^4}{\partial x}\left[i\beta\gamma^2 + \frac{\gamma^2 v}{ic}\right] + \frac{\partial A^1}{\partial t}\left[\frac{\gamma^2 v}{c^2} - \frac{\beta\gamma^2}{c}\right]\\ &= \frac{\partial A^1}{\partial x} + \frac{\partial A^2}{\partial y} + \frac{\partial A^3}{\partial z} + \frac{1}{ic}\frac{\partial A^4}{\partial t}\end{aligned}$$

It proves that

$$\frac{\partial A'^i}{\partial x'^i} = \frac{\partial A^i}{\partial x^i}$$

6. Electromagnetic field tensor

For the four-potential A^i, the electromagnetic field tensor $T^{\alpha\beta}$ is defined as

$$T^{\alpha\beta} = \frac{\partial A^\beta}{\partial x^\alpha} - \frac{\partial A^\alpha}{\partial x^\beta}$$

where the indices α and β can assume, independently, the values 1, 2, 3, 4.

$$T^{\alpha\beta} = \begin{bmatrix} T^{11} & T^{12} & T^{13} & T^{14} \\ T^{21} & T^{22} & T^{23} & T^{24} \\ T^{31} & T^{32} & T^{33} & T^{34} \\ T^{41} & T^{42} & T^{43} & T^{44} \end{bmatrix} \tag{12.56}$$

Obviously, $T^{\alpha\beta}$ is a square matrix (tensor of rank 2) of order 4×4. That is, there are 16 elements. When $\alpha = \beta$ (diagonal elements of matrix), we have $T^{\alpha\alpha} = 0$. That is, four diagonal elements are zero. Further,

$$T^{\beta\alpha} = \frac{\partial A^\alpha}{\partial x^\beta} - \frac{\partial A^\beta}{\partial x^\alpha} = -\left(\frac{\partial A^\beta}{\partial x^\alpha} - \frac{\partial A^\alpha}{\partial x^\beta}\right) = -T^{\alpha\beta}$$

That is, $T^{\alpha\beta}$ is anti-symmetric tensor (matrix). Hence, we have to calculate only half of the off-diagonal elements. That is, we have to calculate only 6 elements. The electromagnetic field vectors $\vec{E}$ and $\vec{B}$ are related to the vector potential $\vec{A}$ and scalar potential ϕ as

$$\vec{E} = -\nabla\phi - \frac{\partial \vec{A}}{\partial t} \qquad \text{and} \qquad \vec{B} = \nabla \times \vec{A} \tag{12.57}$$

The elements of the electromagnetic field tensor can be obtained in the following manner. The x-component equation for the first relation is

$$E^1 = -\frac{\partial \phi}{\partial x^1} - \frac{\partial A^1}{\partial t} \qquad \frac{i}{c}\,E^1 = -\frac{\partial (i\phi/c)}{\partial x^1} - \frac{i}{c}\frac{\partial A^1}{\partial t}$$

$$\frac{i}{c}\,E^1 = -\frac{\partial A^4}{\partial x^1} + \frac{\partial A^1}{\partial x^4} \qquad \frac{i}{c}\,E^1 = \frac{\partial A^1}{\partial x^4} - \frac{\partial A^4}{\partial x^1} = T^{41} \tag{12.58}$$

In the same way for other two components, we get

$$\frac{i}{c}\,E^2 = \frac{\partial A^2}{\partial x^4} - \frac{\partial A^4}{\partial x^2} = T^{42} \tag{12.59}$$

and

$$\frac{i}{c}\,E^3 = \frac{\partial A^3}{\partial x^4} - \frac{\partial A^4}{\partial x^3} = T^{43} \tag{12.60}$$

The second relation $\vec{B} = \nabla \times \vec{A}$ gives the following relation

$$B^1 = \frac{\partial A^3}{\partial x^2} - \frac{\partial A^2}{\partial x^3} = T^{23} \tag{12.61}$$

$$B^2 = \frac{\partial A^1}{\partial x^3} - \frac{\partial A^3}{\partial x^1} = T^{31} \tag{12.62}$$

$$B^3 = \frac{\partial A^2}{\partial x^1} - \frac{\partial A^1}{\partial x^2} = T^{12} \tag{12.63}$$

Equations (12.58) to (12.63) give the 6 off-diagonal elements of the electromagnetic field tensor. Other 6 off-diagonal elements of tensor can be obtained by considering the fact that the tensor is anti-symmetric. The electromagnetic field tensor is obtained as

$$T^{\alpha\beta} = \begin{bmatrix} T^{11} & T^{12} & T^{13} & T^{14} \\ T^{21} & T^{22} & T^{23} & T^{24} \\ T^{31} & T^{32} & T^{33} & T^{34} \\ T^{41} & T^{42} & T^{43} & T^{44} \end{bmatrix} = \begin{bmatrix} 0 & B^3 & -B^2 & -iE^1/c \\ -B^3 & 0 & B^1 & -iE^2/c \\ B^2 & -B^1 & 0 & -iE^3/c \\ iE^1/c & iE^2/c & iE^3/c & 0 \end{bmatrix} \tag{12.64}$$

6.1 Maxwell's equations in terms of electromagnetic field tensor

The electromagnetic field tensor $T^{\alpha\beta}$ in terms of the components of electric and magnetic fields is expressed as

$$T^{\alpha\beta} = = \begin{bmatrix} 0 & B_z & -B_y & -iE_x/c \\ -B_z & 0 & B_x & -iE_y/c \\ B_y & -B_x & 0 & -iE_z/c \\ iE_x/c & iE_y/c & iE_z/c & 0 \end{bmatrix} \tag{12.65}$$

We would like now to express inhomogeneous and homogeneous pairs of Maxwell's field equations in terms of the components of electromagnetic field tensor. The Maxwell's field equations are

$$\left.\begin{array}{ll} \text{(i) } \nabla \,.\, \vec{E} = \frac{\rho}{\epsilon_0} & \text{(ii) } \nabla \,.\, \vec{B} = 0 \\ \text{(iii) } \nabla \times \vec{E} + \frac{\partial \vec{B}}{\partial t} = 0 & \text{(iv) } \nabla \times \vec{B} - \frac{1}{c^2}\frac{\partial \vec{E}}{\partial t} = \mu_0 \vec{J} \end{array}\right\} \tag{12.66}$$

where $c = 1/\sqrt{\mu_0 \epsilon_0}$. Equations 12.66 (i) and (iv) form inhomogeneous pair while 12.66 (ii) and (iii) form homogeneous pair. The component form of equation 12.66 (iv) in terms of four coordinates (x^1, x^2, x^3, x^4) is written as

$$\hat{x}\Big(\frac{\partial B^3}{\partial x^2} - \frac{\partial B^2}{\partial x^3}\Big) + \hat{y}\Big(\frac{\partial B^1}{\partial x^3} - \frac{\partial B^3}{\partial x^1}\Big) + \hat{z}\Big(\frac{\partial B^2}{\partial x^1} - \frac{\partial B^1}{\partial x^2}\Big)$$

$$-\frac{i}{c}\frac{\partial}{\partial x^4}(\hat{x}E^1 + \hat{y}E^2 + \hat{z}E^3) = \mu_0(\hat{x}J^1 + \hat{y}J^2 + \hat{z}J^3) \tag{12.67}$$

Further, the component form of equation 12.66 (i) is written as

$$\frac{\partial E^1}{\partial x^1} + \frac{\partial E^2}{\partial x^2} + \frac{\partial E^3}{\partial x^3} = \frac{ic\rho}{ic\epsilon_0} = \frac{J^4}{ic\epsilon_0} \tag{12.68}$$

Four current $J^i = (\vec{J},\ ic\rho)$. Equations (12.67) and (12.68) are written as

$$\left.\begin{aligned}
&\text{(i)}\ 0 + \frac{\partial B^3}{\partial x^2} - \frac{\partial B^2}{\partial x^3} - \frac{i}{c}\frac{\partial E^1}{\partial x^4} = \mu_0 J^1\\
&\text{(ii)}\ -\frac{\partial B^3}{\partial x^1} + 0 + \frac{\partial B^1}{\partial x^3} - \frac{i}{c}\frac{\partial E^2}{\partial x^4} = \mu_0 J^2\\
&\text{(iii)}\ \frac{\partial B^2}{\partial x^1} - \frac{\partial B^1}{\partial x^2} + 0 - \frac{i}{c}\frac{\partial E^3}{\partial x^4} = \mu_0 J^3\\
&\text{(iv)}\ \frac{i}{c}\frac{\partial E^1}{\partial x^1} + \frac{i}{c}\frac{\partial E^2}{\partial x^2} + \frac{i}{c}\frac{\partial E^3}{\partial x^3} = \mu_0 J^4
\end{aligned}\right\}$$

or

$$\left.\begin{aligned}
&\text{(i)}\ \frac{\partial T^{11}}{\partial x^1} + \frac{\partial T^{12}}{\partial x^2} + \frac{\partial T^{13}}{\partial x^3} + \frac{\partial T^{14}}{\partial x^4} = \mu_0 J^1\\
&\text{(ii)}\ \frac{\partial T^{21}}{\partial x^1} + \frac{\partial T^{22}}{\partial x^2} + \frac{\partial T^{23}}{\partial x^3} + \frac{\partial T^{24}}{\partial x^4} = \mu_0 J^2\\
&\text{(iii)}\ \frac{\partial T^{31}}{\partial x^1} + \frac{\partial T^{32}}{\partial x^2} + \frac{\partial T^{33}}{\partial x^3} + \frac{\partial T^{34}}{\partial x^4} = \mu_0 J^3\\
&\text{(iv)}\ \frac{\partial T^{41}}{\partial x^1} + \frac{\partial T^{42}}{\partial x^2} + \frac{\partial T^{43}}{\partial x^3} + \frac{\partial T^{44}}{\partial x^4} = \mu_0 J^4
\end{aligned}\right\} \quad (12.69)$$

Equation (12.69) shows the four component-wise relationship between the source J^i and components of electromagnetic field tensor. The concise form of equation (12.69) becomes

$$\sum_{\beta=1}^{4} \frac{\partial T^{\alpha\beta}}{\partial x^\beta} = \mu_0 J^\alpha \qquad \alpha = 1, 2, 3, 4 \quad (12.70)$$

For the homogeneous pair 12.66 (ii and iii), the component-wise form for the pair is written as

$$\left.\begin{aligned}
&\text{(i)}\ \frac{\partial E^3}{\partial x^2} - \frac{\partial E^2}{\partial x^3} + ic\,\frac{\partial B^1}{\partial x^4} = 0\\
&\text{(ii)}\ \frac{\partial E^1}{\partial x^3} - \frac{\partial E^3}{\partial x^1} + ic\,\frac{\partial B^2}{\partial x^4} = 0\\
&\text{(iii)}\ \frac{\partial E^2}{\partial x^1} - \frac{\partial E^1}{\partial x^2} + ic\,\frac{\partial B^3}{\partial x^4} = 0\\
&\text{(iv)}\ \frac{\partial B^1}{\partial x^1} + \frac{\partial B^2}{\partial x^2} + \frac{\partial B^3}{\partial x^3} = 0
\end{aligned}\right\}$$

or

$$\left.\begin{aligned}
&\text{(i)}\ ic\,\frac{\partial T^{34}}{\partial x^2} + ic\,\frac{\partial T^{42}}{\partial x^3} + ic\,\frac{\partial T^{23}}{\partial x^4} = 0\\
&\text{(ii)}\ ic\,\frac{\partial T^{14}}{\partial x^3} + ic\,\frac{\partial T^{43}}{\partial x^1} + ic\,\frac{\partial T^{31}}{\partial x^4} = 0\\
&\text{(iii)}\ ic\,\frac{\partial T^{24}}{\partial x^1} + ic\,\frac{\partial T^{41}}{\partial x^2} + ic\,\frac{\partial T^{12}}{\partial x^4} = 0\\
&\text{(iv)}\ \frac{\partial T^{23}}{\partial x^1} + \frac{\partial T^{31}}{\partial x^2} + \frac{\partial T^{12}}{\partial x^3} = 0
\end{aligned}\right\}$$

or

$$\left.\begin{aligned}
&\text{(i)}\quad \frac{\partial T^{34}}{\partial x^2}+\frac{\partial T^{42}}{\partial x^3}+\frac{\partial T^{23}}{\partial x^4}=0\\
&\text{(ii)}\quad \frac{\partial T^{41}}{\partial x^3}+\frac{\partial T^{13}}{\partial x^4}+\frac{\partial T^{34}}{\partial x^1}=0\\
&\text{(iii)}\quad \frac{\partial T^{12}}{\partial x^4}+\frac{\partial T^{24}}{\partial x^1}+\frac{\partial T^{41}}{\partial x^2}=0\\
&\text{(iv)}\quad \frac{\partial T^{23}}{\partial x^1}+\frac{\partial T^{31}}{\partial x^2}+\frac{\partial T^{12}}{\partial x^3}=0
\end{aligned}\right\} \tag{12.71}$$

Equations (12.71) in concise form are written as

$$\frac{\partial T^{\beta\gamma}}{\partial x^\alpha}+\frac{\partial T^{\gamma\alpha}}{\partial x^\beta}+\frac{\partial T^{\alpha\beta}}{\partial x^\gamma}=0 \tag{12.72}$$

where α, β and γ are three numbers out of 1, 2, 3, 4 in sequence. This is tensor form of the homogeneous pair of field equations and it can be transformed to frame S$'$ to establish its covariant form and yield field component relationship between frames S and S$'$.

In arriving at equation (12.72), since matrix given by (12.65) has been used to formulate equation (12.72), a neutral interlink exists between the homogeneous and inhomogeneous pairs of Maxwell's field equations.

7. Four-tensor of second rank

A four-tensor of second rank T^{ik} may be considered as a product of two four-vectors as

$$T^{ik}=\begin{bmatrix}T^1\\T^2\\T^3\\T^4\end{bmatrix}\begin{bmatrix}T^1 & T^2 & T^3 & T^4\end{bmatrix}=\begin{bmatrix}T^1T^1 & T^1T^2 & T^1T^3 & T^1T^4\\ T^2T^1 & T^2T^2 & T^2T^3 & T^2T^4\\ T^3T^1 & T^3T^2 & T^3T^3 & T^3T^4\\ T^4T^1 & T^4T^2 & T^4T^3 & T^4T^4\end{bmatrix}$$

$$=\begin{bmatrix}T^{11} & T^{12} & T^{13} & T^{14}\\ T^{21} & T^{22} & T^{23} & T^{24}\\ T^{31} & T^{32} & T^{33} & T^{34}\\ T^{41} & T^{42} & T^{43} & T^{44}\end{bmatrix}$$

The four tensor T^{ik} is a set of sixteen quantities. Nine quantities T^{11}, T^{12}, T^{13}, T^{21}, T^{22}, T^{23}, T^{31}, T^{32}, T^{33} form a set of purely spatial transformations. The three components T^{14}, T^{24}, T^{34}, and three components T^{41}, T^{42}, T^{43} constitute three-dimensional vectors, while the component T^{44} is a three-dimensional scalar.

The Lorentz transformations are:

$$T^1=\gamma(T'^1-i\beta T'^4);\qquad T^2=T'^2;\qquad T^3=T'^3;\qquad T^4=\gamma(T'^4+i\beta T'^1)$$

7.1 Transformation laws for symmetric four-tensor of rank two

The symmetric tensor is

$$T^{ik} = \begin{bmatrix} T^{11} & T^{12} & T^{13} & T^{14} \\ T^{21} & T^{22} & T^{23} & T^{24} \\ T^{31} & T^{32} & T^{33} & T^{34} \\ T^{41} & T^{42} & T^{43} & T^{44} \end{bmatrix}$$

The components of T^{ik} transform like the product of components of two four-vectors.

$$T^{11} = T^1T^1 = \gamma^2[T'^1 - i\beta T'^4][T'^1 - i\beta T'^4]i$$

$$= \gamma^2[T'^1T'^1 - 2i\beta T'^4T'^1 - \beta^2T'^4T'^4] = \gamma^2[T'^{11} - 2i\beta T'^{41} - \beta^2T'^{44}]$$

$$T^{12} = T^1T^2 = \gamma[T'^1 - i\beta T'^4]T'^2 = \gamma[T'^1T'^2 - i\beta T'^4T'^2] = \gamma[T'^{12} - i\beta T'^{24}]$$

$$T^{13} = T^1T^3 = \gamma[T'^1 - i\beta T'^4]T'^3 = \gamma[T'^1T'^3 - i\beta T'^4T'^3] = \gamma[T'^{13} - i\beta T'^{34}]$$

$$T^{14} = T^1T^4 = \gamma^2[T'^1 - i\beta T'^4][T'^4 + i\beta T'^1]$$

$$= \gamma^2[T'^1T'^4(1+\beta^2) - i\beta T'^4T'^4 + i\beta T'^1T'^1] = \gamma^2[T'^{14}(1+\beta^2) - i\beta T'^{44} + i\beta T'^{11}]$$

$$T^{22} = T^2T^2 = T'^2T'^2 = T'^{22}$$

$$T^{23} = T^2T^3 = T'^2T'^3 = T'^{23}$$

$$T^{24} = T^2T^4 = \gamma T'^2[T'^4 + i\beta T'^1] = \gamma[T'^2T'^4 + i\beta T'^2T'^1] = \gamma[T'^{24} + i\beta T'^{12}]$$

$$T^{33} = T^3T^3 = T'^3T'^3 = T'^{33}$$

$$T^{34} = T^3T^4 = \gamma T'^3[T'^4 + i\beta T'^1] = \gamma[T'^3T'^4 + i\beta T'^3T'^1] = \gamma[T'^{34} + i\beta T'^{13}]$$

$$T^{44} = T^4T^4 = \gamma^2[T'^4 + i\beta T'^1][T'^4 + i\beta T'^1]$$

$$= \gamma^2[T'^4T'^4 + 2i\beta T'^4T'^1 - \beta^2T'^1T'^1] = \gamma^2[T'^{44} + 2i\beta T'^{14} - \beta^2T'^{11}]$$

7.2 Transformation laws for anti-symmetric four-tensor of rank two

The antisymmetric tensor is

$$T^{ik} = \begin{bmatrix} 0 & T^{12} & T^{13} & T^{14} \\ -T^{12} & 0 & T^{23} & T^{24} \\ -T^{13} & -T^{23} & 0 & T^{34} \\ -T^{14} & -T^{24} & -T^{34} & 0 \end{bmatrix}$$

The components of T^{ik} transform like the product of components of two four-vectors.

$$T^{11} = T^1T^1 = \gamma^2[T'^1 - i\beta T'^4][T'^1 - i\beta T'^4]$$

$$= \gamma^2[T'^1T'^1 - i\beta T'^4T'^1 - i\beta T'^1T'^4 - \beta^2 T'^4T'^4] = \gamma^2[T'^{11} - i\beta T'^{41} - i\beta T'^{14} - \beta^2 T'^{44}]$$

$$= \gamma^2[T'^{11} + i\beta T'^{14} - i\beta T'^{14} - \beta^2 T'^{44}] = 0$$

$$T^{12} = T^1T^2 = \gamma[T'^1 - i\beta T'^4]T'^2 = \gamma[T'^1T'^2 - i\beta T'^4T'^2] = \gamma[T'^{12} + i\beta T'^{24}]$$

$$T^{13} = T^1T^3 = \gamma[T'^1 - i\beta T'^4]T'^3 = \gamma[T'^1T'^3 - i\beta T'^4T'^3] = \gamma[T'^{13} + i\beta T'^{34}]$$

$$T^{14} = T^1T^4 = \gamma^2[T'^1 - i\beta T'^4][T'^4 + i\beta T'^1]$$

$$= \gamma^2[T'^1T'^4 - i\beta T'^4T'^4 + i\beta T'^1T'^1 + \beta^2 T'^4T'^1]$$

$$= \gamma^2[T'^{14} - i\beta T'^{44} + i\beta T'^{11} - \beta^2 T'^{14}] = T'^{14}$$

$$T^{22} = T^2T^2 = T'^2T'^2 = T'^{22} = 0$$

$$T^{23} = T^2T^3 = T'^2T'^3 = T'^{23}$$

$$T^{24} = T^2T^4 = \gamma T'^2[T'^4 + i\beta T'^1] = \gamma[T'^2T'^4 + i\beta T'^2T'^1] = \gamma[T'^{24} - i\beta T'^{12}]$$

$$T^{33} = T^3T^3 = T'^3T'^3 = T'^{33} = 0$$

$$T^{34} = T^3T^4 = \gamma T'^3[T'^4 + i\beta T'^1] = \gamma[T'^3T'^4 + i\beta T'^3T'^1] = \gamma[T'^{34} - i\beta T'^{13}]$$

$$T^{44} = T^4T^4 = \gamma^2[T'^4 + i\beta T'^1][T'^4 + i\beta T'^1]$$

$$= \gamma^2[T'^4T'^4 + i\beta T'^4T'^1 + i\beta T'^1T'^4 - \beta^2 T'^1T'^1]$$

$$= \gamma^2[T'^{44} - i\beta T'^{14} + i\beta T'^{14} - \beta^2 T'^{11}] = 0$$

Anti-symmetric field tensor is

$$T^{\alpha\beta} = \begin{bmatrix} 0 & B_z & -B_y & -iE_x/c \\ -B_z & 0 & B_x & -iE_y/c \\ B_y & -B_x & 0 & -iE_z/c \\ iE_x/c & iE_y/c & iE_z/c & 0 \end{bmatrix}$$

7.3 Lorentz transformations of electromagnetic field

The electromagnetic field tensor is antisymmetric tensor. Let us consider the transformation laws for antisymmetric four-tensor of rank two and putting the values of the corresponding components of electromagnetic field. We have

$$T^{12} = \gamma(T'^{12} + i\beta T'^{24}) \qquad B_z = \gamma(B'_z + i\beta(-iE'_y/c)) \qquad B_z = \gamma\left(B'_z + \frac{v}{c^2}\,E'_y\right)$$

$$T^{13} = \gamma(T'^{13} + i\beta T'^{34}) \qquad -B_y = \gamma(-B'_y + i\beta(-iE'_z/c)) \qquad B_y = \gamma\left(B'_y - \frac{v}{c^2}\,E'_z\right)$$

$$T^{14} = T'^{14} \qquad -iE_x/c = -iE'_x/c \qquad E_x = E'_x$$

$$T^{23} = T'^{23} \qquad B_x = B'_x$$

$$T^{24} = \gamma(T'^{24} - i\beta T'^{12}) \qquad -iE_y/c = \gamma(-iE'_y/c - i\beta B'_z) \qquad E_y = \gamma(E'_y + vB'_z)$$

$$T^{34} = \gamma(T'^{34} - i\beta T'^{13}) \qquad -iE_z/c = \gamma(-iE'_z/c + i\beta B'_y)$$

$$E_z = \gamma(E'_z - vB'_y)$$

The formulas of transformation for the electric field are

$$E_x = E'_x \qquad E_y = \gamma(E'_y + vB'_z) \qquad E_z = \gamma(E'_z - vB'_y) \tag{12.73}$$

and for the magnetic field are

$$B_x = B'_x \qquad B_y = \gamma\left(B'_y - \frac{v}{c^2}\,E'_z\right) \qquad B_z = \gamma\left(B'_z + \frac{v}{c^2}\,E'_y\right) \tag{12.74}$$

It shows that the electric and magnetic fields also are relative. That is, their properties are different in different reference systems. In particular, the electric field or magnetic field can be equal to zero in one reference system and at the same time be present in another system.

For low velocities, let us neglect β^2 in comparison to 1. The formulas (12.73), (12.74) simplify considerably and we have

$$E_x = E'_x \qquad E_y = E'_y + vB'_z \qquad E_z = E'_z - vB'_y$$

$$B_x = B'_x \qquad B_y = B'_y - \frac{v}{c^2}\,E'_z \qquad B_z = B'_z + \frac{v}{c^2}\,E'_y$$

7.4 Invariants of electromagnetic field

From the electric and magnetic field intensities we can form invariant quantities, which remain unchanged in the transition from one inertial reference system to another. Let us consider

$$\vec{E}\,.\,\vec{B} = E_xB_x + E_yB_y + E_zB_z$$

Using transformations, we have

$$\vec{E}\,.\,\vec{H} = E'_xB'_x + \gamma^2(E'_y + vB'_z)\left(B'_y - \frac{v}{c^2}\,E'_z\right) + \gamma^2(E'_z - vB'_y)\left(B'_z + \frac{v}{c^2}\,E'_y\right)$$

$$= E'_xB'_x + E'_yB'_y + E'_zB'_z = \vec{E}'\,.\,\vec{B}'$$

Let us consider

$$c^2B^2 - E^2 = c^2B_x^2 + c^2B_y^2 + c^2B_z^2 - (E_x^2 + E_y^2 + E_z^2)$$

Using transformations, we have

$$c^2B^2 - E^2 = c^2B_x'^2 + c^2\gamma^2\left(B'_y - \frac{v}{c^2}E'_z\right)^2 + c^2\gamma^2\left(B'_z + \frac{v}{c^2}E'_y\right)^2$$

$$-\left[E_x'^2 + \gamma^2(E'_y + vB'_z)^2 + \gamma^2(E'_z - vB'_y)^2\right]$$

$$= c^2B_x'^2 + c^2B_y'^2 + c^2B_z'^2 - (E_x'^2 + E_y'^2 + E_z'^2) = c^2B'^2 - E'^2$$

Thus, we have two invariants

$$c^2B^2 - E^2 = \text{inv} \qquad \text{and} \qquad \vec{E}\,.\,\vec{B} = \text{inv}$$

From the invariance of the two expressions presented, we get the following theorems. If the electric and magnetic fields are mutually perpendicular in any reference system, that is $\vec{E}\,.\,\vec{B} = 0$, then they are also perpendicular in every other reference system. If the absolute values of $\vec{E}$ and $c\vec{B}$ are equal to each other in any reference system, then they are the same in any other reference system.

The following inequalities are also clearly valid. If in any reference system $E > cB$ (or $cB < E$), then in every other reference system we will have $E > cB$ (or $cB < E$). If in any reference system $\vec{E}$ and $\vec{B}$ make an acute (or obtuse) angle, then they make an acute (or obtuse) angle in every other reference system.

By means of Lorentz transformation we can always give $\vec{E}$ and $\vec{B}$ any arbitrary values, subject only to the condition that $E^2 - c^2B^2$ and $\vec{E}\,.\,\vec{B}$ have fixed values. In particular, we can always find an inertial system in which the electric and magnetic fields are parallel to each other at a given point. In this system $\vec{E}\,.\,\vec{B} = EB$, and from the two equations

$$E^2 - c^2B^2 = E_0^2 - c^2B_0^2 \qquad EB = \vec{E}_0\,.\,\vec{B}_0$$

we can find the values of $\vec{E}$ and $\vec{B}$ in this system of reference ($\vec{E}_0$ and $\vec{B}_0$ are the electric and magnetic fields in the original system of reference).

8. Maxwell's field equations in terms of vector potential and scalar potential

The Maxwell's field equations are

$$\left.\begin{array}{ll}\text{(i) } \nabla \cdot \vec{E} = \frac{\rho}{\epsilon_0} & \text{(ii) } \nabla \cdot \vec{B} = 0 \\ \text{(iii) } \nabla \times \vec{E} + \frac{\partial \vec{B}}{\partial t} = 0 & \text{(iv) } \nabla \times \vec{B} - \frac{1}{c^2}\frac{\partial \vec{E}}{\partial t} = \mu_0 \vec{J}\end{array}\right\} \tag{12.75}$$

where $c = 1/\sqrt{\mu_0\epsilon_0}$. Eqs. (12.75) (i) and (iv) form inhomogeneous pair while (12.75) (ii) and (iii) form homogeneous pair. For vector potential $\vec{A}$ and scalar potential ϕ, we consider

$$\vec{E} = -\nabla\phi - \frac{\partial \vec{A}}{\partial t} \qquad \text{and} \qquad \vec{B} = \nabla \times \vec{A}$$

homogeneous pair of field equations is satisfied. We now consider inhomogeneous pair of field equations.

From Maxwell's field equation

$$\nabla \times \vec{B} - \frac{1}{c^2}\frac{\partial \vec{E}}{\partial t} = \mu_0 \vec{J}$$

$$\nabla \times (\nabla \times \vec{A}) + \frac{1}{c^2}\frac{\partial \nabla\phi}{\partial t} + \frac{1}{c^2}\frac{\partial^2 \vec{A}}{\partial t^2} = \mu_0 \vec{J}$$

$$\nabla(\nabla \cdot \vec{A}) - \nabla^2\vec{A} + \nabla\Big(\frac{1}{c^2}\frac{\partial \phi}{\partial t}\Big) + \frac{1}{c^2}\frac{\partial^2 \vec{A}}{\partial t^2} = \mu_0 \vec{J}$$

$$\nabla^2\vec{A} - \frac{1}{c^2}\frac{\partial^2 \vec{A}}{\partial t^2} - \nabla\Big[(\nabla \cdot \vec{A}) + \frac{1}{c^2}\frac{\partial \phi}{\partial t}\Big] = -\mu_0 \vec{J}$$

Using Lorentz condition $\nabla \cdot \vec{A} + (1/c^2)\, \partial\phi/\partial t = 0$, we get

$$\nabla^2\vec{A} - \frac{1}{c^2}\frac{\partial^2 \vec{A}}{\partial t^2} = -\mu_0 \vec{J} \qquad \text{or} \qquad \Box\, \vec{A} = -\mu_0 \vec{J} \tag{12.76}$$

where

$$\Box \equiv \nabla^2 - \frac{1}{c^2}\frac{\partial^2}{\partial t^2} = \frac{\partial^2}{\partial x^2} + \frac{\partial^2}{\partial y^2} + \frac{\partial^2}{\partial z^2} + \frac{\partial^2}{\partial (ict)^2}$$

$$= \frac{\partial^2}{\partial (x^1)^2} + \frac{\partial^2}{\partial (x^2)^2} + \frac{\partial^2}{\partial (x^3)^2} + \frac{\partial^2}{\partial (x^4)^2} = \frac{\partial^2}{\partial (x^i)^2}$$

From Gauss' law, we have

$$\nabla \cdot \vec{E} = \frac{\rho}{\epsilon_0}$$

Using the expression for $\vec{E}$, we have

$$\nabla \cdot \Big(-\frac{\partial \vec{A}}{\partial t} - \text{grad}\ \phi\Big) = \frac{\rho}{\epsilon_0}$$

$$-\frac{\partial(\nabla \cdot \vec{A})}{\partial t} - \nabla^2\phi = \frac{\rho}{\epsilon_0}$$

Using Lorentz condition $\nabla \cdot \vec{A} + (1/c^2)\ \partial\phi/\partial t = 0$, we get

$$\nabla^2\phi - \frac{1}{c^2}\frac{\partial^2\phi}{\partial t^2} = -\frac{\rho}{\epsilon_0} \qquad \text{or} \qquad \Box\,\phi = -\frac{\rho}{\epsilon_0} \tag{12.77}$$

Maxwell's field equations (12.75) are equivalent to two equations (12.76) and (12.77) expressed in terms of vector potential $\vec{A}$ and scalar potential ϕ.

8.1 Covariance of electrodynamics in terms of inhomogeneous field equations:

Maxwell's four field equations are reducible to two inhomogeneous field equations either in terms of potential functions $(\vec{A},\ \phi)$ or in terms of field vectors $(\vec{E},\ \vec{H})$ together with the respective sources $(\vec{J},\ \rho)$. Let us consider the following form to investigate the covariance of electrodynamics

$$\Box\,\vec{A} = -\mu_0\vec{J} \qquad\qquad \Box\,\phi = -\frac{\rho}{\epsilon_0}$$

where the terms have usual meaning. If A^i and J^i represent four-vectors of potential and charge, respectively, these two relations reduce to the following form

$$\Box\,A^i = -\mu_0\ J^i \tag{12.78}$$

where

$$A^i = (\vec{A},\ i\phi/c) \qquad\qquad J^i = (\vec{J},\ ic\rho)$$

The Lorentz transformations are

$$\left.\begin{aligned} A'^1 &= \gamma\left(A^1 - \tfrac{v}{c^2}\ \phi\right) \\ A'^2 &= A^2 \\ A'^3 &= A^3 \\ \phi' &= \gamma(\phi - v\ A^1) \end{aligned}\right\} \qquad \left.\begin{aligned} J'^1 &= \gamma(J^1 - v\ \rho) \\ J'^2 &= J^2 \\ J'^3 &= J^3 \\ \rho' &= \gamma\left(\phi - \tfrac{v}{c^2}\ J^1\right) \end{aligned}\right\} \tag{12.79}$$

The transformations given by (12.79) have already been established. The covariance of relation (12.78) demands that it retains its form even on transforming it to frame of reference S′. That is

$$\Box'\ A'^i = -\mu_0\ J'^i \tag{12.80}$$

where

$$\Box' \ \equiv \nabla'^2 - \frac{1}{c^2}\frac{\partial^2}{\partial t'^2} = \frac{\partial^2}{\partial(x'^i)^2}$$

We know $\Box = \Box'$. Let us use component relations to establish the relation (12.80).

$$\Box' A'^1 = \Box A'^1 = -\mu_0 J'^1$$

Using relation (12.79), we get

$$\Box \gamma\left(A^1 - \frac{v}{c^2}\phi\right) = -\mu_0 \gamma(J^1 - v\rho)$$

$$\Box A^1 - \frac{v}{c^2}\Box \phi = -\mu_0 (J^1 - v\rho)$$

Substituting $\Box\phi = -\rho/\epsilon_0$ and $c^2 = 1/\mu_0\epsilon_0$, we get

$$\Box A^1 = -\mu_0 J^1 \tag{12.81}$$

In the same way,

$$\Box' A'^2 = \Box A'^2 = -\mu_0 J'^2$$

But, $A'^2 = A^2$ and $J'^2 = J^2$. Therefore,

$$\Box A^2 = -\mu_0 J^2 \tag{12.82}$$

Similarly, for third component, we have

$$\Box' A'^3 = \Box A'^3 = -\mu_0 J'^3$$

But, $A'^3 = A^3$ and $J'^3 = J^3$. Therefore,

$$\Box A^3 = -\mu_0 J^3 \tag{12.83}$$

The fourth component, we have

$$\Box' A'^4 = \Box A'^4 = -\mu_0 J'^4$$

Substituting $A'^4 = i\phi'/c$ and $J'^4 = ic\rho'$ and using (12.79), we get

$$\Box \frac{i\phi'}{c} = -\mu_0 (ic\rho') \qquad \Box\phi' = -\mu_0 c^2\rho'$$

Now, transformations for ϕ' and ρ', we have

$$\Box \gamma[\phi - vA^1] = -\mu_0 c^2\gamma\left[\rho - \frac{v}{c^2}J^1\right]$$

Using $\Box\phi = -\rho/\epsilon_0$ and $c^2 = 1/\mu_0\epsilon_0$, we have

$$\Box A^1 = -\mu_0 J^1$$

It also shows that the fourth imaginary component arises due to relativistic motion along the first component, *i.e.*, say, x-axis.

9. Principle of least action

The principle of least action is defined as the following. For each mechanical system there exists a certain integral S, called the action, which has a minimum value for the actual path, so that its variant δS is zero.

9.1 Action integral for free particle

To determine the action integral for a free particle, we note that this integral must not depend on a reference system. That is, it must be invariant under Lorentz transformations. Then, it follows that it must depend on a scalar. The only scalar of this kind that one can construct for a free particle is the interval ds, or α ds, where α is some constant. So for a free particle the action must have the form

$$S = -\alpha \int_a^b \mathrm{d}s$$

where the integral is taken along the world line of the particle between two particular events (arrival of particle at the initial position a and at the final position b) at the times t_1 and t_2, respectively. The α is some constant characterizing the particle. It is easy to see that α must be a positive quantity for all particles, as for the minimum value of action S, a negative sign is placed. Using relation, we have

$$S = -\int_{t_1}^{t_2} \alpha c \sqrt{1 - \frac{v^2}{c^2}}\,\mathrm{d}t = \int_{t_1}^{t_2} L\mathrm{d}t$$

where v is the velocity of the particle. The Lagrangian for the particle is

$$L = -\alpha c\sqrt{1 - v^2/c^2}$$

In classical mechanics, each particle is characterized by its mass m_0. (m_0 denotes the rest-mass which is equal to the mass of particle in classical mechanics.) Let us find the relation between α and m_0. We have

$$L = -\alpha c\sqrt{1 - v^2/c^2} = -\alpha c + \frac{\alpha v^2}{2c}$$

Constant term αc in the Lagrangian does not affect the equation of motion and can be omitted. Omitting αc in L and comparing the classical expression $L = m_0 v^2/2$, we find that $\alpha = m_0 c$. Thus, the action for a free particle is

$$S = -m_0 c \int_a^b \mathrm{d}s \tag{12.84}$$

and the Lagrangian is

$$L = -m_0 c^2 \sqrt{1 - v^2/c^2} \tag{12.85}$$

9.2 Momentum of free particle

After knowing the Lagrangian L for a free particle, we can calculate the momentum $\vec{p}$ of free particle as

$$\vec{p} = \frac{\partial L}{\partial \vec{v}} = \frac{m_0 \vec{v}}{\sqrt{1 - v^2/c^2}} = m\vec{v} \tag{12.86}$$

where

$$m = \frac{m_0}{\sqrt{1 - v^2/c^2}}$$

is the mass of particle moving with velocity v and having rest-mass m_0. For $v << c$, we have $m = m_0$ and $\vec{p} = m_0\vec{v}$. The time-derivative of momentum is

$$\frac{\mathrm{d}\vec{p}}{\mathrm{d}t} = \frac{m_0}{(1 - v^2/c^2)^{3/2}} \frac{\mathrm{d}\vec{v}}{\mathrm{d}t} \tag{12.87}$$

9.3 Energy of free particle

The energy $\mathcal{E}$ is expressed as

$$\mathcal{E} = \frac{\partial L}{\partial \vec{v}} \cdot \vec{v} - L = \frac{m_0 \vec{v}}{\sqrt{1 - v^2/c^2}} \cdot \vec{v} + m_0 c^2 \sqrt{1 - v^2/c^2}$$

$$= \frac{m_0 c^2}{\sqrt{1 - v^2/c^2}} \tag{12.88}$$

It shows that in relativistic mechanics, for zero velocity $v = 0$, the energy of a free particle is m_0c^2, which is not zero. This energy is known as the rest-energy. The energy of free particle is

$$\mathcal{E} = mc^2$$

For $v << c$, the energy is

$$\mathcal{E} = m_0 c^2 \left(1 - \frac{v^2}{c^2}\right)^{-1/2} = m_0 c^2 \left(1 + \frac{v^2}{2c^2}\right) = m_0 c^2 + \frac{m_0 v^2}{2}$$

The first expression on right side is the rest-energy and the second is the classical expression for the kinetic energy.

9.4 Relation between energy and momentum of free particle

Squaring (12.88), we have

$$\mathcal{E}^2 = \frac{m_0^2 c^4}{1 - v^2/c^2} \quad \text{or} \quad \frac{\mathcal{E}^2}{c^2} = \frac{m_0^2 c^2}{1 - v^2/c^2} \tag{12.89}$$

Squaring (12.86), we have

$$p^2 = \frac{m_0^2 v^2}{1 - v^2/c^2}$$

Therefore, we have

$$p^2 + m_0^2c^2 = \frac{m_0^2c^2}{1 - v^2/c^2} \tag{12.90}$$

Comparing equations (12.89) and (12.90), we get

$$\frac{\mathcal{E}^2}{c^2} = p^2 + m_0^2c^2 \qquad \text{or} \qquad \mathcal{E}^2 = p^2c^2 + m_0^2c^4$$

This is relation between energy and momentum.

Another relation between energy and momentum can be obtained by dividing equation (12.86) by (12.88) as

$$\frac{\vec{p}}{\mathcal{E}} = \frac{m_0\vec{v}/\sqrt{1 - v^2/c^2}}{m_0c^2/\sqrt{1 - v^2/c^2}} \qquad \text{or} \qquad \vec{p} = \frac{\mathcal{E}\vec{v}}{c^2} \tag{12.91}$$

For $v = c$, the momentum and energy both of a particle become infinite, showing that a particle with $m_0 \neq 0$ cannot move with the velocity of light. Nevertheless, in relativistic mechanics, particles of zero rest-mass moving with the velocity of light can exist.[1] For such particles, the relation between energy and momentum is

$$p = \frac{E}{c} \tag{12.92}$$

9.5 Hamiltonian of free particle

The Hamiltonian $\mathcal{H}$ of a free particle is

$$\mathcal{H} = c\sqrt{p^2 + m_0^2c^2}$$

For low velocity $p << m_0c$ and we have

$$\mathcal{H} = m_0c^2\left(1 + \frac{p^2}{m_0^2c^2}\right)^{1/2} = m_0c^2 + \frac{p^2}{2m_0}$$

The first part is the rest energy and the second the classical expression for Hamiltonian.

Review Questions

1. What are Galilean transformations? Write them. Is the time coupled with the space?

2. Derive Lorentz transformations for coordinates. Is the time coupled with the space? Show that when $v << c$, the Lorentz transformations reduce to the Galilean transformations.

3. Derive inverse Lorentz transformations for coordinates. Is the time coupled with the space?

[1] Examples of such particles are the photons and neutrinos.

4. Write Lorentz transformations for four-vector. Discuss the invariance of $A^i A^i$.
5. Decide if ∇, ∇^2, $\Box$ operators are invariant under Lorentz transformations.
6. Derive electromagnetic field tensor. Express Maxwell's equations in terms of electromagnetic field tensor.
7. Using Lorentz transformations for four-vectors, derive Lorentz transformations for antisymmetric four-tensor of rank two.
8. Derive Lorentz transformations of electromagnetic field.
9. Show that the electric field can be equal to zero in one reference system and at the same time be present in another system.
10. Show that the magnetic field can be equal to zero in one reference system and at the same time be present in another system.
11. Show that in a reference system, the magnetic and electric fields are mutually perpendicular to each other.
12. Show that if the electric and magnetic fields are mutually perpendicular in a reference system, then they are also perpendicular in every other reference system.
13. Show that if the absolute values of electric and magnetic fields are equal to each other in any reference system, then they are also equal in every reference system.
14. Show that if in any reference system $E > cB$ then in every other reference system we have $E > cB$.
15. Show that if in any reference system $E < cB$ then in every other reference system we have $E < cB$.
16. Show that if in any reference system $\vec{E}$ and $\vec{B}$ make an acute then they make an acute angle in every other reference system.
17. Show that if in any reference system $\vec{E}$ and $\vec{B}$ make an obtuse then they make an obtuse angle in every other reference system.
18. Derive an expression for the action and Lagrangian of a free particle.
19. Derive an expression for Lagrangian for a charge particle interacting with a field.
20. Write notes on the following

    (i) Postulates of special theory of relativity
    (ii) Galilean transformations
    (iii) Simultaneity of two events

(iv) Lorentz transformations for coordinates

(v) Four-vectors

(vi) Four-tensor of second rank

(vii) Four-velocity

(viii) Four-momentum

(ix) Four-current

(x) Electromagnetic four-potential

(xi) Invariant form of continuity equation

(xii) Invariant form of Lorentz condition

(xiii) Lorentz invariance of $\partial A^i/\partial x^i$

(xiv) Lagrangian of a free particle

(xv) Action of a free particle

(xvi) Rest-mass and rest-energy

(xvii) Non-uniqueness of scalar potential for constant electric field

(xviii) Non-uniqueness of vector potential for constant magnetic field

Multiple-Choice Questions

1. In which transformations, the speed of light is infinite?

   A. Galilean only B. Lorentz only C. Galilean and Lorentz both

   D. neither in Galilean nor in Lorentz

   Ans. A

2. In which transformations, the speed of light is finite?

   A. Galilean only B. Lorentz only C. Galilean and Lorentz both

   D. neither in Galilean nor in Lorentz

   Ans. B

3. In which transformations, the space and time are not coupled?

   A. Galilean only B. Lorentz only C. Galilean and Lorentz both

   D. neither in Galilean nor in Lorentz

   Ans. A

4. In which transformations, the space and time are coupled?

   A. Galilean only B. Lorentz only C. Galilean and Lorentz both

   D. neither in Galilean nor in Lorentz

   Ans. B

5. For four-space x^i, what is $x^i x^i$?

   A. zero   B. 1   C. $x^2 + y^2 + z^2 - c^2t^2$ D. $x^2 + y^2 + z^2 + c^2t^2$

   Ans. C

6. For four-velocity u^i, what is $u^i u^i$?

   A. $c^2 - v^2$ B. $v^2 - c^2$ C. $-c^2$ D. v^2

   Ans. C

7. What is time-component of four-velocity u^i ?

   A. ct B. $i\gamma c$ C. $\gamma\vec{v}$ D. 1

   Ans. B

8. What is space-component of four-velocity u^i ?

   A. ct B. $i\gamma c$ C. $\gamma\vec{v}$ D. 1

   Ans. C

9. What is time-component of four-momentum p^i?

   A. $m_0\vec{v}$ B. $\gamma m_0\vec{v}$ C. iE/c D. $i\gamma E/c$

   Ans. C

10. What is space-component of four-momentum p^i?

    A. $m_0\vec{v}$ B. $\gamma m_0\vec{v}$ C. iE/c D. $i\gamma E/c$

    Ans. B

11. For four-momentum p^i, what is $p^i p^i$?

    A. $(p^2 - E^2/c^2)$ B. $(p^2 - \gamma^2 E^2/c^2)$ C. $(p^2 + E^2/c^2)$ D. $(p^2 + \gamma^2 E^2/c^2)$

    Ans. A

12. For four-coordinates x^i, what is $x^i x^i$?

    A. $(ct, \vec{r})$ B. $(c^2t^2, \vec{r}^2)$ C. $(\vec{r}, -ct)$ D. $(\vec{r}^2, -c^2t^2)$

    Ans. D

13. What is time-component of four-potential A^i?

    A. ϕ B. ϕ/c C. $i\phi/c$ D. $i\phi c$

    Ans. C

14. What is time-component of four-current J^i?

    A. ρ B. ρc C. $i\rho c$ D. $i\gamma\rho$

    Ans. C

15. For the four-vector A^i, which is not the Lorentz transformation ?

    A. $A'^1 = \gamma(A^1 - i\beta A^4)$ B. $A'^2 = A^2$ C. $A'^3 = A^3$ D. $A'^4 = \gamma(A^4 - i\beta A^1)$

    Ans. A

16. For the four-vector A^i, which is not the inverse Lorentz transformation ?

    A. $A^1 = \gamma(A'^1 + i\beta A'^4)$ B. $A^2 = A'^2$ C. $A^3 = A'^3$ D. $A^4 = \gamma(A'^4 + i\beta A'^1)$

    Ans. A

17. For the four-vector A^i, which is not the Lorentz transformation ?

    A. $A'^1 = \gamma(A^1 + i\beta A^4)$ B. $A'^2 = A^2$ C. $A'^3 = A^3$ D. $A'^4 = \gamma(A^4 + i\beta A^1)$

    Ans. D

18. For the four-vector A^i, which is not the inverse Lorentz transformation ?

    A. $A^1 = \gamma(A'^1 - i\beta A'^4)$ B. $A^2 = A'^2$ C. $A^3 = A'^3$ D. $A^4 = \gamma(A'^4 - i\beta A'^1)$

    Ans. D

19. For the four-space x^i, which is not the Lorentz transformation ?

    A. $x'^1 = \gamma(x^1 + i\beta x^4)$ B. $x'^2 = x^2$ C. $x'^3 = x^3$ D. $x'^4 = \gamma(x^4 + i\beta x^1)$

    Ans. D

20. For the four-space x^i, which is not the inverse Lorentz transformation ?

    A. $x^1 = \gamma(x'^1 - i\beta x'^4)$ B. $x^2 = x'^2$ C. $x^3 = x'^3$ D. $x^4 = \gamma(x'^4 - i\beta x'^1)$

    Ans. D

21. For the four-space x^i, which is not the inverse Lorentz transformation ?

    A. $x^1 = \gamma(x'^1 + i\beta x'^4)$ B. $x^2 = x'^2$ C. $x^3 = x'^3$ D. $x^4 = \gamma(x'^4 + i\beta x'^1)$

    Ans. A

22. For the four-space x^i, which is not the Lorentz transformation ?

    A. $x'^1 = \gamma(x^1 - i\beta x^4)$ B. $x'^2 = x^2$ C. $x'^3 = x^3$ D. $x'^4 = \gamma(x^4 - i\beta x^1)$

    Ans. A

23. In an inertial reference frame, the electric and magnetic fields are perpendicular to each other. After transformation to another inertial reference frame, the angle between the electric and magnetic fields is

    A. $< 90°$ B. $90°$ C. $> 90°$ D. dependent on situation

    Ans. B

24. In an inertial reference frame, the electric and magnetic fields are inclined at an acute angle. After transformation to another inertial reference frame, the angle between the electric and magnetic fields is

    A. acute B. 90° C. obtuse D. depends on situation

    Ans. A

25. In an inertial reference frame, the electric and magnetic fields are inclined at an obtuse angle. After transformation to another inertial reference frame, the angle between the electric and magnetic fields is

    A. acute B. 90° C. obtuse D. depends on situation

    Ans. C

26. In an inertial reference frame, the electric and magnetic fields are equal in magnitude. After transformation to another inertial reference frame, the relation between the magnitudes E and B of electric and magnetic fields, respectively, is

    A. $E < cB$ B. $E = cB$ C. $E > cB$ D. depends on situation

    Ans. B

27. In an inertial reference frame, the magnitude of electric field is smaller than that of the magnetic field. After transformation to another inertial reference frame, the relation between the magnitudes E and B of electric and magnetic fields, respectively, is

    A. $E < cB$ B. $E = cB$ C. $E > cB$ D. depends on situation

    Ans. A

28. In an inertial reference frame, the magnitude of electric field is larger than that of the magnetic field. After transformation to another inertial reference frame, the relation between the magnitudes E and B of electric and magnetic fields, respectively, is

    A. $E < cB$ B. $E = cB$ C. $E > cB$ D. depends on situation

    Ans. C

29. Which of the followings is invariant for the transformation of electromagnetic field.

    A. $\vec{E} + c\vec{B}$ B. $\vec{E} - c\vec{B}$ C. $\vec{E} \,.\, \vec{B}$ D. $E^2 + c^2B^2$

    Ans. C

30. Which of the followings is invariant for the transformation of electromagnetic field.

    A. $\vec{E} + c\vec{B}$ B. $\vec{E} - c\vec{B}$ C. $\vec{E} \,.\, \vec{B}$ D. $E^2 + c^2B^2$

    Ans. C

31. E_x, E_y, E_z are components of electric field and B_x, B_y, B_z are components of magnetic field. Which of the following remains unchanged under the Lorentz transformations of electromagnetic field?

A. E_x B. E_y C. E_z D. B_z

Ans. A

32. E_x, E_y, E_z are components of electric field and B_x, B_y, B_z are components of magnetic field. Which of the following remains unchanged under the Lorentz transformations of electromagnetic field.

A. B_x B. B_y C. B_z D. E_y

Ans. A

33. For vector potential $\vec{A}$ and scalar potential ϕ, steady electric field is determined by

A. $\vec{A}$ only B. ϕ and $\vec{A}$ both C. ϕ only D. neither ϕ nor $\vec{A}$

Ans. C

34. For vector potential $\vec{A}$ and scalar potential ϕ, steady magnetic field is determined by

A. $\vec{A}$ only B. ϕ and $\vec{A}$ both C. ϕ only D. neither ϕ nor $\vec{A}$

Ans. A

35. For vector potential $\vec{A}$ and scalar potential ϕ, the energy depends on

A. $\vec{A}$ only B. ϕ and $\vec{A}$ both C. ϕ only D. neither ϕ nor $\vec{A}$

Ans. C

36. If the action S of free particle is expressed as

$$S = \int_{t_1}^{t_2} Ldt$$

the expression for the Lagrangian L is

A. $-mc^2/\sqrt{1-v^2/c^2}$ B. $-mc^2\sqrt{1-v^2/c^2}$

C. $-mc/\sqrt{1-v/c}$ D. $-mc\sqrt{1-v/c}$

Ans. B

37. In a constant electromagnetic field, the potential energy depends on

A. $\vec{A}$ only B. ϕ and $\vec{A}$ both C. ϕ only D. neither ϕ nor $\vec{A}$

Ans. C

38. In a reference system, we have $\vec{B}' = 0$ and $\vec{E}' \neq 0$. After transformation into another reference system, the electric field $\vec{E}$ and magnetic field $\vec{B}$ can have the values

    A. $\vec{E} = 0$ and $\vec{B} = 0$ B. $\vec{E} \neq 0$ and $\vec{B} = 0$

    C. $\vec{E} = 0$ and $\vec{B} \neq 0$ D. $\vec{E} \neq 0$ and $\vec{B} \neq 0$

    Ans. D

39. In a reference system, we have $\vec{E}' = 0$ and $\vec{B}' \neq 0$. After transformation into another reference system, the electric field $\vec{E}$ and magnetic field $\vec{B}$ can have the values

    A. $\vec{E} = 0$ and $\vec{B} = 0$ B. $\vec{E} \neq 0$ and $\vec{B} = 0$

    C. $\vec{E} = 0$ and $\vec{B} \neq 0$ D. $\vec{E} \neq 0$ and $\vec{B} \neq 0$

    Ans. D

40. Which of the electric field $\vec{E}$ and magnetic field $\vec{B}$ is always steady.

    A. $\vec{E}$ only B. $\vec{E}$ and $\vec{B}$ both C. $\vec{B}$ only D. neither $\vec{E}$ nor $\vec{B}$

    Ans. C

41. Which of the electric field $\vec{E}$ and magnetic field $\vec{B}$ may not be steady.

    A. $\vec{E}$ only B. $\vec{E}$ and $\vec{B}$ both C. $\vec{B}$ only D. neither $\vec{E}$ nor $\vec{B}$

    Ans. A

42. The inertial reference frame S′ is moving with velocity v along x-axis with respect to the inertial reference frame S. Suppose, we have pure electric field $\vec{E}'$ in S′. This field, after transformation, in S, the electric field $\vec{E}$ and magnetic field $\vec{B}$ satisfy the condition

    A. $\vec{E} = 0$ & $\vec{B} = 0$ B. $\vec{E} \neq 0$ & $\vec{B} = 0$

    C. $\vec{E} = 0$ & $\vec{B} \neq 0$ D. $\vec{E} \neq 0$ & $\vec{B} \neq 0$

    Ans. D

43. The inertial reference frame S′ is moving with velocity v along x-axis with respect to the inertial reference frame S. Suppose, we have pure magnetic field $\vec{B}'$ in S′. This field, after transformation, in S, the electric field $\vec{E}$ and magnetic field $\vec{B}$ satisfy the condition

    A. $\vec{E} = 0$ & $\vec{B} = 0$ B. $\vec{E} \neq 0$ & $\vec{B} = 0$

    C. $\vec{E} = 0$ & $\vec{B} \neq 0$ D. $\vec{E} \neq 0$ & $\vec{B} \neq 0$

    Ans. D

44. The inertial reference frame S′ is moving with velocity v along x-axis with respect to the inertial reference frame S. Suppose, we have pure electric field

$\vec{E}'$ in S′. This field, after transformation, in S, the electric field $\vec{E}$ and magnetic field $\vec{B}$ satisfy the condition

A. $\vec{E} = -\vec{v} \times \vec{B}$ B. $\vec{B} = \vec{v} \times \vec{E}$ C. $\vec{v} = -\vec{E} \times \vec{B}$ D. none of these

Ans. B

45. The inertial reference frame S′ is moving with velocity v along x-axis with respect to the inertial reference frame S. Suppose, we have pure magnetic field $\vec{H}'$ in S′. This field, after transformation, in S, the electric field $\vec{E}$ and magnetic field $\vec{H}$ satisfy the condition

A. $\vec{E} = -\vec{v} \times \vec{B}$ B. $\vec{B} = \vec{v} \times \vec{E}$ C. $\vec{v} = -\vec{E} \times \vec{B}$ D. none of these

Ans. A

46. For the special theory of relativity proposed by Einstein, which statement is wrong?

A. Speed of light is finite.

B. Space and time are coupled with each other.

C. Time interval between two events, when measured in two different inertial reference frames, is not equal

D. Time is absolute.

Ans. D

47. Which of the following represents Maxwell's field equation?

A. $\nabla^2\phi - \frac{1}{c^2}\frac{\partial^2\phi}{\partial t^2} = -\frac{\rho}{\epsilon_0}$ B. $\nabla^2\phi + \frac{1}{c^2}\frac{\partial^2\phi}{\partial t^2} = -\frac{\rho}{\epsilon_0}$

C. $\nabla^2\phi - \frac{1}{c^2}\frac{\partial^2\phi}{\partial t^2} = \frac{\rho}{\epsilon_0}$ D. $\nabla^2\phi + \frac{1}{c^2}\frac{\partial^2\phi}{\partial t^2} = \frac{\rho}{\epsilon_0}$

Ans. A

48. Which of the following represents Maxwell's field equation?

A. $\nabla^2\vec{A} - \frac{1}{c^2}\frac{\partial^2\vec{A}}{\partial t^2} = -\mu_0\vec{J}$ B. $\nabla^2\vec{A} + \frac{1}{c^2}\frac{\partial^2\vec{A}}{\partial t^2} = -\mu_0\vec{J}$

C. $\nabla^2\vec{A} - \frac{1}{c^2}\frac{\partial^2\vec{A}}{\partial t^2} = \mu_0\vec{J}$ B. $\nabla^2\vec{A} + \frac{1}{c^2}\frac{\partial^2\vec{A}}{\partial t^2} = \mu_0\vec{J}$

Ans. A

Chapter 13

Retarded Systems

1. Retarded potentials

Vector potential $\vec{A}$ and scalar potential ϕ satisfy the relations

$$\nabla^2\vec{A} - \frac{1}{c^2}\frac{\partial^2\vec{A}}{\partial t^2} = -\mu_0\vec{J} \tag{13.1}$$

and

$$\nabla^2\phi - \frac{1}{c^2}\frac{\partial^2\phi}{\partial t^2} = -\frac{\rho}{\epsilon_0} \tag{13.2}$$

where ρ is volume charge density and $\vec{J}$ the current density.

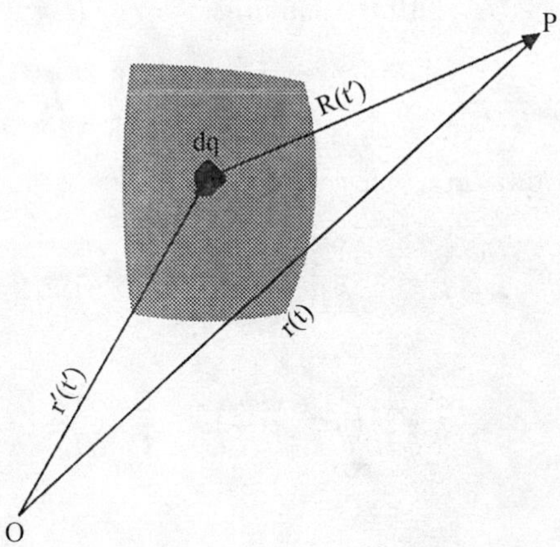

Figure 1: The space is divided into large number of small regions. Here, $t > t'$. This is precisely the meaning of retardation. It means that when the signal is observed at the time t, it was produced at an earlier time $t' < t$.

Equations (13.1) and (13.2) are inhomogeneous linear equations. Solution of a inhomogeneous linear equation has two parts: (i) without the right side (*i.e.*, the solution of homogeneous linear equation), (ii) including the right side, called, the particular integral. The particular integral (including the right-hand side) can be obtained by dividing the space into large number of small regions, and to determine the field produced by the charges located in one of these volume-elements. Because of the linearity of field equations, the actual fields are the sum of the fields produced by all such elements.

Suppose, a given element of volume $\mathrm{d}\tau$ has charge $\mathrm{d}q$ at a given time t. We choose that the element has origin of coordinate system, so that the charge density is $\rho = (\mathrm{d}q(t)/\mathrm{d}\tau)\ \delta(\vec{R})$, where $\vec{R}$ represents the displacement from the origin taken at the element. (In the element, we have $R = 0$ and $\delta(\vec{R}) = 1$). We assume that the charge distribution is stationary in space, but varies harmonically in time. Thus, for this volume element, we must solve the equation

$$\nabla^2\phi - \frac{1}{c^2}\frac{\partial^2\phi}{\partial t^2} = -\frac{1}{\epsilon_0}\frac{\mathrm{d}q(t)}{\mathrm{d}\tau}\ \delta(\vec{R}) \tag{13.3}$$

It is clear that in this case, ϕ is spherically symmetric and therefore, ϕ is a function of R only.

$$\nabla^2 \equiv \frac{1}{r^2}\Big[\frac{\partial}{\partial r}\Big(r^2\frac{\partial}{\partial r}\Big) + \frac{1}{\sin\theta}\frac{\partial}{\partial\theta}\Big(\sin\theta\frac{\partial}{\partial\theta}\Big) + \frac{1}{\sin^2\theta}\frac{\partial^2}{\partial\phi^2}\Big]$$

Therefore, when we express Laplace operator in the spherical coordinates, equation (13.3) reduces to the form

$$\frac{1}{R^2}\frac{\partial}{\partial R}\Big(R^2\frac{\partial\phi}{\partial R}\Big) - \frac{1}{c^2}\frac{\partial^2\phi}{\partial t^2} = -\frac{1}{\epsilon_0}\frac{\mathrm{d}q(t)}{\mathrm{d}\tau}\ \delta(\vec{R}) \tag{13.4}$$

Let us substitute $\phi = \chi/R$. Differentiating $\phi = \chi/R$ with respect to R, we get

$$\frac{\partial\phi}{\partial R} = -\frac{\chi}{R^2} + \frac{1}{R}\frac{\partial\chi}{\partial R} \qquad \text{or} \qquad R^2\frac{\partial\phi}{\partial R} = -\chi + R\frac{\partial\chi}{\partial R}$$

Differentiating this equation with respect to R, we get

$$\frac{\partial}{\partial R}\Big(R^2\frac{\partial\phi}{\partial R}\Big) = -\frac{\partial\chi}{\partial R} + \frac{\partial\chi}{\partial R} + R\frac{\partial^2\chi}{\partial R^2} = R\frac{\partial^2\chi}{\partial R^2}$$

Hence, we have

$$\frac{1}{R^2}\frac{\partial}{\partial R}\Big(R^2\frac{\partial\phi}{\partial R}\Big) = \frac{1}{R}\frac{\partial^2\chi}{\partial R^2} \tag{13.5}$$

Further, we have

$$\frac{\partial\phi}{\partial t} = \frac{1}{R}\frac{\partial\chi}{\partial t} \qquad \text{and} \qquad \frac{\partial^2\phi}{\partial t^2} = \frac{1}{R}\frac{\partial^2\chi}{\partial t^2} \tag{13.6}$$

Using equations (13.5) and (13.6) in (13.4), we get

$$\frac{1}{R}\frac{\partial^2\chi}{\partial R^2} - \frac{1}{c^2}\frac{1}{R}\frac{\partial^2\chi}{\partial t^2} = -\frac{1}{\epsilon_0}\frac{\mathrm{d}q(t)}{\mathrm{d}\tau})\ \delta(\vec{R})$$

or

$$\frac{\partial^2 \chi}{\partial R^2} - \frac{1}{c^2}\frac{\partial^2 \chi}{\partial t^2} = -\frac{R}{\epsilon_0}\frac{dq(t)}{d\tau}\,\delta(\vec{R})$$

Everywhere, except at the origin, the value of $\delta(\vec{R})$ is zero, and we have

$$\frac{\partial^2 \chi}{\partial R^2} - \frac{1}{c^2}\frac{\partial^2 \chi}{\partial t^2} = 0 \tag{13.7}$$

But, this is equation of plane waves, whose solution has the form

$$\chi = f_1\left(t - \frac{R}{c}\right) + f_2\left(t + \frac{R}{c}\right)$$

Since we only want a particular solution of this equation, it is sufficient to choose only one of the functions f_1 and f_2. Usually, it turns out to be convenient to take $f_2 = 0$. That is, we consider a wave moving in the forward direction. Then, everywhere, except at the origin, ϕ has the form

$$\phi = \frac{\chi(t - R/c)}{R} \tag{13.8}$$

So far, the function χ is arbitrary. We now choose it, so that we also obtain the correct value for the potential at the origin. That is, we must have χ so that at the origin equation (13.4) is satisfied. This can be easily done by noting that as $R \to 0$, the potential increases to infinity, and therefore its derivatives with respect to the coordinates increase more rapidly than its time derivative.

Consequently, as $R \to 0$, we can, in equation (13.3), neglect the term $(1/c^2)\, \partial^2\phi/\partial t^2$ as compared with $\nabla^2\phi$. Then equation (13.4) goes over into

$$\nabla^2\phi = -\frac{1}{\epsilon_0}\frac{dq(t)}{d\tau}\,\delta(\vec{R})$$

It corresponds to the potential[1]

$$\nabla^2\phi = -\frac{1}{\epsilon_0}\frac{dq(t)}{d\tau}\,\delta(\vec{R}) \tag{13.9}$$

Except at the origin, we have

$$\nabla^2\phi = 0 \qquad \text{or} \qquad \frac{1}{R^2}\frac{\partial}{\partial R}\left(R^2\frac{\partial \phi}{\partial R}\right) = 0 \tag{13.10}$$

[1]For the electrostatic field (time independent electric field), we have

$$\text{div}\,\vec{E} = \frac{\rho}{\epsilon_0} \qquad \text{and} \qquad \text{curl}\,\vec{E} = 0$$

On expressing $\vec{E} = -\nabla\phi$, we get

$$\nabla^2\phi = -\frac{\rho}{\epsilon_0}$$

For a point-charge dq, the charge density $\rho = (dq/dV)\delta(\vec{R})$ and the potential ϕ at a distance R is (Coulomb potential)

$$\phi = \frac{1}{4\pi\epsilon_0}\frac{dq}{R} \qquad \text{and} \qquad \nabla^2\phi = -\frac{1}{\epsilon_0}\frac{dq(t)}{d\tau}\,\delta(\vec{R})$$

Equation (13.10) is satisfied by

$$\phi = \frac{dq(t)}{R} \tag{13.11}$$

Thus, near the origin, equation (13.8) must go over into the Coulomb law, expressed by equation (13.11). Hence, equation (13.8) should be as

$$\phi = \frac{1}{4\pi\epsilon_0} \frac{dq(t - R/c)}{R}$$

From this, it is easy to get the solution of equation (13.2) for an arbitrary distribution of charges with the charge density $\rho(x, y, z, t)$. To do this, it is sufficient to write $dq = \rho \, d\tau$ ($d\tau$ is the volume element) and to integrate over the whole space. To this solution of the inhomogeneous equation (13.2) we can still add the solution ϕ_0 of the same equation without the right-hand side. Thus, the general solution has the form

$$\phi(\vec{r},\ t) = \frac{1}{4\pi\epsilon_0} \int \frac{1}{R}\, \rho\left(\vec{r},'\ t - \frac{R}{c}\right)\, d\tau' \tag{13.12}$$

This equation says that the potential $\phi(\vec{r},\ t)$ is at the position $\vec{r}$ and at time t whereas the charge density $\rho(\vec{r},'\ t{-}R/c)$ is at the position $r\,'$ at time $t' = t_r = t{-}R/c$. The time t_r is earlier than t and is known as the retarded time and $\phi(\vec{r}, t)$ is known as the retarded potential. We have

$$\vec{R} = \vec{r} - \vec{r}\,' \qquad \text{and} \qquad d\tau' = dx'\ dy'\ dz'$$

where

$$\vec{r} = x\ \hat{x} + y\ \hat{y} + z\ \hat{z} \qquad \text{and} \qquad \vec{r}' = x'\ \hat{x} + y'\ \hat{y} + z'\ \hat{z}$$

and R is the distance from the volume element $d\tau'$ to the field-point P where the potential is determined.

Similarly, we have for the retarded vector potential:

$$\vec{A}(\vec{r},\ t) = \frac{\mu_0}{4\pi} \int \frac{1}{R}\, \vec{J}\left(\vec{r},'\ t - \frac{R}{c}\right) d\tau' \tag{13.13}$$

Example 1: Show that $\nabla R = \hat{R}$ and $\nabla(1/R) = -\hat{R}/R^2$,

Solution: We have

$$\vec{R} = \vec{r} - \vec{r}\,' = (x - x')\hat{x} + (y - y')\hat{y} + (z - z')\hat{z}$$

Thus,

$$R = \sqrt{(x - x')^2 + (y - y')^2 + (z - z')^2}$$

We have

$$\frac{\partial R}{\partial x} = \frac{(x - x')}{\sqrt{(x - x')^2 + (y - y')^2 + (z - z')^2}} = \frac{(x - x')}{R}$$

Similarly,

$$\frac{\partial R}{\partial y} = \frac{(y - y')}{R} \qquad \text{and} \qquad \frac{\partial R}{\partial z} = \frac{(z - z')}{R}$$

Hence

$$\nabla R = \frac{\partial R}{\partial x}\,\hat{x} + \frac{\partial R}{\partial y}\,\hat{y} + \frac{\partial R}{\partial z}\,\hat{z}$$

$$= \frac{(x - x')\,\hat{x} + (y - y')\,\hat{y} + (z - z')\,\hat{z}}{R} = \frac{\vec{R}}{R} = \hat{R}$$

We have

$$\frac{1}{R} = \frac{1}{\sqrt{(x - x')^2 + (y - y')^2 + (z - z')^2}}$$

Therefore,

$$\frac{\partial(1/R)}{\partial x} = -\frac{(x - x')}{[(x - x')^2 + (y - y')^2 + (z - z')^2]^{3/2}} = -\frac{(x - x')}{R^3}$$

Similarly,

$$\frac{\partial(1/R)}{\partial y} = -\frac{(y - y')}{R^3} \qquad \frac{\partial(1/R)}{\partial z} = -\frac{(z - z')}{R^3}$$

Hence

$$\nabla(1/R) = \frac{\partial(1/R)}{\partial x}\,\hat{x} + \frac{\partial(1/R)}{\partial y}\,\hat{y} + \frac{\partial(1/R)}{\partial z}\,\hat{z}$$

$$= -\frac{(x - x')\,\hat{x} + (y - y')\,\hat{y} + (z - z')\,\hat{z}}{R^3} = -\frac{\vec{R}}{R^3} = -\frac{\hat{R}}{R^2}$$

Example 2: An infinite straight wire carries the current

$$I(t) = \begin{cases} 0 & \text{for} \quad t \leq 0 \\ I_0 & \text{for} \quad t > 0 \end{cases}$$

Find the resulting electric and magnetic fields.

Solution: A constant current I_0 is turned on abruptly at $t = 0$. Let the wire lie along the z-axis as shown in Figure. The wire is presumably electrically neutral, so the scalar potential is zero. The retarded vector potential at point P is

$$\vec{A}(s,\ t) = \frac{\mu_0}{4\pi}\int_{-\infty}^{\infty} \frac{I(t_r)\hat{z}}{R}\,\mathrm{d}z$$

For $t < s/c$, the signal has not yet reached P, and the potential is zero. For $t > s/c$, only the segment

$$ct \geq \sqrt{z^2 + s^2} \qquad \text{or} \qquad |z| \leq \sqrt{(ct)^2 - s^2}$$

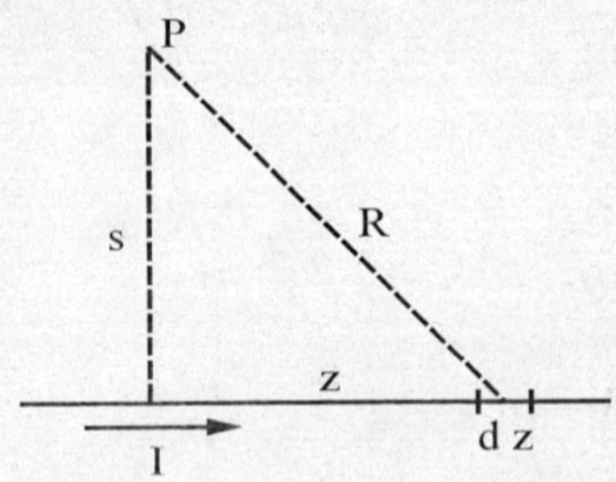

Figure 2: Current flowing through infinite straight wire.

contributes. Here, $t \geq R/c$ and t_r is not negative. Outside this range, t_r is negative, and contribution is zero. Thus,

$$\vec{A}(s,t) = \left(\frac{\mu_0 I_0 \hat{z}}{4\pi}\right) 2 \int_0^{\sqrt{(ct)^2 - s^2}} \frac{dz}{\sqrt{z^2+s^2}}$$

$$= \left(\frac{\mu_0 I_0 \hat{z}}{2\pi}\right) \left[\ln\left(\sqrt{z^2+s^2}+z\right)\right]_0^{\sqrt{(ct)^2-s^2}}$$

$$= \left(\frac{\mu_0 I_0}{2\pi}\right) \ln\left[\frac{ct + \sqrt{(ct)^2 - s^2}}{s}\right] \hat{z}$$

The electric field is

$$\vec{E}(s,\ t) = -\frac{\partial \vec{A}}{\partial t} = -\frac{\mu_0 I_0 c}{2\pi\sqrt{(ct)^2 - s^2}}\ \hat{z}$$

and the magnetic field is

$$\vec{B}(s,\ t) = \nabla \times \vec{A} = -\frac{\partial A_z}{\partial s}\ \hat{\phi} = \frac{\mu_0 I_0}{2\pi s} \frac{ct}{\sqrt{(ct)^2 - s^2}}\ \hat{\phi}$$

Notice that as $t \to \infty$, we get the static case: $\vec{E} = 0$ and $\vec{B} = (\mu_0 I_0/2\pi s)\ \hat{\phi}$

2. Electric and magnetic fields due to retarded potentials

Given the retarded potentials

$$\phi(\vec{r},\ t) = \frac{1}{4\pi\epsilon_0}\int \frac{\rho(\vec{r}',\ t_r)}{R}\, d\tau' \qquad \vec{A}(\vec{r},\ t) = \frac{\mu_0}{4\pi}\int \frac{\vec{J}(\vec{r}',\ t_r)}{R}\, d\tau' \tag{13.14}$$

It is, in principle, a straightforward matter to determine the fields:

$$\vec{E} = -\nabla\phi - \frac{\partial \vec{A}}{\partial t} \qquad \text{and} \qquad \vec{B} = \nabla \times \vec{A} \tag{13.15}$$

But the details are not entirely trivial because, as the integrands depend on $\vec{r}$ both explicitly, through $R = |\vec{r} - \vec{r}\,'|$ in the denominator, and implicitly, through the retarded time $t_r = t - R/c$ in the argument of the numerator. Thus,

$$\nabla\phi = \frac{1}{4\pi\epsilon_0}\int\Big[(\nabla\rho)\,\frac{1}{R} + \rho\,\nabla\Big(\frac{1}{R}\Big)\Big]\mathrm{d}\tau' \tag{13.16}$$

and[2]

$$\nabla\rho = \frac{\partial\rho}{\partial t}\,\nabla t_r = -\frac{1}{c}\,\frac{\partial\rho}{\partial t}\,\nabla R \tag{13.17}$$

Now, $\nabla R = \hat{R}$ and $\nabla(1/R) = -\hat{R}/R^2$, so

$$\nabla\phi = \frac{1}{4\pi\epsilon_0}\int\Big[-\frac{1}{c}\,\frac{\partial\rho}{\partial t}\,\frac{\hat{R}}{R} - \rho\,\frac{\hat{R}}{R^2}\Big]\mathrm{d}\tau' \tag{13.18}$$

The time derivative of $\vec{A}$ is

$$\frac{\partial\vec{A}}{\partial t} = \frac{\mu_0}{4\pi}\int\frac{\partial\vec{J}}{\partial t}\,\frac{1}{R}\,\mathrm{d}\tau' \tag{13.19}$$

Putting them together (and using $c^2 = 1/\mu_0\epsilon_0$):

$$\vec{E}(\vec{r},\,t) = \frac{1}{4\pi\epsilon_0}\int\Big[\frac{\rho(\vec{r},'\,t_r)}{R^2}\hat{R} + \frac{\partial\rho}{\partial t}\,\frac{\hat{R}}{cR} - \frac{\partial\vec{J}}{\partial t}\,\frac{1}{c^2R}\Big]\mathrm{d}\tau' \tag{13.20}$$

This is the time-dependent generalization of Coulomb's law. For static case, it reduces to

$$\vec{E}(\vec{r},t) = \frac{1}{4\pi\epsilon_0}\int\Big[\frac{\rho(\vec{r},'\,t_r)}{R^2}\hat{R}\Big]\mathrm{d}\tau'$$

$$= \frac{1}{4\pi\epsilon_0}\,\frac{\hat{R}}{R^2}\int\rho(\vec{r},'\,t_r)\;\mathrm{d}\tau' = \frac{1}{4\pi\epsilon_0}\,\frac{q}{R^2}\,\hat{R}$$

This is Coulomb's law.

As for $\vec{B}$, the curl of $\vec{A}$ contains two terms:

$$\nabla\times\vec{A} = \frac{\mu_0}{4\pi}\int\Big[\frac{1}{R}(\nabla\times\vec{J}) - \vec{J}\times\nabla\Big(\frac{1}{R}\Big)\Big]\mathrm{d}\tau'$$

Now

$$(\nabla\times\vec{J})_x = \frac{\partial J_z}{\partial y} - \frac{\partial J_y}{\partial z}$$

and

$$\frac{\partial J_z}{\partial y} = \frac{\partial J_z}{\partial t}\,\frac{\partial t_r}{\partial y} = -\frac{1}{c}\,\frac{\partial J_z}{\partial t}\,\frac{\partial R}{\partial y}$$

Similarly,

[2]Note that $\partial/\partial t_r = \partial/\partial t$, since $t_r = t - R/c$ and R is independent of t.

$$\frac{\partial J_y}{\partial z} = -\frac{1}{c}\frac{\partial J_y}{\partial t}\frac{\partial R}{\partial z}$$

so

$$(\nabla \times \vec{J})_x = -\frac{1}{c}\Big(\frac{\partial J_z}{\partial t}\frac{\partial R}{\partial y} - \frac{\partial J_y}{\partial t}\frac{\partial R}{\partial z}\Big)$$

$$= \frac{1}{c}\Big[\frac{\partial \vec{J}}{\partial t} \times (\nabla R)\Big]_x = \frac{1}{c}\Big[\frac{\partial \vec{J}}{\partial t} \times \hat{R}\Big]_x$$

where we have used $\nabla R = \hat{R}$. Similarly,

$$(\nabla \times \vec{J})_y = \frac{1}{c}\Big[\frac{\partial \vec{J}}{\partial t} \times \hat{R}\Big]_y \qquad \text{and} \qquad (\nabla \times \vec{J})_z = \frac{1}{c}\Big[\frac{\partial \vec{J}}{\partial t} \times \hat{R}\Big]_z$$

Therefore,

$$\nabla \times \vec{J} = \frac{1}{c}\frac{\partial \vec{J}}{\partial t} \times \hat{R} \tag{13.21}$$

Meanwhile $\nabla(1/R) = -\hat{R}/R^2$, and hence

$$\vec{B}(\vec{r},\ t) = \frac{\mu_0}{4\pi}\int \Big[\frac{\vec{J}(\vec{r},'\, t_r)}{R^2} + \frac{\partial \vec{J}(\vec{r},'\, t_r)}{\partial t}\ \frac{1}{cR}\Big] \times \hat{R}\ \mathrm{d}\tau' \tag{13.22}$$

This is the time-dependent generalization of the Biot-Savart law. For static case, it reduces to

$$\vec{B}(\vec{r},\ t) = \frac{\mu_0}{4\pi}\int \frac{\vec{J}(\vec{r},'\, t_r)}{R^2} \times \hat{R}\ \mathrm{d}\tau'$$

$$= \frac{\mu_0}{4\pi}\Big(\int \vec{J}(\vec{r},'\, t_r)\ \mathrm{d}\tau'\Big) \times \frac{\hat{R}}{R^2} = \frac{\mu_0}{4\pi} I \mathrm{d}\vec{\ell} \times \frac{\hat{R}}{R^2}$$

This is Biot-Savart law.

3. Lienard-Wiechert potentials

The retarded potentials are

$$\phi(\vec{r},\ t) = \frac{1}{4\pi\epsilon_0}\int \frac{\rho(\vec{r},'\ t_r)}{R}\ \mathrm{d}\tau' \qquad \vec{A}(\vec{r},\ t) = \frac{\mu_0}{4\pi}\int \frac{\vec{J}(\vec{r},'\ t_r)}{R}\ \mathrm{d}\tau'$$

Next, we would like to calculate the (retarded) potentials, $\phi(r,\ t)$ and $\vec{A}(r,\ t)$, of a point-charge q that is moving on a specified trajectory

$$w(t) \equiv \text{position of } q \text{ at time t} \tag{13.23}$$

Consider the retarded potential

$$\phi(\vec{r},\ t) = \frac{1}{4\pi\epsilon_0}\int \frac{\rho(\vec{r},'\, t_r)}{R}\ \mathrm{d}\tau' \tag{13.24}$$

We know that potential due to charge q at a distance R is

$$\frac{1}{4\pi\epsilon_0}\frac{q}{R}$$

But this is wrong, for a very subtle reason. It is true that for a point source the denominator R in equation (13.24) comes outside the integral, but what remains,

$$\int \rho(\vec{r},' t_r)\, d\tau' \qquad (13.25)$$

is not equal to the charge of the particle (and depends, through t_r on the location of the point $\vec{r}$). To calculate the total charge of a configuration, we must integrate ρ over the entire distribution at one *instant of time*, but here the retardation, $t_r = t - R/c$, can be used to evaluate ρ at different times for different parts of the configuration. If the source is moving, this provides a different picture of total charge.

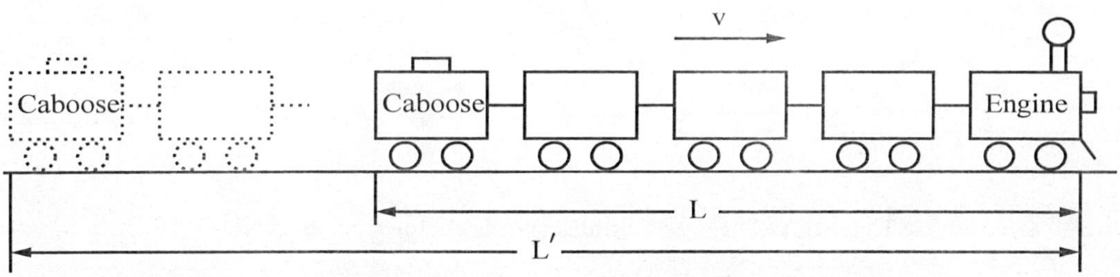

Figure 3: Train moving with velocity v.

In Maxwell's electrodynamics, formulated as it is in terms of charge and current densities, a point charge must be regarded as the limit of an extended charge, when the size goes to zero. And for an extended charge, no matter how small, the retardation in Equation (13.25) throws in a factor $(1 - \hat{R}\,.\,\vec{v}/c)^{-1}$, where $\vec{v}$ is the velocity of charge at the retarded time:

$$\int \rho(\vec{r},' t_r)\, d\tau' = \frac{q}{1 - \hat{R}\,.\,\vec{v}/c} \qquad (13.26)$$

The inclusion of the factor $(1 - \hat{R}\,.\,\vec{v}/c)^{-1}$ is a purely **geometrical effect**, which can be understood in the following manner.

It may not be noticed that a train coming towards the observer appears a little longer than its real length. It is because the light one receives from the caboose (last bogie of train) has left earlier than the light from the engine, received by the same person simultaneously, and at that earlier time the train was farther away (Fig. 3).

In the interval it receives light from the caboose to travel the larger distance L', the train itself moves a distance $(L' - L)$. Thus, we have

$$\frac{L'}{c} = \frac{L' - L}{v} \qquad \text{or} \qquad L' = \frac{L}{1 - v/c}$$

Thus, the approaching trains appear longer, by a factor $(1 - v/c)^{-1}$. By contrast, a train going away from the observer looks shorter,by a factor $(1 + v/c)^{-1}$.

In general, if the train's direction is inclined at an angle θ with the line-of-sight of the observer, the light from the caboose covers a distance, $L' \cos\theta$ (Fig. 4) in the time $L' \cos\theta / c$. In the same time, the train moves a distance $(L' - L)$, and we have

$$\frac{L' \cos\theta}{c} = \frac{L' - L}{v} \qquad \text{or} \qquad L' = \frac{L}{1 - v\cos\theta/c}$$

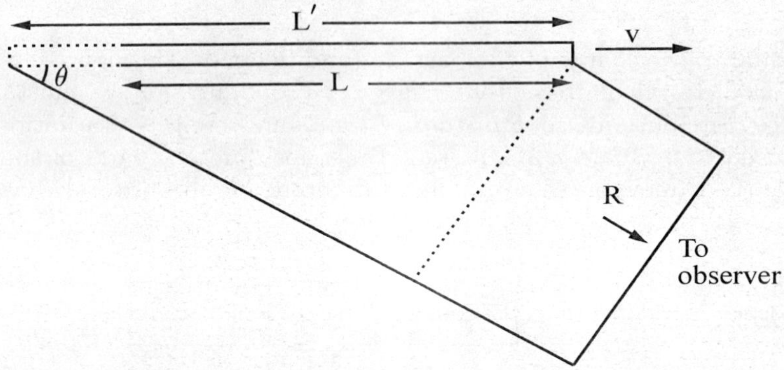

Figure 4: Direction of observer is inclined at angle θ.

It may be noted that this effect does not affect the dimensions perpendicular to the direction of motion (the height and width of the train appear unchanged). It does not matter that the light from the far side is delayed in reaching the observer (relative to light from the near side). As there is no motion in that direction, they still look the same distance apart. The apparent volume τ' of the train, then, is related to the actual volume τ by

$$\tau' = \frac{\tau}{1 - \hat{R} \,.\, \vec{v}/c} \tag{13.27}$$

where $\hat{R}$ is a unit vector from the train to the observer.

A connection between moving trains and retarded potentials can be understood in the following manner.

When we do an integral of the type in equation (13.25), in which the integrand is evaluated at the retarded time, the effective volume is modified by the factor in equation (13.27), just as the apparent volume of the train was. As this correction factor is irrespective of the particle, it is as significant for a point charge as for an extended charge. Further, for a point charge the retarded time can be determined implicitly using the equation:

$$|\vec{r} - \vec{w}(t_r)| = c(t - t_r) \tag{13.28}$$

The left side is the distance the signal must travel, and $(t - t_r)$ is the time it takes to make the trip (Fig. 5); R is the vector from the retarded position to the field point $\vec{r}$:

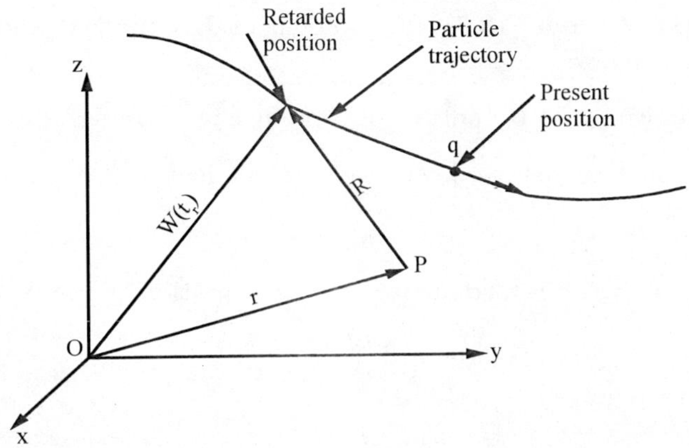

Figure 5: Motion of particle.

$$\vec{R} = \vec{r} - \vec{w}(t_r) \tag{13.29}$$

It is important to note that from a point on the trajectory signal is received at the point P at the position $\vec{r}$ and time t. Let there were two such points, with retarded times t_{r1} and t_{r2}, so that

$$R_1 = c(t - t_{r1}) \qquad \text{and} \qquad R_2 = c(t - t_{r2})$$

Then $R_1 - R_2 = c(t_{r2} - t_{r1})$, therefore the average speed of the particle in the direction of the point $\vec{r}$ is c and that is not counting whatever velocity the charge might have in other directions.

Since no charged particle can travel at the speed of light, it follows that only one retarded point contributes to the potentials, at any given moment.[3]

Then, it follows that

$$\phi(\vec{r},\ t) = \frac{1}{4\pi\epsilon_0}\,\frac{1}{R}\,\frac{q}{1 - \hat{R}\ .\ \vec{v}/c} = \frac{1}{4\pi\epsilon_0}\,\frac{qc}{(Rc - \vec{R}\ .\ \vec{v})} \tag{13.30}$$

where $\vec{v}$ is the velocity of the charge at the retarded time, and $\vec{R}$ the vector from the retarded position to the field point P. Moreover, since the current density is $\rho\,\vec{v}$, the vector potential is

$$\vec{A}(\vec{r},\ t) = \frac{\mu_0}{4\pi}\int \frac{\rho(\vec{r},'\,t_r)\ \vec{v}(t_r)}{R}\,\mathrm{d}\tau' = \frac{\mu_0\vec{v}}{4\pi R}\int \rho(\vec{r},'\,t_r)\ \mathrm{d}\tau'$$

or

$$\vec{A}(\vec{r},\ t) = \frac{\mu_0\vec{v}}{4\pi R}\,\frac{q}{1 - \hat{R}\ .\ \vec{v}/c} = \frac{\mu_0}{4\pi}\,\frac{qc\vec{v}}{(Rc - \vec{R}\ .\ \vec{v})} = \frac{\vec{v}}{c^2}\,\phi(\vec{r},\ t) \tag{13.31}$$

[3]For the same reason, an observer at $\vec{r}$ finds the particle in only one place at a time. By contrast, it is possible to hear an object in two places at once. Consider a bear who growls at you and then runs towards you at the speed of sound and growls again; you hear both growls at the same time, corning from two different locations, but there's only one bear.

Equations (13.30) and (13.31) are known as **Lienard-Wiechert** potentials for a moving point-charge.

Example 3: Find the potentials of a point charge moving with constant velocity.

Solution: For convenience, let us say the particle passes through the origin at time $t = 0$, so that

$$\vec{w}(t) = \vec{v}\ t$$

We first compute the retarded time, using Equation (13.28):

$$|\vec{r} - \vec{v}\ t_r| = c(t - t_r)$$

on squaring, we get

$$r^2 - 2\vec{r}\ .\ \vec{v}\ t_r + v^2 t_r^2 = c^2(t^2 - 2t\ t_r + t_r^2)$$

Solving for t_r by the quadratic formula, we have

$$t_r = \frac{(c^2 t - \vec{r}\ .\ \vec{v}) \pm \sqrt{(c^2 t - \vec{r}\ .\ \vec{v})^2 + (c^2 - v^2)(r^2 - c^2 t^2)}}{(c^2 - v^2)} \tag{13.32}$$

To decide about the sign, consider the limit $v = 0$:

$$t_r = t \pm \frac{r}{c}$$

In this case the charge is at rest at the origin, and the retarded time should be $(t - r/c)$; evidently we want the minus sign. Now, from equations (13.28) and (13.29), we have

$$R = c(t - t_r) \qquad \text{and} \qquad \hat{R} = \frac{\vec{r} - \vec{v}\ t_r}{c(t - t_r)}$$

so

$$\begin{aligned} R(1 - \hat{R}\ .\ \vec{v}/c) &= c(t - t_r)\Big[1 - \frac{\vec{v}}{c}\ .\ \frac{(\vec{r} - \vec{v}\ t_r)}{c(t - t_r)}\Big] \\ &= c(t - t_r) - \frac{\vec{v}\ .\ \vec{r}}{c} + \frac{v^2}{c}\ t_r \\ &= \frac{1}{c}\Big[(c^2 t - \vec{v}\ .\ \vec{r}) - (c^2 - v^2)t_r\Big] \\ &= \frac{1}{c}\sqrt{(c^2 t - \vec{r}\ .\ \vec{v})^2 + (c^2 - v^2)(r^2 - c^2 t^2)} \end{aligned} \tag{13.33}$$

Here, the value of t_r (equation 13.32) with minus sign has been used. Using equation (13.33) in (13.30), we get

$$\phi(\vec{r},\ t) = \frac{1}{4\pi\epsilon_0}\ \frac{qc}{\sqrt{(c^2 t - \vec{r}\ .\ \vec{v})^2 + (c^2 - v^2)(r^2 - c^2 t^2)}} \tag{13.34}$$

Using equation (13.34) in (13.31), we get

$$\vec{A}(\vec{r},\ t) = \frac{\mu_0}{4\pi}\ \frac{qc\vec{v}}{\sqrt{(c^2 t - \vec{r}\ .\ \vec{v})^2 + (c^2 - v^2)(r^2 - c^2 t^2)}} \tag{13.35}$$

Equations (13.34) and (13.35) are the Lienard-Wiechert potentials.

3.1 Fields of a moving point charge

Now, we want to calculate the electric and magnetic fields of a point charge in arbitrary motion, using the Lienard-Wiechert potentials:

$$\phi(\vec{r},t) = \frac{1}{4\pi\epsilon_0}\,\frac{qc}{(Rc - \vec{R}\,.\,\vec{v})}, \qquad \vec{A}(\vec{r},t) = \frac{\vec{v}}{c^2}\,\phi(\vec{r},t) \tag{13.36}$$

For these potentials, the electric field $\vec{E}$ and magnetic field $\vec{B}$ are:

$$\vec{E} = -\nabla\phi - \frac{\partial\vec{A}}{\partial t} \qquad \text{and} \qquad \vec{B} = \nabla\times\vec{A} \tag{13.37}$$

However, the differentiation is tricky, because

$$\vec{R} = \vec{r} - \vec{w}(t_r) \qquad \text{and} \qquad \vec{v} = \frac{\partial\vec{w}}{\partial t_r} \tag{13.38}$$

are both evaluated at the retarded time t_r, and the t_r defined by the equation

$$|\vec{r} - \vec{w}(t_r)| = c(t - t_r) \tag{13.39}$$

is itself a function of $\vec{r}$ and t.

Let us begin with the gradient of ϕ (using equation 13.36):

$$\nabla\phi = \frac{qc}{4\pi\epsilon_0}\,\frac{-1}{(Rc - \vec{R}\,.\,\vec{v})^2}\,\nabla(Rc - \vec{R}\,.\,\vec{v}) \tag{13.40}$$

Since $R = c(t - t_r)$, therefore,

$$\nabla R = -c\nabla t_r \tag{13.41}$$

As for the second term, we have identity:

$$\nabla(\vec{R}\,.\,\vec{v}) = (\vec{R}\,.\,\nabla)\vec{v} + (\vec{v}\,.\,\nabla)\vec{R} + \vec{R}\times(\nabla\times\vec{v}) + \vec{v}\times(\nabla\times\vec{R}) \tag{13.42}$$

Evaluating these terms one at a time:

$$\begin{aligned}(\vec{R}\,.\,\nabla)\vec{v} &= \left(R_x\,\frac{\partial}{\partial x} + R_y\,\frac{\partial}{\partial y} + R_z\,\frac{\partial}{\partial z}\right)\vec{v}(t_r)\\ &= R_x\,\frac{\mathrm{d}\vec{v}}{\mathrm{d}t_r}\frac{\partial t_r}{\partial x} + R_y\,\frac{\mathrm{d}\vec{v}}{\mathrm{d}t_r}\frac{\partial t_r}{\partial y} + R_z\,\frac{\mathrm{d}\vec{v}}{\mathrm{d}t_r}\frac{\partial t_r}{\partial z}\\ &= \frac{\mathrm{d}\vec{v}}{\mathrm{d}t_r}\left(R_x\,\frac{\partial t_r}{\partial x} + R_y\,\frac{\partial t_r}{\partial y} + R_z\,\frac{\partial t_r}{\partial z}\right)\\ &= \vec{a}\,(\vec{R}\,.\,\nabla t_r)\end{aligned} \tag{13.43}$$

where $\vec{a} = \mathrm{d}\vec{v}/\mathrm{d}t_r$ is the acceleration of the particle at the retarded time. Now

$$(\vec{v}\,.\,\nabla)\vec{R} = (\vec{v}\,.\,\nabla)\vec{r} - (\vec{v}\,.\,\nabla)\vec{w} \tag{13.44}$$

and

$$(\vec{v}\,.\,\nabla)\,\vec{r} = \Big(v_x\,\frac{\partial}{\partial x} + v_y\,\frac{\partial}{\partial y} + v_z\,\frac{\partial}{\partial z}\Big)(x\,\hat{x} + y\,\hat{y} + z\,\hat{z})$$

$$= v_x\,\hat{x} + v_y\,\hat{y} + v_z\,\hat{z} = \vec{v} \tag{13.45}$$

while following the procedure of Equation (13.43), we have

$$(\vec{v}\,.\,\nabla)\vec{w} = \vec{v}\,(\vec{v}\,.\,\nabla t_r)$$

where $\vec{v} = \mathrm{d}\vec{w}/\mathrm{d}t_r$ is velocity of particle. Moving on to the third term in Equation (13.42),

$$\nabla\times\vec{v} = \Big(\frac{\partial v_z}{\partial y} - \frac{\partial v_y}{\partial z}\Big)\hat{x} + \Big(\frac{\partial v_x}{\partial z} - \frac{\partial v_z}{\partial x}\Big)\hat{y} + \Big(\frac{\partial v_y}{\partial x} - \frac{\partial v_x}{\partial y}\Big)\hat{z}$$

$$= \Big(\frac{\partial v_z}{\partial t_r}\frac{\partial t_r}{\partial y} - \frac{\partial v_y}{\partial t_r}\frac{\partial t_r}{\partial z}\Big)\hat{x} + \Big(\frac{\partial v_x}{\partial t_r}\frac{\partial t_r}{\partial z} - \frac{\partial v_z}{\partial t_r}\frac{\partial t_r}{\partial x}\Big)\hat{y} + \Big(\frac{\partial v_y}{\partial t_r}\frac{\partial t_r}{\partial x} - \frac{\partial v_x}{\partial t_r}\frac{\partial t_r}{\partial y}\Big)\hat{z}$$

$$= -\vec{a}\times\nabla t_r \tag{13.46}$$

Finally, we have

$$\nabla\times\vec{R} = \nabla\times\vec{r} - \nabla\times\vec{w} \tag{13.47}$$

but $\nabla\times\vec{r} = 0$, while, following the procedure of Equation (13.46), we have

$$\nabla\times\vec{w} = -\vec{v}\times\nabla t_r \tag{13.48}$$

Putting all this in equation (13.42), we have

$$\nabla(\vec{R}\,.\,\vec{v}) = \vec{a}\,(\vec{R}\,.\,\nabla t_r) + \vec{v} - \vec{v}\,(\vec{v}\,.\,\nabla t_r) - \vec{R}\times(\vec{a}\times\nabla t_r) + \vec{v}\times(\vec{v}\times\nabla t_r)$$

$$= \vec{a}\,(\vec{R}\,.\,\nabla t_r) + \vec{v} - \vec{v}\,(\vec{v}\,.\,\nabla t_r) - [\vec{a}(\vec{R}\,.\,\nabla t_r) - \nabla t_r(\vec{R}\,.\,\vec{a})]$$

$$+[\vec{v}(\vec{v}\,.\,\nabla t_r) - \nabla t_r(\vec{v}\,.\,\vec{v})] = \vec{v} + (\vec{R}\,.\,\vec{a} - v^2)\nabla t_r \tag{13.49}$$

Collecting equations (13.41) and (13.49), we have

$$\nabla\phi = \frac{qc}{4\pi\epsilon_0}\,\frac{-1}{(Rc - \vec{R}\,.\,\vec{v})^2}\,[-c^2\nabla t_r - \vec{v} - (\vec{R}\,.\,\vec{a} - v^2)\nabla t_r)]$$

$$= \frac{qc}{4\pi\epsilon_0}\,\frac{1}{(Rc - \vec{R}\,.\,\vec{v})^2}\Big[\vec{v} + (c^2 - v^2 + \vec{R}\,.\,\vec{a})\nabla t_r\Big] \tag{13.50}$$

For the calculation, we need to know ∇t_r. This can be found by taking the gradient of equation (13.39), as we have already done in equation (13.41), and expanding ∇R:

$$-c\nabla t_r = \nabla R = \nabla(\sqrt{\vec{R}\,.\,\vec{R}}) = \frac{1}{2\sqrt{\vec{R}\,.\,\vec{R}}}\,\nabla(\vec{R}\,.\,\vec{R})$$

$$= \frac{1}{R}\,[(\vec{R}\,.\,\nabla)\vec{R} + \vec{R}\times(\nabla\times\vec{R})] \tag{13.51}$$

But, using the idea of derivation of Equation (13.44), we get

$$(\vec{R}\,.\,\nabla)\vec{R} = \vec{R} - \vec{v}(\vec{R}\,.\,\nabla t_r) \tag{13.52}$$

while from equations (13.47) and (13.48), we get

$$\nabla\times\vec{R} = \vec{v}\times\nabla t_r \tag{13.53}$$

Using equations (13.52) and (13.53) in (13.51), we get

$$-c\nabla t_r = \frac{1}{R}\,i\Big[\vec{R} - \vec{v}(\vec{R}\,.\,\nabla t_r) + \vec{R}\times(\vec{v}\times\nabla t_r)\Big]$$

$$= \frac{1}{R}\,\Big[\vec{R} - \vec{v}(\vec{R}\,.\,\nabla t_r) + \vec{v}(\vec{R}\,.\,\nabla t_r) - \nabla t_r(\vec{R}\,.\,\vec{v})\Big]$$

$$= \frac{1}{R}\,[\vec{R} - (\vec{R}\,.\,\vec{v})\nabla t_r]$$

and hence

$$\nabla t_r = \frac{-\vec{R}}{Rc - \vec{R}\,.\,\vec{v}} \tag{13.54}$$

Incorporating this result in equation (13.50), we get

$$\nabla\phi = \frac{1}{4\pi\epsilon_0}\,\frac{qc}{(Rc - \vec{R}\,.\,\vec{v})^3}\Big[(Rc - \vec{R}\,.\,\vec{v})\vec{v} - ((c^2 - v^2) + \vec{R}\,.\,\vec{a})\vec{R}\Big] \tag{13.55}$$

A similar calculation gives

$$\frac{\partial\vec{A}}{\partial t} = \frac{1}{4\pi\epsilon_0}\,\frac{qc}{(Rc - \vec{R}\,.\,\vec{v})^3}\Big[(Rc - \vec{R}\,.\,\vec{v})(-\vec{v}$$

$$+R\,\vec{a}/c) + \frac{R}{c}\,(c^2 - v^2 + \vec{R}\,.\,\vec{a})\vec{v}\Big] \tag{13.56}$$

Using equations (13.55) and (13.56) in (13.37), and introducing the vector

$$\vec{u} \equiv c\hat{R} - \vec{v} \tag{13.57}$$

we get

$$\vec{E}(\vec{r},t) = \frac{q}{4\pi\epsilon_0}\,\frac{R}{(\vec{R}\,.\,\vec{u})^3}[(c^2 - v^2)\vec{u} + \vec{R}\times(\vec{u}\times\vec{a})] \tag{13.58}$$

Further,

$$\nabla\times\vec{A} = \frac{1}{c^2}\,\nabla\times(\phi\vec{v}) = \frac{1}{c^2}\,[\phi(\nabla\times\vec{v}) - \vec{v}\times(\nabla\phi)]$$

We already know $\nabla \times \vec{v}$ (equation 13.46) and $\nabla \phi$ (equation 13.55). Putting these together, we have

$$\nabla \times \vec{A} = -\frac{1}{c}\frac{q}{4\pi\epsilon_0}\frac{1}{(\vec{R}\,.\,\vec{u})^3}\vec{R} \times [(c^2 - v^2)\vec{v} + (\vec{R}\,.\,\vec{a})\vec{v} + (\vec{R}\,.\,\vec{u})\vec{a}]$$

The quantity in brackets is similar to the one in equation (13.58), which can be written, using expression for vector triple product, as $\left[(c^2 - v^2)\vec{u} + (\vec{R}\,.\,\vec{a})\vec{u} + (\vec{R}\,.\,\vec{u})\vec{a}\right]$; the main difference is that we have $\vec{v}$'s instead of $\vec{u}$'s in the first two terms. In fact, since it is all crossed into $\vec{R}$ anyway, we can with impunity change these $\vec{v}$'s into $-\vec{u}$'s; the extra term proportional to $\vec{R}$ disappears in the cross product. It follows that

$$\vec{B}(\vec{r},\, t) = \frac{1}{c}\,\hat{R} \times \vec{E}(\vec{r}, t) \tag{13.59}$$

Evidently the magnetic field of a point charge is always perpendicular to the electric field, and to the vector from the retarded point.

The first term in $\vec{E}$ (the one involving $(c^2 - v^2)\vec{u}$) falls off as the inverse square of the distance from the particle. If the velocity and acceleration are both zero, this term alone survives and reduces to the old electrostatic result

$$\vec{E} = \frac{1}{4\pi\epsilon_0}\frac{q}{R^2}\,\hat{R}$$

For this reason, the first term in E is sometimes called the generalized Coulomb field.

Review Questions

1. Derive expression for retarded potentials.

2. Derive expression for Lienard-Wiechert potentials.

3. Describe the geometrical effect and derive expression for the apparent length of a train moving towards an observer.

4. Describe the geometrical effect and derive expression for the apparent length of a train moving at an angle θ relative to the direction of observer.

5. Show that $\chi = f_1\left(t - \frac{R}{c}\right)$ is solution of equation:

$$\frac{\partial^2 \chi}{\partial R^2} - \frac{1}{c^2}\frac{\partial^2 \chi}{\partial t^2} = 0$$

6. Show that $\chi = f_1\left(t + \frac{R}{c}\right)$ is solution of equation:

$$\frac{\partial^2 \chi}{\partial R^2} - \frac{1}{c^2}\frac{\partial^2 \chi}{\partial t^2} = 0$$

7. Write notes on the following

   (i) Retarded potentials

   (ii) Lienard-Wiechert potentials

   (iii) Geometrical effect

Multiple-Choice Questions

1. Suppose $P = f_1\left(t - \frac{R}{c}\right)$ and $Q = f_1\left(t + \frac{R}{c}\right)$, then which is the solution of following equation?

$$\frac{\partial^2 \chi}{\partial R^2} - \frac{1}{c^2}\frac{\partial^2 \chi}{\partial t^2} = 0$$

   A. P only B. Q only C. P and Q both D. neither P nor Q

   Ans. C

2. A train of length L is moving with velocity v towards an observer. What is apparent length of train observed by the observer?

   A. $L(1 - v/c)$ B. $L(1 + v/c)$ C. $L/(1 - v/c)$ D. $L/(1 + v/c)$

   Ans. C

3. A train of length L is moving with velocity v in a direction inclined at an angle θ with the direction of observer. What is apparent length of train observed by the observer?

   A. $L(1 - v\cos\theta/c)$ B. $L(1 + v\cos\theta/c)$

   C. $L/(1 - v\cos\theta/c)$ D. $L/(1 + v\cos\theta/c)$

   Ans. C

4. A train having actual volume τ is moving with velocity $\vec{v}$. If $\hat{R}$ is unit vector in the direction of observer from the train, what is the apparent volume of train?

   A. $\tau(1-\hat{R}\,.\,\vec{v}/c)$ B. $\tau(1+\hat{R}\,.\,\vec{v}/c$ C. $\tau/(1-\hat{R}\,.\,\vec{v}/c)$ D. $\tau/(1+\hat{R}\,.\,\vec{v}/c)$

   Ans. C

Chapter 14

Charged Particle in the Fields

1. Motion of charge in constant uniform electric field

Suppose, a particle having charge e is moving in a constant uniform electric field $\vec{E}$. The field which does not depend on the space coordinates is known as the uniform field, and which does not depend on time is known as the constant field. We take the direction of field as the x-axis, so that $\vec{E} = E\,\hat{x}$. The motion of particle is obviously in a plane, which we take as the xy plane. Then the equation of motion is

$$\frac{\mathrm{d}\vec{p}}{\mathrm{d}t} = e\vec{E}$$

On resolving this equation into components, we get

$$\frac{\mathrm{d}p_x}{\mathrm{d}t} = eE \qquad \text{and} \qquad \frac{\mathrm{d}p_y}{\mathrm{d}t} = 0$$

Integration of these equations gives

$$p_x = eEt \qquad \text{and} \qquad p_y = p_0 \tag{14.1}$$

We have taken $p_x = 0$ and $p_y = p_0$ at the time $t = 0$. The energy of the particle related to the momentum as $\mathcal{E} = c\sqrt{m_0^2c^2 + p^2}$. Substituting equation (14.1), we find

$$\mathcal{E} = \sqrt{m_0^2c^4 + c^2p_0^2 + (ceEt)^2} = \sqrt{\mathcal{E}_0^2 + (ceEt)^2} \tag{14.2}$$

where $\mathcal{E}_0 = \sqrt{m_0^2c^4 + c^2p_0^2}$ is the energy at $t = 0$. The velocity of the particle can be expressed as $\vec{v} = \vec{p}\,c^2/\mathcal{E}$. On resolving this relation into components, we have

$$v_x = \frac{p_xc^2}{\mathcal{E}} = \frac{c^2eEt}{\sqrt{\mathcal{E}_0^2 + (ceEt)^2}} \qquad v_y = \frac{p_yc^2}{\mathcal{E}} = \frac{p_0c^2}{\sqrt{\mathcal{E}_0^2 + (ceEt)^2}}$$

On integrating these equations, we find

$$x = \frac{1}{eE}\sqrt{\mathcal{E}_0^2 + (ceEt)^2} \qquad \text{and} \qquad y = \frac{p_0c}{eE}\sinh^{-1}\left(\frac{ceEt}{\mathcal{E}_0}\right) \tag{14.3}$$

The constant of integration we set equal to zero. We obtain the equation of trajectory by expressing t in terms of y and substituting t in x.

$$x = \frac{\mathcal{E}_0}{eE} \cosh \frac{eEy}{p_0 c} \tag{14.4}$$

Thus, in a constant uniform electric field a charge moves along a catenary curve.

If the velocity of the particle is $v << c$, then we can set $p_0 = mv_0$, $\mathcal{E}_0 = m_0 c^2$, and expand (14.4) in series in powers of $1/c$. Then we get, to within terms of higher order,

$$x = \frac{eE}{2m_0 v_0^2} y^2 + \text{const}$$

that is, the charge moves along a parabola, a result well known from classical mechanics.

2. Motion of charge in uniform magnetic field

Suppose, a particle having charge e is moving in a uniform magnetic field $\vec{B}$. The field which does not depend on the space coordinates is known as the uniform field. We choose the direction of the field as the z-axis, so that $\vec{B} = B\,\hat{z}$. The equation of motion is

$$\frac{d\vec{p}}{dt} = e(\vec{v} \times \vec{B}) \tag{14.5}$$

We use the relation between momentum and velocity

$$\vec{p} = \frac{\mathcal{E}\vec{v}}{c^2} \tag{14.6}$$

where $\mathcal{E}$ is the energy of the particle, which is constant in a magnetic field. Using equation (14.6) in (14.5), the equation of motion takes the form

$$\frac{\mathcal{E}}{c^2}\frac{d\vec{v}}{dt} = e(\vec{v} \times \vec{B}) \tag{14.7}$$

For $\vec{v} = v_x\,\hat{x} + v_y\,\hat{y} + v_z\,\hat{z}$, we have $\vec{v} \times \vec{B} = v_y B\,\hat{x} - v_x B\,\hat{y}$. Equation (14.8) can be expressed as

$$\frac{d}{dt}(v_x\,\hat{x} + v_y\,\hat{y} + v_z\,\hat{z}) = v_y\omega\,\hat{x} - v_x\omega\,\hat{y} \tag{14.8}$$

where we have introduced the notation

$$\omega = \frac{ec^2 B}{\mathcal{E}} \tag{14.9}$$

On resolving this equation into components, we get

$$\frac{dv_x}{dt} = \omega\, v_y \qquad \frac{dv_y}{dt} = -\omega\, v_x \qquad \frac{dv_z}{dt} = 0 \tag{14.10}$$

By multiplying the second equation of (14.10) by i and add it to the first, we get

$$\frac{\mathrm{d}}{\mathrm{d}t}(v_x + iv_y) = -i\omega(v_x + iv_y)$$

The solution of this equation is

$$v_x + i\ v_y = a\ \mathrm{e}^{-i\ \omega t}$$

where a is a complex constant. This can be written in the form $a = v_{0t}\ \mathrm{e}^{-i\alpha}$ where v_{0t} and α are real. Then, we have

$$v_x + i\ v_y = v_{0t}\ \mathrm{e}^{-i(\omega t+\alpha)}$$

On separating real and imaginary parts of this equation, we find

$$v_x = v_{0t}\ \cos(\omega t + \alpha) \qquad v_y = -v_{0t}\ \sin(\omega t + \alpha) \tag{14.11}$$

From equation (14.11), on squaring and adding, we get

$$v_{0t} = \sqrt{v_x^2 + v_y^2}$$

That is, v_{0t} is the velocity of particle in the xy plane, which remains constant throughout the motion. Integration of equation (14.11) gives

$$x = x_0 + r\ \sin(\omega t + \alpha) \qquad y = y_0 + r\ \cos(\omega t + \alpha) \tag{14.12}$$

where x_0 and y_0 are constants of integration, and

$$r = \frac{v_{0t}}{\omega} = \frac{v_{0t}\mathcal{E}}{e\ c^2 B} = \frac{c\ p_t}{e\ B} \tag{14.13}$$

where p_t is the projection of momentum on the xy plane. Integration of third equation of (14.10), gives $v_z = v_{0z}$ and

$$z = z_0 + v_{0z}t \tag{14.14}$$

Here, v_{0z} and z_0 are constants of integration. Equations (14.12) and (14.14) show that the charge moves in a uniform magnetic field along a helix having its axis along the direction of the applied magnetic field and with a radius r given by equation (14.13). The velocity of the particle is constant. In the special case where $v_{0z} = 0$, that is, the charge has no velocity component along the direction of the field, it moves along a circle in the plane perpendicular to the field.

The quantity ω is the angular frequency of rotation of the particle in the plane perpendicular to the field. When the velocity of the particle is low, then we can approximately set $\mathcal{E} = m_0c^2$. The the frequency ω is changed to

$$\omega = \frac{eB}{m_0} \tag{14.15}$$

3. Particle in uniform electric and magnetic fields

Consider a particle of mass m and charge q moving in a uniform magnetic field $\vec{B}$ and a uniform electric field $\vec{E}$ with velocity $\vec{v}$. The force acting on the particle is $q(\vec{E} + \vec{v} \times \vec{B})$ and the equation of motion of the particle is

$$m \frac{d\vec{v}}{dt} = q(\vec{E} + \vec{v} \times \vec{B}) \tag{14.16}$$

Velocity of the particle is non-relativistic and therefore its mass is assumed to remain constant. Here, we use the Cartesian coordinate system. Let us take z-axis along the direction of the magnetic field $\vec{B}$. That is, $\vec{B} = B\,\hat{z}$, where B is constant magnitude of the field. Now, we take xz-plane so that it contains the applied uniform electric field $\vec{E}$. That is,

$$\vec{E} = E_x\,\hat{x} + E_z\,\hat{z} \tag{14.17}$$

where E_x and E_z are constant components of the electric field along the x-axis and z-axes, respectively. For the velocity

$$\vec{v} = v_x\,\hat{x} + v_y\,\hat{y} + v_z\,\hat{z} \tag{14.18}$$

we have

$$\vec{v} \times \vec{B} = v_y B\,\hat{x} - v_x B\,\hat{y} \tag{14.19}$$

Using equations (14.17), (14.18) and (14.19) in (14.16), and equating the coefficients of $\hat{x}$, $\hat{y}$ and $\hat{z}$ on the two sides of the resulting equation, we get

$$m \frac{dv_x}{dt} = qE_x + qBv_y \tag{14.20}$$

$$m \frac{dv_y}{dt} = -qBv_x \tag{14.21}$$

$$m \frac{dv_z}{dt} = qE_z \tag{14.22}$$

Integration of equation (14.22) gives

$$v_z = \frac{qE_z}{m}\,t + v_{z0} \tag{14.23}$$

where v_{z0} is the z-component of velocity at $t = 0$. Thus, the velocity of the particle along z-direction increases linearly with time t. That is, the particle is accelerated. For the accelerated motion along z-axis, E_z must be non-zero. Further integration of equation (14.23) gives

$$z = \frac{qE_z}{2m}\,t^2 + v_{z0}t + z_0$$

showing that when v_{z0} is non-zero, the particle moves along the z-axis. Here, z_0 is the constant of integration giving z-component of position at $t = 0$. On differentiating equation (14.20) with respect to t and using the value of $\mathrm{d}v_y/\mathrm{d}t$ from equation (14.21), we get

$$\frac{\mathrm{d}^2 v_x}{\mathrm{d}t^2} = \frac{qB}{m}\frac{\mathrm{d}v_y}{\mathrm{d}t} = -\left(\frac{qB}{m}\right)^2 v_x \tag{14.24}$$

Similarly, on differentiating equation (14.21) with respect to t and using the value of $\mathrm{d}v_x/\mathrm{d}t$ from equation (14.20), we get

$$\frac{\mathrm{d}^2 v_y}{\mathrm{d}t^2} = -\frac{qB}{m}\frac{\mathrm{d}v_x}{\mathrm{d}t} = -\frac{qB}{m}\left(\frac{qE_x}{m} + \frac{qB}{m}v_y\right)$$

$$= -\left(\frac{qB}{m}\right)^2\left(\frac{E_x}{B} + v_y\right) \tag{14.25}$$

On defining the cyclotron frequency $\omega_c = |q|B/m$, equation (14.24) can be written as

$$\frac{\mathrm{d}^2 v_x}{\mathrm{d}t^2} = -\omega_c^2\, v_x \tag{14.26}$$

Equation (14.26) is a second order differential equation with constant coefficients and its solution is

$$v_x = v_{x0}\cos(\omega_c t + \phi) \tag{14.27}$$

where v_{x0} is constant. Equation (14.20) can be rearranged as

$$v_y = \frac{m}{qB}\frac{\mathrm{d}v_x}{\mathrm{d}t} - \frac{E_x}{B} = \pm\frac{1}{\omega_c}\frac{\mathrm{d}v_x}{\mathrm{d}t} - \frac{E_x}{B}$$

Using equation (14.27) here, we have

$$v_y = \mp v_{x0}\sin(\omega_c t + \phi) - \frac{E_x}{B} \tag{14.28}$$

Here, the sign $\mp$ represents the sign of charge of the particle. From equations (14.27) and (14.28), we have

$$v_x^2 + \left(v_y + \frac{cE_x}{B}\right)^2 = v_{x0}^2 = v_\perp^2$$

Here, $v_\perp$ represents the velocity of particle in the xy-plane, which is perpendicular to the direction of the applied magnetic field. We have

$$v_\perp = \sqrt{v_x^2 + \left(v_y + \frac{E_x}{B}\right)^2}$$

Integration of equation (14.27) with respect to t gives

$$x = \frac{v_\perp}{\omega_c}\sin(\omega_c t + \phi) + x_0 \tag{14.29}$$

where x_0 is the constant of integration. Integration of equation (14.28) gives

$$y = \pm\frac{v_\perp}{\omega_c}\cos(\omega_c t + \phi) - \frac{E_x t}{B} + y_0 \tag{14.30}$$

where y_0 is the constant of integration. We define the Larmor radius r_L as

$$r_L = \frac{v_\perp}{\omega_c} = \frac{mcv_\perp}{|q|B}$$

so that equations (14.29) and (14.30) can be written as

$$x - x_0 = r_L \sin(\omega_c t + \phi) \tag{14.31}$$

and

$$y - \left(y_0 - \frac{E_x}{B}\, t\right) = \pm r_L \cos(\omega_c t + \phi) \tag{14.32}$$

Equations (14.31) and (14.32) represent a circle:

$$[x - x_0]^2 + \left[y - \left(y_0 - \frac{E_x}{B}\, t\right)\right]^2 = r_L^2$$

of radius r_L and with the center at $(x_0,\, y_0 - cE_x t/B)$. Hence, y-coordinate of the center is changing with time in the negative y direction. It shows that the guiding center drifts with a velocity

$$v_{gc} = \frac{E_x}{B}$$

in the negative y direction, which is perpendicular to both the direction of applied electric and magnetic fields. Thus, we have

$$\vec{v}_{gc} = \frac{\vec{E} \times \vec{B}}{B^2}$$

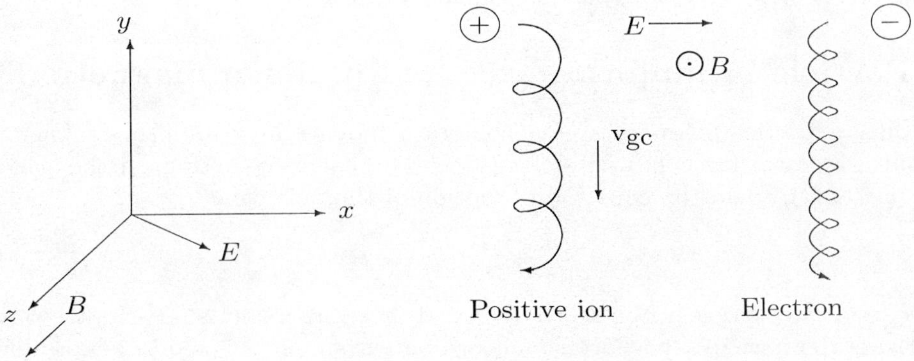

Figure 1: Charged particles drift in the downward direction (negative y direction) when the applied magnetic field is perpendicular to the plane of the paper in the upward direction.

The drift velocity is independent of the mass and charge of the particle. Hence, the positive ions as well as electrons drift with the same velocity in the same direction. Figure 1 shows drift of an ion and an electron. The drift is in the downward direction (negative y direction) when the applied magnetic field is perpendicular to the plane of the paper in the upward direction.

Physical reason for the drift can be understood in the following manner. The charged particles are gyrating about the direction of magnetic field. For a positive ion, in the first half-cycle when the velocity of the ion has component along the direction of the electric field, the ion gains energy from the electric field and thus, $v_\perp$ increases, and in turn the Larmor radius r_L increases.

In the second half-cycle when the velocity of the ion has component opposite to the direction of the electric field, the ion loses energy for opposing the electric field and thus, $v_\perp$ decreases, and in turn r_L decreases. This difference in r_L on the two sides of the orbit causes the drift. An electron gyrates in the opposite direction and also loses and gains energy in the opposite sense, and has a drift.

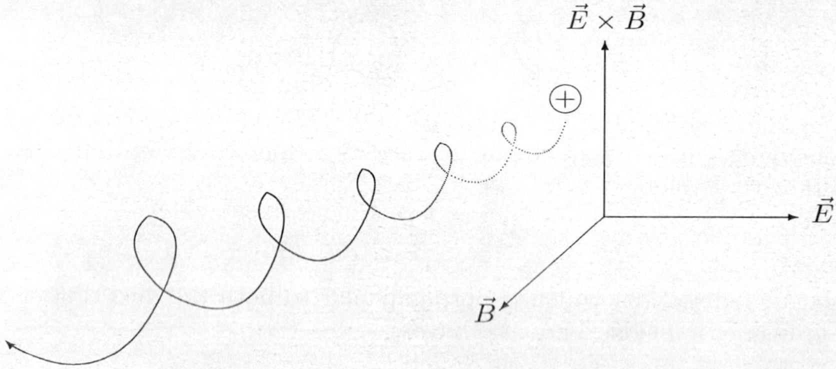

Figure 2: Orbit of a gyrating particle in space.

Further, the velocity of the particle in the direction of z-axis increases with time. Actual orbit of the gyrating particle is shown in Figure 2.

4. Particle in uniform force and uniform magnetic field

Consider a particle of mass m and charge q moving in a uniform magnetic field $\vec{B}$ and a uniform force $\vec{F}$ with velocity $\vec{v}$. The net force acting on the particle is $\vec{F} + q(\vec{v} \times \vec{B})/c$ and the equation of motion of the particle is

$$m\,\frac{\mathrm{d}\vec{v}}{\mathrm{d}t} = \vec{F} + q(\vec{v} \times \vec{B}) \tag{14.33}$$

Velocity of the particle is non-relativistic and therefore its mass is assumed to remain constant. Here, we use the Cartesian coordinate system. Let us take z-axis along the direction of the magnetic field $\vec{B}$. That is $\vec{B} = B\,\hat{z}$, where B is constant magnitude of the field. Now, we take xz-plane so that it contains the applied uniform force $\vec{F}$. That is,

$$\vec{F} = F_x\,\hat{x} + F_z\,\hat{z} \tag{14.34}$$

where F_x and F_z are constant components of the force along the x-axis and z-axis, respectively. For the velocity

$$\vec{v} = v_x\ \hat{x} + v_y\ \hat{y} + v_z\ \hat{z} \tag{14.35}$$

we have

$$\vec{v} \times \vec{B} = v_y B\ \hat{x} - v_x B\ \hat{y} \tag{14.36}$$

Using equations (14.34), (14.35) and (14.36) in (14.33), and equating the coefficients of $\hat{x}$, $\hat{y}$ and $\hat{z}$ on the two sides of the resulting equation, we get

$$m\ \frac{dv_x}{dt} = F_x + qBv_y \tag{14.37}$$

$$m\ \frac{dv_y}{dt} = -qBv_x \tag{14.38}$$

$$m\ \frac{dv_z}{dt} = F_z \tag{14.39}$$

Integration of equation (14.39) gives

$$v_z = \frac{F_z}{m}\ t + v_{z0}$$

where v_{z0} is the z-component of the velocity at $t = 0$. Thus, the velocity of the particle along z-direction increases linearly with t. That is, the particle is accelerated. For the acceleration along the z-axis, F_z must be non-zero. Further integration of this equation gives

$$z = \frac{F_z}{2m}\ t^2 + v_{z0}t + z_0$$

showing that when v_{z0} is non-zero, the particle moves along the z-axis. Here, z_0 is the constant of integration giving z-component of position at $t = 0$. On differentiating equation (14.37) with respect to t and using the value of dv_y/dt from equation (14.38), we get

$$\frac{d^2v_x}{dt^2} = \frac{qB}{m}\frac{dv_y}{dt} = -\left(\frac{qB}{m}\right)^2 v_x \tag{14.40}$$

Similarly, on differentiating equation (14.38) with respect to t and using the value of dv_x/dt from equation (14.37), we get

$$\frac{d^2v_y}{dt^2} = -\frac{qB}{m}\frac{dv_x}{dt} = -\frac{qB}{m}\left(\frac{F_x}{m} + \frac{qB}{m}v_y\right)$$

$$= -\left(\frac{qB}{m}\right)^2\left(\frac{F_x}{qB} + v_y\right) \tag{14.41}$$

On defining the cyclotron frequency $\omega_c = |q|B/m$, equation (14.40) can be written as

$$\frac{d^2v_x}{dt^2} = -\omega_c^2 v_x \tag{14.42}$$

Equation (14.42) is a second order differential equation with constant coefficients and its solution is

$$v_x = v_{x0}\cos(\omega_c t + \phi) \tag{14.43}$$

where v_{x0} is a constant. Equation (14.37) can be rearranged as

$$v_y = \frac{m}{qB}\frac{\mathrm{d}v_x}{\mathrm{d}t} - \frac{F_x}{qB} = \pm\frac{1}{\omega_c}\frac{\mathrm{d}v_x}{\mathrm{d}t} - \frac{F_x}{qB}$$

Using equation (14.43) here, we have

$$v_y = \mp v_{x0}\sin(\omega_c t + \phi) - \frac{F_x}{qB} \tag{14.44}$$

Here, the sign $\pm$ represents the sign of charge of the particle. From equations (14.43) and (14.44), we have

$$v_x^2 + \left(v_y + \frac{F_x}{qB}\right)^2 = v_\perp^2$$

Here, $v_\perp$ represents the velocity of particle in the xy-plane, which is perpendicular to the direction of the applied magnetic field. We have

$$v_\perp = \sqrt{v_x^2 + \left(v_y + \frac{F_x}{qB}\right)^2}$$

Integration of equation (14.43) with respect to t gives

$$x = \frac{v_\perp}{\omega_c}\sin(\omega_c t + \phi) + x_0 \tag{14.45}$$

where x_0 is the constant of integration. Integration of equation (14.44) gives

$$y = \mp\frac{v_\perp}{\omega_c}\cos(\omega_c t + \phi) - \frac{F_x t}{qB} + y_0 \tag{14.46}$$

where y_0 is the constant of integration. We define the Larmor radius r_L as

$$r_L = \frac{v_\perp}{\omega_c} = \frac{mcv_\perp}{|q|\ B}$$

so that equations (14.45) and (14.46) can be written as

$$x - x_0 = r_L\sin(\omega_c t + \phi) \tag{14.47}$$

and

$$y - \left(y_0 - \frac{F_x}{qB}t\right) = \pm r_L\cos(\omega_c t + \phi) \tag{14.48}$$

Equations (14.47) and (14.48) represent a circle

$$[x - x_0]^2 + \left[y - \left(y_0 - \frac{F_x t}{qB}\right)\right]^2 = r_L^2$$

of radius r_L and with the center at $(x_0,\ y_0 - cF_x t/qB)$. Hence, y-coordinate of the center is changing with time in the negative y direction. It shows that the guiding center drifts with a velocity

$$v_f = \frac{F_x}{qB}$$

in the negative y direction, which is perpendicular to both the direction of applied electric and magnetic fields. Thus, we have

$$\vec{v}_f = \frac{\vec{F} \times \vec{B}}{qB^2}$$

The drift velocity depends on the charge of the particle. Thus, the positive ions and electrons drift in the opposite directions, resulting in a current.

Physical reason for the drift can be understood in the following manner. The charged particles are gyrating about the direction of magnetic field. For a particle, in the first half-cycle, when velocity of the particle has component along the direction of the force, the particle gains energy from the force and $v_\perp$ increases, and hence r_L increases.

In the second half-cycle when velocity of the particle has component opposite to the direction of the force, the particle loses energy for opposing the force and $v_\perp$ decreases, and hence r_L decreases. This difference in r_L on the two sides of the orbit causes the drift. An electron gyrates in the opposite direction and also loses and gains energy in the opposite sense, and has a drift.

Here, the drift velocity depends on charge of the particle. The oppositely charged particles move in the opposite directions, and therefore constitute the current. For the hydrogen plasma of protons and electrons, the current density $\vec{J}$ is

$$\vec{J} = 2nc\,\frac{\vec{F} \times \vec{B}}{B^2}$$

where n is the density of protons as well as electron.

4.1 For gravitational force

For the gravitational force $\vec{F} = m\vec{g}$ and the drift velocity is given by

$$\vec{v}_g = \frac{m\vec{g} \times \vec{B}}{qB^2}$$

Here, the drift velocity depends on charge as well as mass of the particle. The oppositely charged particles move in the opposite directions (Figure 3), and therefore they constitute the current. For the hydrogen plasma of protons and electrons the current density $\vec{J}$ is

$$\vec{J} = n(m_p + m_e)\frac{\vec{g} \times \vec{B}}{B^2}$$

where m_p and m_e are the masses of proton and electron, respectively, and n is the density of protons as well as electrons.

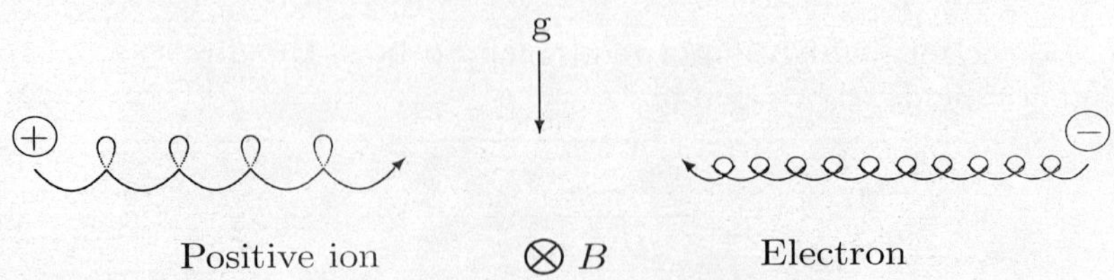

Figure 3: Drifts of gyrating particles in a gravitational field. Direction of the applied magnetic field is perpendicular to the plane of the paper in the inward direction. The acceleration due to gravity is in the downward direction.

5. Curvature drift

Suppose for a constant magnetic field, the field lines are curved in shape with a constant radius R (Figure 4). (The magnetic field is not along a straight line.) The particles moving along the field lines experience a centrifugal force

$$\vec{F}_{cf} = \frac{mv_{\parallel}^2}{R}\hat{R} = \frac{mv_{\parallel}^2\vec{R}}{R^2}$$

where m is mass of the particle and $v_{\parallel}^2$ the average square of the component of velocity of the particle along the magnetic field. The centrifugal force obviously makes an angle with the magnetic field and gives rise to a drift, known as the curvature drift. The guiding center of the particle moves with a velocity

$$\vec{v_R} = \frac{1}{q}\frac{\vec{F}_{cf}\times\vec{B}}{B^2} = \frac{1}{q}\frac{mv_{\parallel}^2\vec{R}}{R^2}\times\frac{\vec{B}}{B^2} = \frac{mv_{\parallel}^2}{qB^2}\frac{\vec{R}\times\vec{B}}{R^2}$$

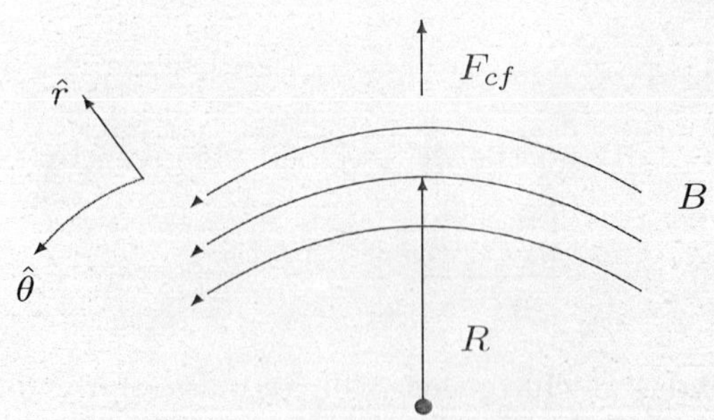

Figure 4: Constant magnetic field in the curved shape with a constant radius R.

When $\vec{R}$ and $\vec{B}$ are perpendicular,the magnitude of velocity of the guiding center is

$$v_R = \frac{mv_{\parallel}^2}{qB^2}\frac{RB}{R^2} = \frac{mv_{\parallel}^2}{qBR}$$

6. Particle in non-uniform magnetic field

We have discussed about the uniform fields where we could get exact expressions for the drifts of the guiding centers. When the inhomogeneity is introduced in a field, the situation becomes quite complicated and exact solution is not feasible. Therefore, we go for an approximate solution where we expand in terms of the small ratio r_L/L, where r_L is the Larmor radius and L the scale length of the inhomogeneity. This type of theory to obtain an approximate solution is known as the *orbit theory.* In the present discussion, we discuss about simple cases where only one type of inhomogeneity is considered for at a time.

6.1 Grad-B drift

For a particle of mass m and charge q moving with velocity $\vec{v}$ in a uniform magnetic field $\vec{B}$, the force $\vec{F}$ acting on the charge is $\vec{F} = q\vec{v}\times\vec{B}/c$. The equation of motion is

$$m\,\frac{\mathrm{d}\vec{v}}{\mathrm{d}t} = q(\vec{v}\times\vec{B})$$

The z-axis is taken along the direction of the magnetic field $\vec{B}$, so that $\vec{B} = B\ \hat{z}$. Taking $\vec{v} = v_x\ \hat{x} + v_y\ \hat{y} + v_z\ \hat{z}$,for uniform magnetic field (as discussed earlier), the particle moves in a helical path

$$x = r_L\sin(\omega_c t + \phi) + x_0 \qquad y = \pm r_L\cos(\omega_c t + \phi) + y_0 \qquad z = v_{z0}t + z_0$$

where x_0, y_0 and z_0 are constants and r_L the Larmor radius. The components of the velocity are

$$v_x = v_\perp\cos(\omega_c t + \phi) \qquad v_y = \mp v_\perp\sin(\omega_c t + \phi) \qquad v_z = v_{z0}$$

where v_{z0} is the z-component of initial velocity. The averages[1] of the components of the force due to the field are

$$\langle F_x\rangle = \Big\langle qBv_y\Big\rangle = \Big\langle qB\ (\mp v_\perp\sin(\omega_c t + \phi))\Big\rangle$$

[1]Average values of some trigonometric functions are as the following. Here, a is an integer and b a constant.

$$\Big\langle\sin(ax+b)\Big\rangle = \frac{\int_0^{2\pi}\sin(ax+b)\ \mathrm{d}x}{\int_0^{2\pi}\mathrm{d}x} = \frac{[-\cos(ax+b)]_0^{2\pi}}{2\pi a} = 0$$

$$\Big\langle\cos(ax+b)\Big\rangle = \frac{\int_0^{2\pi}\cos(ax+b)\ \mathrm{d}x}{\int_0^{2\pi}\mathrm{d}x} = \frac{[\sin(ax+b)]_0^{2\pi}}{2\pi a} = 0$$

$$\Big\langle\sin(ax+b)\ \cos(ax+b)\Big\rangle = \frac{\int_0^{2\pi}\sin(ax+b)\ \cos(ax+b)\ \mathrm{d}x}{\int_0^{2\pi}\mathrm{d}x}$$

$$= \frac{1}{2\pi}\int_0^{2\pi}\frac{1}{2}\ \sin(2ax+2b)\ \mathrm{d}x = \frac{1}{8\pi a}[-\cos(2ax+2b)]_0^{2\pi} = 0$$

$$\Big\langle\cos^2(ax+b)\Big\rangle = \frac{\int_0^{2\pi}\cos^2(ax+b)\ \mathrm{d}x}{\int_0^{2\pi}\mathrm{d}x} = \frac{1}{2\pi}\int_0^{2\pi}\frac{1}{2}\ [1+\cos(2ax+2b)]\ \mathrm{d}x$$

$$= \frac{1}{4\pi}\int_0^{2\pi}\mathrm{d}x + \frac{1}{4\pi}\int_0^{2\pi}\cos(2ax+2b)\ \mathrm{d}x = \frac{1}{2}$$

$$= \mp\left(qBv_{\perp}\right)\,\langle\sin(\omega_c t+\phi)\rangle = 0$$

and

$$\langle F_y\rangle = -\left\langle qBv_x\right\rangle = -\left\langle qBv_{\perp}\,\cos(\omega_c t+\phi)\right\rangle$$

$$= -\left(qBv_{\perp}\right)\,\langle\cos(\omega_c t+\phi)\rangle = 0$$

Hence, there is no net force acting on the particle in case of the uniform magnetic field. Now, we introduce inhomogeneity in the magnetic field by considering that the density of the field lines (also known as the lines of force) increases, say, in the y-direction (Figure 5). Hence, the magnetic field is now a function of y and by using Taylor expansion, we have

$$B(y) = B[y_0 \pm r_L\cos(\omega_c t+\phi)] = B(y_0) \pm r_L\cos(\omega_c t+\phi)\,\frac{\partial B}{\partial y}$$

$$= B_0 \pm r_L\cos(\omega_c t+\phi)\frac{\partial B}{\partial y}$$

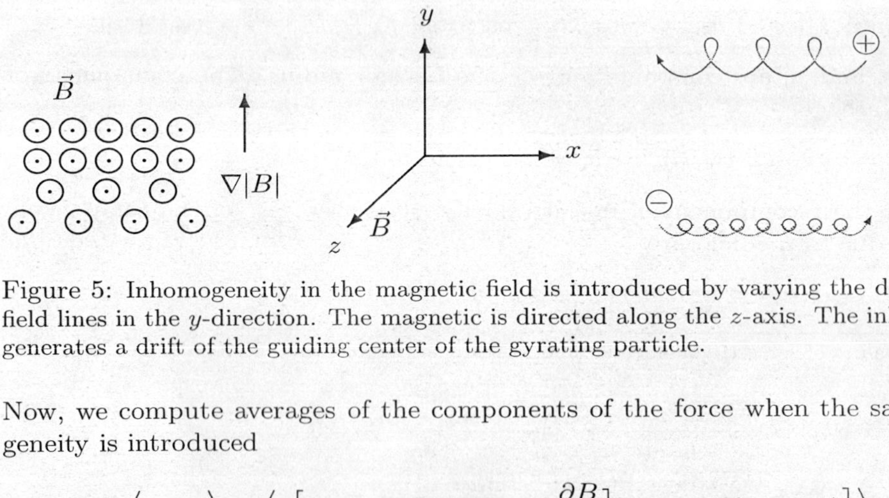

Figure 5: Inhomogeneity in the magnetic field is introduced by varying the density of the field lines in the y-direction. The magnetic is directed along the z-axis. The inhomogeneity generates a drift of the guiding center of the gyrating particle.

Now, we compute averages of the components of the force when the said inhomogeneity is introduced

$$\langle F_x\rangle = \left\langle qBv_y\right\rangle = \left\langle q\Big[B_0 \pm r_L\cos(\omega_c t+\phi)\frac{\partial B}{\partial y}\Big][\mp v_{\perp}\sin(\omega_c t+\phi)]\right\rangle$$

$$\left\langle\sin^2(ax+b)\right\rangle = \frac{\int_0^{2\pi}\sin^2(ax+b)\,\mathrm{d}x}{\int_0^{2\pi}\mathrm{d}x} = \frac{1}{2\pi}\int_0^{2\pi}\frac{1}{2}\,[1-\cos(2ax+2b)]\,\mathrm{d}x$$

$$= \frac{1}{4\pi}\int_0^{2\pi}\mathrm{d}x - \frac{1}{4\pi}\int_0^{2\pi}\cos(2ax+2b)\,\mathrm{d}x = \frac{1}{2}$$

$$= \Big\langle \mp qB_0 v_\perp \ \sin(\omega_c t + \phi)\Big\rangle - \Big\langle q r_L v_\perp \ \sin(\omega_c t + \phi) \ \cos(\omega_c t + \phi)\frac{\partial B}{\partial y}\Big\rangle$$

$$= \mp qB_0 v_\perp \langle \sin(\omega_c t + \phi)\rangle - q r_L v_\perp \ \frac{\partial B}{\partial y}\langle \sin(\omega_c t + \phi)\cos(\omega_c t + \phi)\rangle = 0$$

and

$$\langle F_y\rangle = -\Big\langle qBv_x\Big\rangle = -\Big\langle q\Big[B_0 \pm r_L \cos(\omega_c t + \phi)\frac{\partial B}{\partial y}\Big] v_\perp \cos(\omega_c t + \phi)\Big\rangle$$

$$= \Big\langle - qB_0 v_\perp \ \cos(\omega_c t + \phi)\Big\rangle \mp \Big\langle q r_L v_\perp \ \cos^2(\omega_c t + \phi)\frac{\partial B}{\partial y}\Big\rangle$$

$$= -qB_0 v_\perp \ \langle \cos(\omega_c t + \phi)\rangle \mp q r_L v_\perp \ \frac{\partial B}{\partial y}\langle \cos^2(\omega_c t + \phi)\rangle$$

$$= \mp \frac{q r_L v_\perp}{2}\ \frac{\partial B}{\partial y}$$

It shows that the average of the x-component of the force is still zero. It is because the inhomogeneity is considered in the y-direction. The average of the y-component of the force is non-zero, showing that besides the applied magnetic field in the direction of z-axis, there is a net force $\langle F_y\rangle$ working in the direction of y-axis. This generates the drift of the guiding center with a velocity

$$\vec{v}_{gc} = \frac{\langle F_y\rangle \hat{y} \times B\hat{z}}{qB^2} = \frac{\langle F_y\rangle \hat{x}}{qB} = \mp \frac{r_L v_\perp}{2B}\frac{\partial B}{\partial y}\hat{x}$$

The choice of y-axis for inhomogeneity of the field is arbitrary. For general variation of magnetic field in space, the drift velocity of the guiding center due to ∇B is

$$\vec{v}_{\nabla B} = \pm \frac{r_L v_\perp}{2}\ \frac{\vec{B} \times \nabla B}{B^2}$$

This drift is known as the grad-B drift and is independent of the charge of the particle.

6.2 Curvature and grad-B drifts

In the curvature drift, discussed earlier, we accounted for a curved magnetic field with constant magnitude. However, for a curved magnetic field, it is difficult to have a constant magnitude. Hence, for a curved magnetic field, both the curvature drift and grad-B drift arise together. Hence, both of them are considered together. Let us consider a magnetic field (Figure 4) where the field lines are along the z-axis of the cylindrical coordinates (ρ, ϕ, z).

Density of the field lines varies with ρ so that the field strength decreases with the increase of r. The magnetic field obviously does not depend on ϕ and z coordinates, and the component B_ϕ is zero. For a magnetic field $\vec{B}$, we have

$$\nabla \, . \, \vec{B} = 0 \qquad \text{or} \qquad \frac{1}{\rho}\frac{\partial}{\partial \rho}(\rho B_\rho) = 0$$

Thus,

$$B_\rho\ \rho = \text{constant} = b \qquad \text{or} \qquad B_\rho = \frac{b}{\rho}$$

It shows that the radial component of the magnetic field decreases with the increase of ρ. Gradient of the magnetic field is

$$\nabla B = \frac{\partial B_\rho}{\partial \rho}\ \hat{\rho} = -\frac{b}{\rho^2}\ \hat{\rho} = -\frac{B_\rho}{\rho}\ \hat{\rho}$$

Thus, the grad-B drift velocity is

$$\vec{v}_{\nabla B} = \pm\frac{v_\perp r_L}{2B^2}\ \vec{B}\times\nabla B = \pm\frac{v_\perp r_L}{2B^2}\ \vec{B}\times\left(-\frac{B_\rho}{\rho}\ \hat{\rho}\right)$$

Taking $\vec{B} = B_z\ \hat{z}$, we have

$$\vec{v}_{\nabla B} = \mp\frac{v_\perp r_L B_r}{2\rho B_z}\ \hat{\phi} = \mp\frac{v_\perp^2 B_r}{2\omega_c \rho B_z}\ \hat{\phi} = \mp\frac{m v_\perp^2 B_r}{2|q|\rho B_z^2}\ \hat{\phi} = -\frac{m v_\perp^2 B_r}{2q\rho B_z^2}\ \hat{\phi} \tag{14.49}$$

where $v_\perp^2$ is the average square of the velocity component, perpendicular to the magnetic field, and the Larmor radius r_L and cyclotron frequency ω_c are

$$r_L = \frac{v_\perp}{\omega_c} \qquad \text{and} \qquad \omega_c = \frac{|q|B}{m}$$

where m and q, respectively, are the mass and charge of the particle. The field lines are curved with a radius R (Figure 4), and the particles moving along the field lines experience a centrifugal force. Thus, the curvature drift is

$$\vec{v}_R = \frac{m v_\parallel^2}{qB^2}\ \frac{\vec{R}\times\vec{B}}{R^2}$$

where $v_\parallel^2$ is the average square of the velocity-component, along the magnetic field. Using $\vec{B} = B_z\ \hat{z}$ and $\vec{R} = R\ \hat{\rho}$, we have

$$\vec{v}_R = -\frac{m v_\parallel^2}{qB_z R}\ \hat{\phi} \tag{14.50}$$

Combined drift due to grad-B and curvature of the field lines can be obtained by adding the equations (14.49) and (14.50)

$$\vec{v}_{R+\nabla B} = \vec{v}_R + \vec{v}_{\nabla B} = -\left(\frac{m v_\perp^2 B_r}{2q\rho B_z^2}\ \hat{\phi} + \frac{m v_\parallel^2}{qB_z R}\ \hat{\phi}\right) \tag{14.51}$$

From the technology point of view, it is a discouraging situation as both the drifts have the same sign and they add up. For confinement of the thermonuclear plasma, when the magnetic field is bent into a torus, the particles would drift out of the torus due to the term $v_\perp$.

For a Maxwellian distribution, the average energy of a particle is $3KT/2$. Velocity of a particle can be resolved into two components, parallel and perpendicular to the magnetic field. The parallel component $v_{\|}$ involves one degree of freedom whereas the perpendicular component $v_{\perp}$ involves two degrees of freedom. Thus, we have

$$\frac{1}{2}\, m\langle v_{\|}^2\rangle = \frac{1}{2}\, KT \qquad \text{and} \qquad \frac{1}{2}\, m\langle v_{\perp}^2\rangle = 2\,\frac{1}{2}\, KT$$

$$\langle v_{\|}^2\rangle = \frac{KT}{m} \qquad \text{and} \qquad \langle v_{\perp}^2\rangle = \frac{2KT}{m} \tag{14.52}$$

Using equations (14.52) in (14.51), we get

$$\vec{v}_{R+\nabla B} = -\Big(\frac{mv_{\perp}^2 B_\rho}{2q\rho B_z^2}\,\hat{\phi} + \frac{mv_{\|}^2}{qB_zR}\,\hat{\phi}\Big) = -\Big(\frac{KT}{m}\frac{mB_\rho}{q\rho B_z^2}\,\hat{\phi} + \frac{KT}{m}\frac{m}{qB_zR}\,\hat{\phi}\Big)$$

$$= -\frac{KT}{qB_z}\Big(\frac{B_\rho}{\rho B_z} + \frac{1}{R}\Big)\,\hat{\phi}$$

Defining the thermal velocity $v_{th} = \sqrt{2KT/m}$, we have

$$\vec{v}_{R+\nabla B} = -\frac{mv_{th}^2}{2qB_z}\Big(\frac{B_\rho}{\rho B_z} + \frac{1}{R}\Big)\,\hat{\phi} = \mp\frac{v_{th}^2}{2\omega_c}\Big(\frac{B_\rho}{\rho B_z} + \frac{1}{R}\Big)\,\hat{\phi}$$

Here, the $\vec{v}_{R+\nabla B}$ depends on the charge of the particle, but does not depend on its mass.

7. Particle in non-uniform electric field and uniform magnetic field

Consider a particle of mass m and charge q moving with a velocity $\vec{v}$ in a uniform magnetic field $\vec{B}$ and a nonuniform electric field $\vec{E}$ is expressed as

$$\vec{E} = E_0\cos(kx)\,\hat{x} = E_x\,\hat{x} \tag{14.53}$$

Such sinusoidal electric field with wavelength $\lambda = 2\pi/k$ can arise in a plasma during the motion of a wave motion through it. The force acting on the particle is $q[\vec{E} + \vec{v}\times\vec{B}]$ and the equation of motion of particle is

$$m\,\frac{d\vec{v}}{dt} = q(\vec{E} + \vec{v}\times\vec{B}) \tag{14.54}$$

Velocity of the particle is non-relativistic and therefore its mass is assumed to remain constant. Here, we use the Cartesian coordinate system. Let us take z-axis along the direction of the magnetic field $\vec{B}$, so that $\vec{B} = B\,\hat{z}$, where B is constant. For the velocity

$$\vec{v} = v_x\,\hat{x} + v_y\,\hat{y} + v_z\,\hat{z} \tag{14.55}$$

we have

$$\vec{v} \times \vec{B} = v_y B\,\hat{x} - v_x B\,\hat{y} \tag{14.56}$$

Using equations (14.53), (14.55) and (14.56) in (14.54), and equating the coefficients of $\hat{x}$, $\hat{y}$ and $\hat{z}$ on the two sides of the resultant equation, we get

$$m\,\frac{\mathrm{d}v_x}{\mathrm{d}t} = qE_x + qBv_y \tag{14.57}$$

$$m\,\frac{\mathrm{d}v_y}{\mathrm{d}t} = -qBv_x \tag{14.58}$$

$$m\,\frac{\mathrm{d}v_z}{\mathrm{d}t} = 0 \tag{14.59}$$

Integration of equation (14.59) gives

$$v_z = v_{z0}$$

where v_{z0} is the z-component of the initial velocity. Thus, the motion of the particle along the direction of the magnetic field is uniform. Integration of the above equation gives

$$z = v_{z0}t + z_0$$

where z_0 is the constant of integration. This equation shows that the distance along z-axis changes linearly with time provided the particle has z-component of the initial velocity. On differentiating equation (14.57) and using the value of $\mathrm{d}v_y/\mathrm{d}t$ from equation (14.58), we get (mind that E_x is independent of t)

$$\frac{\mathrm{d}^2 v_x}{\mathrm{d}t^2} = \frac{qB}{m}\,\frac{\mathrm{d}v_y}{\mathrm{d}t} = -\left(\frac{qB}{m}\right)^2 v_x = -\omega_c^2 v_x \tag{14.60}$$

where $\omega_c = |q|B/m$ is the cyclotron frequency. On differentiating equation (14.58) and using the value of $\mathrm{d}v_x/\mathrm{d}t$ from equation (14.57), we get

$$\frac{\mathrm{d}^2 v_y}{\mathrm{d}t^2} = -\frac{qB}{m}\,\frac{\mathrm{d}v_x}{\mathrm{d}t} = -\frac{qB}{m}\left(\frac{qE_0}{m}\cos(kx) + \frac{qB}{m}\,v_y\right)$$

$$= -\frac{c\omega_c^2 E_0}{B}\cos(kx) - \omega_c^2 v_y \tag{14.61}$$

For solving equation (14.61), knowledge of x (*i.e.*, the knowledge of the particle's orbit) is essentially needed. If the electric field is weak, as an approximation, we can use x which was obtained in absence of the electric field (case of uniform magnetic field)

$$x = x_0 + r_L \sin(\omega_c t + \phi) \tag{14.62}$$

Since we are not interested in a gyration at ω_c, it can be taken out by averaging over a cycle. The oscillatory term $\mathrm{d}^2 v_y/\mathrm{d}t^2$ clearly averages to zero so that

$$\left\langle \frac{\mathrm{d}^2 v_y}{\mathrm{d}t^2} \right\rangle = -\frac{c\omega_c^2 E_0}{B}\langle \cos\left[k\{x_0 + r_L \sin(\omega_c t + \phi)\}\right]\rangle - \omega_c^2 \langle v_y \rangle \tag{14.63}$$

where we have used equation (14.62). Expanding the cosine function, we have

$$\cos\left[k\{x_0 + r_L \sin(\omega_c t + \phi)\}\right] = \cos(kx_0)\cos[kr_L \sin(\omega_c t + \phi)]$$

$$- \sin(kx_0)\sin[kr_L \sin(\omega_c t + \phi)] \qquad (14.64)$$

For small Larmor radius, we have $kr_L << 1$ and the Taylor series expansions are

$$\cos[kr_L \sin(\omega_c t + \phi)] = 1 - \frac{1}{2}(kr_L \sin(\omega_c t + \phi))^2 - \ldots$$

and

$$\sin[kr_L \sin(\omega_c t + \phi)] = kr_L \sin(\omega_c t + \phi) - \ldots$$

after neglecting the higher order terms in the Taylor expansions, and using them in equation (14.64), we have

$$\cos\left[k\{x_0 + r_L \sin(\omega_c t + \phi)\}\right] = \cos(kx_0)\left[1 - \frac{k^2 r_L^2}{2}\sin^2(\omega_c t + \phi)\right]$$

$$- \sin(kx_0)\ kr_L \sin(\omega_c t + \phi)$$

thus,

$$\langle \cos\left[k\{x_0 + r_L \sin(\omega_c t + \phi)\}\right]\rangle = \cos(kx_0)\left[1 - \frac{k^2 r_L^2}{2}\langle \sin^2(\omega_c t + \phi)\rangle\right]$$

$$- \sin(kx_0) kr_L \Big\langle \sin(\omega_c t + \phi)\Big\rangle = \cos(kx_0)\left[1 - \frac{1}{4}\ k^2 r_L^2\right] \qquad (14.65)$$

Using equation (14.65) in (14.63), we have

$$\langle v_y \rangle = -\frac{E_0}{B}\cos(kx_0)\left[1 - \frac{1}{4}\ k^2 r_L^2\right] = -\frac{E_x(x_0)}{B}\left[1 - \frac{1}{4}\ k^2 r_L^2\right] \qquad (14.66)$$

Let us now evaluate the average of the oscillatory terms $\mathrm{d}^2 v_x/\mathrm{d}t^2$ to zero so that

$$\left\langle \frac{\mathrm{d}^2 v_x}{\mathrm{d}t^2} \right\rangle = -\omega_c^2 \langle v_x \rangle = 0 \qquad \text{or} \qquad \langle v_x \rangle = 0$$

Thus, the drift is along the y-direction (equation 14.66) and the usual drift, discussed earlier, is modified to

$$\vec{v}_E = \frac{\vec{E} \times \vec{B}}{B^2}\left[1 - \frac{1}{4}\ k^2 r_L^2\right]$$

The correction term depends on k^2 and hence on the second derivative[2] of the electric field $\vec{E}$. For the assumed $\vec{E}$, the second derivative is always negative. For an arbitrary variation of $\vec{E}$, we need only to replace (ik) by ∇, and the drift velocity is

$$v_{\vec{E}} = \left[1 + \frac{1}{4}\ r_L^2 \nabla^2\right]\frac{\vec{E} \times \vec{B}}{B^2}$$

[2]For the electric field $\vec{E}$ we have

$$\vec{E} = E_0 \cos(kx)\ \hat{x} \qquad \frac{\partial \vec{E}}{\partial x} = -kE_0 \sin(kx)\ \hat{x} \qquad \frac{\partial^2 \vec{E}}{\partial x^2} = -k^2 E_0 \cos(kx)\hat{x} = -k^2 \vec{E}$$

Figure 6 shows the drift of a gyrating particle in a nonuniform electric field $E_0 \cos(kx)\hat{x}$ and uniform magnetic field.

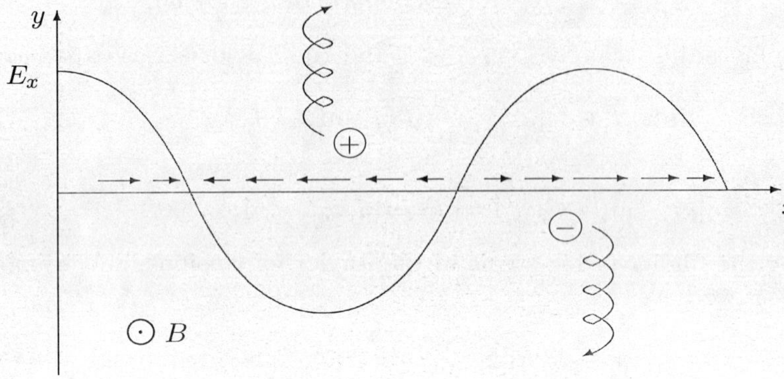

Figure 6: Drift of a gyrating particle in a nonuniform electric field $E_0 \cos(kx)\hat{x}$ and uniform magnetic field.

8. Particle in time-varying electric field and uniform magnetic field

Consider a particle of mass m and charge q moving in a uniform magnetic field $\vec{B}$ and a time varying electric field $\vec{E}$

$$\vec{E} = E_0 \cos(\omega t + \phi')\ \hat{x} \tag{14.67}$$

with velocity $\vec{v}$. The force acting on the particle is $q(\vec{E} + \vec{v} \times \vec{B})$ and the equation of motion of the particle is

$$m\ \frac{\mathrm{d}\vec{v}}{\mathrm{d}t} = q(\vec{E} + \vec{v} \times \vec{B}) \tag{14.68}$$

Velocity of the particle is non-relativistic and therefore its mass is assumed to remain constant. Here, we use the Cartesian coordinate system. Let us take z-axis along the direction of the magnetic field $\vec{B}$, so that $\vec{B} = B\ \hat{z}$, where B is constant. For the velocity

$$\vec{v} = v_x\ \hat{x} + v_y\ \hat{y} + v_z\ \hat{z} \tag{14.69}$$

we have

$$\vec{v} \times \vec{B} = v_y B\ \hat{x} - v_x B\ \hat{y} \tag{14.70}$$

Using equations (14.67), (14.69) and (14.70) in (14.68), and equating the coefficients of $\hat{x}$, $\hat{y}$ and $\hat{z}$ on the two sides of the resulting equation, we get

$$m\ \frac{\mathrm{d}v_x}{\mathrm{d}t} = qE_0 \cos(\omega t + \phi') + qBv_y \tag{14.71}$$

$$m\,\frac{\mathrm{d}v_y}{\mathrm{d}t} = -qBv_x \tag{14.72}$$

$$m\,\frac{\mathrm{d}v_z}{\mathrm{d}t} = 0 \tag{14.73}$$

Equation (14.73) shows that z-component of the velocity (*i.e.*, along the direction of the applied magnetic field) is constant.

$$v_z = v_{z0}$$

where v_{z0} is the z-component of initial velocity. Thus, the motion of the particle along the direction of the magnetic field is uniform. Integration of the above equation gives

$$z = v_{z0}t + z_0$$

where z_0 is the constant of integration. It shows that the distance along z-axis changes linearly with time provided the particle has z-component of initial velocity. On differentiating equation (14.71) and using the value of $\mathrm{d}v_y/\mathrm{d}t$ from equation (14.72), we get

$$\begin{aligned}\frac{\mathrm{d}^2v_x}{\mathrm{d}t^2} &= -\frac{\omega qE_0\sin(\omega t+\phi')}{m} + \frac{qB}{m}\frac{\mathrm{d}v_y}{\mathrm{d}t} \\ &= -\frac{\omega qE_0\sin(\omega t+\phi')}{m} - \left(\frac{qB}{m}\right)^2 v_x \\ &= \mp\frac{\omega E_0\omega_c\sin(\omega t+\phi')}{B} - \omega_c^2 v_x \end{aligned} \tag{14.74}$$

where $\omega_c = |q|B/m$ is the cyclotron frequency. Similarly, on differentiating equation (14.72) and using the value of $\mathrm{d}v_x/\mathrm{d}t$ from equation (14.71), we get

$$\begin{aligned}\frac{\mathrm{d}^2v_y}{\mathrm{d}t^2} &= -\frac{qB}{m}\frac{\mathrm{d}v_x}{\mathrm{d}t} \\ &= -\frac{qB}{mc}\left(\frac{qE_0\cos(\omega t+\phi')}{m} + \frac{qB}{m}v_y\right) \\ &= -\frac{c\omega_c^2E_0\cos(\omega t+\phi')}{B} - \omega_c^2 v_y \end{aligned} \tag{14.75}$$

On defining

$$\overline{v}_P = \mp\frac{\omega E_0\sin(\omega t+\phi')}{\omega_c B} \qquad \text{and} \qquad \overline{v}_E = -\frac{E_0\cos(\omega t+\phi')}{B}$$

equations (14.74) and (14.75) can be expressed as

$$\frac{\mathrm{d}^2v_x}{\mathrm{d}t^2} = -\omega_c^2(v_x - \overline{v}_P) \tag{14.76}$$

and

$$\frac{d^2 v_y}{dt^2} = -\omega_c^2(v_y - \overline{v}_E) \tag{14.77}$$

When $\overline{v}_P$ and $\overline{v}_E$ are zero, *i.e.*, there is only uniform magnetic field, the solutions are discussed earlier. Let us try the solution of equations (14.76) and (14.77), which is sum of a drift and a gyration, in the following form

$$v_x = v_\perp \cos(\omega_c t + \phi) + \overline{v}_P \tag{14.78}$$

and

$$v_y = \mp v_\perp \sin(\omega_c t + \phi) + \overline{v}_E \tag{14.79}$$

Two times differentiation of equations gives

$$\frac{d^2 v_x}{dt^2} = -\omega_c^2 v_x + (\omega_c^2 - \omega^2)\overline{v}_P \tag{14.80}$$

and

$$\frac{d^2 v_y}{dt^2} = -\omega_c^2 v_y + (\omega_c^2 - \omega^2)\overline{v}_E \tag{14.81}$$

Equations (14.80) and (14.81) are not the same as (14.76) and (14.77), respectively, unless $\omega^2 << \omega_c^2$. This condition can be obtained if we assume that $\vec{E}$ is so slowly varying such that $\omega^2 << \omega_c^2$. Then equations (14.78) and (14.79), respectively, are the solutions of equations (14.74) and (14.75). Integration of equation (14.78) gives

$$x = r_L \sin(\omega_c t + \phi) \pm \frac{E_0 \cos(\omega t + \phi')}{\omega_c B} \tag{14.82}$$

where $r_L = v_\perp/\omega_c$ is the Larmor radius. Integration of equation (14.79) gives

$$y = \pm r_L \cos(\omega_c t + \phi) - \frac{E_0 \sin(\omega t + \phi')}{\omega B} \tag{14.83}$$

From equations (14.82) and (14.83), we have

$$\left[x \mp \frac{E_0 \cos(\omega t + \phi')}{\omega_c B}\right]^2 + \left[y + \frac{E_0 \sin(\omega t + \phi')}{\omega B}\right]^2 = r_L^2$$

It shows that the guiding center is at (x_0, y_0), where

$$x_0 = \pm\frac{cE_0 \cos(\omega t + \phi')}{\omega_c B} \qquad y_0 = -\frac{cE_0 \sin(\omega t + \phi')}{\omega B}$$

Both the coordinates x_0 and y_0 are varying with time. The velocity of the guiding center along the y-axis is

$$\frac{dy_0}{dt} = -\frac{\omega E_0 \cos(\omega t + \phi')}{\omega B} = \overline{v}_E$$

The velocity of the guiding center along the x-axis is

$$\frac{dx_0}{dt} = \mp\frac{\omega E_0 \sin(\omega t + \phi')}{\omega_c B} = \overline{v}_P$$

The y-component is the usual drift

$$\vec{v}_E = -\frac{E_0 \cos(\omega t + \phi')}{B} i\, \hat{y} = \frac{c\vec{E} \times \vec{B}}{B^2}$$

except that $\vec{v}_E$ now oscillates slowly with a frequency ω. The x-component, along the direction of the electric field $\vec{E}$, known as *polarization drift*, is

$$\vec{v}_P = \mp \frac{\omega E_0 \sin(\omega t + \phi')}{\omega_c B} \hat{x} = \pm \frac{1}{\omega_c B} \frac{\mathrm{d}\vec{E}}{\mathrm{d}t} = \frac{m}{qB^2} \frac{\mathrm{d}\vec{E}}{\mathrm{d}t}$$

The polarization drift depends on the charge of the particle. Thus, the opposite charges move in opposite directions, and they constitute a current. For the hydrogen plasma of protons and electrons the current density $\vec{J}$ is

$$\vec{J}_p = ne(v_{ip} - v_{ep}) = \frac{ne}{eB^2}(m_p + m_e)\frac{\mathrm{d}\vec{E}}{\mathrm{d}t} = \frac{\rho}{B^2} \frac{\mathrm{d}\vec{E}}{\mathrm{d}t}$$

where m_p and m_e are masses of proton and electron, respectively, n is the density of protons as well as electron, and ρ the mass density.

9. Particle in time-varying magnetic field

Consider a particle of mass m and charge q moving in a time-varying magnetic field $\vec{B}$ with velocity $\vec{v}$. As the Lorentz force $\vec{F}(= q\ \vec{v} \times \vec{B})$ due to the magnetic field $\vec{B}$ is perpendicular to the direction of the field, the magnetic field does not do any work on the particle and therefore, it does not change any energy of the particle. However, when $\vec{B}$ changes with time, it has an associated electric field $\vec{E}$ given by the Maxwell's equation

$$\nabla \times \vec{E} = -\frac{\partial \vec{B}}{\partial t} \tag{14.84}$$

This produced electric field $\vec{E}$ is perpendicular to the direction of the magnetic field $\vec{B}$, and it can accelerate/decelerate the charged particles. Now, velocity of a particle is also changing and thus, the Lorentz force is changing. Owing to the electric field, velocity of the particle is almost in the direction perpendicular to the magnetic field. Let

$$\vec{v}_\perp = \frac{\mathrm{d}\vec{\ell}}{\mathrm{d}t}$$

be the transverse velocity of the particle (with $v_\parallel$ negligibly small). Here, $\vec{\ell}$ is the element of the path along the particle trajectory. For the electric field $\vec{E}$ and magnetic field $\vec{B}$, the equation of motion of the particle is

$$m\frac{\mathrm{d}\vec{v}}{\mathrm{d}t} = q(\vec{E} + \vec{v} \times \vec{B})$$

We have $\vec{v} = \vec{v}_\perp$, and therefore,

$$m\ \frac{\mathrm{d}\vec{v}_\perp}{\mathrm{d}t} = q(\vec{E} + \vec{v}_\perp \times \vec{B}) \tag{14.85}$$

Taking scalar product of $\vec{v}_\perp$ with equation (14.85), we have

$$m\vec{v}_\perp \cdot \frac{d\vec{v}_\perp}{dt} = q\vec{v}_\perp \cdot (\vec{E} + \vec{v}_\perp \times \vec{B})$$

$$\frac{d}{dt}\left(\frac{1}{2}m\, v_\perp^2\right) = q\vec{v}_\perp \cdot \vec{E} = q\vec{E} \cdot \frac{d\vec{\ell}}{dt}$$

The change in one gyration is obtained by integrating over one period

$$\left[\frac{1}{2}\, mv_\perp^2\right]_{t=0}^{2\pi/\omega_c} = \int_{t=0}^{2\pi/\omega_c} q\,\vec{E} \cdot \frac{d\vec{\ell}}{dt}\, dt$$

$$\delta\left(\frac{1}{2}\, mv_\perp^2\right) = \int_{t=0}^{2\pi/\omega_c} q\vec{E} \cdot \frac{d\vec{\ell}}{dt}\, dt \tag{14.86}$$

For slow variation of the magnetic field, the time integral in equation (14.86) can be replaced by a line integral

$$\delta\left(\frac{1}{2}\, mv_\perp^2\right) = \oint q\vec{E} \cdot d\vec{l} \tag{14.87}$$

Using Stokes' theorem in equation (14.87), we have

$$\delta\left(\frac{1}{2}\, mv_\perp^2\right) = q\int_S (\nabla \times \vec{E}) \cdot d\vec{S} = -q\int_S \dot{\vec{B}} \cdot d\vec{S} \tag{14.88}$$

where we have used equation(14.84). Here, $\vec{S}$ is the surface enclosed by the Larmor orbit and is in the direction of the thumb when the fingers of the right hand point in the direction of $\vec{v}$ (right-hand rule). Since a plasma is diamagnetic, we have $\vec{B} \cdot d\vec{S} < 0$ for ions and $\vec{B} \cdot d\vec{S} > 0$ for electrons. Further, for slowing varying field, $\dot{\vec{B}}$ can be taken out in equation (14.88) and we have

$$\delta\left(\frac{1}{2}\, mv_\perp^2\right) = \pm q\frac{\partial B}{\partial t}\, \pi r_L^2 = \pm q\pi\, \frac{\partial B}{\partial t}\frac{v_\perp^2}{\omega_c^2}$$

$$= \pm q\pi\, \frac{\partial B}{\partial t}\frac{v_\perp^2}{\omega_c}\frac{m}{(\pm qB)} = \frac{\frac{1}{2}mv_\perp^2}{B}\frac{2\pi(\partial B/\partial t)}{\omega_c}$$

The quantity $[2\pi(\partial B/\partial t)/\omega_c] = [(\partial B/\partial t)/f_c]$ is the change δB during one rotation and thus, we have

$$\delta\left(\frac{1}{2}\, mv_\perp^2\right) = \mu\, \delta B \tag{14.89}$$

where the magnetic moment μ is defined as

$$\mu = \frac{mv_\perp^2/2}{B} \tag{14.90}$$

Using equation (14.90) in (14.89), we have

$$\delta(\mu B) = \mu\, \delta B \qquad \mu\, \delta B + B\, \delta\mu = \mu\, \delta B \qquad B\, \delta\mu = 0$$

As the magnetic field is non-zero, we have

$$\delta\mu = 0 \tag{14.91}$$

Hence, in the slowly varying magnetic field, the magnetic moment is invariant. As the magnetic field varies in strength, $v_\perp$ (from equation 14.89) and hence the Larmor orbit expands and contacts, and the particles lose and gain transverse energy. This exchange of energy between the particles and the field is described by equation (14.91), that the magnetic moment remains conserved.

Theorem: The magnetic flux through a Larmor orbit is conserved.

Proof: The flux Φ is given by $\Phi = BS$, where $S = \pi r_L^2$ is the area enclosed by Larmor orbit. Thus,

$$\Phi = B\pi r_L^2 = B\pi \frac{v_\perp^2}{\omega_c^2} = B\pi \frac{v_\perp^2 m^2}{q^2\ B^2}$$

$$= \frac{2\pi m}{q^2} \frac{\frac{1}{2}\ mv_\perp^2}{B} = \frac{2\pi m}{q^2} \mu$$

Since μ is constant, therefore, the flux through a Larmor orbit is constant.

Review Questions

1. On a charged particle moving with constant velocity, both the electric and magnetic fields are applied. The direction of the magnetic field is along the z-axis and the electric field is in the xz-plane. Describe the drift velocity of the particle.

2. On a charged particle moving with constant velocity, a uniform force and magnetic field are applied. The direction of the magnetic field is along the z-axis and the force is in the xz-plane. Describe the drift velocity of the particle.

3. A charged particle is moving in a non-uniform magnetic field. Describe the drift velocity of the particle.

4. Describe the curvature drift.

5. A charged particle is moving in a non-uniform electric field and uniform magnetic field. Describe the motion of the particle.

6. A charged particle is moving in a time varying electric field and uniform magnetic field. Describe the drift velocity of the particle.

7. Write notes on the following

   (i) Curvature drift

Multiple-Choice Questions

1. A particle having positive charge $+q$ and mass m is moving along the $+x$-axis. On application of a magnetic field along the $+z$-axis, the direction of force on the particle is along the

   A. $+z$-axis B. $-z$-axis C. $+y$-axis D. $-y$-axis

   Ans. D

2. An electron is moving along the $+x$-axis. On application of a magnetic field along the $+z$-axis, the direction of force on the electron is along the

   A. $+z$-axis B. $-z$-axis C. $+y$-axis D. $-y$-axis

   Ans. C

3. A charged particle moving in the uniform electric and magnetic fields moves in a circular path when (v_{z0} is the z-component of velocity at $t = 0$)

   A. E_x and v_{z0} are zero B. E_y and v_{z0} are zero

   C. E_z and v_{z0} are zero D. E_x, E_y and E_z are zero

   Ans. C

4. A charged particle is moving in the uniform electric field $\vec{E}$ and uniform magnetic field $\vec{B}$. The guiding center moves

   A. along the direction of $\vec{B}$ B. along the direction of $\vec{E}$

   C. perpendicular to both $\vec{B}$ and $\vec{E}$ D. None of them

   Ans. C

5. A charged particle moving in the uniform force $\vec{F}$ and uniform magnetic field moves in a circular path when (v_{z0} is the z-component of velocity at $t = 0$)

   A. F_x and v_{z0} are zero B. F_y and v_{z0} are zero

   C. F_z and v_{z0} are zero D. F_x, F_y and F_z are zero

   Ans. C

6. A charged particle is moving in the uniform electric field $\vec{E}$ and uniform magnetic field $\vec{B}$. The drift velocity of electron v_{ge} i and of ion v_{gi} satisfy the relation

   A. $v_{gi} > v_{ge}$ B. $v_{gi} < v_{ge}$

   C. $v_{gi} = v_{ge}$ D. depends on situation

   Ans. C

7. A charged particle is moving in the uniform force $\vec{F}$ and uniform magnetic field $\vec{B}$. The drift velocity of electron v_{ge} and of ion v_{gi} satisfy the relation

   A. $v_{gi} > v_{ge}$ B. $v_{gi} < v_{ge}$ C. $v_{gi} = v_{ge}$ D. depends on situation

   Ans. C

8. On a stationary charged particle, both the electric and magnetic fields are applied. The direction of magnetic field is along the z-axis and the electric field is in the xz-plane. Along the z-axis, the particle moves with

   A. zero velocity. B. constant velocity

   C. acceleration. D. arbitrary velocity

   Ans. C

9. On a stationary charged particle, both the electric and magnetic fields are applied. The direction of both the electric and magnetic fields is along the z-axis. Along the z-axis, the particle moves with

   A. zero velocity. B. constant velocity

   C. acceleration. D. arbitrary velocity

   Ans. C

10. On a charged particle moving with constant velocity, both the electric and magnetic fields are applied. The direction of the magnetic field is along the z-axis and the electric field along the x-axis. Along the z-axis, the particle moves with

    A. zero velocity. B. constant velocity

    C. acceleration. D. arbitrary velocity

    Ans. B

11. On a stationary charged particle, both the electric and magnetic fields are applied. The direction of the magnetic field is along the z-axis and the electric field is in the xz-plane. The drift velocity of the guiding-center along

    A. $+z$-axis B. $-z$-axis C. $+y$-axis D. $-y$-axis

    Ans. D

12. On a stationary charged particle, having mass m and charge q, both the electric and magnetic fields are applied. The direction of the magnetic field is along the z-axis and the electric field is in the xz-plane. The drift velocity of the guiding-center depends

    A. on m only B. on both m and q

    C. on q only D. neither on m nor q

    Ans. D

13. On a stationary charged particle, having mass m and charge q, both the electric and magnetic fields are applied. The direction of the magnetic field is along the z-axis and the electric field is in the xz-plane. Which of the following statements is correct for the motion of positive ions and negatively charged electrons.

    A. same direction and same velocity

    B. different direction but same velocity

C. same direction but different velocity

D. different direction and different velocity

Ans. A

14. On a stationary charged particle, having mass m and charge q, both the electric and magnetic fields are applied. The direction of the magnetic field is along the z-axis and the electric field is in the xz-plane. The value of the Larmor radius depends

A. on m only B. on both m and q

C. on q only D. neither on m nor q

Ans. B

15. On a stationary charged particle, a uniform force and magnetic field are applied. The direction of magnetic field is along the z-axis and the force is in the xz-plane. Along the z-axis, the particle moves with

A. zero velocity. B. constant velocity

C. acceleration. D. arbitrary velocity

Ans. C

16. On a stationary charged particle, a uniform force and magnetic field are applied. The direction of both the force and magnetic fields is along the z-axis. Along the z-axis, the particle moves with

A. zero velocity. B. constant velocity

C. acceleration. D. arbitrary velocity

Ans. C

17. On a charged particle moving with constant velocity, a uniform force and magnetic field are applied. The direction of the magnetic field is along the z-axis and the force along the x-axis. Along the z-axis, the particle moves with

A. zero velocity. B. constant velocity

C. acceleration. D. arbitrary velocity

Ans. B

18. Path of a charged particle moving with an acute angle with the direction of a uniform magnetic field is

A. straight line B. parabola C. helix D. ellipse

Ans. C

19. Path of a charged particle moving with an obtuse angle with the direction of a uniform magnetic field is

A. straight line B. parabola C. helix D. ellipse

Ans. C

20. Path of a charged particle moving with a right angle with the direction of a uniform magnetic field is

A. straight line B. parabola C. circle D. ellipse

Ans. C

21. Initial motion of a charged particle lies in a plane and a uniform magnetic field is applied in a direction perpendicular to the plane. The path of the particle is

A. circle B. parabola C. helix D. ellipse

Ans. A

Chapter 15

Plasma State

Three states of matter, called, solid, liquid and gas, are generally known to a common person. One of these states can, in general, be converted into others through the exchange of energy. In day-to-day life, H_2O molecule is a remarkable example, as it is found in all the three states, as ice (solid), water (liquid) and steam (gas). Obviously, by supplying energy to matter, solid state can be converted into liquid one, and liquid state into gaseous one; for the reverse process, the energy is extracted out from the matter. In some cases, it is possible to convert solid state into gaseous state directly without going through the liquid state. NH_4Cl is example of such case.

Further supply of energy to matter in the gaseous state, breaks molecules into its constituent atoms, and then atoms are stripped off their electrons producing, positively charged ions and negatively charged electrons. The amount of energy required to liberate an electron from atom is known as *ionization potential.*[1] This supplied energy may be in the form of heat, radiation, due to collision or due to chemical reaction. The ionization due to heat occurs at very high temperature, of the order of million Kelvin which can be produced in laboratories. This state of matter where the charged as well as neutral particles exist simultaneously is generally known as plasma. This term 'plasma' was first coined by Langmuir in 1923. The plasma may be defined as the following:

Plasma is quasi-neutral gas of charged and neutral particles which exhibit collective behaviour.

As the plasma is found in a natural form in a number of cosmic objects and in the upper atmosphere of Earth, therefore, it is sometimes called *fourth state of matter.*

When the gas is ionized, its dynamical behaviour is influenced by external electric and magnetic fields. Moreover, the separated charged particles within plasma give rise to new forces between the constituent particles. Thus, the properties of plasma become quite different from those of neutral atoms and molecules. In this chapter we have discussed some processes and properties of plasma.

[1]From an atom, having more than one electron, electrons may be liberated one by one. Thus, an atom can have several ionization potentials; each subsequent ionization of atom has larger value of ionization potential.

1. Natural plasma

Natural plasma exists in some cosmic objects, such as interiors and atmospheres of hot stars, planetary nebulae, regions of ionized hydrogen in interstellar medium, and upper atmosphere of Earth. On the Earth, plasma can however be produced in laboratories. The reason for finding natural plasma in cosmic objects and not at the Earth is due to large differences in density and temperature in the two regions. On the Earth, the density is very large and the temperature is very low as compared to those in the cosmic objects. Some regions where natural plasma is found are as the following.

1.1 Ionosphere

Streams of charged particles, known as the solar wind, are being continuously emitted by our Sun. Some of these charged particles reach up to the upper atmosphere of Earth. Moreover, intense radiations (γ-rays, x-rays, UV radiations) coming from the outer space ionize the upper atmosphere of Earth. This upper atmosphere of Earth is known as **ionosphere**. It ranges from about 50 km to about 1000 km above Earth's surface. Since the hazardous radiations (γ-rays, x-rays, UV radiations) are absorbed in the upper atmosphere of Earth. The Earth's atmosphere thus plays important role in our life by shielding us from hazardous radiations coming from the outer space. This ionosphere had been in use for communication purposes, as the radiations of frequency less than plasma frequency are reflected back by plasma in the ionosphere.

1.2 van Allen belts

Investigation made by satellites has found two regions, known as van Allen radiation belts, which envelop Earth. One belt is at a distance of about 9700 km and second at 22,500 km from the surface of Earth. The thickness of the inner and outer belts is, respectively, about 4800 km and 8000 km. The belts contain charged particles (ions) trapped between the magnetic lines of force.

1.3 Aurorae

Above the Earth's magnetic poles, the charged particles have free access to Earth's surface, as the lines of magnetic field are concentrated towards the surface. These charged particles interact with the molecules of atmosphere, causing a glow from time to time. These glows are aurorae which are also known as northern lights and southern lights.

1.4 Solar corona

Two strong emission lines at the wavelengths 5303 Å and 6374 Å have been found in solar atmosphere. Later on some more lines were found in the spectra of Sun. These lines could not be assigned to any of the known atoms or their singly ionized ions. Since the estimated temperature of photosphere of Sun is about 5800 K, scientists could expect either atoms or their singly ionized ions in the solar atmosphere. In absence of any atom to which these lines could be assigned, scientist coined a name coronium to some unknown atom.

No one knew about physical properties of this coronium, except to say that it was responsible for the generation of those unassigned lines. Later on, through laboratory studies at very high temperatures, it was found that these unknown lines could be generated by highly ionized ions. For example, the lines at λ 5303 Å and 6374 Å are generated by Fe XIV (thirteen times ionized iron, Fe^{+13}) and Fe X (nine times ionized iron, Fe^{+9}), respectively. Since these highly ionized ions are produced at temperature of million Kelvin, there was no alternative but to accept that the temperature in corona around the Sun is of million Kelvin.

In spite of energy losses due to the radiation as well as conduction, the temperature of corona is found maintained at million Kelvin. It is indeed a challenging task before scientists to find out source for solar coronal heating. Such coronae are found around a number of stars.

1.5 Core of Sun

For understanding source of energy inside Sun, it is now well established that in the core of Sun, temperature is of the order of 15 million K and the process of fusion of four hydrogen nuclei into a helium nucleus is going on. Thus, the core of Sun is so hot that the matter there is in the plasma state. Besides the fusion of hydrogen nuclei in helium nucleus, in some stars, CNO cycle or triple α reactions are going on. Consequently, in the cores of shining stars, temperature is sufficiently high to maintain the nuclear reactions. Hence, the material in the cores of stars is in plasma state.

1.6 HII regions

In some parts of interstellar medium, temperature is so high that hydrogen gas is in the ionized state. These regions are generally known as HII regions, and are either associated with evolved stars (stars in the late stage) or they form a big cloud in which the process of star formation is going on.

2. Concept of temperature

Let us first understand the meaning of 'temperature' of gas whose constituent particles may be neutral or may have charge. In a gas in thermal equilibrium, the constituent particles interact (collide) constantly with one another and move with various velocities. The most probable distribution of velocities of these particles is known as Maxwell-Boltzmann (MB) distribution. Though this velocity distribution is three-dimensional distribution, we however first consider one-dimensional MB distribution expressed as

$$f(u) = A \exp\Big(-\frac{mu^2}{2KT}\Big) \tag{15.1}$$

where u is velocity a particle, $f(u)\,du$ the number of particles per unit length having velocities in the range from u to $u + du$, m the mass of a particle in gas, K the Boltzmann constant[2] and T is a physical parameter of gas, termed as 'temperature'. The density n, the number of particles per unit length, is

$$n = \int_{-\infty}^{\infty} f(u)\,du = \int_{-\infty}^{\infty} A\,\exp\Big(-\frac{mu^2}{2KT}\Big)\,du$$

On defining

$$v_{th} = \sqrt{2KT/m} \qquad \text{and} \qquad y = u/v_{th}$$

we have

$$n = A\int_{-\infty}^{\infty} e^{-y^2} v_{th}\,dy = Av_{th}\int_{-\infty}^{\infty} e^{-y^2}dy = Av_{th}\sqrt{\pi} = A\sqrt{\frac{2\pi KT}{m}}$$

Here, A is assumed as independent of velocity of particles. Thus,

$$A = n\sqrt{\frac{m}{2\pi KT}}$$

A is found to depend on the temperature of gas, and thus, the distribution function (equation 15.1) is

$$f(u) = n\sqrt{\frac{m}{2\pi KT}}\,\exp\Big(-\frac{mu^2}{2KT}\Big)$$

Variation of distribution function $f(u)$ with velocity u is shown in Figure 1 for three temperatures, 100, 120 and 140 K. The figure shows that with the increase of temperature, the number of particles having average velocity (corresponding to $u = 0$) decreases, but the width of plot increases, such that total area under curve remains constant, giving the density of particles (*i.e.*, the number of particles per unit length). In order to understand the meaning of T, we compute mean kinetic energy of particles averaged over the MB-distribution (15.1) as

$$\begin{aligned} E_{av} &= \frac{\int_{-\infty}^{\infty} \frac{1}{2}mu^2 f(u)\,du}{\int_{-\infty}^{\infty} f(u)\,du} \\ &= \frac{\int_{-\infty}^{\infty} \frac{1}{2}mu^2 A\,\exp(-mu^2/2KT)\,du}{\int_{-\infty}^{\infty} A\,\exp(-mu^2/2KT)\,du} \end{aligned} \tag{15.2}$$

Here, we have assumed that interaction between the constituent particles is very weak, so that the potential energy is neglected. However, the collisions between particles take place momentarily. Using the relations

$$v_{th} = \sqrt{2KT/m} \qquad \text{and} \qquad y = u/v_{th}$$

equation (15.2) gives

$$E_{av} = \frac{\frac{1}{2}\,mv_{th}^2\int_{-\infty}^{\infty} y^2 e^{-y^2}dy}{\int_{-\infty}^{\infty} e^{-y^2}dy} = \frac{\frac{1}{2}\,mv_{th}^2\,(\sqrt{\pi}/2)}{\sqrt{\pi}} = \frac{1}{4}\,mv_{th}^2 = \frac{1}{2}\,KT$$

[2]We have used K for Boltzmann constant because k is used for wavevector.

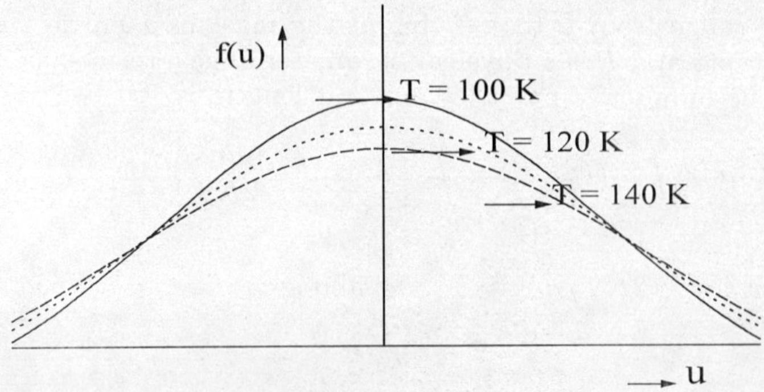

Figure 1: Variation of $f(u$ (in arbitrary units) versus u for three temperatures, 100, 120 and 140 K.

Thus, for one-dimensional MB distribution, the average kinetic energy is $KT/2$. Hence, the temperature may be defined as the average kinetic energy per degree of freedom divided by $(K/2)$. Let us now take three-dimensional MB distribution

$$f(u,v,w) = B\ \exp\Big[-m(u^2+v^2+w^2)/2KT\Big] \tag{15.3}$$

where u, v and w are three mutually perpendicular components of velocity. The density n of gas is

$$\begin{aligned} n &= \int_{-\infty}^{\infty}\int_{-\infty}^{\infty}\int_{-\infty}^{\infty} f(u,v,w)\ du\ dv\ dw \\ &= \int_{-\infty}^{\infty}\int_{-\infty}^{\infty}\int_{-\infty}^{\infty} B\ \exp\Big[-m(u^2+v^2+w^2)/2KT\Big]\ du\ dv\ dw \\ &= B\int_{-\infty}^{\infty}\exp(-mu^2/2KT)\ du\int_{-\infty}^{\infty}\exp(-mv^2/2KT)\ dv\int_{-\infty}^{\infty}\exp(-mw^2/2KT)\ dw \\ &= B\sqrt{\frac{2\pi KT}{m}}\sqrt{\frac{2\pi KT}{m}}\sqrt{\frac{2\pi KT}{m}} = B\Big(\frac{2\pi KT}{m}\Big)^{3/2} \end{aligned}$$

Here, B is assumed to be independent of velocity of particles. Thus,

$$B = n\Big(\frac{m}{2\pi KT}\Big)^{3/2}$$

B is found to depend on the temperature of gas, and the distribution function (equation 15.3) is

$$f(u,v,w) = n\Big(\frac{m}{2\pi KT}\Big)^{3/2}\exp\Big[-m(u^2+v^2+w^2)/2KT\Big] \tag{15.4}$$

We compute mean kinetic energy of the particles averaged over this distribution function (equation 15.4)

$$E_{av} = \frac{\int_{-\infty}^{\infty}\int_{-\infty}^{\infty}\int_{-\infty}^{\infty}\frac{1}{2}m(u^2+v^2+w^2)f(u,v,w)\ du\ dv\ dw}{\int_{-\infty}^{\infty}\int_{-\infty}^{\infty}\int_{-\infty}^{\infty}f(u,v,w)\ du\ dv\ dw}$$

$$= \frac{\int_{-\infty}^{\infty}\int_{-\infty}^{\infty}\int_{-\infty}^{\infty}\frac{1}{2}mu^2\exp[-m(u^2+v^2+w^2)/2KT]\ du\ dv\ dw}{\int_{-\infty}^{\infty}\int_{-\infty}^{\infty}\int_{-\infty}^{\infty}\exp[-m(u^2+v^2+w^2)/2KT]\ du\ dv\ dw}$$

$$+ \frac{\int_{-\infty}^{\infty}\int_{-\infty}^{\infty}\int_{-\infty}^{\infty}\frac{1}{2}mv^2\exp[-m(u^2+v^2+w^2)/2KT]\ du\ dv\ dw}{\int_{-\infty}^{\infty}\int_{-\infty}^{\infty}\int_{-\infty}^{\infty}\exp[-m(u^2+v^2+w^2)/2KT]\ du\ dv\ dw}$$

$$+ \frac{\int_{-\infty}^{\infty}\int_{-\infty}^{\infty}\int_{-\infty}^{\infty}\frac{1}{2}mw^2\exp[-m(u^2+v^2+w^2)/2KT]\ du\ dv\ dw}{\int_{-\infty}^{\infty}\int_{-\infty}^{\infty}\int_{-\infty}^{\infty}\exp[-m(u^2+v^2+w^2)/2KT]du\ dv\ dw} \tag{15.5}$$

Let us compute one of the three parts on right side of equation (15.5)

$$z = \frac{\int_{-\infty}^{\infty}\int_{-\infty}^{\infty}\int_{-\infty}^{\infty}\frac{1}{2}mu^2\exp[-\frac{1}{2}m(u^2+v^2+w^2)/KT]\ du\ dv\ dw}{\int_{-\infty}^{\infty}\int_{-\infty}^{\infty}\int_{-\infty}^{\infty}\exp[-\frac{1}{2}m(u^2+v^2+w^2)/KT]\ du\ dv\ dw}$$

$$= \frac{\int_{-\infty}^{\infty}\frac{1}{2}mu^2\exp(-\frac{1}{2}\frac{mu^2}{KT})\ du}{\int_{-\infty}^{\infty}\exp(-\frac{1}{2}\frac{mu^2}{KT})\ du} = \frac{\int_{-\infty}^{\infty}\frac{1}{2}mv_{th}^2y^2\exp(-y^2)\ v_{th}\ dy}{\int_{-\infty}^{\infty}\exp(-y^2)v_{th}\ dy}$$

$$= \frac{\frac{1}{2}mv_{th}^2\sqrt{\pi}/2}{\sqrt{\pi}} = \frac{1}{4}mv_{th}^2 = \frac{1}{2}\ K\ T$$

Similarly, other two parts would also have the value $KT/2$. Thus,

$$E_{av} = \frac{1}{2}\ KT + \frac{1}{2}\ KT + \frac{1}{2}\ KT = \frac{3}{2}\ KT$$

Hence, a general result is that the average kinetic energy per degree of freedom is $(KT/2)$. As earlier, the temperature can be defined as the average kinetic energy per degree of freedom divided by $(K/2)$. Temperature and energy in plasma are so closely related that it is customary to express temperature in the units of energy. 1 eV energy corresponds to a temperature of 11,605 K.

2.1 Simultaneous existence of several temperatures

It is interesting to notice that plasma may have various kinetic temperatures at the same time. It may happen as the ions and electrons may have independent MB distributions corresponding to different temperatures T_i and T_e of ions and electrons, respectively. This can be understood as the rates of collisions among electrons and among ions may be larger than those among ions and electrons. Thus, each of the species (ions as well as electrons) may have their independent thermal equilibrium states. But, this situation of two different temperatures may not last long and they equalize fast.

When the velocities of the ions are much smaller than those of the electron, the temperature in plasma is controlled by electrons, and hence the MB distribution is given by the electron temperature T_e. Further, when there is magnetic field, even a single species can have two temperatures. It is because velocity of a particle has two components, one along the magnetic field and the other perpendicular to the magnetic field. The parallel and perpendicular components of velocities may have their different MB distributions, corresponding to the temperatures, $T_{||}$ and $T_{\perp}$, respectively.

3. Debye shielding

A fundamental characteristic of plasma is that it can shield out electric potentials that are applied to it. In order to understand it, let us put an electric field inside plasma by inserting two charged balls connected to a battery (Figure 2). Here, we assume that a layer of dielectric restricts the plasma from recombining on the surface or the battery is large enough to maintain the potential in spite of the recombination. Now, the balls almost immediately would attract particles of opposite charge and a cloud of positive ions would be formed around the negative ball and a cloud of electrons would be formed around the positive ball.

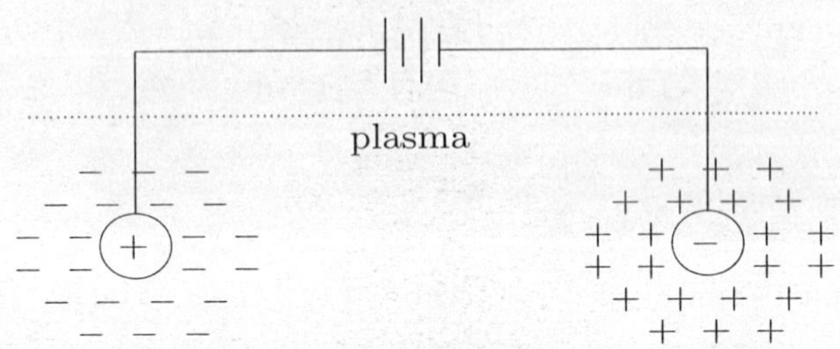

Figure 2: Figure shows two charged balls connected to a battery and inserted inside plasma

In case the plasma is cold, there are no thermal motions of charged particles, and the surrounding cloud in each case has as many charges as there are in the ball. Thus, in the plasma, outside the clouds, there is no electric field and the shielding is prefect. On the other hand, if the plasma temperature is finite, the particles at the edge of cloud (there electric field is weak), have enough thermal energy to escape from the electrostatic potential well. The edge of the cloud is estimated to occur at the radial distance where the potential energy is approximately equal to the thermal kinetic energy KT of particles. Consequently, the shielding is not complete.

Let us calculate approximate thickness of such a charge cloud. Suppose, the potential ϕ on the plane $x = 0$ is kept at a value ϕ_0 by a perfectly transparently grid (Figure 3). Here, x is the distance measured radially outward. Now, our object is to compute $\phi(x)$. For simplicity, we assume that the ratio M/m_e (ratio of mass of positively charged ion and that of election) is very high (infinite in the mathematical language), so that the positively charged ions do not move but form a uniform background of positive charge through which negatively charged electrons are moving in the gas. Since the situation is spherically symmetric, we can account for the radial variation. Thus, we consider one-dimensional Poisson equation

$$\nabla^2\phi = -\frac{q}{\epsilon_0} \qquad\qquad \frac{d^2\phi}{dx^2} = -\frac{e(n_i - n_e)}{\epsilon_0} \tag{15.6}$$

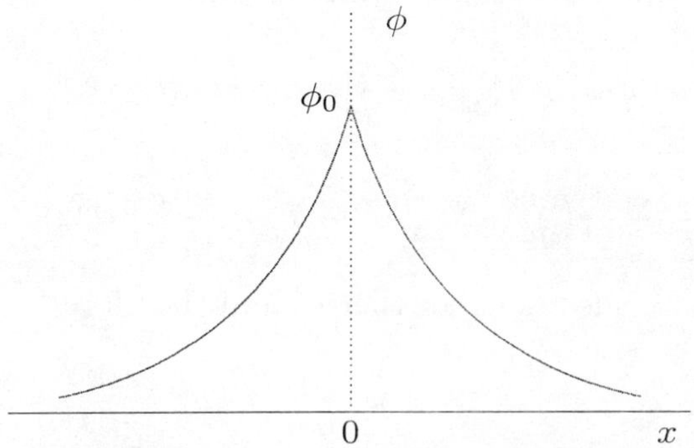

Figure 3: Variation of potential as a function of distance.

Here, we have considered hydrogen plasma, and hence the positively charged ions are protons. At a large distance, density of protons is equal to that of the electrons and is denoted by n. As the protons are not moving, the density of protons n_i is n everywhere. Thus,

$$n_i = n \tag{15.7}$$

In the presence of a potential energy $q\phi(\equiv -e\phi)$, the electron distribution function is

$$f(u) = A\,\exp\Big[-\Big(\frac{1}{2}m_e u^2 - e\phi\Big)/KT_e\Big] \tag{15.8}$$

where T_e is electron temperature, u the velocity of electron and ϕ the potential. Electron density is

$$n_e = \int_{-\infty}^{\infty} f(u)\,du = \int_{-\infty}^{\infty} A\,\exp\Big[-\Big(\frac{1}{2}m_e u^2 - e\phi\Big)/KT_e\Big]\,du$$

$$= A\,\exp\Big(\frac{e\phi}{KT_e}\Big)\int_{-\infty}^{\infty}\exp\Big(-\frac{m_e u^2}{2KT_e}\Big)\,du$$

$$= A\,\exp(e\phi/KT_e)\sqrt{\frac{2\pi KT_e}{m_e}} \tag{15.9}$$

At large distance

$$x \to \infty \qquad\qquad \phi \to 0 \qquad\qquad n_e \to n$$

and therefore, we have

$$n = A\sqrt{\frac{2\pi KT_e}{m_e}}$$

Thus, equation (15.9) becomes

$$n_e = n\ \exp(e\phi/KT_e) \tag{15.10}$$

Using equations (15.7), (15.8), and (15.10) in (15.6), we get

$$\frac{\mathrm{d}^2\phi}{\mathrm{d}x^2} = -\frac{e}{\epsilon_0}\Big[n - n\ \exp(e\phi/KT_e)\Big] = \frac{en}{\epsilon_0}\Big[\exp(e\phi/KT_e) - 1\Big]$$

$$= \frac{en}{\epsilon_0}\Big[1 + \frac{e\phi}{KT_e} + \frac{1}{2!}\Big(\frac{e\phi}{KT_e}\Big)^2 + \ldots - 1\Big] = \frac{e^2 n\phi}{\epsilon_0 KT_e}\Big[1 + \frac{1}{2!}\ \frac{e\phi}{KT_e} + \ldots\Big]$$

Since $e\phi < KT_e$, we can neglect smaller therms and therefore

$$\frac{\mathrm{d}^2\phi}{\mathrm{d}x^2} = \frac{ne^2}{\epsilon_0 KT_e}\phi \qquad \text{or} \qquad \frac{\mathrm{d}^2\phi}{\mathrm{d}x^2} = \frac{\phi}{\lambda_D^2} \tag{15.11}$$

where

$$\lambda_D = \Big(\frac{\epsilon_0 KT_e}{ne^2}\Big)^{1/2}$$

Solution of equation (15.11) is

$$\phi = C\ \exp(-|x|/\lambda_D) + D\ \exp(|x|/\lambda_D)$$

where C and D are constants. Second part on right side of this equation is not feasible as it shows unphysical situation of increase of potential with the increase of distance. Hence, the constant D is substituted zero, and we have

$$\phi = C\ \exp(-|x|/\lambda_D) \tag{15.12}$$

At $x = 0$, we have $\phi = \phi_0$. Thus, $C = \phi_0$, and equation (15.12) becomes

$$\phi = \phi_0\ \exp(-|x|/\lambda_D)$$

The parameter λ_D is known as Debye length and is a measure of shielding distance or of the thickness of the sheath. Thus, the Debye length is the distance at which the potential drops by a factor e.

4. Dielectric constant of plasma

Suppose, plasma is subject to an oscillating electric field $\vec{E} = \vec{E_0}\ \mathrm{e}^{i\omega t}$, where ω is frequency of applied field. Due to this field, a current flows in plasma and when collisions of charged particles with neutral atoms are neglected, the equation of motion is

$$m\ \frac{\mathrm{d}\vec{v}}{\mathrm{d}t} = e\ \vec{E_0}\ \mathrm{e}^{i\omega t}$$

For non-relativistic motion, mass of particle remains constant and integration of this equation gives

$$\vec{v} = \frac{e\vec{E}_0}{mi\omega}\, e^{i\omega t} = \frac{e\vec{E}}{mi\omega} \tag{15.13}$$

For convenience, the constant of integration is assumed to be zero. If n is number of particles per unit volume, the current density is

$$\vec{J} = ne\vec{v} = \frac{ne^2\vec{E}}{m\ i\ \omega} \tag{15.14}$$

From Maxwell's equation, we have

$$\nabla \times \vec{B} = \mu_0\ \vec{J} + \frac{1}{c^2}\ \frac{d\vec{E}}{dt}$$

Using expression for $\vec{E}$ and equation (15.14) here, we get

$$\nabla \times \vec{B} = \frac{\mu_0 ne^2\vec{E}}{m\ i\ \omega} + \frac{i\omega\vec{E}}{c^2} = \frac{i\omega\vec{E}}{c^2}\left[1 - \frac{\omega_p^2}{\omega^2}\right] \tag{15.15}$$

where $\omega_p = \sqrt{ne^2/m\epsilon_0}$ is electron plasma frequency. If ϵ is permittivity, Maxwell's equation is (there is no current)

$$\nabla \times \vec{B} = \epsilon\mu\ \frac{d\vec{E}}{dt}$$

Using the expression for $\vec{E}$, we get

$$\nabla \times \vec{B} = \epsilon\ \mu\ i\ \omega\vec{E} \tag{15.16}$$

Since permeability of mediums is almost constant, from equations (15.15) and (15.16), the dielectric constant $\epsilon_r = \epsilon/\epsilon_0$ is

$$\epsilon_r = 1 - \frac{\omega_p^2}{\omega^2}$$

This equation shows that when $\omega_p < \omega$, the dielectric constant is positive and the wave can propagate through the medium. But when $\omega_p > \omega$, the dielectric constant is negative and the wave cannot propagate through the medium. The cut off takes place when $\omega_p = \omega$.

4.1 Refractive index of plasma

The refractive index n is equal to the square root of dielectric constant. For a wave having frequency ω, the refractive index n of plasma is

$$n = \sqrt{1 - \frac{\omega_p^2}{\omega^2}}$$

where ω_p is plasma frequency. When ω_p is smaller than ω, the radiation flows through plasma. As the refractive index is less than one, the velocity of radiation in plasma exceeds the speed of light c. This is the phase velocity of wave. When ω_p is larger than ω, the refractive index becomes imaginary and therefore the incident radiation is reflected back from plasma. When $\omega_p = \omega$, the refractive index is zero, which is true only in case of an ideal plasma where collisions are not considered.

5. Magnetic mirrors

Let us discuss the case when grad-B is along the direction of $\vec{B}$. Consider magnetic field which is primarily pointed in z-direction of cylindrical coordinates (ρ, ϕ, z) and its magnitude varies in z-direction (Figure 4). The field obviously does not depend on ϕ, and the component B_ϕ is zero. The density of field lines varies radially so that the field strength varies with ρ. For magnetic field $\vec{B}$, in cylindrical polar coordinates (ρ, ϕ, z), we have

$$\nabla \cdot \vec{B} = 0 \qquad \text{or} \qquad \frac{1}{\rho}\frac{\partial\,(\rho B_\rho)}{\partial \rho} + \frac{\partial B_z}{\partial z} = 0 \tag{15.17}$$

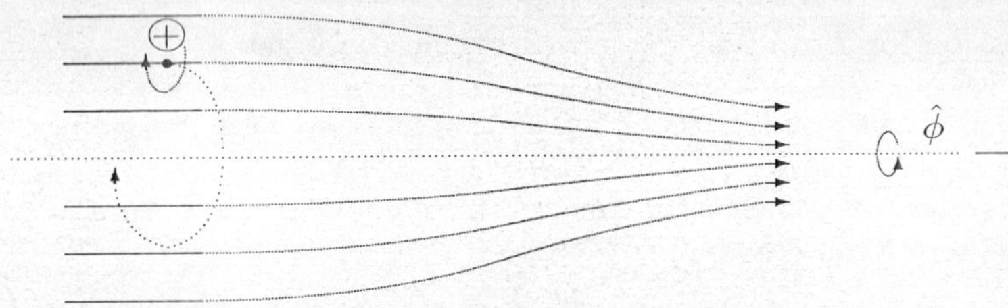

Figure 4: Magnetic field is primarily pointed in z-direction and the magnitude of field varies with ρ and z.

Suppose, the variation of the magnetic field is known on the z-axis (*i.e.*, $\partial B_z/\partial z$ is known at $\rho = 0$), and it does not vary much with ρ, then equation (15.17) can be integrated as

$$\rho B_\rho = -\int_0^\rho \rho \left[\frac{\partial B_z}{\partial z}\right]_{\rho=0} \mathrm{d}\rho = -\left[\frac{\partial B_z}{\partial z}\right]_{\rho=0} \int_0^\rho \rho\ \mathrm{d}\rho = -\left[\frac{\partial B_z}{\partial z}\right]_{\rho=0} \frac{\rho^2}{2}$$

Thus, we have

$$B_\rho = -\frac{\rho}{2}\left[\frac{\partial B_z}{\partial z}\right]_{\rho=0} \tag{15.18}$$

Hence, the variation of magnetic field with r creates grad-B

$$\nabla B = \frac{\partial B_\rho}{\partial \rho}\ \hat{\rho} = -\frac{1}{2}\left[\frac{\partial B_z}{\partial z}\right]_{\rho=0} \hat{\rho}$$

This ∇B generates drift of guiding center with velocity

$$\vec{v}_{\nabla B} = \pm\frac{1}{2}\ r_L v_\perp\ \frac{\vec{B}\times\nabla B}{B^2} = \mp\frac{1}{4}\ \frac{r_L v_\perp}{B}\left[\frac{\partial B_z}{\partial z}\right]_{\rho=0} \hat{\phi}$$

For the magnetic field $\vec{B} = B_\rho\ \hat{\rho} + B_z\ \hat{z}$ and velocity $\vec{v} = v_\rho\ \hat{\rho} + v_\phi\ \hat{\phi} + v_z\ \hat{z}$, the force is $\vec{F}(= q\vec{v}\times\vec{B})$. Expressing $\vec{F} = F_\rho\ \hat{\rho} + F_\phi\ \hat{\phi} + F_z\ \hat{z}$, we get

$$F_\rho = qv_\phi B_z \qquad\qquad F_\phi = qv_z B_\rho - qv_\rho B_z \qquad\qquad F_z = -qv_\phi B_\rho$$

The terms $qv_\phi B_z$ and $-qv_\rho B_z$ give rise to the usual Larmor gyration about z-direction. The term $qv_z B_\rho$ vanishes on the z-axis. When it does not vanish, it causes a drift with velocity

$$\vec{v_f} = \frac{1}{q}\frac{qv_z B_\rho\, \hat{\phi}\times\vec{B}}{B^2} = \frac{v_z B_\rho}{B}\,\hat{\rho}$$

Hence, the drift is in the radial direction and it makes the guiding centers to follow the field lines. The last term $-qv_\phi B_\rho$ is of great interest in the present context. Using equation (15.18), we have

$$F_z = -qv_\phi B_\rho = \frac{qv_\phi \rho}{2}\frac{\partial B_z}{\partial z}$$

Let us now compute average of F_z over one gyration. For simplicity, let us consider a particle whose guiding center lies on the z-axis. Then because of the symmetry about z-axis, v_ϕ is constant during gyration, and depending on the sign of charge, we have $v_\theta = \mp v_\perp$. Further, for the gyration, ρ is Larmor radius r_L. Thus, the average of F_z is

$$\langle F_z\rangle = \frac{q(\mp v_\perp)r_L}{2}\frac{\partial B_z}{\partial z} = \mp\frac{qv_\perp v_\perp}{2\omega_c}\frac{\partial B_z}{\partial z}$$

$$= \mp\frac{qv_\perp^2 m}{2|q|B}\frac{\partial B_z}{\partial z} = -\frac{mv_\perp^2}{2B}\frac{\partial B_z}{\partial z}$$

A gyrating particle creates a magnetic field, and we define the magnetic moment[3] of the particle as

$$\mu = \frac{1}{2}\frac{mv_\perp^2}{B} \tag{15.19}$$

so that

$$\langle F_z\rangle = -\mu\,\frac{\partial B_z}{\partial z}$$

The negative sign shows that it is an example of force acting on a diamagnetic particle. This force along the field lines can be written as

$$\vec{F_\parallel} = -\mu\,\frac{\partial B}{\partial s}\,\hat{s} = -\mu\,\nabla_\parallel B \tag{15.20}$$

where s is length measured along the field line. It is interesting to notice that as the particle moves in the regions of stronger and weaker magnetic field, the magnetic

[3]When a particle of charge $|q|$ makes $\omega_c/2\pi$ rounds per second in an orbit (loop) of radius r_L, we have current I and area of cross section A as

$$I = |q|\omega_c/2\pi \qquad\qquad A = \pi r_L^2 = \pi v_\perp^2/\omega_c^2$$

The magnetic moment μ of current loop is

$$\mu = AI = \frac{\pi v_\perp^2}{\omega_c^2}\frac{|q|\omega_c}{2\pi} = \frac{1}{2}\frac{v_\perp^2|q|}{\omega_c} = \frac{1}{2}\frac{mv_\perp^2}{B}$$

This shows that the definition of magnetic moment is the usual one.

moment μ remains constant, which can be verified as follows. Equation (15.20) can be expressed as

$$m\,\frac{dv_{\|}}{dt} = -\mu\,\frac{\partial B}{\partial s}$$

Multiplying this equation by $v_{\|}(= ds/dt)$, we have

$$mv_{\|}\,\frac{dv_{\|}}{dt} = -\mu\,\frac{\partial B}{\partial s}\frac{ds}{dt} \qquad \text{or} \qquad \frac{d}{dt}\left(\frac{1}{2}\,mv_{\|}^2\right) = -\mu\,\frac{dB}{dt} \tag{15.21}$$

For conservation of energy of particle, we have

$$\frac{d}{dt}\left(\frac{1}{2}\,mv_{\|}^2 + \frac{1}{2}\,mv_{\perp}^2\right) = 0 \tag{15.22}$$

Using equation (15.19) in (15.22), we have

$$\frac{d}{dt}\left(\frac{1}{2}\,mv_{\|}^2 + \mu B\right) = 0 \qquad \text{or} \qquad \frac{d}{dt}\left(\frac{1}{2}\,mv_{\|}^2\right) + \frac{d}{dt}(\mu\,B) = 0 \tag{15.23}$$

Using equation (15.21) in (15.23) we have

$$-\mu\,\frac{dB}{dt} + \frac{d}{dt}(\mu B) = 0 \qquad \text{or} \qquad B\,\frac{d\mu}{dt} = 0$$

Since the magnitude of magnetic field is not zero, we have

$$\frac{d\mu}{dt} = 0$$

showing that μ remains constant. As m and μ for a particle are constant, equation (15.19) gives

$$v_{\perp}^2 \propto B$$

Thus, when a particle moves from a weak-field region to a strong-field region, perpendicular component of the velocity $v_{\perp}$ increases fast as it has degree 2. For conservation of energy, with the increase of $v_{\perp}$, the parallel component of velocity $v_{\|}$ necessarily decreases fast. If B is high enough in the 'throat' of the field, the $v_{\|}$ becomes practically zero, and the particle is reflected back towards a weak-field region where $v_{\|}$ increases fact.

On the other side also there is another 'throat' where magnetic field is again high and the $v_{\|}$ becomes practically zero and the particle is reflected back. Thus, the particle moves back and forth between the two throats (called mirrors) of magnetic field, and is confined between the mirrors. Figure 5 shows a simple pair of coils producing nonuniform magnetic field which behaves like two magnetic mirrors between which plasma can be trapped. Obviously, a large value of $v_{\perp}$ is required for trapping of particles in the magnetic field. This effect works on both the ions and electrons.

This method of trapping plasma is however not perfect. For example for the particles with $v_{\perp} = 0$, there is no magnetic moment (equation 15.19) and it would not experience any magnetic force. Further, for example, at the mid-plane ($B = B_0$), the magnetic field is the lowest, and following equation (15.19), $v_\perp$ is the minimum and therefore $v_\parallel$ is the maximum. The particles can escape from these regions. Let us obtain a criterion for the escape of particles.

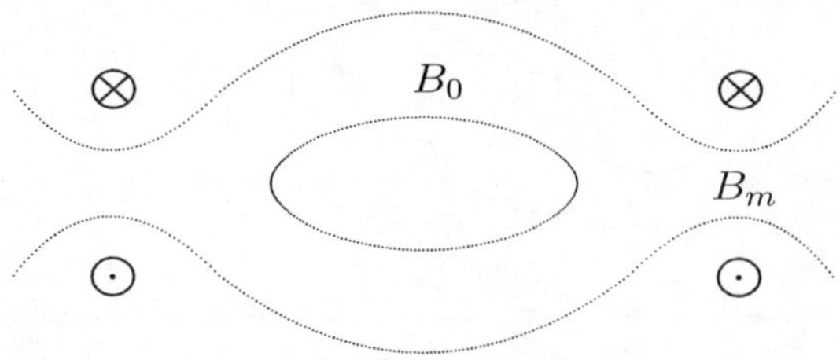

Figure 5: A simple pair of coils produces nonuniform magnetic field which behaves like two magnetic mirrors between which plasma may be trapped.

5.1 Criterion for escape of particles

Suppose, at the mid-plane ($B = B_0$) a particle has $v_\perp = v_{\perp 0}$ and $v_\parallel = v_{\parallel 0}$. When this particle reaches a turning point ($B = B_t$), let $v_\perp = v_{\perp t}$ and $v_\parallel = 0$. Then the invariance of magnetic moment μ gives

$$\frac{1}{2}\frac{mv_{\perp 0}^2}{B_0} = \frac{1}{2}\frac{mv_{\perp t}^2}{B_t} \tag{15.24}$$

Since in a magnetic field, there is no loss or gain of energy of a particle. Hence, the conservation of energy of a particle gives

$$\frac{1}{2}m(v_{\perp 0}^2 + v_{\parallel 0}^2) = \frac{1}{2}mv_{\perp t}^2 \qquad \text{or} \qquad v_{\perp t}^2 = v_{\perp 0}^2 + v_{\parallel 0}^2 = v_0^2 \tag{15.25}$$

where v_0 is magnitude of velocity of particle. From equations (15.24) and (15.25), we have

$$\frac{B_0}{B_t} = \frac{v_{\perp 0}^2}{v_{\perp t}^2} = \frac{v_{\perp 0}^2}{v_0^2} = \sin^2\theta \tag{15.26}$$

where θ is the pitch angle at the mid-plane. For a particle, smaller value of θ gives smaller value of $v_{\perp 0}$ and in turn smaller value of magnetic moment μ and thus the particle has larger probability of escaping away.

The particles which have low value of θ (low value of $v_\perp$) in the weak-field region have larger value of $v_\perp$ in the high-field region and is reflected back. If θ is too small, a high-field region does not have sufficiently high value of $v_\perp$ and therefore no reflection occurs at all.

At a throat replacing B_t by B_m (which is the maximum field strength) in equation (15.26), we have the smallest value of pitch angle θ_m at the mid-plane as

$$\sin^2\theta_m = \frac{B_0}{B_m} = \frac{1}{R_m}$$

where R_m is the ratio of magnetic field in a throat and at the mid-plane of the magnetic field and is known as the *mirror ratio.*

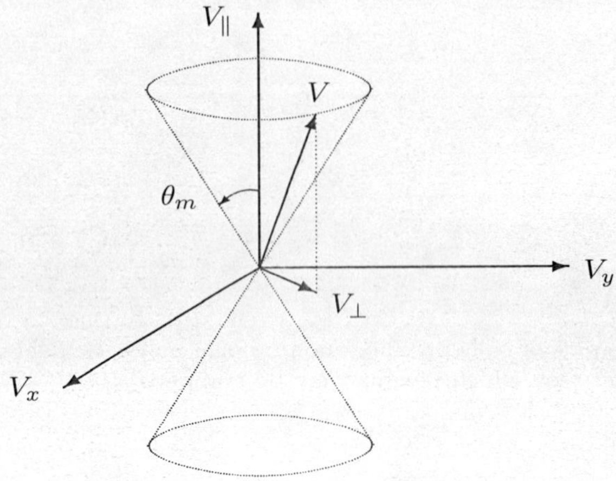

Figure 6: Velocity-space where θ_m is the minimum value of pitch angle at the mid-plane.

If we consider velocity-space as shown in Figure 6, θ_m is the angle of cone, called the 'loss-cone' in the space. This loss-cone is independent of mass m and charge q of particle and depends on the magnetic fields B_0 and B_m. The particles lying within the cone are not confined by the magnetic field and escape out. Electrons are generally lost more easily because they have a higher collision frequency.

The idea of magnetic mirror was first envisaged by Fermi as a mechanism for acceleration of cosmic rays. He proposed that the protons bouncing between the magnetic mirrors could gain energy in each bounce. Another example of the magnetic confinement of plasma are the van Allen belts around the Earth. The magnetic field is strong at the pole and weak at the equator and thus it makes a natural mirror with rather large R_m.

Review Questions

1. Describe plasma. Write the name of some places in nature where plasma is found.

2. Show that in Maxwellian distribution, the average kinetic energy per degree of freedom is $(KT/2)$, where K is Boltzmann constant and T the temperature. Discuss the concept of temperature. How a plasma can have more than one temperature simultaneously.

3. What is Debye shielding? Derive an expression for Debye length.

4. Describe dielectric constant of plasma and derive an expression for it. Discuss refractive index of plasma.

5. Describe a method for confinement of plasma with the help of magnetic field. Discuss criterion of escape of particles.

6. Write short notes on the following

   (i) Plasma in ionosphere
   (ii) Plasma in van Allen belts
   (iii) Plasma in aurorae
   (iv) Plasma in solar corona
   (v) Plasma in the core of Sun
   (vi) Debye length
   (vii) Debye shielding
   (viii) Dielectric constant of plasma
   (ix) Refractive index of plasma
   (x) Magnetic mirrors

Multiple-Choice Questions

1. In which of the following objects, natural plasma is not found?

   A. Ionosphere B. van Allen belts
   C. Solar corona D. Molecular cloud

   Ans. D

2. In which of the following objects, natural plasma is not found?

   A. Core of Sun B. van Allen belts
   C. Solar corona D. Molecular cloud

   Ans. D

3. An ideal plasma contains

   A. negatively charged electrons B. positively charged ions
   C. neutral particles D. All of the three

   Ans. D

4. At temperature T, for one-dimensional case, Maxwell-Boltzmann distribution function for velocity is expressed as

   A. $A\exp\left(-\dfrac{mu^2}{2KT}\right)$ B. $A\exp\left(-\dfrac{m^2u^2}{2KT}\right)$

   C. $A\exp\left(\dfrac{mu^2}{2KT}\right)$ D. $A\exp\left(\dfrac{m^2u^2}{2KT}\right)$

   Ans. A

5. Which of the following is the value of definite integral

$$\int_{-\infty}^{\infty} e^{-y^2} dy$$

A. $\sqrt{\pi}$ B. $\sqrt{\pi/2}$ C. π D. $\sqrt{2\pi}$

Ans. A

6. Which of the following is the value of definite integral

$$\int_{0}^{\infty} e^{-y^2} dy$$

A. $\sqrt{\pi}/2$ B. $\sqrt{\pi/2}$ C. π D. $\sqrt{2\pi}$

Ans. A

7. Expression for the Debye length is

A. $\left(\frac{KT_e}{4\pi ne^2}\right)^{1/2}$ B. $\left(\frac{KT_e}{4\pi ne^2}\right)^{-1/2}$ C. $\left(\frac{KT_e}{4\pi n^2e^2}\right)^{1/2}$ D. $\left(\frac{KT_e}{4\pi n^2e^2}\right)^{-1/2}$

Ans. A

8. Plasma frequency is ω_p. For a wave of frequency ω propagating through a dielectric medium, the dielectric constant is

A. $\left(1-\frac{\omega^2}{\omega_p^2}\right)$ B. $\left(1+\frac{\omega^2}{\omega_p^2}\right)$ C. $\left(1-\frac{\omega_p^2}{\omega^2}\right)$ D. $\left(1+\frac{\omega_p^2}{\omega^2}\right)$

Ans. C

9. If ω_p is plasma frequency of ionosphere, the ionosphere can be used for transmission of signal having frequency ω such that

A. $\omega < \omega_p$ B. $\omega = \omega_p$ C. $\omega > \omega_p$ D. $\omega = \omega_p^2$

Ans. A

10. If ω_p is plasma frequency of ionosphere, the ionosphere is opaque for the signal having frequency ω such that

A. $\omega < \omega_p$ B. $\omega = \omega_p$ C. $\omega > \omega_p$ D. $\omega = \omega_p^2$

Ans. A

11. If ω_p is plasma frequency of ionosphere, the ionosphere is transparent for signal having frequency ω when

A. $\omega < \omega_p$ B. $\omega = \omega_p$ C. $\omega > \omega_p$ D. $\omega = \sqrt{\omega_p}$

Ans. C

12. If ω_p is plasma frequency of ionosphere, a geo-stationary satellite is used for the transmission of signal having frequency ω such that

A. $\omega < \omega_p$ B. $\omega = \omega_p$ C. $\omega > \omega_p$ D. $\omega = \sqrt{\omega_p}$

Ans. C

13. If ω_p is plasma frequency of ionosphere, a geo-stationary satellite cannot be used for the transmission of signal having frequency ω such that

A. $\omega < \omega_p$ B. $\omega > \omega_p^2$ C. $\omega > \omega_p$ D. $\omega > \omega_p^3$

Ans. A

14. Plasma frequency is ω_p. The refractive index of plasma for a wave having frequency ω, such that $\omega > \omega_p$, is

A. real B. imaginary C. complex D . any of these

Ans. A

15. The shape of magnetic field used for confinement of plasma is

A. spherical B. paraboloidal C. cylindrical D. ellipsoidal

Ans. C

16. In magnetic confinement of plasma, $v_\perp$ and $v_\parallel$ denote the components of velocity in the directions perpendicular and parallel to the applied magnetic field B, respectively. The magnetic field B is proportional to

A. $v_\perp$ B. $v_\parallel$ C. $v_\perp^2$ D. $v_\parallel^2$

Ans. C

17. Suppose, $v_\perp$ and $v_\parallel$ denote the components of velocity in the directions perpendicular and parallel to applied magnetic field B, respectively. The magnetic moment is equal to

A. $mv_\perp^2/2B$ B. $mv_\parallel^2/2B$ C. $mv_\perp/2B$ D. $mv_\parallel/2B$

Ans. A

Chapter 16

Waves in Fluid Plasma

1. Phase velocity and group velocity

A wave traveling along positive z-direction may be expressed as

$$y = A\ \cos(kz - \omega t) \tag{16.1}$$

where A is amplitude, $\omega(= 2\pi\nu)$ is angular frequency and $k(= 2\pi/\lambda)$ is wave number. Here, ν is frequency and λ is wavelength of wave. **Phase velocity** of wave is defined as the ratio of angular frequency ω to wave number k. Thus, the phase velocity v_p is

$$v_p = \frac{\omega}{k}$$

The quantity $(kz - \omega t)$ in equation (16.1) is known as the phase of wave motion. The particle of constant phase travels such that $(kz - \omega t)$ = constant with time. Thus, we have

$$\frac{\mathrm{d}}{\mathrm{d}t}(kz - \omega t) = 0 \qquad \text{or} \qquad k\,\frac{\mathrm{d}z}{\mathrm{d}t} - \omega = 0$$

Therefore, we get

$$\frac{\mathrm{d}z}{\mathrm{d}t} = \frac{\omega}{k} = v_p$$

Hence, the phase velocity of a wave is the same as the velocity of wave. We can thus write phase velocity

$$v_p = \nu\lambda \tag{16.2}$$

For electromagnetic wave

$$E = h\nu \qquad \text{or} \qquad \nu = \frac{E}{h} = \frac{mc^2}{h}$$

Here, we have used the energy-mass equivalence relation, $E = mc^2$. According to de Broglie hypothesis, we have

$$\lambda = \frac{h}{p} = \frac{h}{mv}$$

where v is velocity of particle. Using expressions for ν and λ in equation (16.2), we get

$$v_p = \frac{mc^2}{h}\frac{h}{mv} = \frac{c^2}{v} \tag{16.3}$$

Since the velocity of particle is always smaller than the speed of light ($v < c$), the phase velocity is always larger than the speed of light. Therefore, the phase velocity does not correspond to the velocity of a particle. This difficulty of phase velocity being larger than the speed of light can be overcome by assuming that each moving particle is associated with a group of waves or a wave packet rather than a single wave. Thus, a particle is represented by a wave packet and we have group velocity associated with a particle.

To understand the concept of group velocity, let us consider a combination of two waves having equal amplitude A as

$$y_1 = A\ \cos(k_1 z - \omega_1 t) \qquad \text{and} \qquad y_2 = A\ \cos(k_2 z - \omega_2 t)$$

where k_1 and k_2 are wave numbers, and ω_1 and ω_2 are angular frequencies of waves. The phase velocities of waves are ω_1/k_1 and ω_2/k_2, respectively.

The superposition of waves gives

$$y = y_1 + y_2 = A\ \cos(k_1 z - \omega_1 t) + A\ \cos(k_2 z - \omega_2 t)$$

Using trigonometric relation

$$\cos C + \cos D = 2\cos\left[\frac{C-D}{2}\right]\ \cos\left[\frac{C+D}{2}\right]$$

we have

$$y = 2A\ \cos\left[\frac{(k_1 z - \omega_1 t) - (k_2 z - \omega_2 t)}{2}\right]\ \cos\left[\frac{(k_1 z - \omega_1 t) + (k_2 z - \omega_2 t)}{2}\right]$$

$$= 2A\ \cos\left[\frac{\Delta k}{2}\,z - \frac{\Delta\omega}{2}\,t\right]\ \cos(kz - \omega t) \tag{16.4}$$

where we have

$$\omega = \frac{\omega_1 + \omega_2}{2} \qquad k = \frac{k_1 + k_2}{2} \qquad \Delta\omega = (\omega_1 - \omega_2) \qquad \Delta k = (k_1 - k_2)$$

The resultant wave (16.4) has two segments:

(i) A wave having angular frequency ω and wave number k so that its velocity (phase velocity) is

$$v_p = \frac{\omega}{k}$$

(ii) Another wave having angular frequency $\Delta\omega/2$ and wave number $\Delta k/2$, and corresponding velocity is

$$v_g = \frac{\Delta\omega/2}{\Delta k/2} = \frac{\Delta\omega}{\Delta k} = \frac{\mathrm{d}\omega}{\mathrm{d}k}$$

The v_g is known as the **group velocity**.

1.1 Relation between v_p and v_g

The phase velocity v_p and group velocity v_g are expressed as

$$v_p = \frac{\omega}{k} \qquad \text{and} \qquad v_g = \frac{d\omega}{dk}$$

Now, we have

$$v_g = \frac{d\omega}{dk} = \frac{d(v_p\, k)}{dk} = v_p + k\,\frac{dv_p}{dk} = v_p + k\,\frac{dv_p}{d\lambda}\frac{d\lambda}{dk}$$

We know

$$\lambda = \frac{2\pi}{k} \qquad \text{and} \qquad \frac{d\lambda}{dk} = -\frac{2\pi}{k^2}$$

Thus, we have

$$v_g = v_p + k\,\frac{dv_p}{d\lambda}\left(-\frac{2\pi}{k^2}\right) = v_p - \frac{2\pi}{k}\frac{dv_p}{d\lambda} = v_p - \lambda\,\frac{dv_p}{d\lambda}$$

It is the required relation.

Example 1. Show that group velocity v_g of a wave-packet is equal to the velocity of a material particle. Derive the relation $v_g v_p = c^2$, where v_p is phase velocity of a wave in the wave-packet and c is speed of light.

Solution: For the relations

$$E = \hbar\omega \qquad \text{and} \qquad p = \hbar k$$

we have

$$dE = \hbar\, d\omega \qquad \text{and} \qquad dp = \hbar dk$$

group velocity

$$v_g = \frac{d\omega}{dk} = \frac{dE}{dp}$$

(i) For a free non-relativistic particle, potential energy is zero and we have

$$E = \frac{p^2}{2m} \qquad \text{and thus,} \qquad \frac{dE}{dp} = \frac{p}{m} = v$$

(ii) For a relativistic particle, we have[1]

$$E^2 = c^2p^2 + m_0^2c^4 \qquad \text{and thus,} \qquad \frac{dE}{dp} = \frac{c^2 p}{E} = \frac{c^2 mv}{mc^2} = v$$

---

[1]We have

$$E = mc^2 \qquad \text{and} \qquad p = mc$$

Therefore,

$$E^2 - p^2c^2 = m^2c^4 - c^2m^2c^2 = m^2c^4\left(1 - \frac{c^2}{c^2}\right)$$

Using the relation

$$m = \frac{m_0}{\sqrt{1 - c^2/c^2}}$$

we have

$$E^2 - p^2c^2 = m_0^2c^4$$

This shows that velocity of a particle and group velocity of corresponding wave packet are the same, *i.e.*, $v_g = v$. The phase velocity expressed as

$$v_p = \frac{\omega}{k} = \frac{E}{p} = \frac{mc^2}{mv} = \frac{c^2}{v}$$

has no physical significance and it is not a measurable quantity. The product of group velocity and phase velocity is

$$v_g v_p = v\,\frac{c^2}{v} = c^2$$

2. Fluid equations

As a simple case, let us consider plasma comprising two species: positive ions and electrons. For more species, the extension of following equations is straight forward. When ions and electrons in a plasma have charges q_i and q_e, particle densities n_i and n_e, and are moving with velocities $\vec{v_i}$ and $\vec{v_e}$, respectively, the charge density ρ and current density $\vec{J}$ are

$$\rho = n_i q_i + n_e q_e \qquad \text{and} \qquad \vec{J} = n_i q_i\,\vec{u_i} + n_e q_e\,\vec{u_e}$$

In the plasma (fluid), we have a large number of particles. Hence, we no more consider motion of individual particle and therefore, we use the velocity $\vec{u}$ for fluid instead of $\vec{v}$ for individual particle. When we neglect collisions (*i.e.*, fully ionized plasma)[2] and viscosity (*i.e.*, isotropic plasma), a set of equations is as the following.

$$\nabla \cdot \vec{E} = \frac{1}{\epsilon_0}\,(n_i q_i + n_e q_e) \tag{16.5}$$

$$\nabla \times \vec{E} = -\frac{\partial \vec{B}}{\partial t} \tag{16.6}$$

$$\nabla \cdot \vec{B} = 0 \tag{16.7}$$

$$\nabla \times \vec{B} = \frac{1}{c^2}\frac{\partial \vec{E}}{\partial t} + \mu_0(n_i q_i \vec{u_i} + n_e q_e \vec{u_e}) \tag{16.8}$$

$$m_j n_j \Big[\frac{\partial \vec{u_j}}{\partial t} + (\vec{u_j}\cdot\nabla)\vec{u_j}\Big] = q_j n_j\,[\vec{E} + \vec{u_j}\times\vec{B}] - \nabla p_j \tag{16.9}$$

$$\frac{\partial n_j}{\partial t} + \nabla\cdot(n_j\,\vec{u_j}) = 0 \tag{16.10}$$

$$p_j = C_j n_j^{\gamma_j} \tag{16.11}$$

[2] As the collisions between charged particles and neutral particles are quite frequent whereas the collisions between charged particles themselves are rare.

where $c = 1/\sqrt{\epsilon_0 \mu_0}$ is the velocity of electromagnetic wave propagating through a medium having permittivity ϵ_0 and permeability μ_0, and j denotes the ion i or electron e.

The equations (16.5), (16.6), (16.7), (16.8) are Maxwell's equations. In a plasma, the ions and electrons comprising the plasma are the equivalent of the **bound** charges and currents. Since these charges move in a complicated way, it is impractical to try to lump their effects into two constants, ϵ and μ. Consequently, in the plasma physics, one generally works with the vacuum equations, in which ρ and $\vec{J}$ include all the charges and currents, both external and internal.

Equation (16.9) is the equation of motion where electromagnetic field is included and p denotes the pressure. Equations (16.10) (16.11) are the continuity equation and equation of state, respectively. Equations (16.9), (16.10), (16.11) depend on the species (ion or electron). There are 16 unknowns: n_i, n_e, p_i, p_e, $\vec{u_i}$, $\vec{u_e}$, $\vec{E}$ and $\vec{H}$ and 18 equations when we account each vector equation as three scalar equations. Two Maxwell's equations are superfluous as (16.5) and (16.7) can be obtained from the divergence of (16.8) and (16.6).[3] A simultaneous solution of 16 equations in 16 unknowns gives a self-consistent values of fields and motions in the fluid approximation.

2.1 Equation of state of plasma

A fluid where the exchange of energy from the outide is not allowed, the pressure p and density ρ of the fluid are related through the equation of state

$$p = C\rho^\gamma \tag{16.12}$$

where C is constant and γ the ratio of specific heats C_p/C_v of fluid. For the particle density n, equation (16.12) can be written as

$$p = C(mn)^\gamma = C_1\, n^\gamma \tag{16.13}$$

[3]Taking divergence of equation (16.6), we have

$$\nabla \,.\, (\nabla \times \vec{E}) = 0 = -\nabla \,.\, \frac{\partial \vec{B}}{\partial t} \qquad\qquad \frac{\partial}{\partial t}(\nabla \,.\, \vec{B}) = 0$$

Thus, $\nabla \,.\, \vec{B} = 0$ when it is initially zero. This is equation (16.7). Taking divergence of equation (16.8), we have

$$\nabla \,.\, (\nabla \times \vec{B}) = 0 = \mu_0[q_i \nabla \,.\, (n_i \vec{u_i}) + q_e \nabla \,.\, (n_e \vec{u_e}) + \frac{1}{c^2}\, \nabla \,.\, \frac{\partial \vec{E}}{\partial t}$$

Now, for the equation of continuity, we have

$$\nabla \,.\, (n_i \vec{u_i}) = -\frac{\partial n_i}{\partial t} \qquad\qquad \nabla \,.\, (n_e \vec{u_e}) = -\frac{\partial n_e}{\partial t}$$

Using these relations here, we get

$$\mu_0\left[-q_i \frac{\partial n_i}{\partial t} - q_e \frac{\partial n_e}{\partial t}\right] + \frac{1}{v^2}\nabla \,.\, \frac{\partial \vec{E}}{\partial t} = 0 \qquad\qquad \frac{\partial}{\partial t}\left[\nabla \,.\, \vec{E} - \mu_0\, c^2 (n_i q_i + n_e q_e)\right] = 0$$

Thus, $\nabla \,.\, \vec{E} - (n_i q_i + n_e q_e)/\epsilon_0 = 0$ when it is initially zero. This is equation (16.5).

where m is mass of a particle, and $C_1 (= Cm^\gamma)$ is constant. Thus, we have

$$\nabla p = C_1 \ \gamma \ n^{\gamma-1} \ \nabla n$$

and therefore,

$$\frac{\nabla p}{p} = \gamma \ \frac{\nabla n}{n} \tag{16.14}$$

For an isothermal system, we have $p = nKT$. Thus,

$$\nabla p = \nabla(nKT) = KT \ \nabla n$$

and therefore,

$$\frac{\nabla p}{p} = \frac{\nabla n}{n} \tag{16.15}$$

Comparison of equation (16.15) with (16.14) tells that γ is 1 for an isothermal fluid. For an adiabatic compression, T also changes and therefore, the value of γ is larger than 1. For a system with N degrees of freedom, the γ is

$$\gamma = 1 + \frac{2}{N}$$

For a fluid, we have $\nabla p = \gamma \ KT \ \nabla n$ and the equation of motion is expressed as

$$mn\Big[\frac{\partial \vec{u}}{\partial t} + (\vec{u} \ . \ \nabla)\vec{u}\Big] = qn \ [\vec{E} + \vec{u} \times \vec{B}] - \gamma KT \nabla n$$

3. Linearisation of equations

Consider a plasma where a particle has mass m, charge q and average velocity is u in an electric field $\vec{E}$ and magnetic field $\vec{B}$. Let the plasma be homogeneous and isotropic with the particle density n and kinetic temperature T. We have given a set of equations. This set of equations can be solved with the help of the process of linearisation of equations.

Here, we have discussed the procedure of linearisation of some equations.

(a) Equation of motion of particle

$$mn\Big[\frac{\partial \vec{u}}{\partial t} + (\vec{u} \ . \ \nabla)\vec{u}\Big] = qn \ [\vec{E} + \vec{u} \times \vec{B}] - \gamma KT \nabla n \tag{16.16}$$

where K is Boltzmann constant, and γ the ratio of specific heats at constant pressure and constant volume.

(b) Equation of continuity

$$\frac{\partial n}{\partial t} + \nabla \ . \ (n\vec{u}) = 0 \tag{16.17}$$

(c) Poisson equation

$$\nabla \ . \ \vec{E} = \frac{e(n_i - n_e)}{\epsilon_0} \tag{16.18}$$

where n_i and n_e are densities of positive ions and electrons, respectively, and an ion (proton) and an electron carry equal magnitude e of charge.

The procedure of linearisation is applicable only when the amplitude of each oscillating quantity is small, so that the terms having square or higher degrees of amplitude can be neglected. For linearisation, the first step is to separate the variable parameters into two parts: (i) an equilibrium part, which represents the state of matter in absence of oscillations and is indicated by the subscript 0, and (ii) the perturbation part due to oscillations, indicated by the subscript 1:

$$n = n_0 + n_1 \qquad \vec{u} = \vec{u_0} + \vec{u_1} \qquad \phi = \phi_0 + \phi_1$$

$$\vec{E} = \vec{E_0} + \vec{E_1} \qquad \vec{B} = \vec{B_0} + \vec{B_1} \tag{16.19}$$

Before perturbations, the plasma is homogeneous and at rest, so that

$$\nabla n_0 = \nabla \phi_0 = \vec{u_0} = \vec{E_0} = 0 \qquad \frac{\partial n_0}{\partial t} = 0 \tag{16.20}$$

3.1 Linearisation of equation of motion

In equilibrium, we have

$$mn_0 \left[\frac{\partial \vec{u}_0}{\partial t} + (\vec{u}_0 \, . \, \nabla)\vec{u}_0 \right] = qn_0 \, [\vec{E_0} + \vec{u}_0 \times \vec{B_0}] - \gamma KT \nabla n_0$$

Using equations (16.19) and (16.20) in (16.16) we get

$$m(n_0 + n_1)\left[\frac{\partial}{\partial t}(\vec{u_0} + \vec{u_1}) + \left\{ (\vec{u_0} + \vec{u_1}) \, . \, \nabla \right\} (\vec{u_0} + \vec{u_1}) \right]$$

$$= q(n_0 + n_1)[(\vec{E_0} + \vec{E_1}) + (\vec{u_0} + \vec{u_1}) \times (\vec{B_0} + \vec{B_1})] - \gamma KT \nabla (n_0 + n_1)$$

so that

$$m(n_0 + n_1)\left[\frac{\partial \vec{u_1}}{\partial t} + (\vec{u_1} \, . \, \nabla)\vec{u_1} \right] = q(n_0 + n_1)[\vec{E_1} + \vec{u_1} \times (\vec{B_0} + \vec{B_1})]$$

$$-\gamma KT \nabla n_1 \tag{16.21}$$

Neglecting the square and higher degree terms in equation (16.21), we have

$$mn_0 \frac{\partial \vec{u_1}}{\partial t} = qn_0(\vec{E_1} + \vec{u_1} \times \vec{B_0}) - \gamma KT \nabla n_1$$

3.2 Linearisation of equation of continuity

In equilibrium, we have

$$\frac{\partial n_0}{\partial t} + \nabla \, . \, (n_0 \vec{u}_0) = 0$$

Using equations (16.19) and (16.20) in (16.17) we get

$$\frac{\partial}{\partial t}(n_0 + n_1) + \nabla \, . \, [(n_0 + n_1)(\vec{u_0} + \vec{u_1})] = 0 \qquad \frac{\partial n_1}{\partial t} + n_0 \nabla \, . \, \vec{u_1} = 0$$

3.3 Linearisation of Poisson equation

We have

$$\nabla \,.\, \vec{E} = \frac{e(n_i - n_e)}{\epsilon_0}$$

Using equations (16.19) and (16.20) in (16.18) we get

$$\nabla \,.\, (\vec{E_0} + \vec{E_1}) = \frac{e[n_{i0} - (n_{e0} + n_{e1})]}{\epsilon_0} \qquad\qquad \nabla \,.\, \vec{E_1} = -\frac{en_{e1}}{\epsilon_0}$$

where we have used $n_i = n_{i0}$ as ions are stationary, $n_e = n_{e0} + n_{e1}$, and $n_{i0} = n_{e0}$ (as in the equilibrium, the neutrality is maintained).

3.4 Linearisation of electron current density

Electron current density is expressed as

$$\vec{J} = -n_e e \vec{u}_e$$

Expressing $\vec{J} = \vec{J_0} + \vec{J_1}$, $\vec{u}_e = \vec{u}_{e0} + \vec{u}_{e1}$, $n_e = n_{e0} + n_{e1}$ and consider that for plasma, $\vec{u}_{e0} = 0$, linearized form of the equation of current density is

$$\vec{J_1} = -n_{e0} e \vec{u}_{e1}$$

3.5 Perturbed quantities

Perturbed quantities are considered oscillating, behaving sinusoidally as $e^{i(\vec{k}\,.\,\vec{r}-\omega t)}$. Then, the time derivative $(\partial/\partial t)$ can be replaced by $-i\omega$, and the gradient ∇ by $i\vec{k}$.

$$\frac{\partial}{\partial t}\, e^{i(\vec{k}\,.\,\vec{r}-\omega t)} = -i\omega\; e^{i(\vec{k}\,.\,\vec{r}-\omega t)} \qquad\qquad \frac{\partial}{\partial t} \longrightarrow -i\omega$$

We have $\vec{k} = k_x\;\hat{x} + k_y\;\hat{y} + k_z\;\hat{z}$ and $\vec{r} = x\;\hat{x} + y\;\hat{y} + z\;\hat{z}$, so that $\vec{k}\;.\;\vec{r} = k_x\;x + k_y\;y + k_z\;z$.

$$\nabla e^{i(\vec{k}\,.\,\vec{r}-\omega t)} = \Big(\frac{\partial}{\partial x}\;\hat{x} + \frac{\partial}{\partial y}\;\hat{y} + \frac{\partial}{\partial z}\;\hat{z}\Big) e^{i(k_x x + k_y y + k_z z - \omega t)}$$

$$= i\Big(k_x\;\hat{x} + k_y\;\hat{y} + k_z\;\hat{z}\Big) e^{i(k_x x + k_y y + k_z z - \omega t)}$$

$$= i\vec{k}\; e^{i(k_x x + k_y y + k_z z - \omega t)}$$

$$\nabla \longrightarrow i\vec{k} \qquad\qquad \frac{\partial}{\partial x} \longrightarrow ik_x$$

4. Plasma oscillations

In equilibrium position, charged particles in plasma are uniformly distributed in such a manner that neutrality is maintained everywhere. When the electrons in plasma are displaced relative to uniform background of ions, an electric field is developed in such a direction that it tries to pull the electrons back to restore neutrality. While returning back, owing to inertia, the electrons overshoot and the electric field is developed in reverse direction which tries to pull back the electrons to their positions of equilibrium. Thus, the electrons oscillate about their equilibrium positions with a characteristic frequency, known as the **plasma frequency**.

Because of low mass of electrons, these oscillations are so fast that massive ions are not capable to respond the oscillating field, generated by the oscillations of electrons. Hence, the ions may be treated as fixed. Expression for plasma frequency ω_p can be derived in a simple manner using following assumptions:

(i) The plasma has a quite large extent.

(ii) There is no magnetic field, so that from Maxwell's equation, we have

$$\nabla \times \vec{E} = -\frac{\partial \vec{B}}{\partial t} = 0 \qquad \text{and hence,} \qquad \vec{E} = -\nabla\phi$$

where ϕ is scalar quantity.

(iii) The ions are fixed and form uniform background in space.

(iv) There are no thermal motions.

(v) Due to electric field generated in plasma, motions of electrons are along one direction, say, the x direction, so that we have

$$\nabla = \frac{\partial}{\partial x}\,\hat{x} \qquad \text{and} \qquad \vec{E} = E\,\hat{x}$$

The equations of motion and continuity for electron are

$$m_e n_e \Big[\frac{\partial \vec{u_e}}{\partial t} + (\vec{u_e}\,.\,\nabla)\vec{u_e}\Big] = -en_e\vec{E} \tag{16.22}$$

and

$$\frac{\partial n_e}{\partial t} + \nabla\,.\,(n_e\vec{u_e}) = 0 \tag{16.23}$$

Here, we are assuming collision-less, homogeneous and isotropic plasma. These are high frequency oscillations, generated due to deviation from neutrality and therefore, we can use Poisson equation for one-dimensional motion

$$\frac{\partial E}{\partial x} = \frac{e(n_i - n_e)}{\epsilon_0} \tag{16.24}$$

where n_i and n_e are densities of ions and electrons, respectively. When amplitude of each oscillating quantity is small, the set of equations (16.22) – (16.24) can be solved by using the procedure of linearisation, where the terms having square and higher powers of amplitude are neglected. For this, let us first separate the variable parameters into two parts: (i) an equilibrium part, which represents the state of plasma in absence of oscillations, and is indicated by subscript 0 and (ii) perturbation part due to oscillations, indicated by subscript 1:

$$n_e = n_0 + n_1 \qquad \vec{u_e} = \vec{u_0} + \vec{u_1} \qquad \vec{E} = \vec{E_0} + \vec{E_1} \tag{16.25}$$

Before displacement of electrons, the plasma is homogeneous and at rest, so that

$$\nabla n_0 = \vec{u_0} = \vec{E_0} = 0 \qquad \text{and} \qquad \frac{\partial n_0}{\partial t} = 0 \tag{16.26}$$

Using equations (16.25) and (16.26) in (16.22) - (16.24) and on linearisation, we get

$$m_e \frac{\partial \vec{u_1}}{\partial t} = -e\vec{E_1}; \qquad \frac{\partial n_1}{\partial t} + n_0 \nabla \, . \, \vec{u_1} = 0; \qquad \frac{\partial E_1}{\partial x} = -\frac{en_1}{\epsilon_0} \tag{16.27}$$

For convenience, we consider one-dimensional case and assume that oscillating quantities behave sinusoidally as $e^{i(kx-\omega t)}$. Then time derivative $(\partial/\partial t)$ can be replaced by $-i\omega$, and gradient ∇ by ik. From equations (16.27), we get

$$-im_e\omega u_1 = -eE_1 \tag{16.28}$$

$$-i\omega n_1 = -n_0 iku_1 \tag{16.29}$$

$$ikE_1 = -\frac{en_1}{\epsilon_0} \tag{16.30}$$

Equations (16.28) – (16.30) can be rearranged as

$$-im_e\omega u_1 = -e\left(\frac{-e}{ik\epsilon_0}\right)n_1 = \left(\frac{e^2}{ik\epsilon_0}\right)\left(\frac{n_0 k}{\omega}\right)u_1 = -i\left(\frac{n_0 e^2}{\omega\epsilon_0}\right)u_1$$

so that

$$\left(m_e\omega - \frac{n_0 e^2}{\omega\epsilon_0}\right)u_1 = 0$$

Since u_1 is not zero, we have

$$\omega^2 = \frac{n_0 e^2}{m_e\epsilon_0}$$

Thus, plasma frequency ω_p is

$$\omega_p = \left(\frac{n_0 e^2}{m_e\epsilon_0}\right)^{1/2} \tag{16.31}$$

The plasma frequency depending on plasma density is also a fundamental parameter for plasma. Since the mass of electron is very low, plasma frequency is very high. The frequency f_p is

$$f_p = \frac{\omega_p}{2\pi} = \left(\frac{n_0 e^2}{4\pi^2 m_e\epsilon_0}\right)^{1/2}$$

In the SI system, charge of electron $e = 1.6 \times 10^{-19}$ C, mass of electron $m_e = 9.1 \times 10^{-31}$ kg, and $(1/4\pi\epsilon_0) = 8.99 \times 10^9$ N m^2/C^2. Thus, we have

$$f_p = \frac{1.6 \times 10^{-19}\sqrt{8.99 \times 10^9}}{\sqrt{\pi}\; 9.1 \times 10^{-31}}\sqrt{n_0} = 8.975\ \sqrt{n_0}\ \text{Hz}$$

where n_0 is electron density per m^3. For plasma having electron density $n_0 = 10^{18}$ m^{-3}, we have

$$f_p = 8.975\left(10^{18}\right)^{1/2} = 9 \times 10^9\ \text{Hz} = 9\ \text{GHz}$$

Hence, the plasma frequency generally lies in the microwave region. Equation (16.31) shows that ω_p does not depend on k, and therefore, the group velocity $d\omega_p/dk$ is zero. Thus, no information transmits through these waves from one region to another.

Electromagnetic radiation of frequency smaller than plasma frequency ($\omega < \omega_p$), on incidence on plasma are reflected back, whereas the radiation of frequency larger than plasma frequency ($\omega > \omega_p$) transmit through the plasma. This property of plasma in ionosphere around the earth has been exploited for communication purpose. Obviously, in order to communicate with geostationary satellites, frequency of signal must be larger than the plasma frequency of ionosphere.

5. Electron plasma waves

In the above, we have assumed no thermal motions and found that electrons oscillate about their equilibrium positions with frequency ω_p which is independent of k. Thus, the group velocity is zero, and hence, there is no transmission of information from one place to another. When thermal motions are included, they provide propagation of plasma oscillations. Electrons moving with their thermal velocities into adjacent layers of plasma can carry information regarding the happening in the previous oscillating region.

Propagation of such information is known as **electron plasma waves**. Such waves can be obtained through inclusion of term $-\nabla p_e$, due to thermal motions of electrons, in the equation of motion. If we take all assumptions of preceding section except the assumption (iv), (*i.e.*, here we include thermal motions), the equations of motion and continuity for electron are

$$m_e n_e\left[\frac{\partial \vec{u_e}}{\partial t} + (\vec{u_e}.\nabla)\vec{u_e}\right] = -en_e\vec{E} - \nabla p_e \tag{16.32}$$

and

$$\frac{\partial n_e}{\partial t} + \nabla.\ (n_e\vec{u_e}) = 0 \tag{16.33}$$

Here, we are assuming collision-less, homogeneous and isotropic plasma. These are high frequency oscillations which are generated due to deviation from neutrality and therefore, we can use Poisson equation for one-dimensional motion

$$\frac{\partial E}{\partial x} = \frac{e(n_i - n_e)}{\epsilon_0} \tag{16.34}$$

where n_i and n_e are densities of ions and electrons, respectively. When amplitude of each oscillating quantity is small, the set of equations (16.32) – (16.34) can be solved by using the procedure of linearisation, where the terms having square and higher powers of amplitude are neglected. For this, let us first separate variable parameters into two parts (one-dimensional case): (i) an equilibrium part, which represents the state of plasma in absence of oscillations, and is indicated by subscript 0 and (ii) perturbation part due to oscillations, indicated by subscript 1:

$$n_e = n_0 + n_1 \qquad u_e = u_0 + u_1 \qquad E = E_0 + E_1 \tag{16.35}$$

Before displacement of electrons, plasma is homogeneous and at rest, so that

$$\frac{\partial n_0}{\partial x} = u_0 = E_0 = 0 \qquad \text{and} \qquad \frac{\partial n_0}{\partial t} = 0 \tag{16.36}$$

The term ∇p_e for one-dimensional space can be written as

$$\frac{\partial p_e}{\partial x} = \frac{\gamma p_e}{n_e}\frac{\partial n_e}{\partial x} = \frac{\gamma n_e K T_e}{n_e}\frac{\partial n_e}{\partial x}$$

$$= \gamma K T_e \frac{\partial}{\partial x}(n_0 + n_1) = \gamma K T_e \frac{\partial n_1}{\partial x} \tag{16.37}$$

For one-dimensional space we have

$$\gamma = 1 + \frac{2}{N} = 1 + \frac{2}{1} = 3$$

Using equations (16.35) – (16.37) in (16.32) – (16.34) and on linearisation, we get

$$m_e n_0 \frac{\partial u_1}{\partial t} = -e n_0 E_1 - 3KT_e \frac{\partial n_1}{\partial x} \tag{16.38}$$

$$\frac{\partial n_1}{\partial t} + n_0 \frac{\partial u_1}{\partial x} = 0 \tag{16.39}$$

$$\frac{\partial E_1}{\partial x} = -\frac{e n_1}{\epsilon_0} \tag{16.40}$$

For convenience, we consider one-dimensional case and assume that oscillating quantities behave sinusoidally as $e^{i(kx-\omega t)}$. Then time derivative $(\partial/\partial t)$ can be replaced by $-i\omega$, and gradient $\partial/\partial x$ by ik. From the equations (16.38) – (16.40), we get

$$-i m_e n_0 \omega u_1 = -e n_0 E_1 - 3KT_e i k n_1 \tag{16.41}$$

$$-i\omega n_1 = -n_0 i k u_1 \tag{16.42}$$

$$ikE_1 = -\frac{e n_1}{\epsilon_0} \tag{16.43}$$

Equations (16.41) – (16.43) can be rearranged as

$$-im_e n_0 \omega u_1 = -en_0\left(\frac{-e}{ik\epsilon_0}\right)n_1 - 3KT_e ikn_1$$

$$= -i\left(\frac{n_0 e^2}{k\epsilon_0} + 3KT_e k\right)n_1 = -i\left(\frac{n_0 e^2}{k\epsilon_0} + 3KT_e k\right)\frac{n_0 k}{\omega}\; u_1$$

so that

$$\left[\omega^2 - \frac{n_0 e^2}{m_e \epsilon_0} - \frac{3KT_e k^2}{m_e}\right]u_1 = 0$$

Since u_1 is not zero, we have

$$\omega^2 = \frac{n_0 e^2}{m_e\epsilon_0} + \frac{3KT_e k^2}{m_e} = \omega_p^2 + \frac{3}{2}\; k^2 u_{th}^2$$

where $\omega_p (= \sqrt{n_0 e^2/m_e\epsilon_0})$ is the plasma frequency and $u_{th} (= \sqrt{2KT_e/m_e})$ is thermal velocity in one-dimensional space. Figure 1 shows a plot of the dispersion relation – a variation of ω versus k.

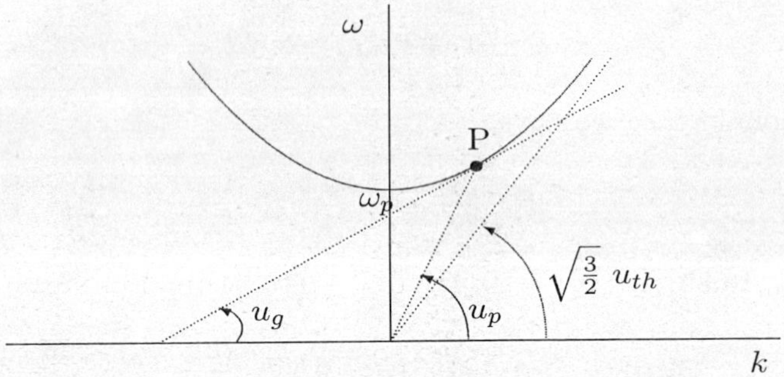

Figure 1: Variation of ω versus k for electron plasma waves

At a point P on the curve, the slope of line joining the point and origin gives the phase velocity u_p, whereas tangent to the curve at P gives the group velocity u_g. We can write

$$\frac{\omega^2}{k^2} = \frac{\omega_p^2}{k^2} + \frac{3}{2}\; u_{th}^2 \qquad \text{or} \qquad \frac{\omega}{k} = \sqrt{\frac{3}{2}}\; u_{th}\left(1 + \frac{2\;\omega_p^2}{3k^2 u_{th}^2}\right)^{1/2} \equiv u_p$$

It shows that the phase velocity u_p is never smaller than $\sqrt{3/2}\; u_{th}$. When $k \to \infty$, the phase velocity $u_p \to \sqrt{3/2}\; u_{th}$. The group velocity u_g is obtained in the following manner

$$2\omega\; \mathrm{d}\omega = \frac{3}{2}\; u_{th}^2 2k\; \mathrm{d}k$$

Hence, the group velocity is

$$u_g = \frac{\mathrm{d}\omega}{\mathrm{d}k} = \frac{3}{2}\frac{k}{\omega}u_{th}^2 = \sqrt{\frac{3}{2}}\; u_{th}\left(1 + \frac{2\omega_p^2}{3k^2 u_{th}^2}\right)^{-1/2}$$

It shows that group velocity u_g is never larger than $\sqrt{3/2}\, u_{th}$. When $k \to \infty$, group velocity $u_g \to \sqrt{3/2}\, u_{th}$.

When we move along the dispersion curve from a large value of k towards the point $k = 0$, the phase velocity varies from $\sqrt{3/2}\, u_{th}$ to infinite value whereas the group velocity varies from $\sqrt{3/2}\, u_{th}$ to zero value. Thus, the range of phase velocity is from $\sqrt{3/2}\, u_{th}$ to infinite whereas the range of group velocity is from zero to $\sqrt{3/2}\, u_{th}$. We can write

$$k = \sqrt{\frac{2}{3}}\frac{1}{u_{th}}\left(\omega^2 - \omega_p^2\right)^{1/2}$$

For $\omega < \omega_p$, k becomes imaginary. It shows that there is cut-off frequency $\omega = \omega_p$ above which an electron plasma wave can propagate. For imaginary k, the spatial dependence $\exp(ikx)$ shows an exponential attenuation. The depth δ through which an electron wave with $\omega < \omega_p$ can propagate can be obtained as

$$e^{ikx} = e^{-|k|x} = e^{-k/\delta} \qquad \text{or} \qquad \delta = |k|^{-1} = \sqrt{\frac{3}{2}}\frac{u_{th}}{(\omega_p^2 - \omega^2)^{1/2}}$$

6. Sound waves

After discussion of electron waves, we want to discuss about ion waves. Before the discussion of ion waves, however, let us first discuss about the sound waves, which take place in an ordinary medium. Neglecting the viscosity, equation of motion is

$$\rho\left[\frac{\partial \vec{u}}{\partial t} + (\vec{u}\ .\ \nabla)\vec{u}\right] = -\nabla p \tag{16.44}$$

and equations of continuity and state are

$$\frac{\partial \rho}{\partial t} + \nabla\ .\ (\rho\vec{u}) = 0 \qquad \text{and} \qquad p = C\rho^\gamma \tag{16.45}$$

where ρ is mass density. We have

$$\nabla p = \frac{\gamma p}{\rho}\ \nabla\rho \tag{16.46}$$

When amplitude of each oscillating quantity is small, the set of equations (16.44) – (16.46) can be solved by using the procedure of linearisation, where the terms having square and higher powers of amplitude are neglected. For this, let us first separate variable parameters into two parts: (i) an equilibrium part, which represents the state of matter in absence of oscillations, and is indicated by subscript 0 and (ii) perturbation part due to oscillations, indicated by subscript 1:

$$\rho = \rho_0 + \rho_1 \qquad \vec{u} = \vec{u_0} + \vec{u_1} \qquad p = p_0 + p_1 \tag{16.47}$$

Before displacement of particles, medium is homogeneous and at rest, so that

$$\nabla\rho_0 = \vec{u_0} = \nabla p_0 = 0 \qquad \text{and} \qquad \frac{\partial \rho_0}{\partial t} = \frac{\partial \vec{u_0}}{\partial t} = 0 \tag{16.48}$$

Using equations (16.46) – (16.48) in (16.44) and after linearisation, we get

$$\rho_0 \frac{\partial \vec{u_1}}{\partial t} = -\frac{\gamma p_0}{\rho_0} \nabla \rho_1 \tag{16.49}$$

Using equations (16.47) – (16.48) in first of (16.45) and after linearisation, we get

$$\frac{\partial \rho_1}{\partial t} + \rho_0 \nabla \cdot \vec{u_1} = 0 \tag{16.50}$$

For convenience, we consider one-dimensional case and assume that oscillating quantities behave sinusoidally as $e^{i(kx-\omega t)}$. Then time derivative $(\partial/\partial t)$ can be replaced by $-i\omega$, and the gradient $(\partial/\partial x)$ by ik. Now, equations (16.49) and (16.50) become

$$-i\omega \rho_0 u_1 = -\frac{\gamma p_0}{\rho_0} ik\rho_1 \tag{16.51}$$

and

$$-\omega \rho_1 + \rho_0 k u_1 = 0 \tag{16.52}$$

On elimination of ρ, equations (16.51) and (16.52) can be rearranged as

$$\left[\omega^2 - \frac{\gamma p_0}{\rho_0} k^2\right] u_1 = 0$$

Since u_1 is not zero, we have

$$\omega^2 = \frac{\gamma\, p_0}{\rho_0} k^2 \qquad \text{or} \qquad \frac{\omega}{k} = \left(\frac{\gamma\, p_0}{\rho_0}\right)^{1/2}$$

Using $p_0 = n_0 KT$ and $\rho_0 = n_0 m_p$, where m_p is mass of proton, we get

$$\frac{\omega}{k} = \left(\frac{\gamma KT}{m_p}\right)^{1/2} \equiv c_s$$

where c_s is velocity of sound wave in neutral gas. The waves are propagating from one layer to another through collisions among the gas molecules, and hence they are **pressure waves**. For one-dimensional case, $\gamma = 3$, and thus, the phase velocity v_p and group velocity v_g of wave are

$$v_p = \frac{\omega}{k} = \sqrt{\frac{3}{2}}\, u_{th} \qquad \text{and} \qquad v_g = \frac{d\omega}{dk} = \sqrt{\frac{3}{2}}\, u_{th}$$

where $u_{th} = \sqrt{2KT/m_p}$ is average thermal velocity of atoms in gas. It is interesting to find that in a fully ionized plasma (*i.e.*, in absence of collisions), analogous waves, called *ion acoustic waves*, or simply, *ion waves* are found.

7. Ion waves

In a fully ionized plasma, in absence of collisions, sound waves do not occur. Ions, being charged particles, can transmit vibrations to each other. Since the ions are massive particles, these vibrations are low frequency oscillations. In absence of magnetic field $\vec{B}$, from Maxwell's equation, we have

$$\nabla \times \vec{E} = -\frac{\partial \vec{B}}{\partial t} = 0 \qquad \text{and hence,} \qquad \vec{E} = -\nabla\phi$$

where ϕ is scalar quantity. The equations of motion and continuity for an ion are

$$m_i n_i \left[\frac{\partial \vec{u_i}}{\partial t} + (\vec{u_i} \cdot \nabla)\vec{u_i}\right] = e n_i \vec{E} - \nabla p_i \tag{16.53}$$

and

$$\frac{\partial n_i}{\partial t} + \nabla \cdot (n_i \vec{u_i}) = 0 \tag{16.54}$$

Using the equation of state $p_i = C\rho_i^{\gamma_i}$, we have

$$\nabla p_i = \gamma_i p_i \frac{\nabla n_i}{n_i} = \frac{\gamma_i n_i K T_i}{n_i} \nabla n_i = \gamma_i K T_i \nabla n_i$$

Here, we have used $p_i = n_i K T_i$. Hence, equation (16.53) of motion can be written as

$$m_i n \left[\frac{\partial \vec{u_i}}{\partial t} + (\vec{u_i} \cdot \nabla)\vec{u_i}\right] = -en\nabla\phi - \gamma_i K T_i \nabla n \tag{16.55}$$

where we have used the plasma approximation $n_i = n_e = n$, and therefore, we do not use Poisson equation. When amplitude of each oscillating quantity is small, let us linearize equation (16.55) by neglecting the terms having square and higher powers of amplitude. For this, let us first separate variable parameters into two parts: (i) an equilibrium part, which represents the state of plasma in absence of oscillations, and is indicated by subscript 0 and (ii) perturbation part due to oscillations, indicated by subscript 1:

$$n = n_0 + n_1 \qquad \vec{u_i} = \vec{u_0} + \vec{u_1} \qquad \phi = \phi_0 + \phi_1 \tag{16.56}$$

Before displacement of electrons, plasma is homogeneous and at rest, so that

$$\nabla n_0 = \vec{u_0} = \nabla\phi_0 = 0 \qquad \text{and} \qquad \frac{\partial n_0}{\partial t} = 0 \tag{16.57}$$

Using equations (16.56) and (16.57) in (16.55) and on linearisation, we get

$$m_i n_0 \frac{\partial \vec{u_1}}{\partial t} = -e n_0 \nabla\phi_1 - \gamma_i K T_i \nabla n_1 \tag{16.58}$$

Fluid drift parallel to $\vec{B}$ is the same as in absence of $\vec{B}$. For electric field $\vec{E} = -\nabla\phi = -\nabla(\phi_0 + \phi_1) = -\nabla\phi_1$, density of mass-less electrons is (mass of electron may be considered as negligible in comparison to that of proton)

$$n_e = n = n_0 \exp\left(\frac{e\phi_1}{KT_e}\right) = n_0\left(1 + \frac{e\phi_1}{KT_e} + \ldots\right) = n_0\left(1 + \frac{e\phi_1}{KT_e}\right)$$

We have neglected higher order terms, and we have

$$n - n_0 = n_0 \frac{e\phi_1}{KT_e} \qquad \text{or} \qquad n_1 = n_0 \frac{e\phi_1}{KT_e} \tag{16.59}$$

Equation (16.59) can be written as

$$m_i n_0 \frac{\partial \vec{u_1}}{\partial t} = -en_0 \nabla \phi_1 - \frac{\gamma_i n_0 e T_i}{T_e} \nabla \phi_1 \tag{16.60}$$

Using equations (16.56) – (16.57) in (16.54) and on linearisation, we get

$$\frac{\partial n_1}{\partial t} + n_0 \nabla \ . \ \vec{u_1} = 0 \tag{16.61}$$

For convenience, we consider one-dimensional case and assume that oscillating quantities behave sinusoidally as $e^{i(kx-\omega t)}$. Then time derivative $(\partial/\partial t)$ can be replaced by $-i\omega$, and gradient ∇ by ik. From equations (16.59) – (16.61), we get

$$n_1 = n_0 \frac{e\phi_1}{KT_e} \tag{16.62}$$

$$-im_i n_0 \omega u_1 = -en_0 ik\phi_1 - \frac{\gamma_i n_0 e T_i}{T_e} ik\phi_1 \tag{16.63}$$

$$-i\omega n_1 + n_0 iku_1 = 0 \tag{16.64}$$

Equations (16.62) – (16.64) can be rearranged as

$$\begin{aligned} -im_i n_0 \omega u_1 &= -iken_0 \Big(1 + \frac{\gamma_i T_i}{T_e}\Big)\phi_1 \\ &= -iken_0 \Big(1 + \frac{\gamma_i T_i}{T_e}\Big)\frac{KT_e}{n_0 e} n_1 = -ikK(T_e + \gamma_i T_i)\frac{n_0 k}{\omega} u_1 \end{aligned}$$

so that

$$\Big[\omega^2 - \frac{k^2 K}{m_i}\Big(T_e + \gamma_i T_i\Big)\Big] u_1 = 0$$

Since u_1 is not equal to zero, we have

$$\omega^2 = \frac{k^2 K}{m_i}\Big(T_e + \gamma_i T_i\Big) \qquad \text{or} \qquad \frac{\omega}{k} = \Big(\frac{KT_e + \gamma_i KT_i}{m_i}\Big)^{1/2} \equiv v_s \tag{16.65}$$

This is dispersion relation for ion acoustic waves. Here, v_s is sound speed in plasma. Obviously, v_s is equal to the phase velocity as well as group velocity of wave. When ion temperature T_i tends to zero, ion wave still exist and its velocity is

$$v_s = \Big(\frac{KT_e}{m_i}\Big)^{1/2}$$

This does happen in neutral gas.

8. Validity of plasma approximation

In preceding section, in derivation of the velocity of ion waves, we have used neutrality condition $n_i = n_e$, and have taken finite value for electric field $\vec{E}$. It is in violation of Poisson equation. For estimating error introduced in the process, let us assume $n_i \neq n_e$ and consider Poisson equation for hydrogen plasma

$$\nabla \cdot \vec{E} = \frac{e(n_i - n_e)}{\epsilon_0}$$

In absence of $\vec{B}$, we have $\vec{E} = -\nabla\phi$, and Poisson equation is

$$-\nabla^2\phi = \frac{e(n_i - n_e)}{\epsilon_0} \tag{16.66}$$

When amplitude of each oscillating quantity is small, let us linearize equation (16.66) by neglecting the terms having square and higher powers of amplitude. For this, let us first separate variable parameters into two parts: (i) an equilibrium part, which represents the state of plasma in absence of oscillations, and is indicated by subscript 0 and (ii) perturbation part due to oscillations, indicated by subscript 1:

$$n_i = n_{i0} + n_{i1} \qquad n_e = n_{e0} + n_{e1} \qquad \phi = \phi_0 + \phi_1 \tag{16.67}$$

Using equation (16.67) in (16.66), we have

$$-\nabla^2(\phi_0 + \phi_1) = \frac{e(n_{i0} + n_{i1} - n_{e0} - n_{e1})}{\epsilon_0} \tag{16.68}$$

Before displacement of electrons, plasma is homogeneous and at rest, so that

$$-\nabla^2\phi_0 = \frac{e(n_{i0} - n_{e0})}{\epsilon_0} \tag{16.69}$$

Subtracting equation (16.69) from (16.68), we get

$$-\nabla^2\phi_1 = \frac{e(n_{i1} - n_{e1})}{\epsilon_0} \tag{16.70}$$

Fluid drift parallel to $\vec{B}$ is the same as in absence of $\vec{B}$. The density of mass-less electrons is (mass of an electron may be considered as negligible in comparison to that of proton)

$$n_e = n_0 \exp\left(\frac{e\phi}{KT_e}\right)$$

$$n_{e0} + n_{e1} = n_0 \exp\left(\frac{e(\phi_0 + \phi_1)}{KT_e}\right) \tag{16.71}$$

We have

$$n_{e0} = n_0 \exp\left(\frac{e\phi_0}{KT_e}\right) \tag{16.72}$$

Subtracting equation (16.72) from (16.71), we get

$$n_{e1} = n_0 \exp\Big(\frac{e(\phi_0+\phi_1)}{KT_e}\Big) - n_0 \exp\Big(\frac{e\phi_0}{KT_e}\Big)$$

$$= n_0 \exp\Big(\frac{e\phi_0}{KT_e}\Big)\Big[1 + \frac{e\phi_1}{KT_e} + \ldots - 1\Big] = n_{e0}\,\frac{e\phi_1}{KT_e}$$

Using this expression in equation (16.70), we get

$$-\nabla^2\phi_1 + \frac{n_{e0}}{\epsilon_0}\,\frac{e^2\phi_1}{KT_e} = \frac{en_{i1}}{\epsilon_0} \tag{16.73}$$

For convenience, we consider one-dimensional case and assume that oscillating quantities behave sinusoidally as $e^{i(kx-\omega t)}$. Then time derivative $(\partial/\partial t)$ can be replaced by $-i\omega$, and gradient $\partial/\partial x$ by ik. From equation (16.73), we get

$$k^2\phi_1 + \frac{en_{e0}}{\epsilon_0}\,\frac{e\phi_1}{KT_e} = \frac{en_{i1}}{\epsilon_0} \qquad \phi_1\Big(k^2 + \frac{n_{e0}e^2}{KT_e\epsilon_0}\Big) = \frac{en_{i1}}{\epsilon_0}$$

$$\phi_1\Big(k^2 + \frac{1}{\lambda_D^2}\Big) = \frac{en_{i1}}{\epsilon_0} \qquad \phi_1\Big(k^2\lambda_D^2 + 1\Big) = \frac{en_{i1}\lambda_D^2}{\epsilon_0} \tag{16.74}$$

Equations (16.63), (16.64) and (16.74) can be rearranged as

$$-im_i n_0 \omega u_1 = -iken_0\Big[1 + \frac{\gamma_i T_i}{T_e}\Big]\phi_1 = -iken_0\Big[1 + \frac{\gamma_i T_i}{T_e}\Big]\Big[\frac{e\lambda_D^2}{(1+k^2\lambda_D^2)\epsilon_0}\Big]n_{i1}$$

$$= -iken_0\Big[1 + \frac{\gamma_i T_i}{T_e}\Big]\Big[\frac{e}{(1+k^2\lambda_D^2)\epsilon_0}\Big]\Big[\frac{KT_e\epsilon_0}{n_{e0}e^2}\Big]\Big[\frac{n_0 k}{\omega}\Big]u_1$$

so that

$$\Big[\omega^2 - \frac{k^2K}{m_i}\Big(T_e + \gamma_i T_i\Big)\frac{1}{(1+k^2\lambda_D^2)}\Big]u_1 = 0$$

Since u_1 is not equal to zero, we have

$$\omega^2 = \frac{k^2K}{m_i}\Big(T_e + \gamma_i T_i\Big)\frac{1}{(1+k^2\lambda_D^2)}$$

or

$$\frac{\omega}{k} = \Big(\frac{KT_e + \gamma_i KT_i}{m_i}\Big)^{1/2}\frac{1}{\sqrt{(1+k^2\lambda_D^2)}} \tag{16.75}$$

Comparison of equations (16.75) and (16.65) shows that in our plasma approximation $(n_i = n_e)$, an error of order of $k^2\lambda_D^2$ $(= \{2\pi\lambda_D/\lambda\}^2)$ is introduced. Since λ_D in comparison to λ is very small in most of the cases, plasma approximation is valid, except for the cases of shortest wavelengths.

9. Some definitions for waves

In the discussion so far, we did not account for the magnetic field $\vec{B}$. In presence of $\vec{B}$, several other types of waves are possible. Let us first introduce some terminology for waves.

(i) The words 'parallel' and 'perpendicular' are used to denote the direction of $\vec{k}$ relative to unperturbed magnetic field $\vec{B}_0$.

(ii) The words 'longitudinal' and 'transverse' are used to denote the direction of $\vec{k}$ relative to perturbed (oscillating) electric field $\vec{E}_1$.

(iii) The wave is termed 'electrostatic' when oscillating magnetic field $\vec{B}_1$ is zero.

(iv) The wave is termed 'electromagnetic' when oscillating magnetic field $\vec{B}_1$ is finite (non-zero).

Let us consider Maxwell's equation

$$\nabla \times \vec{E} = -\frac{\partial \vec{B}}{\partial t} \tag{16.76}$$

and resolve $\vec{E}$ and $\vec{B}$ into two parts: (i) an equilibrium part, which represents the state of plasma in absence of oscillations, and indicated by subscript 0 and (ii) perturbation part due to oscillations, indicated by subscript 1:

$$\vec{E} = \vec{E}_0 + \vec{E}_1 \qquad \text{and} \qquad \vec{B} = \vec{B}_0 + \vec{B}_1 \tag{16.77}$$

Using equation (16.77) in (16.76), we have

$$\nabla \times (\vec{E}_0 + \vec{E}_1) = -\frac{\partial \vec{B}_0}{\partial t} - \frac{\partial \vec{B}_1}{\partial t}$$

Since the equilibrium parts do not depend on time and space coordinates, we have

$$\nabla \times \vec{E}_1 = -\frac{\partial \vec{B}_1}{\partial t} \tag{16.78}$$

Let us assume that perturbation parts behave sinusoidally as $e^{i(\vec{k}\,\cdot\,\vec{r}-\omega t)}$, then we replace time derivative $(\partial/\partial t)$ by $-i\omega$, and gradient ∇ by $i\vec{k}$. Thus, we have

$$i\vec{k} \times \vec{E}_1 = -(-i\omega)\,\vec{B}_1 \qquad \text{or} \qquad \vec{k} \times \vec{E}_1 = \omega \vec{B}_1$$

When a wave is longitudinal, *i.e.*, $\vec{k}$ is parallel to $\vec{E}_1$, we have $\vec{k} \times \vec{E}_1 = 0$, and $\vec{B}_1 = 0$ showing that the wave is electrostatic also. When a wave is transverse, *i.e.*, $\vec{k}$ is perpendicular to $\vec{E}_1$, we have $\vec{k} \times \vec{E}_1 \neq 0$, and $\vec{B}_1 \neq 0$ showing that the wave is electromagnetic also.

In general, $\vec{k}$ may have an arbitrary angle to $\vec{B}_0$ and $\vec{E}_1$, then the wave has a combination of principal modes discussed above in (i) – (iv).

10. Electrostatic electron waves perpendicular to $\vec{B}$

In equilibrium position, charged particles in plasma are uniformly distributed in such a manner that neutrality is maintained. When electrons in a plasma are displaced relative to uniform background of ions, an electric field is developed in such a direction that it tries to pull electrons back to restore the neutrality of plasma. While returning back, owing to inertia, the electrons overshoot and now electric field is developed in reverse direction which tries to pull back electrons to their equilibrium positions.

Thus, the electrons oscillate about their equilibrium positions. Because of low mass of electrons, their oscillations are so fast that massive ions are not capable to respond the oscillating field, generated by oscillations of electrons. Hence, the ions may be treated as fixed forming uniform background. Further, owing to the presence of magnetic field, the motions of electrons are not linear. We assume also that there are no thermal motions ($T_e = 0$). The equations of motion and continuity for electrons are

$$m_e n_e\left[\frac{\partial \vec{u_e}}{\partial t} + (\vec{u_e} \,.\, \nabla)\vec{u_e}\right] = -en_e(\vec{E} + \vec{u}_e \times \vec{B}) \tag{16.79}$$

and

$$\frac{\partial n_e}{\partial t} + \nabla \,.\, (n_e\vec{u_e}) = 0 \tag{16.80}$$

Here, we are assuming collision-less, homogeneous, isotropic and cold plasma. These are high frequency oscillations which are generated due to deviation from neutrality and therefore, we can use Poisson equation

$$\nabla \,.\, \vec{E} = \frac{e(n_i - n_e)}{\epsilon_0} \tag{16.81}$$

where n_i and n_e are densities of ions and electrons, respectively. When amplitude of each oscillating quantity is small, the set of equations (16.79) – (16.81) can be solved by using the procedure of linearisation, where the terms having square and higher powers of amplitude are neglected. For this, let us first separate variable parameters into two parts: (i) an equilibrium part, which represents the state of plasma in absence of oscillations, and is indicated by subscript 0 and (ii) perturbation part due to oscillations, indicated by subscript 1:

$$n_e = n_0 + n_1 \qquad \vec{u_e} = \vec{u_0} + \vec{u_1} \qquad \vec{E} = \vec{E_0} + \vec{E_1} \tag{16.82}$$

Since we want to address an electrostatic wave, perturbation of $\vec{B}$ is not considered (*i.e.*, $\vec{B} = \vec{B_0}$). Before displacement of electrons, plasma is homogeneous and at rest, so that

$$\nabla n_0 = \vec{u_0} = \vec{E_0} = 0 \qquad \text{and} \qquad \frac{\partial n_0}{\partial t} = 0 \tag{16.83}$$

Using equations (16.82) and (16.83) in (16.79) – (16.81) and on linearisation, we get

$$m_e \frac{\partial \vec{u_1}}{\partial t} = -e(\vec{E_1} + \vec{u_1} \times \vec{B_0}) \tag{16.84}$$

$$\frac{\partial n_1}{\partial t} + n_0 \nabla \cdot \vec{u_1} = 0 \tag{16.85}$$

$$\nabla \cdot \vec{E_1} = -\frac{en_1}{\epsilon_0} \tag{16.86}$$

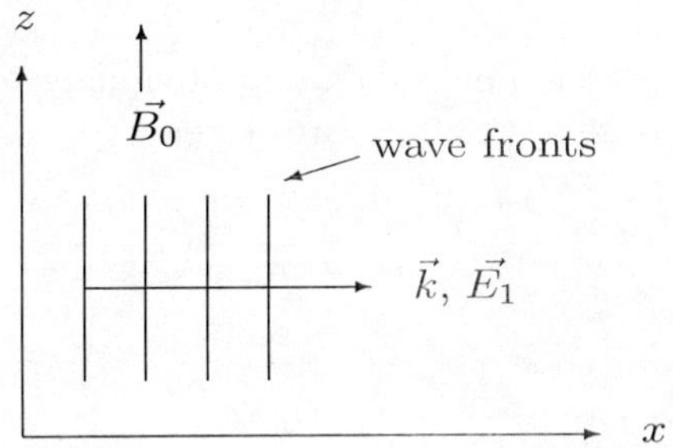

Figure 2: A longitudinal plane wave propagating at right angle to $\vec{B_0}$

Here, we consider only longitudinal waves with $\vec{k}$ parallel $\vec{E_1}$. Let us consider x-axis along $\vec{k}$ and $\vec{E_1}$, and z-axis along $\vec{B_0}$ (Figure 2). Hence, $\vec{k} = k\ \hat{x}$; $\vec{E_1} = E_1 \hat{x}$; $\vec{B_0} = B_0\ \hat{z}$; $\vec{u_1} = u_x\ \hat{x} + u_y\ \hat{y} + u_z\ \hat{z}$. Equation (16.84) can be expressed as

$$m_e \frac{\partial u_x}{\partial t} = -eE_1 - eu_y B_0 \tag{16.87}$$

and

$$m_e \frac{\partial u_y}{\partial t} = eu_x B_0 \tag{16.88}$$

We assume that oscillating quantities behave sinusoidally as $e^{i(\vec{k}\ \cdot\ \vec{r} - \omega t)}$. Then time derivative $(\partial/\partial t)$ can be replaced by $-i\omega$, and gradient ∇ by $i\vec{k}$. From equations (16.85) – (16.88), we get

$$-i\omega n_1 + n_0 iku_x = 0 \tag{16.89}$$

$$ikE_1 = -4\pi en_1 \tag{16.90}$$

$$-im_e\omega u_x = -eE_1 - eu_y B_0 \tag{16.91}$$

$$-im_e\omega u_y = eu_x B_0 \tag{16.92}$$

Using u_y from equation (16.92) in (16.91), we get

$$-im_e\omega u_x = -eE_1 - eB_0\Big(\frac{eB_0}{-im_e\omega}\Big)u_x$$

Thus,

$$u_x = \frac{eE_1}{im_e\omega} + \frac{e^2B_0^2}{\omega^2 m_e^2}\,u_x = \frac{eE_1}{im_e\omega} + \frac{\omega_c^2}{\omega^2}\,u_x$$

giving,

$$u_x\left(1 - \frac{\omega_c^2}{\omega^2}\right) = \frac{eE_1}{im_e\omega} \qquad \text{or} \qquad u_x = \frac{eE_1}{im_e\omega}\left(1 - \frac{\omega_c^2}{\omega^2}\right)^{-1} \tag{16.93}$$

where $\omega_c = eB_0/m_e$ is cyclotron frequency. The u_x obviously tends to infinite when $\omega \to \omega_c$. Equations (16.89), (16.90), (16.93) can be rearranged as

$$ikE_1 = -e\left(\frac{n_0 k}{\omega\epsilon_0}\right)u_x = -\left(\frac{n_0 ek}{\omega\epsilon_0}\right)\frac{eE_1}{im_e\omega}\left(1 - \frac{\omega_c^2}{\omega^2}\right)^{-1}$$

so that

$$\left[1 - \frac{\omega_c^2}{\omega^2} - \frac{n_0 e^2}{m_e\epsilon_0}\frac{1}{\omega^2}\right]E_1 = 0$$

Since E_1 is not zero, we have

$$1 - \frac{\omega_c^2}{\omega^2} = \frac{4\pi n_0 e^2}{m_e}\frac{1}{\omega^2} \qquad \text{or} \qquad \omega^2 = \omega_p^2 + \omega_c^2 \equiv \omega_h^2 \tag{16.94}$$

where $\omega_p = (n_0 e^2/m_e\epsilon_0)^{1/2}$ is plasma frequency. The frequency ω_h is known as **upper hybrid frequency**. Electrostatic electron waves perpendicular to $\vec{B}$ have this frequency ω_h whereas electron waves parallel to $\vec{B}$ have frequency ω_p. The group velocity is zero, as the thermal motion is not considered. When magnetic field goes to zero, ω_c goes to zero and we have plasma oscillations with frequency ω_p. When electron density goes to zero, ω_p goes to zero and we have a simple Larmor gyration. Existence of upper hybrid frequency has been verified experimentally.

11. Electrostatic ion waves perpendicular to $\vec{B}$

Let us investigate ion acoustic wave where $\vec{k}$ is perpendicular to $\vec{B}_0$. Here, we consider that angle between $\vec{k}$ and $\vec{B}_0$ is almost $\pi/2$ so that $\vec{k}\,.\,\vec{B}_0 \neq 0$. The geometry is shown in Figure 3.

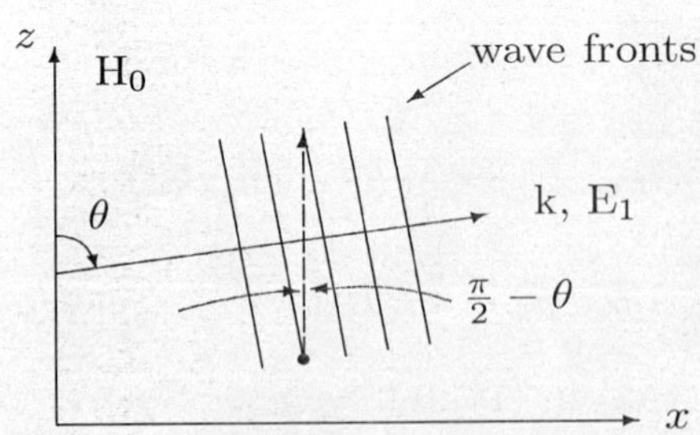

Figure 3: Electrostatic ion cyclotron wave propagating nearly at right angle to $\vec{B}_0$

We consider electrostatic waves with $\vec{B}_1 = 0$, and therefore, $\vec{k}\times\vec{E}_1 = 0$. The wave is longitudinal. Further, $\nabla\times\vec{E}_1 = 0$ and therefore, $\vec{E}_1 = -\nabla\phi = -(\partial\phi/\partial x)\,\hat{x}$, where

ϕ is scalar quantity. We assume also that there are no thermal motions ($T_i = 0$). The equations of motion and continuity for ions are

$$m_i n_i\left[\frac{\partial \vec{u_i}}{\partial t} + (\vec{u_i}.\nabla)\vec{u_i}\right] = en_i\left(-\frac{\partial \phi}{\partial x}\,\hat{x} + \vec{u_i} \times \vec{B_0}\right) \tag{16.95}$$

and

$$\frac{\partial n_i}{\partial t} + \nabla\,.\,(n_i\vec{u_i}) = 0 \tag{16.96}$$

Here, we have used plasma approximation $n_i = n_e = n$, and therefore, we do not use Poisson equation. When amplitude of each oscillating quantity is small, let us linearize equations by neglecting the terms having square and higher powers of amplitude. For this, let us first separate variable parameters into two parts: (i) an equilibrium part, which represents the state of plasma in absence of oscillations, and is indicated by subscript 0 and (ii) perturbation part due to oscillations, indicated by subscript 1:

$$\vec{u_i} = \vec{u_0} + \vec{u_1} \qquad \phi = \phi_0 + \phi_1 \qquad n_i = n_0 + n_1 \tag{16.97}$$

At equilibrium, we have

$$\nabla n_0 = \vec{u_0} = \nabla\phi_0 = 0 \qquad \text{and} \qquad \frac{\partial n_0}{\partial t} = 0 \tag{16.98}$$

Using equations (16.97) and (16.98) in (16.95) and on linearisation, we get

$$m_i\,\frac{\partial \vec{u_1}}{\partial t} = -e\,\frac{\partial \phi_1}{\partial x}\,\hat{x} + e\,\vec{u_1} \times \vec{B_0} \tag{16.99}$$

For $\vec{u_1} = u_x\,\hat{x} + u_y\,\hat{y} + u_z\,\hat{z}$, equation (16.99) can be expressed as following.

$$m_i\,\frac{\partial u_x}{\partial t} = -e\,\frac{\partial \phi_1}{\partial x} + eu_y B_0 \tag{16.100}$$

and

$$m_i\,\frac{\partial u_y}{\partial t} = -eu_x B_0 \tag{16.101}$$

For electric field $\vec{E} = -\nabla\phi = -\nabla(\phi_0 + \phi_1) = -\nabla\phi_1$, density of mass-less electrons is (mass of an electron may be considered as negligible in comparison to that of proton)

$$n_e = n = n_0\,\exp\left(\frac{e\phi_1}{KT_e}\right)$$

$$= n_0\left(1 + \frac{e\phi_1}{KT_e} + \ldots\right) = n_0\left(1 + \frac{e\phi_1}{KT_e}\right)$$

Neglecting higher order terms, we have

$$n - n_0 = n_0\,\frac{e\phi_1}{KT_e} \qquad \text{or} \qquad n_{e1} = n_0\,\frac{e\phi_1}{KT_e}$$

Under plasma approximation, we have $n_e = n_i$. Further, we have $n_{e0} = n_{i0}$ and therefore $n_{e1} = n_{i1}$. So, under plasma approximation, we get

$$n_1 = n_{i1} = n_0 \, \frac{e\phi_1}{KT_e} \tag{16.102}$$

Using equations (16.97) and (16.98) in (16.96) and on linearisation, we get

$$\frac{\partial n_1}{\partial t} + n_0 \nabla \, . \, \vec{u_1} = 0 \tag{16.103}$$

We consider only longitudinal waves with $\vec{k}$ parallel to $\vec{E_1}$. Let us consider x-axis along $\vec{k}$ and $\vec{E_1}$, and z-axis along $\vec{B_0}$. Hence, $\vec{k} = k \, \hat{x}$; $\vec{B_0} = B_0 \, \hat{z}$. We assume that oscillating quantities behave sinusoidally as $e^{i(\vec{k}.\vec{r} - \omega t)}$. Then time derivative $(\partial/\partial t)$ can be replaced by $-i\omega$, and gradient ∇ by $i\vec{k}$. From equations (16.100) – (16.103), we get

$$-i\omega m_i u_x = -eik\phi_1 + eu_y B_0 \tag{16.104}$$

$$-i\omega m_i u_y = -eu_x B_0 \tag{16.105}$$

$$n_1 = n_0 \, \frac{e\phi_1}{KT_e} \tag{16.106}$$

$$-i\omega n_1 + n_0 iku_x = 0 \tag{16.107}$$

Using u_y from equation (16.105) in (16.104), we get

$$-i\omega m_i u_x = -eik\phi_1 + eB_0\Big(\frac{eB_0}{i\omega m_i}\Big)u_x$$

Hence,

$$u_x = \frac{ek\phi_1}{\omega m_i} + \Big(\frac{e^2 B_0^2}{\omega^2 m_i^2}\Big)u_x = \frac{ek\phi_1}{\omega m_i} + \frac{\Omega_c^2}{\omega^2}u_x$$

giving,

$$u_x\Big(1 - \frac{\Omega_c^2}{\omega^2}\Big) = \frac{ek\phi_1}{\omega m_i} \qquad \text{or} \qquad u_x = \frac{ek\phi_1}{\omega m_i}\Big(1 - \frac{\Omega_c^2}{\omega^2}\Big)^{-1} \tag{16.108}$$

where $\Omega_c = eB_0/m_i$ is ion cyclotron frequency. Equations (16.106) – (16.108) can be rearranged as

$$u_x = \frac{ek}{\omega m_i}\Big(1 - \frac{\Omega_c^2}{\omega^2}\Big)^{-1}\Big(\frac{KT_e}{n_0 e}\Big)\, n_1 \qquad u_x\Big(1 - \frac{\Omega_c^2}{\omega^2}\Big) = \frac{ek}{\omega m_i}\Big(\frac{KT_e}{n_0 e}\Big)\Big(\frac{n_0 k}{\omega}\Big)\, u_x$$

Since u_x is not zero, we have

$$\Big(1 - \frac{\Omega_c^2}{\omega^2}\Big) = \frac{k^2 KT_e}{\omega^2 m_i} \qquad \text{or} \qquad \omega^2 - \Omega_c^2 = k^2 v_s^2$$

where $v_s = \sqrt{KT_e/m_i}$ is velocity of ion waves. This is dispersion relation for electrostatic ion waves. Electrostatic ion waves, which also have been detected experimentally.

12. Lower hybrid frequency

In the preceding section, we have considered that angle between $\vec{k}$ and $\vec{B_0}$ is almost $\pi/2$ so that $\vec{k}\ .\ \vec{B_0} \neq 0$. When angle between $\vec{k}$ and $\vec{B_0}$ is exactly $\pi/2$, the electrons are not allowed to preserve neutrality by flowing along the lines of force, and hence they also obey the equation of motion. The equations of motion and continuity for ions and electrons are

$$m_i n_i \Big[\frac{\partial \vec{u_i}}{\partial t} + (\vec{u_i}\ .\ \nabla)\vec{u_i}\Big] = e n_i(-\nabla\phi + \vec{u_i} \times \vec{B_0}) \tag{16.109}$$

$$\frac{\partial n_i}{\partial t} + \nabla\ .\ (n_i \vec{u_i}) = 0 \tag{16.110}$$

$$m_e n_e \Big[\frac{\partial \vec{v_e}}{\partial t} + (\vec{v_e}\ .\ \nabla)\vec{u_e}\Big] = -e n_e(-\nabla\phi + \vec{v_e} \times \vec{B_0}) \tag{16.111}$$

$$\frac{\partial n_e}{\partial t} + \nabla\ .\ (n_e\ \vec{u_e}) = 0 \tag{16.112}$$

Equations (16.109) and (16.110) are solved in the preceding section and we have

$$u_{ix} = \frac{ek\phi_1}{\omega m_i}\Big(1 - \frac{\Omega_c^2}{\omega^2}\Big)^{-1} \qquad \text{and} \qquad n_{i1} = \frac{n_0 k}{\omega}\ u_{ix} \tag{16.113}$$

Solution of the equations (16.111) and (16.112) can be obtained in a similar manner by replacing e by $-e$, m_i by m_e, Ω_c by $-\omega_c$ in equations (16.113) as

$$u_{ex} = -\frac{ek\phi_1}{\omega m_e}\Big(1 - \frac{\omega_c^2}{\omega^2}\Big)^{-1} \qquad \text{and} \qquad n_{e1} = \frac{n_0 k}{\omega}\ u_{ex} \tag{16.114}$$

The plasma approximation $n_i = n_e = n_0$ requires $n_{i1} = n_{e1}$ and therefore, we have (using second equations of 16.113 and 16.114)

$$\frac{n_0 k}{\omega}\ u_{ix} = \frac{n_0 k}{\omega}\ u_{ex} \qquad \text{or} \qquad u_{ix} = u_{ex}$$

Using first equations of 16.113 and 16.114 here, we have

$$\frac{ek\phi_1}{\omega m_i}\Big(1 - \frac{\Omega_c^2}{\omega^2}\Big)^{-1} = -\frac{ek\phi_1}{\omega m_e}\Big(1 - \frac{\omega_c^2}{\omega^2}\Big)^{-1} \qquad m_i\Big(1 - \frac{\Omega_c^2}{\omega^2}\Big) = -m_e\Big(1 - \frac{\omega_c^2}{\omega^2}\Big)$$

$$\omega^2(m_i + m_e) = m_e\omega_c^2 + m_i\Omega_c^2 = \frac{e^2 B_0^2}{c^2}\Big(\frac{1}{m_i} + \frac{1}{m_e}\Big)$$

giving,

$$\omega^2 = \frac{e^2 B_0^2}{c^2 m_i m_e} = \Omega_c\omega_c \qquad \text{or} \qquad \omega = \sqrt{\Omega_c\omega_c} \equiv \omega_l$$

This frequency ω_l is known as **lower hybrid frequency**.

13. Electromagnetic waves with $\vec{B}_0 = 0$

Let us consider Maxwell's equations

$$\nabla \times \vec{E} = -\frac{\partial \vec{B}}{\partial t} \qquad \text{and} \qquad \nabla \times \vec{B} = \frac{1}{c^2}\frac{\partial \vec{E}}{\partial t} + \mu_0 \vec{J}$$

For linearization of these equations, let us first separate variable parameters into two parts: (i) an equilibrium part, which represents the state of plasma in absence of oscillations, and is indicated by subscript 0 and (ii) perturbation part due to oscillations, indicated by subscript 1:

$$\vec{B} = \vec{B}_0 + \vec{B}_1 \qquad \vec{E} = \vec{E}_0 + \vec{E}_1 \qquad \vec{J} = \vec{J}_0 + \vec{J}_1$$

In plasma in equilibrium state, we have $\vec{E_0} = 0$ and $\vec{J_0} = 0$. Thus, for $\vec{B_0} = 0$, the Maxwell's equations reduce to

$$\nabla \times \vec{E_1} = -\frac{\partial \vec{B_1}}{\partial t} \tag{16.115}$$

and

$$\nabla \times \vec{B_1} = \frac{1}{c^2}\frac{\partial \vec{E_1}}{\partial t} + \mu_0 \vec{J_1} \tag{16.116}$$

Differentiating equation (16.116) with respect to time, we have

$$\nabla \times \frac{\partial \vec{B_1}}{\partial t} = \frac{1}{c^2}\frac{\partial^2 \vec{E_1}}{\partial t^2} + \mu_0 \frac{\partial \vec{J_1}}{\partial t} \tag{16.117}$$

Taking curl of equation (16.115), we have

$$\nabla \times (\nabla \times \vec{E_1}) = -\nabla \times \frac{\partial \vec{B_1}}{\partial t}$$

$$\nabla(\nabla \,.\, \vec{E_1}) - \nabla^2 \vec{E_1} = -\nabla \times \frac{\partial \vec{B_1}}{\partial t} \tag{16.118}$$

Using equation (16.117) in (16.118), we have

$$\nabla(\nabla \,.\, \vec{E_1}) - \nabla^2 \vec{E_1} = -\mu_0 \frac{\partial \vec{J_1}}{\partial t} - \frac{1}{c^2}\frac{\partial^2 \vec{E_1}}{\partial t^2} \tag{16.119}$$

We assume that the oscillating quantities behave sinusoidally as $e^{i(\vec{k}.\vec{r}-\omega t)}$. Then time derivative $(\partial/\partial t)$ can be replaced by $-i\omega$, and gradient ∇ by $i\vec{k}$. From equation (16.119), we get

$$i\vec{k}(i\vec{k} \,.\, \vec{E_1}) - (i\vec{k})^2 \vec{E_1} = i\omega\mu_0 \, \vec{J_1} - \frac{(-i\omega)^2}{c^2}\vec{E_1}$$

$$-\vec{k}(\vec{k} \,.\, \vec{E_1}) + k^2 \vec{E_1} = i\omega\mu_0 \, \vec{J_1} + \frac{\omega^2}{c^2}\vec{E_1}$$

For transverse wave, we have $\vec{k}\ .\ \vec{E_1} = 0$. Thus,

$$k^2 \vec{E_1} = i\omega\mu_0\ \vec{J_1} + \frac{\omega^2}{c^2}\ \vec{E_1} \qquad \text{or} \qquad (\omega^2 - c^2k^2)\vec{E_1} = -i\omega\mu_0 c^2\ \vec{J_1} \quad (16.120)$$

For high frequency waves, ions may be considered as fixed. Then current density is due to the motion of electrons only

$$\vec{J_1} = -n_0 e\ \vec{u}_{e1} \quad (16.121)$$

The equation of motion for electrons is

$$m_e n_e \left[\frac{\partial \vec{u}_e}{\partial t} + (\vec{u}_e\ .\ \nabla)\vec{u}_e\right] = -en_e\ (\vec{E} + \vec{u}_e \times \vec{B})$$

Here, we are considering that electron temperature T_e is zero (cold plasma). We express as $\vec{u}_e = \vec{u}_0 + \vec{u}_1$. After linearisation with $\vec{u_0} = 0$, $\vec{E_0} = 0$ and $\vec{B_0} = 0$, the equation of motion reduces to

$$m_e\ \frac{\partial \vec{u}_{e1}}{\partial t} = -e\vec{E_1}$$

We assume that oscillating quantities behave sinusoidally as $e^{i(\vec{k}\ .\ \vec{r} - \omega t)}$. Then time derivative $(\partial/\partial t)$ can be replaced by $-i\omega$ and we get

$$m_e(-i\omega)\vec{u}_{e1} = -e\vec{E_1} \qquad\qquad \vec{u}_{e1} = \left(\frac{e}{im_e\omega}\right)\vec{E_1} \quad (16.122)$$

Using equation (16.122) in (16.121), we have

$$\vec{J_1} = -\left(\frac{n_0 e^2}{im_e\omega}\right)\vec{E_1}$$

Using $\vec{J_1}$ in equation (16.120), we get

$$(\omega^2 - c^2k^2)\vec{E_1} = \omega_p^2\ \vec{E_1}$$

Since $\vec{E_1} \neq 0$, we have

$$\omega^2 - c^2k^2 = \omega_p^2 \qquad\qquad \omega^2 = \omega_p^2 + c^2k^2$$

where $\omega_p = (n_0e^2/m_e\epsilon_0)^{1/2}$ is plasma frequency. This is dispersion relation for transverse electromagnetic waves propagating in plasma with $\vec{B_0} = 0$. For electromagnetic waves in the medium, we have[4] $\omega^2 = c^2k^2$. It shows that there is additional term ω_p^2. The phase velocity is obtained as

$$\frac{\omega^2}{k^2} = \frac{\omega_p^2}{k^2} + c^2 \qquad \text{or} \qquad \frac{\omega}{k} = c\sqrt{1 + \frac{\omega_p^2}{c^2k^2}} \equiv v_p$$

[4]Maxwell's equations for electromagnetic waves in plasma are

$$\nabla \times \vec{E_1} = -\frac{\partial \vec{B_1}}{\partial t} \quad (16.123)$$

$$\nabla \times \vec{B_1} = \frac{1}{c^2}\ \frac{\partial \vec{E_1}}{\partial t} \quad (16.124)$$

The phase velocity v_p is greater than c. The group velocity is obtained as

$$2\omega\, \mathrm{d}\omega = 2c^2 k\, \mathrm{d}k \qquad \frac{\mathrm{d}\omega}{\mathrm{d}k} = \frac{c^2}{\omega/k} = c\left(1 + \frac{\omega_p^2}{c^2\,k^2}\right)^{-1/2} \equiv v_g$$

The group velocity v_g is smaller than c. When $k \to \infty$, the phase velocity $v_p \to c$, and also the group velocity $v_g \to c$. The dispersion curve is shown in Figure 4.

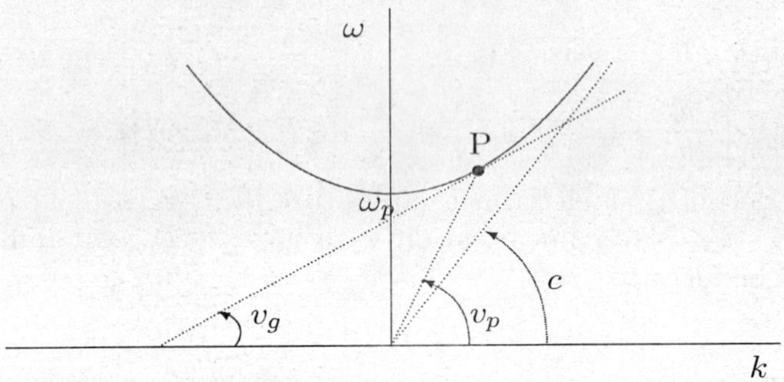

Figure 4: Variation of ω with k for the electromagnetic waves in a plasma with $\vec{B}_0 = 0$

This figure is similar to that in case of electron plasma waves, but behaviour is

---

Since in vacuum we have $\vec{J} = 0$. Differentiating equation (16.123) with respect to time t, we have

$$\nabla \times \frac{\partial \vec{E}_1}{\partial t} = -\frac{\partial^2 \vec{B}_1}{\partial t^2} \tag{16.125}$$

Taking curl of equation (16.124), we have

$$\nabla \times (\nabla \times \vec{B}_1) = \frac{1}{c^2}\, \nabla \times \frac{\partial \vec{E}_1}{\partial t} \tag{16.126}$$

Using equation (16.125) in equation (16.126), we have

$$\nabla \times (\nabla \times \vec{B}_1) = -\frac{1}{c^2}\frac{\partial^2 \vec{B}_1}{\partial t^2}$$

$$\nabla(\nabla \cdot \vec{B}_1) - \nabla^2 \vec{B}_1 = -\frac{1}{c^2}\frac{\partial^2 \vec{B}_1}{\partial t^2}$$

$$\nabla^2 \vec{B}_1 = \frac{1}{c^2}\frac{\partial^2 \vec{B}_1}{\partial t^2}$$

where we have used $\nabla \cdot \vec{B}_1 = 0$. We assume that oscillating quantities behave sinusoidally as $e^{i(\vec{k}\,\cdot\,\vec{r} - \omega t)}$. Then time derivative $(\partial/\partial t)$ can be replaced by $-i\omega$, and gradient ∇ by $i\vec{k}$, and we have

$$(i\vec{k})^2 \vec{B}_1 = \frac{(-i\omega)^2}{c^2}\,\vec{B}_1 \qquad c^2 k^2 \vec{B}_1 = \omega^2\, \vec{B}_1$$

Since $\vec{B}_1 \neq 0$, we have

$$c^2 k^2 = \omega^2 \qquad \frac{\omega}{k} = c \qquad \frac{\mathrm{d}\omega}{\mathrm{d}k} = c$$

Thus, the phase velocity and group velocity of the light waves in the plasma are c.

quite different as the asymptotic velocity in present case is v and in that case was $\sqrt{3/2}\ v_{th}$.

14. Electromagnetic waves perpendicular to $\vec{B_0}$

Earlier, we have considered propagation of electromagnetic waves when there is no magnetic field. Now, we consider propagation of electromagnetic waves when magnetic field $\vec{B_0}$ is present. In the present section, we consider the case of perpendicular propagation, $\vec{k} \perp \vec{B_0}$. If we take transverse wave with $\vec{k} \perp \vec{E_1}$, we have two options for the direction of $\vec{E_1}$: (i) $\vec{E_1}$ is parallel to $\vec{B_0}$ – case of ordinary wave and (ii) $\vec{E_1}$ is perpendicular to $\vec{B_0}$ – case of extraordinary wave. Let us consider one case at a time.

14.1 Ordinary wave

In this case, $\vec{E_1}$ is parallel to $\vec{B_0}$ and $\vec{k}$ is perpendicular to $\vec{B_0}$ as shown in Figure 5.

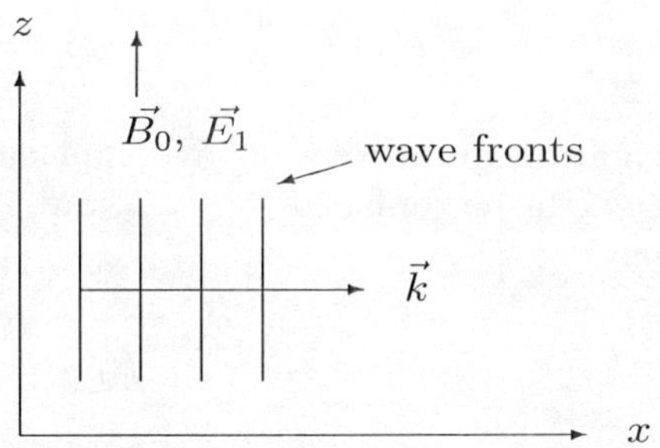

Figure 5: Propagation of an ordinary electromagnetic wave perpendicular to $\vec{B_0}$

We have $\vec{B_0} = B_0\ \hat{z}$, $\vec{k} = k\ \hat{x}$, and $\vec{E_1} = E_1\ \hat{z}$. Let us consider Maxwell's equations

$$\nabla \times \vec{E} = -\frac{\partial \vec{B}}{\partial t} \qquad \text{and} \qquad \nabla \times \vec{B} = \frac{1}{c^2}\frac{\partial \vec{E}}{\partial t} + \mu_0\ \vec{J}$$

For linearization of these equations, let us first separate variable parameters into two parts: (i) an equilibrium part, which represents the state of plasma in absence of oscillations, and is indicated by subscript 0 and (ii) perturbation part due to oscillations, indicated by subscript 1:

$$\vec{B} = \vec{B_0} + \vec{B_1} \qquad \vec{E} = \vec{E_0} + \vec{E_1} \qquad \vec{J} = \vec{J_0} + \vec{J_1}$$

In plasma, we have $\vec{E_0} = 0$ and $\vec{J_0} = 0$. Equilibrium parts do not depend on time and space coordinates and we have Maxwell's equations

$$\nabla \times \vec{E_1} = -\frac{\partial \vec{B_1}}{\partial t} \tag{16.127}$$

and

$$\nabla \times \vec{B}_1 = \frac{1}{c^2} \frac{\partial \vec{E}_1}{\partial t} + \mu_0 \vec{J}_1 \tag{16.128}$$

Differentiating equation (16.128) with respect to time t, we have

$$\nabla \times \frac{\partial \vec{B}_1}{\partial t} = \frac{1}{c^2} \frac{\partial^2 \vec{E}_1}{\partial t^2} + \mu_0 \frac{\partial \vec{J}_1}{\partial t} \tag{16.129}$$

Taking curl of equation (16.127), we have

$$\nabla \times (\nabla \times \vec{E}_1) = -\nabla \times \frac{\partial \vec{B}_1}{\partial t}$$

$$\nabla(\nabla \cdot \vec{E}_1) - \nabla^2 \vec{E}_1 = -\nabla \times \frac{\partial \vec{B}_1}{\partial t} \tag{16.130}$$

Using equation (16.129) in (16.130), we have

$$\nabla(\nabla \cdot \vec{E}_1) - \nabla^2 \vec{E}_1 = -\frac{1}{c^2} \frac{\partial^2 \vec{E}_1}{\partial t^2} - \mu_0 \frac{\partial \vec{J}_1}{\partial t} \tag{16.131}$$

We assume that oscillating quantities behave sinusoidally as $e^{i(\vec{k} \cdot \vec{r} - \omega t)}$. Then, the time derivative $(\partial/\partial t)$ can be replaced by $-i\omega$, and the gradient ∇ by $i\vec{k}$. From equation (16.131), we get

$$i\vec{k}(i\vec{k} \cdot \vec{E}_1) - (i\vec{k})^2 \vec{E}_1 = -\frac{(-i\omega)^2}{c^2} \vec{E}_1 + \mu_0 i\omega \vec{J}_1$$

$$-\vec{k}(\vec{k} \cdot \vec{E}_1) + k^2 \vec{E}_1 = \frac{\omega^2}{c^2} \vec{E}_1 + \mu_0 i\omega \vec{J}_1$$

Let us take $\vec{J}_1 = J_x \hat{x} + J_y \hat{y} + J_z \hat{z}$. For a transverse wave, we have $\vec{k} \cdot \vec{E}_1 = 0$ and thus, along the z-axis, we have

$$c^2 k^2 E_1 = \omega^2 E_1 + \mu_0 i\omega c^2 J_z \qquad (\omega^2 - c^2 k^2) E_1 = -\mu_0 i\omega c^2 J_z \tag{16.132}$$

For the high frequency waves, the ions may be considered as fixed. Then the current density is due to motion of electrons along z-axis as

$$J_z = -n_{e0}\, e u_z \tag{16.133}$$

The equation of motion for electrons is

$$m_e n_e \left[\frac{\partial \vec{u}_e}{\partial t} + (\vec{u}_e \cdot \nabla)\vec{u}_e\right] = -e n_e (\vec{E} + \vec{u}_e \times \vec{B})$$

We express $\vec{u}_e = \vec{u}_{e0} + \vec{u}_{e1}$. Here, we are assuming that electron temperature T_e is zero (cold plasma). After linearisation with $\vec{u_{e0}} = 0$, and $\vec{E}_0 = 0$, we have

$$m_e \frac{\partial \vec{u}_{e1}}{\partial t} = -e\, \vec{E}_1 - e\, \vec{u}_{e1} \times \vec{B}_0$$

Since $\vec{E_1}$ is parallel to z-axis, we are interested in z-component only and therefore, the term $\vec{u}_{e1} \times \vec{B_0}$ does not contribute.

We assume that oscillating quantities behave sinusoidally as $e^{i(\vec{k} \cdot \vec{r} - \omega t)}$. Then, time derivative $(\partial/\partial t)$ can be replaced by $-i\omega$ and on linearisation along the z-axis, we get

$$m_e(-i\omega)u_z = -eE_1 \qquad u_z = \left(\frac{e}{im_e\omega}\right)E_1 \tag{16.134}$$

From equations (16.133) and (16.134), we have

$$J_z = -\left(\frac{n_{e0}e^2}{im_e\omega}\right)E_1$$

Using this value of J_z in equation (16.132), we have

$$(\omega^2 - c^2k^2)E_1 = \omega_p^2 E_1$$

Since $E_1 \neq 0$, we have

$$\omega^2 - c^2k^2 = \omega_p^2 \qquad \omega^2 = \omega_p^2 + c^2k^2$$

where $\omega_p = (n_{e0}e^2/m_e\epsilon_0)^{1/2}$ is the plasma frequency. This is the dispersion relation for ordinary, transverse, electromagnetic waves propagating perpendicular to $\vec{B_0}$. The dispersion curve is shown in Figure 6.

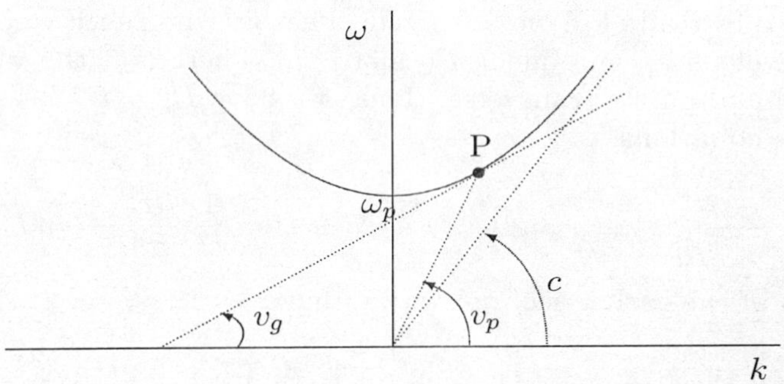

Figure 6: Variation of ω versus k for ordinary, transverse, electromagnetic wave perpendicular to $\vec{B_0}$

The phase velocity is

$$\frac{\omega^2}{k^2} = \frac{\omega_p^2}{k^2} + c^2 \qquad \text{or} \qquad \frac{\omega}{k} = c\sqrt{1 + \frac{\omega_p^2}{c^2\,k^2}} \equiv v_p$$

It shows that the phase velocity is always greater than c. The group velocity is

$$2\omega\, d\omega = 2c^2k\, dk \qquad \frac{d\omega}{dk} = \frac{c^2}{v_p} = c\left(1 + \frac{\omega_p^2}{c^2k^2}\right)^{-1/2} \equiv v_g$$

It shows that the group velocity is always smaller than c. When $k \to \infty$, the phase velocity $v_p \to c$, and thus, the group velocity $v_g \to c$.

Example 2: For electromagnetic, perpendicular, transverse, ordinary wave, derive dispersion relation, and show that the phase velocity is always larger than the speed of light c and the group velocity is always smaller than c.

14.2 Extraordinary wave

In this case, $\vec{E_1}$ is perpendicular to $\vec{B_0}$, and $\vec{k}$ is perpendicular to $\vec{B_0}$ as shown in Figure 7.

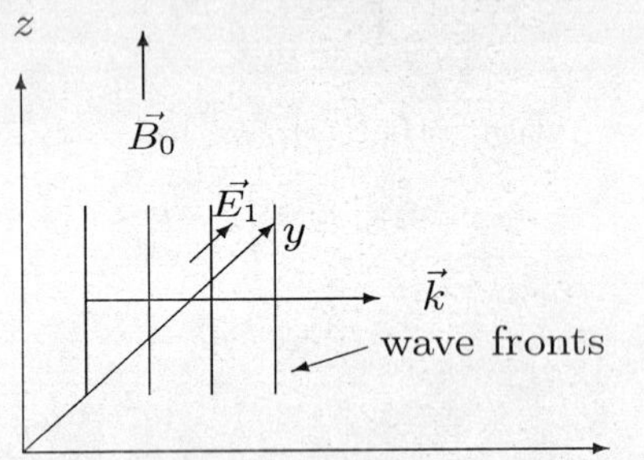

Figure 7: Propagation of extraordinary electromagnetic wave perpendicular to $\vec{B_0}$

Here, $\vec{B_0}$ being perpendicular to $\vec{E_1}$ affects the motion of electrons. We have $\vec{B_0} = B_0\ \hat{z}$, $\vec{k} = k\ \hat{x}$, and $\vec{E_1} = E_1\ \hat{y}$. However, it turns out that waves with $\vec{E_1} \perp \vec{B_0}$ are elliptically polarized instead of plane polarized. That is, when such wave propagates into plasma, it develops a component E_x along $\vec{k}$. Therefore, the wave becomes partly longitudinal and partly transverse. Thus, we have $\vec{E_1} = E_x\ \hat{x} + E_y\ \hat{y}$. Let us consider Maxwell's equations

$$\nabla \times \vec{E} = -\frac{\partial \vec{B}}{\partial t} \qquad \text{and} \qquad \nabla \times \vec{B} = \frac{1}{c^2}\frac{\partial \vec{E}}{\partial t} + \mu_0\ \vec{J}$$

Following the step of the earlier section, we can have

$$-\vec{k}(\vec{k}\ .\ \vec{E_1}) + k^2\vec{E_1} = \frac{\omega^2}{c^2}\ \vec{E_1} + \mu_0 i\omega\ \vec{J_1}$$

Now, we have $\vec{k}\ .\ \vec{E_1} = kE_x$. Thus,

$$-c^2kE_x\vec{k} + \ c^2k^2\vec{E_1} = \omega^2\vec{E_1} + \mu_0 i\omega c^2\ \vec{J_1}$$

$$(\omega^2 - c^2k^2)\vec{E_1} + \ c^2kE_x\vec{k} = -\mu_0 i\omega c^2\ \vec{J_1} \tag{16.135}$$

For high frequency waves, ions may be considered as fixed. Then the current density is due to the motion of electrons as

$$\vec{J_1} = -n_{e0}e\vec{u}_{e1} \tag{16.136}$$

Using equation (16.136) in (16.135), we have

$$(\omega^2 - c^2k^2)\vec{E_1} + c^2kE_x\vec{k} = \mu_0 i\omega c^2 n_{e0} e\, \vec{u}_{e1}$$

The x and y components of this equation are

$$(\omega^2 - c^2k^2)E_x + c^2k^2E_x = \mu_0 i\omega c^2 n_{e0} e\, u_x$$

$$\omega^2 E_x = \mu_0 i\omega c^2 n_{e0} e\, u_x \qquad (16.137)$$

and

$$(\omega^2 - c^2k^2)E_y = \mu_0 i\omega c^2 n_{e0} e\, u_y \qquad (16.138)$$

The equation of motion for electrons is

$$m_e n_e\Big[\frac{\partial \vec{u}_e}{\partial t} + (\vec{u}_e \,.\, \nabla)\vec{u}_e\Big] = -en_e(\vec{E} + \vec{u}_e \times \vec{B})$$

We express $\vec{u}_e = \vec{u}_{e0} + \vec{u}_{e1}$. Here, we are assuming that electron temperature T_e is zero (cold plasma). After linearisation with $\vec{u_{e0}} = 0$ and $\vec{E_0} = 0$, we have

$$m_e\,\frac{\partial \vec{u}_{e1}}{\partial t} = -e(\vec{E_1} + \vec{u}_{e1} \times \vec{B_0})$$

We assume that oscillating quantities behave sinusoidally as $e^{i(\vec{k}\,.\,\vec{r} - \omega t)}$. Then time derivative $(\partial/\partial t)$ can be replaced by $-i\omega$ and we get

$$m_e(-i\omega)\vec{u}_{e1} = -e(\vec{E_1} + \vec{u}_{e1} \times \vec{B_0})$$

Writing $\vec{u}_{e1} = u_x\,\hat{x} + u_y\,\hat{y} + u_z\,\hat{z}$, and considering for non-trivial x and y components, we have

$$im_e\omega u_x = e(E_x + u_yB_0) \qquad \text{and} \qquad im_e\omega u_y = e(E_y - u_xB_0)$$

We have

$$u_x = \frac{e}{im_e\omega}(E_x + u_yB_0) = \frac{eB_0}{im_e\omega}\Big(\frac{E_x}{B_0} + u_y\Big) \qquad (16.139)$$

and

$$u_y = \frac{e}{im_e\omega}(E_y - u_xB_0) = \frac{eB_0}{im_e\omega}\Big(\frac{E_y}{B_0} - u_x\Big) \qquad (16.140)$$

Using equation (16.140) in (16.139), we have

$$u_x = \frac{eB_0}{im_e\omega}\Big[\frac{E_x}{B_0} + \frac{eB_0}{im_e\omega}\Big(\frac{E_y}{B_0} - u_x\Big)\Big]$$

$$= \frac{eE_x}{im_e\omega} - \frac{e^2E_yB_0}{m_e^2\omega^2} + \frac{e^2B_0^2}{m_e^2\omega^2}\,u_x$$

Therefore,

$$u_x\Big(1-\frac{\omega_c^2}{\omega^2}\Big) = \frac{e}{m_e\omega}\Big[-iE_x - \frac{\omega_c}{\omega}\,E_y\Big]$$

$$u_x = \frac{e}{m_e\omega}\Big[-iE_x - \frac{\omega_c}{\omega}\,E_y\Big]\Big(1-\frac{\omega_c^2}{\omega^2}\Big)^{-1} \tag{16.141}$$

where $\omega_c = eB_0/m_e$ is cyclotron frequency. Using equation (16.139) in (16.140), we have

$$u_y = \frac{eB_0}{im_e\omega}\Big[\frac{E_y}{B_0} - \frac{eB_0}{im_e\omega}\Big(\frac{E_x}{B_0}+u_y\Big)\Big] = \frac{eE_y}{im_e\omega} + \frac{e^2E_xB_0}{m_e^2\omega^2} + \frac{e^2B_0^2}{m_e^2\omega^2}\,u_y$$

Therefore

$$u_y\Big(1-\frac{\omega_c^2}{\omega^2}\Big) = \frac{e}{m_e\omega}\Big[-iE_y + \frac{\omega_c}{\omega}E_x\Big]$$

$$u_y = \frac{e}{m_e\omega}\Big[-iE_y + \frac{\omega_c}{\omega}\,E_x\Big]\Big(1-\frac{\omega_c^2}{\omega^2}\Big)^{-1} \tag{16.142}$$

Using equation (16.141) in (16.137), we have

$$\omega^2E_x = \frac{\mu_0ic^2ne^2}{m_e}\Big[-iE_x - \frac{\omega_c}{\omega}\,E_y\Big]\Big(1-\frac{\omega_c^2}{\omega^2}\Big)^{-1}$$

$$\omega^2\Big(1-\frac{\omega_c^2}{\omega^2}\Big)E_x = -i\omega_p^2\Big[iE_x + \frac{\omega_c}{\omega}E_y\Big] = \omega_p^2E_x - \frac{i\omega_p^2\omega_c}{\omega}\,E_y$$

$$(\omega^2-\omega_h^2)E_x + \frac{i\omega_p^2\omega_c}{\omega}\,E_y = 0 \tag{16.143}$$

where $\omega_p = \sqrt{n_{e0}e^2/\epsilon_0m_e}$ is plasma frequency, $\omega_h = \sqrt{\omega_c^2+\omega_p^2}$ is the upper hybrid frequency. Using equation (16.142) in (16.138), we have

$$(\omega^2-c^2k^2)E_y = \frac{\mu_0ic^2n_{e0}e^2}{m_e}\Big[-iE_y + \frac{\omega_c}{\omega}\,E_x\Big]\Big(1-\frac{\omega_c^2}{\omega^2}\Big)^{-1}$$

$$(\omega^2-c^2k^2)\Big(1-\frac{\omega_c^2}{\omega^2}\Big)E_y = \omega_p^2E_y + \frac{i\omega_p^2\omega_c}{\omega}E_x$$

$$-\frac{i\omega_p^2\omega_c}{\omega}E_x + \Big[(\omega^2-c^2k^2)\Big(1-\frac{\omega_c^2}{\omega^2}\Big) - \omega_p^2\Big]E_y = 0 \tag{16.144}$$

For convenience, let us express equations (16.143) and (16.144) as

$$AE_x + BE_y = 0 \qquad \text{and} \qquad CE_x + DE_y = 0$$

These simultaneous equations are satisfied when

$$\begin{vmatrix} \mathrm{A} & \mathrm{B} \\ \mathrm{C} & \mathrm{D} \end{vmatrix} = 0$$

$$AD - CB = 0 \qquad\qquad AD = CB$$

Using A, B, C and D, we get

$$(\omega^2 - \omega_h^2)\Big[(\omega^2 - c^2k^2)\Big(1 - \frac{\omega_c^2}{\omega^2}\Big) - \omega_p^2\Big] = \frac{i\omega_p^2\omega_c}{\omega}\Big(-\frac{i\omega_p^2\omega_c}{\omega}\Big)$$

$$(\omega^2 - \omega_h^2)\Big[\omega^2 - \omega_h^2 - c^2k^2\Big(1 - \frac{\omega_c^2}{\omega^2}\Big)\Big] = \frac{\omega_p^4\omega_c^2}{\omega^2}$$

$$c^2k^2\Big(1 - \frac{\omega_c^2}{\omega^2}\Big) = (\omega^2 - \omega_h^2) - \frac{\omega_p^4\omega_c^2/\omega^2}{\omega^2 - \omega_h^2}$$

Thus,

$$\frac{c^2k^2}{\omega^2} = \frac{(\omega^2 - \omega_h^2) - \Big[(\omega_p^2\omega_c/\omega)^2/(\omega^2 - \omega_h^2)\Big]}{\omega^2 - \omega_c^2} = \frac{(\omega^2 - \omega_h^2)^2 - (\omega_p^4\omega_c^2/\omega^2)}{(\omega^2 - \omega_c^2)(\omega^2 - \omega_h^2)}$$

$$= \frac{(\omega^2 - \omega_h^2)(\omega^2 - \omega_c^2 - \omega_p^2) - (\omega_p^4\omega_c^2/\omega^2)}{(\omega^2 - \omega_c^2)(\omega^2 - \omega_h^2)}$$

$$= \frac{(\omega^2 - \omega_h^2)(\omega^2 - \omega_c^2)}{(\omega^2 - \omega_c^2)(\omega^2 - \omega_h^2)} - \frac{\omega_p^2(\omega^2 - \omega_h^2) + (\omega_p^4\omega_c^2/\omega^2)}{(\omega^2 - \omega_c^2)(\omega^2 - \omega_h^2)}$$

$$= 1 - \frac{\omega_p^2(\omega^2 - \omega_h^2) + (\omega_p^4\omega_c^2/\omega^2)}{(\omega^2 - \omega_c^2)(\omega^2 - \omega_h^2)}$$

$$= 1 - \frac{\omega_p^2}{\omega^2}\Big[\frac{\omega^2(\omega^2 - \omega_c^2) - \omega_p^2(\omega^2 - \omega_c^2)}{(\omega^2 - \omega_c^2)(\omega^2 - \omega_h^2)}\Big] = 1 - \frac{\omega_p^2}{\omega^2}\Big[\frac{\omega^2 - \omega_p^2}{(\omega^2 - \omega_h^2)}\Big]$$

Therefore, we have

$$\frac{c^2\,k^2}{\omega^2} = \frac{c^2}{v_p^2} = 1 - \frac{\omega_p^2}{\omega^2}\Big[\frac{\omega^2 - \omega_p^2}{(\omega^2 - \omega_h^2)}\Big] \tag{16.145}$$

This is dispersion relation for extraordinary waves. It is considerably complicated.

Cut-offs and resonances

In the dispersion relation for extraordinary waves, one can find cut-offs and resonances. When refractive index $\widetilde{n} = c/v_p$ tends to zero, we have a cut-off. At the cut-off, wavelength tends to have value infinite and the wave is reflected back. When refractive index $\widetilde{n}$ tends to infinite, we have a resonance. At the resonance, wavelength tends to have value zero, and the wave is absorbed. When a wave propagates through plasma in which ω_p and ω_c are varying, it may encounter cut-offs as well as resonances. Equation (16.145) shows that at $\omega = \omega_p$, we have $v_p = c$.

Resonances for extraordinary waves can be obtained from equation (16.145) by putting $k = \infty$. It occurs when $\omega \to \omega_h$. Hence, at $\omega = \omega_h = \sqrt{\omega_p^2 + \omega_c^2}$, wave encounters a resonance. When the wave of frequency ω approaches the resonance point, its phase velocity and group velocity both approach to zero, and the wave energy is converted into upper hybrid oscillations. Cut-offs for extraordinary waves can be obtained from equation (16.145) by putting $k = 0$. Thus, we have

$$1 - \frac{\omega_p^2(\omega^2 - \omega_p^2)}{\omega^2(\omega^2 - \omega_h^2)} = 0; \qquad \frac{\omega_p^2}{\omega^2} = \frac{\omega^2 - \omega_p^2 - \omega_c^2}{\omega^2 - \omega_p^2}; \qquad 1 - \frac{\omega_p^2}{\omega^2} = \frac{\omega_c^2/\omega^2}{1 - \omega_p^2/\omega^2}$$

$$\left(1 - \frac{\omega_p^2}{\omega^2}\right)^2 = \frac{\omega_c^2}{\omega^2}; \qquad 1 - \frac{\omega_p^2}{\omega^2} = \pm\frac{\omega_c}{\omega} \qquad \omega^2 \mp \omega\omega_c - \omega_p^2 = 0$$

Thus, the cut-off frequencies are

$$\omega = \frac{\pm\omega_c \pm \sqrt{\omega_c^2 + 4\omega_p^4}}{2}$$

For convention that ω is always positive, the roots are

$$\omega_R = \frac{\omega_c + \sqrt{\omega_c^2 + 4\omega_p^4}}{2} \qquad \text{and} \qquad \omega_L = \frac{-\omega_c + \sqrt{\omega_c^2 + 4\omega_p^4}}{2}$$

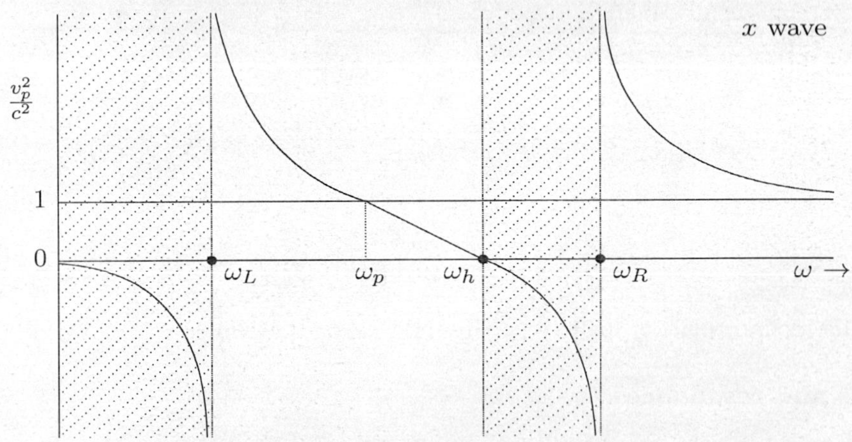

Figure 8: Variation of phase velocity with frequency for extraordinary wave. The wave does not propagate in shaded regions.

The cut-off frequencies ω_R and ω_L are known as right-hand and left-hand cutoffs, respectively. Cut-off and resonance frequencies divide the dispersion diagram into regions of propagation and of non-propagation. Instead of ω versus k diagram, a plot of (ω^2/c^2k^2) versus ω is shown in Figure 8, where the wave does not propagate in the shaded region.

15. Electromagnetic waves parallel to $\vec{B_0}$

Earlier , we have considered the case of perpendicular propagation, $\vec{k} \perp \vec{B_0}$. Now, we take the case of $\vec{k} \parallel \vec{B_0}$, as shown in Figure 9.

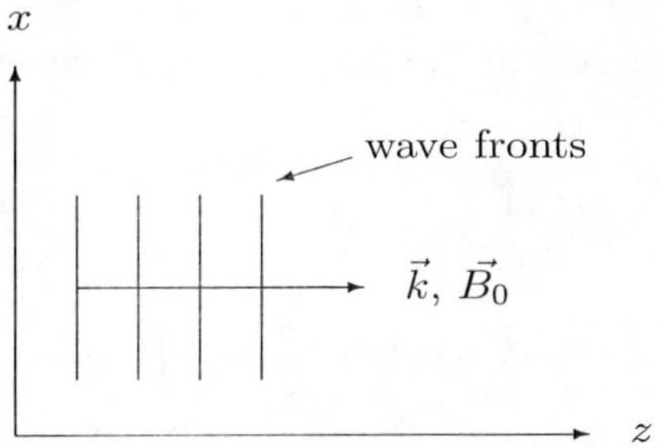

Figure 9: Propagation of electromagnetic wave parallel to $\vec{B_0}$

We have $\vec{B_0} = B_0\ \hat{z}$, $\vec{k} = k\ \hat{z}$. For transverse propagation, we take $\vec{E_1} = E_x\ \hat{x} + E_y\ \hat{y}$. Let us consider Maxwell's equations

$$\nabla \times \vec{E} = -\frac{\partial \vec{B}}{\partial t} \qquad \text{and} \qquad \nabla \times \vec{B} = \frac{\partial \vec{E}}{\partial t} + \mu_0\ \vec{J}$$

Following the step of the earlier section, we can have

$$-\vec{k}(\vec{k}\ .\ \vec{E_1}) + k^2\vec{E_1} = \frac{\omega^2}{c^2}\ \vec{E_1} + \mu_0 i\omega\ \vec{J_1}$$

Now, we have $\vec{k}\ .\ \vec{E_1} = 0$. Thus, we have

$$c^2k^2\vec{E_1} = \omega^2\vec{E_1} + c^2\mu_0 i\omega\ \vec{J_1} \qquad\qquad (\omega^2 - c^2k^2)\vec{E_1} = -c^2\mu_0 i\omega\ \vec{J_1} \tag{16.146}$$

For high frequency waves, ions may be considered as fixed. Then current density is due to the motion of electrons as

$$\vec{J_1} = -n_{e0}e\vec{u}_{e1} \tag{16.147}$$

Using equation (16.147) in (16.146), we have

$$(\omega^2 - c^2k^2)\vec{E_1} = c^2\ \mu_0\ i\ \omega\ n_{e0}e\ \vec{u}_{e1} \tag{16.148}$$

Writing $\vec{u}_{e1} = u_x\ \hat{x} + u_y\ \hat{y} + u_z\ \hat{z}$ and $\vec{E_1} = E_x\ \hat{x} + E_y\ \hat{y}$, the x and y components of this equation are

$$(\omega^2 - c^2k^2)E_x = c^2\ \mu_0\ i\ \omega\ n_{e0}\ e\ u_x \tag{16.149}$$

and

$$(\omega^2 - c^2k^2)E_y = c^2\ \mu_0\ i\ \omega\ n_{e0}\ e\ u_y \tag{16.150}$$

The equation of motion for electrons is

$$m_e n_e\Big[\frac{\partial \vec{u}_e}{\partial t} + (\vec{u}_e\ .\ \nabla)\vec{u}_e\Big] = -en_e(\vec{E} + \vec{u}_e \times \vec{B})$$

Here, we are assuming that electron temperature T_e is zero. We express $\vec{u}_e = \vec{u}_{e0} + \vec{u}_{e1}$. After linearisation with $\vec{u_{e0}} = 0$, and $\vec{E_0} = 0$, we have

$$m_e \frac{\partial \vec{u}_{e1}}{\partial t} = -e(\vec{E_1} + \vec{u}_{e1} \times \vec{B_0})$$

Following the steps discussed in the earlier section, we get

$$u_x = \frac{e}{m_e \omega}\Big[-iE_x - \frac{\omega_c}{\omega}\,E_y\Big]\Big(1 - \frac{\omega_c^2}{\omega^2}\Big)^{-1} \tag{16.151}$$

and

$$u_y = \frac{e}{m_e \omega}\Big[-iE_y + \frac{\omega_c}{\omega}\,E_x\Big]\Big(1 - \frac{\omega_c^2}{\omega^2}\Big)^{-1} \tag{16.152}$$

where $\omega_c = eB_0/m_e$ is cyclotron frequency. Using equation (16.151) in (16.149), we have

$$(\omega^2 - c^2k^2)E_x = \frac{c^2\ \mu_0\ i\ n_{e0}\ e^2}{m_e}\Big[-iE_x - \frac{\omega_c}{\omega}\,E_y\Big]\Big(1 - \frac{\omega_c^2}{\omega^2}\Big)^{-1}$$

$$(\omega^2 - c^2k^2)\Big(1 - \frac{\omega_c^2}{\omega^2}\Big)E_x = \omega_p^2\ E_x - \frac{i\omega_p^2\omega_c}{\omega}\ E_y$$

$$\Big[(\omega^2 - c^2k^2)\Big(1 - \frac{\omega_c^2}{\omega^2}\Big) - \omega_p^2\Big]E_x + \frac{i\omega_p^2\omega_c}{\omega}\ E_y = 0 \tag{16.153}$$

where $\omega_p = \sqrt{n_{e0}e^2/\epsilon_0 m_e}$ is plasma frequency. Using equation (16.152) in (16.150), we have

$$(\omega^2 - c^2k^2)E_y = \frac{c^2\ \mu_0\ i\ n_{e0}\ e^2}{m_e}\Big[-iE_y + \frac{\omega_c}{\omega}\,E_x\Big]\Big(1 - \frac{\omega_c^2}{\omega^2}\Big)^{-1}$$

$$(\omega^2 - c^2k^2)\Big(1 - \frac{\omega_c^2}{\omega^2}\Big)E_y = \omega_p^2\ E_y + \frac{i\omega_p^2\omega_c}{\omega}\ E_x$$

$$-\frac{i\omega_p^2\omega_c}{\omega}\ E_x + \Big[(\omega^2 - c^2k^2)\Big(1 - \frac{\omega_c^2}{\omega^2}\Big) - \omega_p^2\Big]E_y = 0 \tag{16.154}$$

Defining a parameter $\alpha = \omega_p^2/(1 - \omega_c^2/\omega^2)$, equations (16.153) and (16.154) can be written as

$$\Big(\omega^2 - c^2k^2 - \alpha\Big)E_x + \frac{i\alpha\omega_c}{\omega}\ E_y = 0 \tag{16.155}$$

and

$$-\frac{i\alpha\omega_c}{\omega}\ E_x + (\omega^2 - c^2k^2 - \alpha)E_y = 0 \tag{16.156}$$

For convenience, let us express equations (16.155) and (16.156) as

$$AE_x + BE_y = 0 \qquad \text{and} \qquad CE_x + DE_y = 0$$

These simultaneous equations are satisfied when

$$\begin{vmatrix} A & B \\ C & D \end{vmatrix} = 0$$

$$AD - CB = 0 \qquad\qquad AD = CB$$

Using A, B, C and D, we get

$$\left(\omega^2 - c^2k^2 - \alpha\right)^2 = -\frac{i\alpha\omega_c}{\omega}\frac{i\alpha\omega_c}{\omega} = \left(\frac{\alpha\omega_c}{\omega}\right)^2$$

$$\omega^2 - c^2k^2 - \alpha = \pm\alpha\omega_c/\omega$$

Thus,

$$\omega^2 - c^2k^2 = \alpha\left(1 \pm \frac{\omega_c}{\omega}\right) = \frac{\omega_p^2}{1-(\omega_c^2/\omega^2)}\left(1 \pm \frac{\omega_c}{\omega}\right)$$

$$= \omega_p^2 \frac{1 \pm (\omega_c/\omega)}{[1+(\omega_c/\omega)][1-(\omega_c/\omega)]} = \frac{\omega_p^2}{1 \mp (\omega_c/\omega)}$$

The sign $\mp$ shows that there are two possible solutions corresponding to two different waves which propagate along $\vec{B}_0$. The dispersion relations are

$$\frac{c^2k^2}{\omega^2} = 1 - \frac{\omega_p^2/\omega^2}{1-(\omega_c/\omega)} \qquad \text{R wave}$$

and

$$\frac{c^2k^2}{\omega^2} = 1 - \frac{\omega_p^2/\omega^2}{1+(\omega_c/\omega)} \qquad \text{L wave}$$

The R and L waves are circularly polarized. R stands for right-hand circular polarization whereas L stands for left-hand circular polarization.

16. Hydromagnetic waves

In the preceding sections, we have accounted for high frequency electron waves. Now, we discuss for low frequency ion waves. Out of several possible modes, we consider here two of them: (i) hydromagnetic waves along $\vec{B}_0$, called Alfvén waves and (ii) magnetosonic waves across $\vec{B}_0$. Here, we discuss about Alfvén waves.

In a plane geometry, Alfvén wave has $\vec{k}$ along $\vec{B_0}$; $\vec{E_1}$ and $\vec{J_1}$ both perpendicular to $\vec{B_0}$; and both $\vec{B_1}$ and $\vec{u_1}$ perpendicular to both $\vec{B_0}$ and $\vec{E_1}$ as shown in Figure 10.

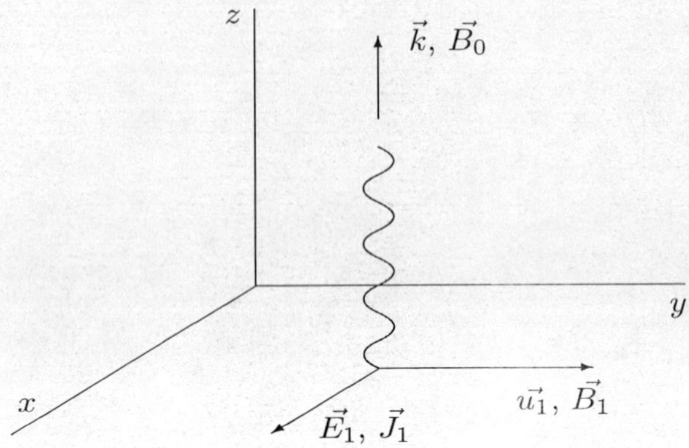

Figure 10: Alfvén wave propagating along $\vec{B_0}$.

Let us consider Maxwell's equations

$$\nabla \times \vec{E} = -\frac{\partial \vec{B}}{\partial t} \qquad \text{and} \qquad \nabla \times \vec{B} = \frac{1}{c^2}\frac{\partial \vec{E}}{\partial t} + \mu_0 \vec{J}$$

Following the step of the earlier section, we can have

$$-\vec{k}(\vec{k} \,.\, \vec{E_1}) + k^2\vec{E_1} = \frac{\omega^2}{c^2}\vec{E_1} + \mu_0 i\omega \vec{J_1}$$

Since $\vec{E_1}$ and $\vec{k}$ are perpendicular to each other, we have $\vec{k} \,.\, \vec{E_1} = 0$, and therefore,

$$(\omega^2 - c^2k^2)\vec{E_1} = -c^2\mu_0 i\omega \vec{J_1} \tag{16.157}$$

Further, we have taken $\vec{E_1}$ along x-axis, and therefore only x-component of this equation is non-trivial. For low frequency waves, current density $\vec{J}$ has contributions from both electrons and ions as

$$\vec{J} = ne(\vec{u}_i - \vec{u}_e)$$

Considering $\vec{J} = \vec{J_0} + \vec{J_1}$, $\vec{u_i} = \vec{u}_{i0} + \vec{u}_{i1}$, $\vec{u}_e = \vec{u}_{e0} + \vec{u}_{e1}$, the linearization process provides

$$\vec{J_1} = ne(\vec{u}_{i1} - \vec{u}_{e1}) \tag{16.158}$$

Considering $\vec{u}_{i1} = \vec{u}_{ix}\,\hat{x} + \vec{u}_{iy}\,\hat{y} + \vec{u}_{iz}\,\hat{z}$, $\vec{u}_{e1} = \vec{u}_{ex}\,\hat{x} + \vec{u}_{ey}\,\hat{y} + \vec{u}_{ez}\,\hat{z}$, and using equation (16.158) in (16.157) and taking x-component of resultant equation, we have

$$(\omega^2 - c^2k^2)E_1 = -c^2\mu_0 i\omega ne(u_{ix} - u_{ex}) \tag{16.159}$$

We assume also that there are no thermal motions ($T_i = 0$). The equations of motion is

$$m_i n_i \Big[\frac{\partial \vec{u_i}}{\partial t} + (\vec{u_i}\ .\ \nabla)\vec{u_i}\Big] = e n_i(\vec{E} + \vec{u_i} \times \vec{B_0})$$

On linearisation, this equation reduces to

$$m_i\ \frac{\partial \vec{u}_{i1}}{\partial t} = e\vec{E_1} + e\ \vec{u}_{i1} \times \vec{B_0} \tag{16.160}$$

We assume that oscillating quantities behave sinusoidally as $e^{i(\vec{k}\ .\ \vec{r}-\omega t)}$. Then time derivative $(\partial/\partial t)$ can be replaced by $-i\omega$, and gradient ∇ by $i\vec{k}$. The x and y components of equation (16.160) are

$$-i\omega m_i u_{ix} = eE_1 + eu_{iy}B_0 \qquad \text{and} \qquad i\omega m_i u_{iy} = eu_{ix}B_0$$

Using the value of u_{iy} from second in the first, we get

$$-i\omega m_i u_{ix}\Big[1 - \Big(\frac{eB_0}{\omega m_i}\Big)^2\Big] = eE_1$$

Thus, we have

$$u_{ix} = \frac{ie}{m_i\omega}\Big(1 - \frac{\Omega_c^2}{\omega^2}\Big)^{-1} E_1 \tag{16.161}$$

where $\Omega_c = eB_0/m_i$ is ion cyclotron frequency. For the equation of motion for electrons, u_{ex} can be written from this expression by replacing m_i by m_e, Ω_c by ω_c, e by $-e$, and thus, we have

$$u_{ex} = -\frac{ie}{m_e\omega}\Big(1 - \frac{\omega_c^2}{\omega^2}\Big)^{-1} E_1$$

Obviously, $\omega_c >> \Omega_c$. For the situation, $\omega^2 << \omega_c^2$, we have

$$u_{ex} = \frac{ie}{m_e\omega}\frac{\omega^2}{\omega_c^2}\ E_1 \to 0 \tag{16.162}$$

For $\omega^2 << \omega_c^2$, Larmor gyrations of electrons are neglected and electrons have $\vec{E}\times\vec{B}$ drift in y-direction. Using equations (16.161) and (16.162) in (16.159), we get

$$(\omega^2 - c^2k^2)E_1 = \frac{c^2\mu_0 n e^2}{m_i}\Big(1 - \frac{\Omega_c^2}{\omega^2}\Big)^{-1} E_1 \tag{16.163}$$

Since $E_1 \neq 0$, and defining the ion plasma frequency $\Omega_p^2 = ne^2/m_i\epsilon_0$, we get

$$\omega^2 - c^2k^2 = \Omega_p^2\Big(1 - \frac{\Omega_c^2}{\omega^2}\Big)^{-1}$$

Further, we assume that hydromagnetic waves have frequencies well below the ion cyclotron frequency $(\omega^2 << \Omega_c^2)$, so that

$$\omega^2 - c^2k^2 = \Omega_p^2\Big(-\frac{\omega^2}{\Omega_c^2}\Big) = -\omega^2\ \frac{\Omega_p^2}{\Omega_c^2}$$

$$= -\omega^2 \frac{ne^2}{m_i \epsilon_0} \frac{m_i^2}{e^2 B_0^2} = -\omega^2 \frac{\rho}{\epsilon_0 B_0^2}$$

where $\rho = nm_i$ is mass density. Hence,

$$\omega^2 \left[1 + \frac{\rho}{B_0^2 \epsilon_0}\right] = c^2 k^2 \qquad \text{or} \qquad \frac{\omega^2}{k^2} = \frac{c^2}{1 + (\rho/\epsilon_0 B_0^2)} \tag{16.164}$$

The denominator on right side of equation (16.164) may be interpreted as relative dielectric constant for low frequency perpendicular motions. Equation (16.164) may be approximated as

$$\frac{\omega^2}{k^2} = \frac{c^2}{(\rho/\epsilon_0 B_0^2)} = \frac{B_0^2}{\mu_0 \rho} \qquad \frac{\omega}{k} = v_p = \frac{B_0}{(\mu_0 \rho)^{1/2}}$$

These hydromagnetic waves travel along $\vec{B_0}$ at a constant velocity v_A, called Alfvén velocity

$$v_A = \frac{B_0}{(\mu_0 \rho)^{1/2}}$$

17. Magnetosonic waves

Here, we consider low frequency electromagnetic waves propagating across the magnetic field $\vec{B_0}$ (Figure 11).

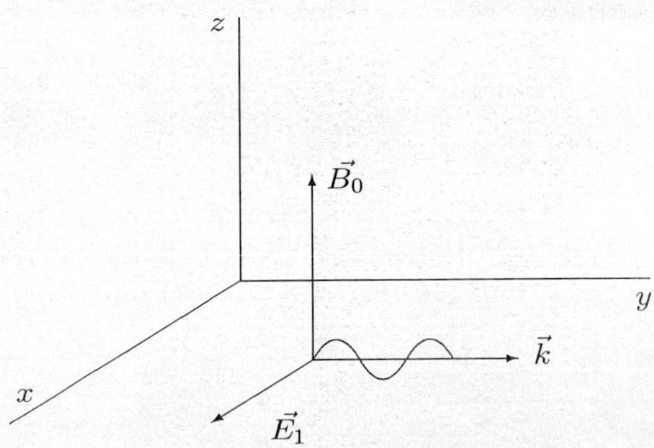

Figure 11: Magnetosonic wave propagating at right angle to $\vec{B_0}$.

We consider $\vec{B_0} = B_0 \, \hat{z}$ and $\vec{E_1} = E_1 \, \hat{x}$, but now let $\vec{k} = k \, \hat{y}$, so that the $\vec{E_1} \times \vec{B_0}$ drifts lie along $\vec{k}$. Hence, the plasma is compressed and released along y-axis during the course of oscillations. Therefore, we have to include pressure term ∇p along y-axis in the equation of motion.

$$\nabla \times \vec{E} = -\frac{\partial \vec{B}}{\partial t} \qquad \text{and} \qquad \nabla \times \vec{B} = \frac{1}{c^2} \frac{\partial \vec{E}}{\partial t} + \mu_0 \vec{J}$$

Following the step of the earlier section, we can have

$$-\vec{k}(\vec{k}\ .\ \vec{E_1}) + k^2\vec{E_1} = \frac{\omega^2}{c^2}\ \vec{E_1} + \mu_0 i\omega\ \vec{J_1}$$

Since $\vec{E_1}$ and $\vec{k}$ are perpendicular to each other, we have $\vec{k}\ .\ \vec{E_1} = 0$, and therefore,

$$(\omega^2 - c^2k^2)E_1 = -c^2\mu_0 i\omega\ \vec{J_1} \tag{16.165}$$

Further, we have taken $\vec{E_1}$ along x-axis, and therefore only x-component of this equation is non-trivial. For low frequency waves, current density $\vec{J_1}$ has contributions from both electrons and ions as

$$\vec{J_1} = ne(\vec{v_{i1}} - \vec{v_{e1}}) \tag{16.166}$$

Considering $\vec{u}_{i1} = \vec{u}_{ix}\ \hat{x} + \vec{u}_{iy}\ \hat{y} + \vec{u}_{iz}\ \hat{z}$, $\vec{u}_{e1} = \vec{u}_{ex}\ \hat{x} + \vec{u}_{ey}\ \hat{y} + \vec{u}_{ez}\ \hat{z}$, and using equation (16.166) in (16.165) and taking x-component of resultant equation, we have

$$(\omega^2 - c^2k^2)E_1 = -c^2\mu_0 i\omega ne(u_{ix} - u_{ex}) \tag{16.167}$$

The equations of motion and continuity are

$$m_i n_i\left[\frac{\partial\vec{u_i}}{\partial t} + (\vec{u_i}\ .\ \nabla)\vec{u_i}\right] = en_i(\vec{E} + \vec{u_i}\times\vec{B_0}) - \gamma_i KT_i\nabla n_i$$

and

$$\frac{\partial n_i}{\partial t} + \nabla\ .\ (n_i\vec{u_i}) = 0$$

On linearisation, these equations reduce to

$$m_i n\ \frac{\partial\vec{u_{i1}}}{\partial t} = en(\vec{E_1} + \vec{u}_{i1}\times\vec{B_0}) - \gamma_i KT_i\nabla n_{i1} \tag{16.168}$$

and

$$\frac{\partial n_{i1}}{\partial t} + n\nabla\ .\ \vec{u_{i1}} = 0 \tag{16.169}$$

We assume that oscillating quantities behave sinusoidally as $e^{i(\vec{k}\ .\ \vec{r}-\omega t)}$. Then time derivative $(\partial/\partial t)$ can be replaced by $-i\omega$, and gradient ∇ by $i\vec{k}$. The x and y components of equation (16.168) are

$$-i\omega m_i n u_{ix} = en(E_1 + u_{iy}\ B_0)$$

and

$$i\omega m_i n u_{iy} = enu_{ix}B_0 - \gamma_i KT_i ikn_{i1}$$

Remember that density variation is along y-axis. Thus, we have

$$u_{ix} = \frac{ie}{m_i\omega}(E_1 + u_{iy}B_0) \tag{16.170}$$

and

$$u_{iy} = \frac{ie}{m_i\omega}(-u_{ix}B_0) + \frac{k}{\omega}\frac{\gamma_i KT_i}{m_i}\frac{n_{i1}}{n} \tag{16.171}$$

Linearisation of equation (16.169) gives

$$-i\omega n_{i1} = -niku_{iy} \qquad \frac{n_{i1}}{n} = \frac{k}{\omega}\,u_{iy} \tag{16.172}$$

Using equation (16.172) in (16.171), we get

$$u_{iy} = -\frac{ieB_0}{m_i\omega}\,u_{ix} + \frac{k^2}{\omega^2}\frac{\gamma_i KT_i}{m_i}\,u_{iy} = -\frac{i\Omega_c}{\omega}u_{ix} + \alpha_i u_{iy}$$

where

$$\Omega_c = \frac{eB_0}{m_i} \qquad \text{and} \qquad \alpha_i = \frac{k^2}{\omega^2}\,\frac{\gamma_i KT_i}{m_i}$$

$$u_{iy}(1-\alpha_i) = -\frac{i\Omega_c}{\omega}u_{ix} \qquad u_{iy} = -\frac{i\Omega_c}{\omega}\,u_{ix}(1-\alpha_i)^{-1} \tag{16.173}$$

Using equation (16.173) in (16.170), we get

$$u_{ix} = \frac{ie}{m_i\omega}\,E_1 + \frac{i\Omega_c}{\omega}\Big(-\frac{i\Omega_c}{\omega}\Big)(1-\alpha_i)^{-1}u_{ix}$$

$$= \frac{ie}{m_i\omega}\,E_1 + \frac{\Omega_c^2}{\omega^2}(1-\alpha_i)^{-1}u_{ix}$$

$$u_{ix}\Big(1 - \frac{\Omega_c^2/\omega^2}{1-\alpha_i}\Big) = \frac{ie}{m_i\omega}E_1 \qquad u_{ix}\Big(\frac{1-\alpha_i-\Omega_c^2/\omega^2}{1-\alpha_i}\Big) = \frac{ie}{m_i\omega}E_1 \tag{16.174}$$

Equation for electron, corresponding to equation (16.174) can be obtained by replacing e by $-e$, m_i by m_e, Ω_c by ω_c and α_i by α_e as

$$u_{ex}\Big(\frac{1-\alpha_e-\omega_c^2/\omega^2}{1-\alpha_e}\Big) = -\frac{ie}{m_e\omega}E_1 \tag{16.175}$$

where

$$\omega_c = \frac{eB_0}{m_e} \qquad \text{and} \qquad \alpha_e = \frac{k^2}{\omega^2}\frac{\gamma_e KT_e}{m_e}$$

for $\omega^2 << \omega_c^2$ and $\omega^2 << k^2\, u_{th}^2$, ($u_{th} = \sqrt{2KT_e/m_e}$) we have

$$u_{ex}\Big(\frac{-\omega_c^2/\omega^2}{1-\alpha_e}\Big) = -\frac{ie}{m_e\omega}\,E_1 \qquad u_{ex} = \frac{ie}{m_e\omega}\frac{\omega^2}{\omega_c^2}(1-\alpha_e)E_1$$

$$u_{ex} = \frac{ie}{m_e\omega}\frac{\omega^2}{\omega_c^2}\Big(1 - \frac{k^2}{\omega^2}\frac{\gamma_e KT_e}{m_e}\Big)E_1 \to -\frac{ik^2}{\omega B_0^2}\frac{\gamma_e KT_e}{e}E_1 \tag{16.176}$$

Using equations (16.174) and (16.176) in (16.167), we have

$$(\omega^2 - c^2k^2)E_1 = -c^2\mu_0 i\omega ne\Big[\frac{ie}{m_i\omega}\Big(\frac{1-\alpha_i}{1-\alpha_i-\Omega_c^2/\omega^2}\Big) + \frac{ik^2}{\omega B_0^2}\frac{\gamma_e KT_e}{e}\Big]E_1$$

For $\omega^2 << \Omega_c^2$, we can neglect $1-\alpha_i$ in comparison to Ω_c^2/ω^2 and for $E_1 \neq 0$, we have

$$\omega^2 - c^2k^2 = -c^2\mu_0 i\omega ne\Big[\frac{ie}{m_i\omega}\Big(\frac{1-\alpha_i}{-\Omega_c^2/\omega^2}\Big) + \frac{ik^2\gamma_e KT_e}{\omega B_0^2 e}\Big]$$

$$= -\frac{ne^2}{m_i\epsilon_0}\frac{\omega^2}{\Omega_c^2}(1-\alpha_i) + \frac{nk^2 m_i\gamma_e KT_e}{B_0^2 m_i\epsilon_0}$$

$$= -\frac{\Omega_p^2\,\omega^2}{\Omega_c^2}(1-\alpha_i) + \frac{\rho k^2\gamma_e KT_e}{B_0^2 m_i\epsilon_0}$$

$$= -\frac{\Omega_p^2\omega^2}{\Omega_c^2}(1-\alpha_i) + \frac{c^2k^2\gamma_e KT_e}{v_A^2 m_i}$$

Thus, we have

$$\omega^2 - c^2k^2 - \frac{c^2k^2}{v_A^2}\frac{\gamma_e KT_e}{m_i} + \frac{\Omega_p^2}{\Omega_c^2}\Big(\omega^2 - \omega^2\alpha_i\Big) = 0$$

$$\omega^2 - c^2k^2\Big(1 + \frac{\gamma_e KT_e}{m_i v_A^2}\Big) + \frac{\Omega_p^2}{\Omega_c^2}\Big(\omega^2 - k^2\frac{\gamma_i KT_i}{m_i}\Big) = 0$$

Now,

$$\frac{\Omega_p^2}{\Omega_c^2} = \frac{ne^2}{m_i\epsilon_0}\,\frac{m_i^2}{e^2B_0^2} = \frac{c^2}{v_A^2}$$

Thus, we have

$$\omega^2\Big(1 + \frac{c^2}{v_A^2}\Big) = c^2k^2\Big(1 + \frac{\gamma_e KT_e}{m_i v_A^2} + \frac{\gamma_i KT_i}{m_i v_A^2}\Big)$$

$$= c^2k^2\Big(1 + \frac{\gamma_e KT_e + \gamma_i KT_i}{m_i v_A^2}\Big) = c^2k^2\Big(1 + \frac{v_s^2}{v_A^2}\Big)$$

where $v_s^2 = (\gamma_e KT_e + \gamma_i KT_i)/m_i$. Therefore,

$$\omega^2\Big(\frac{v_A^2 + c^2}{v_A^2}\Big) = c^2k^2\Big(\frac{v_A^2 + v_s^2}{v_A^2}\Big) \qquad \frac{\omega^2}{k^2} = c^2\,\frac{v_s^2 + v_A^2}{c^2 + v_A^2}$$

This is dispersion relation for magnetosonic waves propagating perpendicular to $\vec{B}_0$. In this acoustic wave, compressions and rarefactions are produced not by the motions along $\vec{E}$, but perpendicular to $\vec{E}$ by $\vec{E}i \times \vec{B}_0$ drift.

When $\vec{B_0} \rightarrow 0$, $v_A \rightarrow 0$, and we have $(\omega/k) = v_s$. Hence, magnetosonic wave turns into ordinary ion acoustic wave. When $KT_i \rightarrow 0$ and $K\ T_e \rightarrow 0$, $v_s \rightarrow 0$, and we have

$$\frac{\omega^2}{k^2} = \frac{c^2 v_A^2}{c^2 + v_A^2} \qquad \text{or} \qquad \frac{\omega}{k} = \frac{v_A}{\sqrt{1 + v_A^2/c^2}}$$

Thus, magnetosonic wave becomes a modified Alfvén wave, whose phase velocity is smaller than that of Alfvén velocity. When v_s is finite, phase velocity of wave is

$$\frac{\omega}{k} = \sqrt{\frac{v_s^2 + v_A^2}{1 + v_A^2/c^2}}$$

It is generally larger than Alfvén velocity, and therefore, the wave is often called fast hydromagnetic wave.

Review Questions

1. Describe phase velocity of sinusoidal wave. Show that phase velocity is always larger than the speed of light.

2. Describe phase velocity and group velocity. Derive relation between them.

3. Show that velocity of a particle is the same as group velocity of corresponding wave packet.

4. For phase velocity u_p and group velocity u_g, derive following relations:

$$u_g = u_p - \lambda \frac{\mathrm{d}u_p}{\mathrm{d}\lambda} \qquad \text{and} \qquad u_p u_g = c^2$$

5. Describe procedure of linearisation of an equation for plasma. Assuming that perturbation part of parameters behave sinusoidally as $e^{i(kx-\omega t)}$ linearize following equation.

$$mn\left[\frac{\partial \vec{u}}{\partial t} + (\vec{u}\ .\ \nabla)\vec{u}\right] = qn(\vec{E} + \vec{u} \times \vec{B}) - \gamma KT\nabla n$$

6. Describe procedure of linearisation of an equation for plasma. Assuming that perturbation part of parameters behave sinusoidally as $e^{i(kx-\omega t)}$ linearize following equations.

$$\frac{\partial n}{\partial t} + \nabla.\ (n\vec{u}) = 0 \qquad \text{and} \qquad \nabla\ .\ \vec{E} = \frac{e(n_i - n_e)}{\epsilon_0}$$

7. Describe procedure of linearisation of an equation for plasma. Assuming that perturbation part of parameters behave sinusoidally as $e^{i(kx-\omega t)}$ linearize following equations.

$$\nabla\ .\ \vec{E} = \frac{e(n_i - n_e)}{\epsilon_0} \qquad \text{and} \qquad \nabla \times \vec{E} = -\frac{\partial \vec{B}}{\partial t}$$

8. Describe procedure of linearisation of an equation for plasma. Assuming that perturbation part of parameters behave sinusoidally as $e^{i(kx-\omega t)}$ linearize following equations.

$$\nabla \times \vec{E} = -\frac{\partial \vec{B}}{\partial t} \qquad \text{and} \qquad \nabla \times \vec{B} = \frac{1}{c^2}\frac{\partial \vec{E}}{\partial t} + \mu_0 \vec{J}$$

9. What is plasma frequency? Derive an expression for the plasma frequency. What happens when thermal motions in plasma are considered.

10. For homogeneous, isotropic, collision-less, cold plasma, derive an expression for plasma frequency.

11. For homogeneous, isotropic, collision-less, hot plasma, derive an expression for dispersion relation.

12. Show that a longitudinal wave is electrostatic and a transverse wave is electromagnetic.

13. Derive an expression for dispersion relation for electron plasma waves. Show that group velocity is never greater than the speed of light, and at a large value of wavenumber k, information travels faster than that at a small value of k.

14. Derive an expression for velocity of sound waves in ordinary gas.

15. Show that in electron plasma waves, the phase velocity is always greater than or equal to $\sqrt{3/2}\, u_{th}$ whereas the group velocity is always less than or equal to $\sqrt{3/2}\, u_{th}$, where u_{th} is thermal velocity of electrons.

16. Derive an expression for velocity of ion waves in fully ionized plasma. Estimate error introduced by plasma approximation.

17. Discuss validity of plasma approximation.

18. Describe propagation of electrostatic electron waves perpendicular to the direction of applied magnetic field.

19. Describe propagation of electrostatic ion waves perpendicular to the direction of applied magnetic field.

20. Describe about lower hybrid frequency.

21. Derive dispersion relation for transverse electromagnetic waves propagating in plasma with $\vec{B_0} = 0$.

22. Derive dispersion relation for ordinary electromagnetic wave propagating perpendicular to $\vec{B_0}$.

23. Derive dispersion relation for extraordinary electromagnetic wave propagating perpendicular to $\vec{B_0}$.

24. Discuss about cut-offs and resonances for electromagnetic waves propagating perpendicular to $\vec{B_0}$.

25. Derive dispersion relation for electromagnetic waves propagating parallel to $\vec{B_0}$.

26. Derive an expression for velocity of Alfvén waves.

27. Derive an expression for velocity of Magnetosonic waves.

28. For electromagnetic, perpendicular, transverse, extraordinary wave, derive the dispersion relation, and show that the phase velocity is always larger than the speed of light c.

29. Write short notes on the following

    (i) Phase velocity

    (ii) Group velocity

    (iii) Plasma frequency

    (iv) Plasma approximation

Multiple-Choice Questions

1. For dispersion relation $\omega^2 = c^2k^2 + c$, group velocity v_g and phase velocity v_p satisfy the relation

   A. $v_g v_p = c$ B. $v_g v_p = c^2$ C. $v_g = v_p$ D. None of these

   Ans. B

2. For dispersion relation $\omega^2 = c^2k^2 + c^3$, group velocity v_g and phase velocity v_p satisfy the relation

   A. $v_g v_p = c$ B. $v_g v_p = c^2$ C. $v_g < v_p$ D. $v_g v_p = c^3$

   Ans. B

3. For dispersion relation $\omega^2 = ck^2 + c^3$, group velocity v_g and phase velocity v_p satisfy the relation

   A. $v_g v_p = c$ B. $v_g v_p = c^2$ C. $v_g < v_p$ D. $v_g v_p = c^3$

   Ans. A

4. Plasma frequency is ω_p. The speed v of wave propagating through plasma and having frequency ω, such that $\omega > \omega_p$, is

   A. $v = c$ B. $v > c$ C. $v < c$ D. $v = \sqrt{c}$

   Ans. B

5. If v is velocity of particle, which of the followings is not expression for group velocity.

   A. $\dfrac{d\omega}{dk}$ B. $\dfrac{dE}{dp}$ C. $\dfrac{\omega}{k}$ D. v

   Ans. C

6. Angular frequency ω and wave number k are related through the relation $\omega = ck$, where c is constant. Which of the following relations between phase velocity v_p and group velocity v_g is correct.

   A. $v_p < v_g$ B. $v_p = v_g$ C. $v_p > v_g$ D. None of them

   Ans. B

7. Angular frequency ω and wave number k are related through the relation $\omega^2 = c^2k^2$, where c is constant. Which of the following relations between phase velocity v_p and group velocity v_g is correct.

   A. $v_p < v_g$ B. $v_p = v_g$ C. $v_p > v_g$ D. None of them

   Ans. B

8. The relation between the phase velocity v_p, group velocity v_g and speed of light c is

   A. $v_pv_g = c$ B. $v_pv_g = c^2$ C. $v_p/v_g = c$ D. $v_p/v_g = c^2$

   Ans. B

9. For a wave expressed as $y = 5\cos(6\pi t - 12x)$, the phase velocity is

   A. 1.2π B. $\pi/2$ C. 2.4 D. $2/\pi$

   Ans. B

10. For a wave expressed as $y = 4\sin(6\pi t - 12x)$, the phase velocity is

    A. 1.5π B. $\pi/2$ C. 3 D. $2/\pi$

    Ans. B

11. For a wave expressed as $y = 10\sin(6\pi t - 12x)$, the phase velocity is

    A. 0.6π B. $\pi/2$ C. 1.2 D. $2/\pi$

    Ans. B

12. Suppose, ions and electrons in plasma have charges q_i and q_e, particle densities n_i and n_e, and are moving with velocities $\vec{}v_i$ and $\vec{v_e}$, respectively. The charge density ρ is

    A. $n_iq_i+n_eq_e$ B. $n_iq_i\vec{u_i}+n_eq_e\vec{u_e}$ C. $n_iq_i-n_eq_e$ D. $n_iq_i\vec{u_i}-n_eq_e\vec{u_e}$

    Ans. A

13. Suppose, ions and electrons in a plasma have charges q_i and q_e, particle densities n_i and n_e, and are moving with velocities $\vec{v_i}$ and $\vec{v_e}$, respectively. The current density $\vec{J}$ is

A. $n_i q_i + n_e q_e$ B. $n_i q_i \vec{u_i} + n_e q_e \vec{u_e}$ C. $n_i q_i - n_e q_e$ D. $n_i q_i \vec{u_i} - n_e q_e \vec{u_e}$

Ans. B

14. For fluid plasma, the following equation is known as

$$mn\left[\frac{\partial \vec{u}}{\partial t} + (\vec{u}\ .\ \nabla)\vec{u}\right] = qn(\vec{E} + \vec{u} \times \vec{B}) - \nabla p$$

A. equation of motion B. continuity equation

C. equation of state D. Maxwell's equation

Ans. A

15. Suppose, n, e, m denote the density, charge and mass of electrons in plasma. The expression for plasma frequency is

A. $\left(\frac{ne^2}{\epsilon_0 m}\right)^{1/2}$ B. $\left(\frac{ne^2}{\epsilon_0 m}\right)^{-1/2}$ C. $\left(\frac{n^2e^2}{\epsilon_0 m}\right)^{1/2}$ D. $\left(\frac{n^2e^2}{\epsilon_0 m}\right)^{-1/2}$

Ans. A

16. For fluid plasma, the following equation is known as

$$mn\left[\frac{\partial \vec{u}}{\partial t} + (\vec{u}\ .\ \nabla)\vec{u}\right] = qn(\vec{E} + \vec{u} \times \vec{B}) - \gamma KT\nabla n$$

A. equation of motion B. continuity equation

C. equation of state D. Maxwell's equation

Ans. A

17. For fluid plasma, the following equation is known as

$$\frac{\partial n}{\partial t} + \nabla\ .\ (n\ \vec{u}) = 0$$

A. equation of motion B. continuity equation

C. equation of state D. Maxwell's equation

Ans. B

18. For fluid plasma, equation $p = Cn^{\gamma}$ is the

A. equation of motion B. continuity equation

C. equation of state D. Maxwell's equation

Ans. C

19. In discussion of plasma, we assume that oscillating quantities behave as $e^{i(\vec{k}\,.\,\vec{r}-\omega t)}$. Then time derivative $(\partial/\partial t)$ is replaced by

A. $-i\omega$ B. $i\omega$ C. $i\vec{k}$ D. $-i\vec{k}$

Ans. A

20. In discussion of plasma, we assume that oscillating quantities behave as $e^{i(\vec{k}\,.\,\vec{r}-\omega t)}$. Then gradient ∇ is replaced by

A. $-i\omega$ B. $i\omega$ C. $i\vec{k}$ D. $-i\vec{k}$

Ans. C

21. In discussion of plasma, we assume that oscillating quantities behave as $e^{i(\vec{k}\,.\,\vec{r}+\omega t)}$. Then time derivative $(\partial/\partial t)$ is replaced by

A. $-i\omega$ B. $i\omega$ C. $i\vec{k}$ D. $-i\vec{k}$

Ans. B

22. In discussion of plasma, we assume that oscillating quantities behave as $e^{i(\vec{k}\,.\,\vec{r}+\omega t)}$. Then gradient ∇ is replaced by

A. $-i\omega$ B. $i\omega$ C. $i\vec{k}$ D. $-i\vec{k}$

Ans. C

23. In discussion of plasma, we assume that oscillating quantities behave as $e^{i(\omega t-\vec{k}\,.\,\vec{r})}$. Then time derivative $(\partial/\partial t)$ is replaced by

A. $-i\omega$ B. $i\omega$ C. $i\vec{k}$ D. $-i\vec{k}$

Ans. B

24. In discussion of plasma, we assume that oscillating quantities behave as $e^{i(\omega t-\vec{k}\,.\,\vec{r})}$. Then gradient ∇ is replaced by

A. $-i\omega$ B. $i\omega$ C. $i\vec{k}$ D. $-i\vec{k}$

Ans. D

25. For adiabatic plasma, pressure p and particle density n are related as (γ is ratio of two specific heats of plasma-gas)

A. $\frac{\nabla p}{p} = \gamma\,\frac{\nabla n}{n}$ B. $\frac{\nabla p}{n} = \gamma\,\frac{\nabla n}{p}$ C. $\frac{\nabla p}{p} = \frac{\nabla n}{n}$ D. $\frac{\nabla p}{n} = \frac{\nabla n}{p}$

Ans. A

26. For adiabatic plasma, pressure p and particle density n are related as (γ is ratio of two specific heats of plasma-gas)

A. $\frac{\nabla p}{p} = \gamma\,\frac{\nabla n}{n}$ B. $\frac{\nabla p}{n} = \gamma\,\frac{\nabla n}{p}$ C. $\frac{\nabla p}{p} = \frac{\nabla n}{n}$ D. $\frac{\nabla p}{n} = \gamma^2\frac{\nabla n}{p}$

Ans. A

27. For isothermal plasma, pressure p and particle density n are related as (γ is ratio of two specific heats of plasma-gas)

A. $\frac{\nabla p}{p} = \gamma \frac{\nabla n}{n}$ B. $\frac{\nabla p}{n} = \gamma \frac{\nabla n}{p}$ C. $\frac{\nabla p}{p} = \frac{\nabla n}{n}$ D. $\frac{\nabla p}{n} = \frac{\nabla n}{p}$

Ans. C

28. For isothermal plasma, pressure p and particle density n are related as (γ is ratio of two specific heats of plasma-gas)

A. $\frac{\nabla p}{p} = \gamma^2 \frac{\nabla n}{n}$ B. $\frac{\nabla p}{n} = \gamma \frac{\nabla n}{p}$ C. $\frac{\nabla p}{p} = \frac{\nabla n}{n}$ D. $\frac{\nabla p}{n} = \frac{\nabla n}{p}$

Ans. C

29. When thermal motions are not considered, the equation of motion of electron in homogeneous, isotropic and collision-less plasma is

A. $m_e n_e \Big[\frac{\partial \vec{u_e}}{\partial t} + (\vec{u_e}.\nabla)\vec{u_e}\Big] = -e n_e \vec{E}$

B. $m_e n_e \Big[\frac{\partial \vec{u_e}}{\partial t} + (\vec{u_e}.\nabla)\vec{u_e}\Big] = -e n_e \vec{E} - \nabla p_e$

C. $m_e n_e \Big[\frac{\partial \vec{u_e}}{\partial t} + (\vec{u_e}.\nabla)\vec{u_e}\Big] = -\nabla p_e$

D. $m_e n_e \Big[\frac{\partial \vec{u_e}}{\partial t} + (\vec{u_e}.\nabla)\vec{u_e}\Big] = -e n_e \vec{E} - -\gamma K T \nabla n$

Ans. A

30. Linearized form of following equation is

$$m_e \Big[\frac{\partial \vec{u_e}}{\partial t} + (\vec{u_e} \, . \, \nabla)\vec{u_e}\Big] = -e\vec{E}$$

A. $m_e \Big[\frac{\partial \vec{u_1}}{\partial t} + (\vec{u_{e1}} \, . \, \nabla)\vec{u_{e0}}\Big] = -e\vec{E}$ B. $m_e \frac{\partial \vec{u_1}}{\partial t} = e\vec{E_1}$

C. $m_e \Big[\frac{\partial \vec{u_1}}{\partial t} + (\vec{u_{e1}} \, . \, \nabla)\vec{u_{e0}}\Big] = -e\vec{E_1}$ D. $m_e \frac{\partial \vec{u_1}}{\partial t} = -e\vec{E_1}$

Ans. D

31. Linearized form of following equation is

$$m_e\left[\frac{\partial \vec{u_e}}{\partial t} + (\vec{u_e} \,.\, \nabla)\vec{u_e}\right] = -e\vec{E}$$

A. $m_e\left[\frac{\partial \vec{u_1}}{\partial t} + (\vec{u_{e0}} \,.\, \nabla)\vec{u_{e1}}\right] = -e\vec{E}$ B. $m_e\,\frac{\partial \vec{u_1}}{\partial t} = e\vec{E_1}$

C. $m_e\left[\frac{\partial \vec{u_1}}{\partial t} + (\vec{u_{e0}} \,.\, \nabla)\vec{u_{e1}}\right] = -e\vec{E_1}$ D. $m_e\,\frac{\partial \vec{u_1}}{\partial t} = -e\vec{E_1}$

Ans. D

32. Linearized form of following equation is

$$m_e\left[\frac{\partial \vec{u_e}}{\partial t} + (\vec{u_e} \,.\, \nabla)\vec{u_e}\right] = e\vec{E}$$

A. $m_e\left[\frac{\partial \vec{u_1}}{\partial t} + (\vec{u_{e1}} \,.\, \nabla)\vec{u_{e0}}\right] = -e\vec{E}$ B. $m_e\,\frac{\partial \vec{u_1}}{\partial t} = e\vec{E_1}$

C. $m_e\left[\frac{\partial \vec{u_1}}{\partial t} + (\vec{u_{e1}} \,.\, \nabla)\vec{u_{e0}}\right] = -e\vec{E_1}$ D. $m_e\,\frac{\partial \vec{u_1}}{\partial t} = -e\vec{E_1}$

Ans. B

33. Linearized form of following equation is

$$\frac{\partial n}{\partial t} + \nabla \,.\, (n\vec{u}) = 0$$

A. $\frac{\partial n_1}{\partial t} + n_0\nabla \,.\, \vec{u_1} = 0$ B. $\frac{\partial n_1}{\partial t} = n_0\nabla \,.\, \vec{u_1}$

C. $\frac{\partial n_1}{\partial t} + \vec{u_0}\nabla n_1 = 0$ D. $\frac{\partial n_1}{\partial t} = \vec{u_0}\nabla n_1$

Ans. A

34. Linearized form of following equation is

$$\frac{\partial n}{\partial t} - \nabla \,.\, (n\vec{u}) = 0$$

A. $\frac{\partial n_1}{\partial t} + n_0\nabla \,.\, \vec{u_1} = 0$ B. $\frac{\partial n_1}{\partial t} = n_0\nabla \,.\, \vec{u_1}$

C. $\frac{\partial n_1}{\partial t} + \vec{u_0}\nabla n_1 = 0$ D. $\frac{\partial n_1}{\partial t} = \vec{u_0}\nabla n_1$

Ans. B

35. Linearized form of following equation is

$$\frac{\partial E}{\partial x} = \frac{e(n_i - n_e)}{\epsilon_0}$$

A. $\frac{\partial E_1}{\partial x} = -\frac{en_{e1}}{\epsilon_0}$ B. $\frac{\partial E_1}{\partial x} = \frac{en_{e1}}{\epsilon_0}$

C. $\frac{\partial E_1}{\partial x} = -\frac{en_{i1}}{\epsilon_0}$ D. $\frac{\partial E_1}{\partial x} = \frac{en_{i1}}{\epsilon_0}$

Ans. A

36. Linearized form of following equation is

$$\frac{\partial E}{\partial x} = -\frac{en_e}{\epsilon_0}$$

A. $\frac{\partial E_1}{\partial x} = -\frac{en_{e1}}{\epsilon_0}$ B. $\frac{\partial E_1}{\partial x} = \frac{en_{e1}}{\epsilon_0}$

C. $\frac{\partial E_1}{\partial x} = -\frac{en_{e0}}{\epsilon_0}$ D. $\frac{\partial E_1}{\partial x} = \frac{en_{e0}}{\epsilon_0}$

Ans. A

37. Linearized form of following equation is

$$\nabla \times \vec{E} = -\frac{\partial \vec{B}}{\partial t}$$

A. $\nabla \times \vec{E_1} = \frac{\partial \vec{B_1}}{\partial t}$ B. $\nabla \times \vec{E_1} = -\frac{\partial \vec{B_0}}{\partial t} + \frac{1}{c}\frac{\partial \vec{B_1}}{\partial t}$

C. $\nabla \times \vec{E_1} = -\frac{\partial \vec{B_1}}{\partial t}$ D. $\nabla \times \vec{E_1} = -\frac{\partial \vec{B_0}}{\partial t} - \frac{1}{c}\frac{\partial \vec{B_1}}{\partial t}$

Ans. C

38. Linearized form of following equation is

$$\nabla \times \vec{E} = -\frac{\partial \vec{B}}{\partial t} + \vec{B}$$

A. $\nabla \times \vec{E_1} = -\frac{\partial \vec{B_1}}{\partial t} + \vec{B_1}$ B. $\nabla \times \vec{E_1} = \frac{\partial \vec{B_0}}{\partial t} - \frac{\partial \vec{B_1}}{\partial t}$

C. $\nabla \times \vec{E_1} = -\frac{\partial \vec{B_1}}{\partial t}$ D. $\nabla \times \vec{E_1} = -\frac{\partial \vec{B_1}}{\partial t} + \vec{B}$

Ans. D

39. Linearized form of following equation is

$$\nabla \times \vec{B} = -\frac{\partial \vec{E}}{\partial t} + \vec{E}$$

A. $\nabla \times \vec{B_1} = -\frac{\partial \vec{E_1}}{\partial t} + \vec{E_1}$ B. $\nabla \times \vec{B_1} = \frac{\partial \vec{E_0}}{\partial t} - \frac{\partial \vec{E_1}}{\partial t}$

C. $\nabla \times \vec{B_1} = -\frac{\partial \vec{E_1}}{\partial t}$ D. $\nabla \times \vec{B_1} = -\frac{\partial \vec{E_1}}{\partial t} + \vec{E}$

Ans. A

40. Linearized form of following equation is

$$\nabla \times \vec{E} = -\frac{\partial \vec{B}}{\partial t}$$

A. $\nabla \times \vec{E_1} = \frac{\partial \vec{B_1}}{\partial t}$ B. $\nabla \times \vec{E_1} = -\frac{\partial \vec{B_0}}{\partial t} - \frac{1}{c}\frac{\partial \vec{B_1}}{\partial t}$

C. $\nabla \times \vec{E_1} = -\frac{\partial \vec{B_1}}{\partial t}$ D. $\nabla \times \vec{E_1} = \frac{\partial \vec{B_0}}{\partial t} - \frac{\partial \vec{B_1}}{\partial t}$

Ans. C

41. Linearized form of following equation is

$$\nabla \times \vec{B} = \frac{\partial \vec{E}}{\partial t}$$

A. $\nabla \times \vec{B_1} = \frac{\partial \vec{E_1}}{\partial t}$ B. $\nabla \times \vec{B_1} = -\frac{\partial \vec{E_0}}{\partial t} - \frac{\partial \vec{E_1}}{\partial t}$

C. $\nabla \times \vec{B_1} = -\frac{\partial \vec{E_1}}{\partial t}$ D. $\nabla \times \vec{B_1} = \frac{\partial \vec{E_0}}{\partial t} - \frac{\partial \vec{E_1}}{\partial t}$

Ans. A

42. Linearized form of following equation for electron is

$$m\left[\frac{\partial \vec{u}}{\partial t} + (\vec{u}.\nabla)\vec{u}\right] = -e(\vec{E} + \vec{u} \times \vec{B})$$

A. $m\,\frac{\partial \vec{u_1}}{\partial t} = e(\vec{E_1} + \vec{u_1} \times \vec{B_0})$ B. $m\,\frac{\partial \vec{u_1}}{\partial t} = -e(\vec{E_1} + \vec{u_0} \times \vec{B_1})$

C. $m\,\frac{\partial \vec{u_1}}{\partial t} = -e(\vec{E_1} + \vec{u_1} \times \vec{B_0})$ D. $m\,\frac{\partial \vec{u_1}}{\partial t} = e(\vec{E_1} - \vec{u_1} \times \vec{B_0})$

Ans. C

43. In plasma, for a perpendicular wave

A. $\vec{k} \perp \vec{B_0}$ B. $\vec{k} \parallel \vec{B_0}$ C. $\vec{k} \perp \vec{E_1}$ D. $\vec{k} \parallel \vec{E_1}$

Ans. A

44. In plasma, for a parallel wave

A. $\vec{k} \perp \vec{B_0}$ B. $\vec{k} \parallel \vec{B_0}$ C. $\vec{k} \perp \vec{E_1}$ D. $\vec{k} \parallel \vec{E_1}$

Ans. B

45. In plasma, for a transverse wave

A. $\vec{k} \perp \vec{B_0}$ B. $\vec{k} \parallel \vec{B_0}$ C. $\vec{k} \perp \vec{E_1}$ D. $\vec{k} \parallel \vec{E_1}$

Ans. C

46. In plasma, for a longitudinal wave

A. $\vec{k} \perp \vec{B_0}$ B. $\vec{k} \parallel \vec{B_0}$ C. $\vec{k} \perp \vec{E_1}$ D. $\vec{k} \parallel \vec{E_1}$

Ans. D

47. In plasma, for an electrostatic wave, the requirement is

A. $\vec{B_1} = 0$ B. $\vec{B_1} \neq 0$ C. $\vec{E_1} = 0$ D. $\vec{E_1} \neq 0$

Ans. A

48. In plasma, for an electromagnetic wave, the requirement is

A. $\vec{B_1} = 0$ B. $\vec{B_1} \neq 0$ C. $\vec{E_1} = 0$ D. $\vec{E_1} \neq 0$

Ans. B

Appendix

A. Useful constants

		SI system	cgs system
c	speed of light	3×10^{8} m/s	3×10^{10} cm/s
e	electron charge	1.602×10^{-19} C	4.806×10^{-10} esu
m_e	electron mass	9.109×10^{-31} kg	9.109×10^{-28} gm
m_p	mass of proton	1.673×10^{-27} kg	1.673×10^{-24} gm
m_p/m_e		1837	1837
k	Boltzmann's constant	1.38×10^{-23} J/K	1.38×10^{-16} erg/K
h	Planck's constant	6.626×10^{-34} Js	6.626×10^{-27} ergs
N	Avogadro number	6.022×10^{23} mol^{-1}	
ϵ_0	permittivity of free space	8.854×10^{-12} F/m	
μ_0	permeability of free space	$4\pi \times 10^{-7}$ H/m	
eV	electron volt	1.602×10^{-19} J	1.602×10^{-12} erg
Å	Angstrom	10^{-10} m	10^{-8} cm

Bibliography

Capri, A.Z. and Panat, P.V. (2002) Introduction to Electrodynamics, Narosa Publishing House, New Delhi.

Chandra, S. (2010) Textbook of Plasma Physics, CBS Publishers & Distributors Pvt Ltd, New Delhi.

Chen, F.F. (1984) Introduction to Plasma Physics and Controlled Fusion, Second Edition, Plenum Press, New York.

Griffiths, D.J. (2013) Introduction to Electrodynamics, Fourth Edition, Pearson.

Gupta, S.L., Kumar, V. and Singh, S.P. (2007) Electrodynamics, Pragati Prakashan, Meerut.

Jackson, J.D. (2008) Classical Electrodynamics, Wiley.

Krishan, V. (1999) Astrophysical Plasmas and Fluids, Kluwer Academic Publishers, Dordrecht.

Landau, L.D. and Lifshitz, E.M. (1984) Electrodynamics of Continuous Media, Pergamon Press, Oxford.

Sen, S.N. (2007) Plasma Physics - Plasma State of Matter, Sixth Edition, Pragati Prakashan, Meerut.

Index